STUDENT'S SOLUTIONS MANUAL

COLLEGE ALGEBRA

SEVENTH EDITION

Lial • Hornsby • Schneider

Prepared with the assistance of

Abby Tanenbaum
Sandra Morris
August Zarcone

College of DuPage

Irene Doo

Austin Community College

ADDISON-WESLEY

An imprint of Addison Wesley Longman, Inc.

Reading, Massachusetts • Menlo Park, California • New York • Harlow, England
Don Mills, Ontario • Sydney • Mexico City • Madrid • Amsterdam

ISBN   0-673-98336-6

5 6 7 8 9 10 CRS 050403020100

PREFACE

This book provides complete solutions for many of the exercises in *College Algebra*, Seventh Edition, by Margaret L. Lial, E. John Hornsby, Jr., and David I. Schneider. Solutions are provided for all of the following exercises except for the writing exercises.

- Odd–numbered section exercises and chapter review exercises

- All exercises in the chapter tests

- All "Discovering Connections" exercises

In addition, three sets of Cumulative Review Exercises are included and complete solutions to these exercises are provided. You may find these exercises helpful in preparing for examinations.

This book should be used as an aid as you work to master your coursework. Try to solve the exercises that your instructor assigns before you refer to the solutions in the book. Then, if you have difficulty, read these solutions to guide you in solving the exercises. The solutions have been written so that they are consistent with the methods used in the textbook.

Solutions to textbook exercises that require graphs will refer to the answer section of the textbook. These graphs are not included in this book.

In addition to solutions, you will find a list of suggestions on how to be successful in mathematics. A careful reading will be helpful to many students.

The following people have made valuable contributions to the production of this *Student's Solution Manual*: Abby Tanenbaum, editor; Judy Martinez, typist; and Carmen Eldersveld, proofreader. Artwork has been provided by Precision Graphics, Therese Brown, Charles Sullivan, and August Zarcone.

We also want to thank Tommy Thompson of Cedar Valley Community College for his suggestions for the essay "To the Student: Success in Mathematics."

CONTENTS

TO THE STUDENT: SUCCESS IN MATHEMATICS

The main reason students have difficulty with mathematics is that they don't know how to study it. Studying mathematics *is* different from studying subjects like English or history. The key to success is regular practice.

This should not be surprising. After all, can you learn to play the piano or to ski well without a lot of regular practice? The same thing is true for learning mathematics. Working problems nearly every day is the key to becoming successful. Here is a list of things you can do to help you succeed in studying algebra.

1. *Attend class regularly.* Pay attention in class to what your instructor says and does, and make careful notes. In particular, note the problems the instructor works on the board and copy the complete solutions. Keep these notes separate from your homework to avoid confusion when you read them over later.

2. Don't hesitate to ask questions in class. It is not a sign of weakness, but of strength. There are always other students with the same question who are too shy to ask.

3. *Read your text carefully.* Many students read only enough to get by, usually only the examples. Reading the complete section will help you to be successful with the homework problems. Most exercises are keyed to specific examples or objectives that will explain the procedures for working them.

4. Before you start on your homework assignment, rework the problems the instructor worked in class. This will reinforce what you have learned. Many students say, "I understand it perfectly when you do it, but I get stuck when I try to work the problem myself."

5. Do your homework assignment only *after* reading the text and reviewing your notes from class. Check your work with the answers in the back of the book. If you get a problem wrong and are unable to see why, mark that problem and ask your instructor about it. Then practice working additional problems of the same type to reinforce what you have learned.

6. Work as neatly as you can. Write your symbols clearly, and make sure the problems are clearly separated from each other. Working neatly will help you to think clearly and also make it easier to review the homework before a test.

7. After you have completed a homework assignment, look over the text again. Try to decide what the main ideas are in the lesson. Often they are clearly highlighted or boxed in the text.

8. Use the chapter test at the end of each chapter as a practice test. Work through the problems under test conditions, without referring to the text or the answers until you are finished. You may want to time yourself to see how long it takes you. When you have finished, check your answers against those in the back of the book and study those problems that you missed. Answers are referenced to the appropriate sections of the text.

9. Keep any quizzes and tests that are returned to you and use them when you study for future tests and the final exam. These quizzes and tests indicate what your instructor considers most important. Be sure to correct any problems on these tests that you missed, so you will have the corrected work to study.

10. Don't worry if you do not understand a new topic right away. As you read more about it and work through the problems, you will gain understanding. Each time you look back at a topic you will understand it a little better. No one understands each topic completely right from the start.

CHAPTER 1 ALGEBRAIC EXPRESSIONS

Section 1.1

1. 1 and 3 are natural numbers.

3. -6, $\dfrac{-12}{4}$ (or -3), 0, 1, and 3 are integers.

5. $-\sqrt{3}$, 2π, and $\sqrt{12}$ are irrational numbers.

7. 29 is a natural number, a whole number, an integer, a rational number, and a real number.

9. $-\dfrac{5}{6}$ is a rational number and a real number.

11. $\sqrt{13}$ is an irrational number and a real number.

15. $-3^5 = -(3 \cdot 3 \cdot 3 \cdot 3 \cdot 3) = -243$

17. $(-3)^4 = (-3)(-3)(-3)(-3) = 81$

19. $(-3)^5 = (-3)(-3)(-3)(-3)(-3) = -243$

21. A negative base raised to an odd exponent is *negative*. A negative base raised to an even exponent is *positive*.

23. $8^2 - (-4) + 11 = 64 - (-4) + 11$
$$= 64 + 4 + 11$$
$$= 79$$

25. $-2 \cdot 5 + 12 \div 3 = -10 + 4 = -6$

27. $-4(9 - 8) + (-7)(2)^3$
$$= -4(1) + (-7)(2)^3$$
$$= -4(1) + (-7)(8)$$
$$= -4 + (-56)$$
$$= -60$$

29. $(4 - 2^3)(-2 + \sqrt{25})$
$$= (4 - 8)(-2 + 5)$$
$$= (-4)(3)$$
$$= -12$$

31. $\left(-\dfrac{2}{9} - \dfrac{1}{4}\right) - \left[-\dfrac{5}{18} - \left(-\dfrac{1}{2}\right)\right]$
$$= \left(-\dfrac{8}{36} - \dfrac{9}{36}\right) - \left(-\dfrac{10}{36} + \dfrac{18}{36}\right)$$
$$= \left(-\dfrac{17}{36}\right) - \left(\dfrac{8}{36}\right)$$
$$= -\dfrac{25}{36}$$

33. $\dfrac{-8 + (-4)(-6) \div 12}{4 - (-3)}$
$$= \dfrac{-8 + 24 \div 12}{4 + 3}$$
$$= \dfrac{-8 + 2}{4 + 3}$$
$$= -\dfrac{6}{7}$$

35. The calculator screen shows the values $x = 5$, $y = 2$, and $z = 2$. Substitute these values into the given expression and use the order of operations.

$$x^2 + 2y - 3z = 5^2 + 2(2) - 3(2)$$
$$= 25 + 4 - 6$$
$$= 23$$

Using a graphing calculator, start with the entries shown on the screen in the textbook. After "z", enter ":$x^2 + 2y - 3z$." The calculator will return 23 as the result.

37. $2(q - r) = 2[8 - (-10)]$
    Let $q = 8$ and $r = -10$
$$= 2(8 + 10)$$
$$= 2(18) = 36$$

39. $\dfrac{q + r}{q + p} = \dfrac{8 + (-10)}{8 + (-4)}$ *Let $p = -4$, $q = 8$, and $r = -10$*
$$= \dfrac{-2}{4}$$
$$= -\dfrac{1}{2}$$

41. $\dfrac{3q}{r} - \dfrac{5}{p} = \dfrac{3(8)}{-10} - \dfrac{5}{-4}$ *Let $p = -4$, $q = 8$, and $r = -10$*
$$= -\dfrac{24}{10} + \dfrac{5}{4}$$
$$= -\dfrac{48}{20} + \dfrac{25}{20}$$
$$= -\dfrac{23}{20}$$

43. Use the magic number formula given in the text.
$$M = W_2 + N_2 - W_1 + 1$$
Find the values to be entered into this formula.

W_2 = number of current wins for Royals
    = 61

N_2 = number of remaining games for Royals
    = 144 − (61 + 58)
    = 25

W_1 = number of current wins for Indians
    = 82

Then
$$M = W_2 + N_2 - W_1 + 1$$
$$= 61 + 25 - 82 + 1$$
$$M = 5.$$

The Indians' magic number on September 6, 1995 was 5.

45. $6 \cdot 12 + 6 \cdot 15 = 6(12 + 15)$

Distributive property

47. $(x + 6) \cdot \left(\dfrac{1}{x + 6}\right) = 1$ if $x + 6 \neq 0$

Inverse property

49. $(7 + y) + 0 = 7 + y$

Identity property

53. $8p - 14p = (8 - 14)p$
$$= -6p$$

55. $18y + 6 = 6(3y + 1)$

57. $-3(z - y) = (-3)(z) + (-3)(-y)$
$$= -3z + 3y$$

59. $a(r + s - t)$
$$= a \cdot r + a \cdot s + a(-t)$$
$$= ar + as - at$$

61. $\dfrac{10}{11}(22z) = \left(\dfrac{10}{11} \cdot 22\right)z$ *Associative property*
$$= 20z$$

63. $\left(-\frac{5}{8}p\right)(-24) = (-24)\left(-\frac{5}{8}p\right)$ *Commutative property*

$= \left[-24\left(-\frac{5}{8}\right)\right]p$ *Associative property*

$= 15p$

65. $-\frac{1}{4}(20m + 8y - 32z)$

$= -\frac{1}{4}(20m) + \left(-\frac{1}{4}\right)(8y) - \left(-\frac{1}{4}\right)(32z)$

Distributive property

$= -5m - 2y + 8z$

67. $\dfrac{64 + 71 + 69 + 67}{4} = 67.75$

His average score per round was 67.75.

69. The ball would travel at an average speed of

$$\frac{35,840 \text{ ft}}{64 \text{ min}} = 560 \text{ ft/min}$$

$$= 33,600 \text{ ft/hr}$$

$$\approx 6.36 \text{ mi/hr.}$$

71. In a sample of 15,000 homeowners, approximately 16% of them, or

$$.16(15,000) = 2400$$

have expenditures between $1000 and $4999.

73. **(a)** Finding a wind speed of 46.0 mph and then moving across to a 30 hour wind duration yields a wave height of 31 feet.

(b) No. It is possible for two different wind speeds and durations to produce the same wave height. For example, a 17.3–mph wind for 30 hours will produce a wave 5 feet high, as will a 23.0–mph wind blowing for 20 hours.

(c) Stronger winds produce higher waves. The duration of the wind will increase the wave height as the wind becomes stronger. At 11.5 mph the wind duration does not affect the wave height, whereas at 57.5 mph the wind duration has a dramatic effect on the wave height.

Section 1.2

1. If $a > b$, then the absolute value of $b - a$ is $a - b$.

If $a > b$, then $b - a < 0$, so

$|b - a| = -(b - a) = -b + a = a - b$.

The given statement is true.

3. If a and b are both positive, $|a + b| = a + b$.

If a and b are both positive, $a + b$ is also positive, so

$$|a + b| = a + b.$$

The given statement is true.

5. If a graphing calculator returns a 0 for a false statement and a 1 for a true statement, then a 0 would be returned for the statement in the screen.

The screen in the textbook shows the statement

$$abs(5 + 6) = abs(5 - 6),$$

which represents the equation

$$|5 + 6| = |5 - 6|.$$

Since

$$|5 + 6| = |11| = 11$$

and

$$|5 - 6| = |-1| = 1,$$

the statement shown in the screen is false, so a 0 will be returned. Therefore, the statement given in this exercise is true.

7. Since $|-8| = 8$, $-|9| = -9$, and $-|-6| = -6$, the order is

$$-|9|, \; -|-6|, \; |-8|.$$

9. Since $\sqrt{8} \approx 2.83$, $-\sqrt{3} \approx -1.73$, and $\sqrt{6} \approx 2.45$, the order is

$$-5, \; -4, \; -2, \; -\sqrt{3}, \; \sqrt{6}, \; \sqrt{8}, \; 3.$$

11. Since $\frac{3}{4} = .75$, $\sqrt{2} \approx 1.414$,

$\frac{7}{5} = 1.4$, $\frac{8}{5} = 1.6$, and $\frac{22}{15} \approx 1.47$,

the order is

$$\frac{3}{4}, \; \frac{7}{5}, \; \sqrt{2}, \; \frac{22}{15}, \; \frac{8}{5}.$$

15. $-2 < 5$

$$3(-2) < 3(5)$$

$$-6 < 15$$

17. $-11 \geq -22$

$$\left(-\frac{1}{11}\right)(-11) \leq \left(-\frac{1}{11}\right)(-22) \quad \begin{array}{l} \textit{Reverse} \\ \textit{inequality} \\ \textit{symbol} \end{array}$$

$$1 \leq 2$$

19. $3x < 9$

$$\frac{1}{3} \cdot 3x < \frac{1}{3} \cdot 9$$

$$x < 3$$

21. $-9k > 63$

$$\left(-\frac{1}{9}\right)(-9k) < \left(-\frac{1}{9}\right)(63) \quad \begin{array}{l} \textit{Reverse} \\ \textit{inequality} \\ \textit{symbol} \end{array}$$

$$k < -7$$

23. $|2x| = |2(-4)| \quad \textit{Let } x = -4$

$$= |-8|$$

$$= 8$$

25. $|x - y| = |-4 - 2| \quad \begin{array}{l} \textit{Let } x = -4 \\ \textit{and } y = 2 \end{array}$

$$= |-6|$$

$$= 6$$

27. $|3x + 4y| = |3(-4) + 4(2)|$
$$\textit{Let } x = -4 \textit{ and } y = 2$$

$$= |-12 + 8|$$

$$= |-4|$$

$$= 4$$

29. $\dfrac{|-8y + x|}{-|x|} = \dfrac{|-8(2) + (-4)|}{-|-4|}$

$$\textit{Let } x = -4 \textit{ and } y = 2$$

$$= \frac{|-16 + (-4)|}{-|-4|}$$

$$= \frac{|-20|}{-(4)}$$

$$= \frac{20}{-4} = -5$$

31. $\left|\dfrac{2x - 3}{2z}\right| = \left|\dfrac{2(-2) - 3}{2(-6)}\right| \quad \begin{array}{l} \textit{Let } x = -2 \\ \textit{and } z = -6 \end{array}$

$$= \left|\frac{-7}{-12}\right| = \frac{7}{12}$$

A calculator will return the value .583333... (with as many decimal places as the calculator displays). This represents the repeating decimal $.58\overline{3}$, which is equal to the fraction $7/12$, so the two answers are equivalent.

33. $|\pi - 3|$

Since $\pi > 3$, $\pi - 3 > 0$, so

$$|\pi - 3| = \pi - 3.$$

35. $|x - 4|$, if $x > 4$

If $x > 4$, $x - 4 > 0$, so

$$|x - 4| = x - 4.$$

37. $|2k - 8|$, if $k < 4$

If $k < 4$, $2k < 8$, and $2k - 8 < 0$, so

$$
\begin{aligned}
|2k - 8| &= -(2k - 8) \\
&= -2k + 8 \\
&= 8 - 2k.
\end{aligned}
$$

39. $|-8 - 4m|$, if $m > -2$

If $m > -2$, $4m > -8$, and $-8 - 4m < 0$, so

$$
\begin{aligned}
|-8 - 4m| &= -(-8 - 4m) \\
&= 8 + 4m.
\end{aligned}
$$

41. $|x - y|$, if $x < y$

If $x < y$, $x - y < 0$, so

$$
\begin{aligned}
|x - y| &= -(x - y) \\
&= -x + y \\
&= y - x.
\end{aligned}
$$

43. $|3 + x^2|$

$(3 + x^2) > 0$, so

$$|3 + x^2| = 3 + x^2.$$

45. The screen shows the statement

$$|x - y| \geq 0.$$

Since the absolute value of *any* number is greater than or equal to 0, the statement is true for any values of x and y. Therefore, the calculator will return a 1 (to indicate a true statement) no matter what values of x and y are stored in the calculator.

47. If $x + 8 < 15$, then $x < 7$.

$$
\begin{aligned}
x + 8 &< 15 \\
x + 8 + (-8) &< 15 + (-8) \\
x &< 7
\end{aligned}
$$

Addition property of order

49. If $x < 5$ and $5 < m$, then $x < m$.

Transitive property of order

51. If $k > 0$, then $8 + k > 8$.

Addition property of order

53. $|k - m| \leq |k| + |-m|$

Triangle inequality,
$|a + b| \leq |a| + |b|$

55. $|12 + 11r| \geq 0$

Property of absolute value, $|a| \geq 0$

57. $P_d = |P - 125|$

$\quad = |116 - 125|$

$\quad = |-9| = 9$

The P_d value for a woman whose actual systolic pressure is 116 and whose normal value should be 125 is 9.

59. The absolute value of the difference in wind–chill factors for wind at 15 mph with a 30°F temperature and wind at 10 mph with a −10°F temperature is

$$|9° - (-33°)| = |-42°| = 42°F.$$

61. The absolute value of the difference in wind–chill factors for wind at 30 mph with a −30°F temperature and wind at 15 mph with a −20°F temperature is

$$|-94° - (-58°)| = |-36°| = 36°F.$$

63. For 1990, the graphs show revenue of 1032 billion dollars and expenditures of 976 billion dollars. The absolute value of the difference between revenue and expenditures is

$$|1032 - 976| = |56|$$
$$= 56 \text{ billion dollars.}$$

Since revenue was greater than expenditures, the governments were "in the black."

65. For 1992, the graphs show revenue of 1185 billion dollars and expenditures of 1355 billion dollars. The absolute value of the difference between revenue and expenditures is

$$|1185 - 1355| = |-170|$$
$$= 170 \text{ billion dollars.}$$

Since revenue was less than expenditures, the governments were "in the red."

Section 1.3

1. $(-4)^5(-4)^2 = 16^7$

When we use the product rule to multiply powers of the same base, the product has the common base and the exponents are added. A correct statement is

$$(-4)^5(-4)^2 = (-4)^7.$$

The given statement is false.

3. $k^r k^{2r+1} = k^{3r+1}$

When we use the product rule, we keep the common base and add the exponents. Since

$$r + (2r + 1) = 3r + 1,$$

the given statement is true.

5. $(-3)^0 = 1$

Any nonzero number raised to the zero power is 1, so the given statement is true.

7. $(2^2)^5 = 2^{2 \cdot 5}$ *Power rule*

   $= 2^{10}$

9. $(2x^5y^4)^3$

   $= 2^3(x^5)^3(y^4)^3$ *Power rule*

   $= 2^3x^{5 \cdot 3}y^{4 \cdot 3}$ *Power rule*

   $= 2^3x^{15}y^{12}$

11. $-\left(\dfrac{p^4}{q}\right)^2 = -\dfrac{(p^4)^2}{q^2}$ *Power rule*

    $= -\dfrac{p^{4 \cdot 2}}{q^2}$ *Power rule*

    $= -\dfrac{p^8}{q^2}$

13. $-5x^{11}$ is a polynomial. It is a monomial since it has one term. It has degree 11 since 11 is the highest exponent.

15. $18p^5q + 6pq$ is a polynomial. It is a binomial since it has two terms. It has degree 6 because 6 is the sum of the exponents in the term $18p^5q$, and this term has a higher degree than the term $6pq$.

17. $\sqrt{2}x^2 + \sqrt{3}x^6$ is a polynomial. It is a binomial since it has two terms. It has degree 6 since 6 is the highest exponent.

19. $\dfrac{1}{3}r^2s^2 - \dfrac{3}{5}r^4s^2 + rs^3$ is a polynomial. It is a trinomial since it has three terms. It has degree 6 because the sum of the exponents in the term $-\dfrac{3}{5}r^4s^2$ is 6, and this term has the highest degree.

21. $\dfrac{5}{p} + \dfrac{2}{p^2} + \dfrac{5}{p^3}$ is not a polynomial since positive exponents in the denominator are equivalent to negative exponents in the numerator.

23. $(3x^2 - 4x + 5) + (-2x^2 + 3x - 2)$

    $= (3x^2 - 2x^2) + (-4x + 3x)$

    $\quad + (5 - 2)$ *Remove parentheses and group like terms*

    $= x^2 - x + 3$ *Combine like terms*

25. $(12y^2 - 8y + 6) - (3y^2 - 4y + 2)$

    $= 12y^2 - 8y + 6 - 3y^2 + 4y - 2$

    Remove parentheses

    $= (12y^2 - 3y^2) + (4y - 8y)$

    $\quad + (6 - 2)$ *Group like terms*

    $= 9y^2 - 4y + 4$

    Combine like terms

27. $(6m^4 - 3m^2 + m) - (2m^3 + 5m^2 + 4m)$

    $\quad + (m^2 - m)$

    $= (6m^4 - 3m^2 + m)$

    $\quad + (-2m^3 - 5m^2 - 4m) + (m^2 - m)$

    $= 6m^4 - 2m^3 - 3m^2 - 5m^2 + m^2 + m$

    $\quad - 4m - m$

    $= 6m^4 - 2m^3 + (-3 - 5 + 1)m^2$

    $\quad + (1 - 4 - 1)m$

    $= 6m^4 - 2m^3 - 7m^2 - 4m$

29. $(4r - 1)(7r + 2)$

    $= 4r(7r) + 4r(2) - 1(7r) - 1(2)$

    FOIL

    $= 28r^2 + 8r - 7r - 2$

    $= 28r^2 + r - 2$ *Combine like terms*

31. $\left(3x - \frac{2}{3}\right)\left(5x + \frac{1}{3}\right)$

$$= 3x(5x) + 3x\left(\frac{1}{3}\right) + \left(-\frac{2}{3}\right)(5x)$$

$$+ \left(-\frac{2}{3}\right)\left(\frac{1}{3}\right) \qquad FOIL$$

$$= 15x^2 + \frac{3}{3}x - \frac{10}{3}x - \frac{2}{9}$$

$$= 15x^2 - \frac{7}{3}x - \frac{2}{9}$$

33. $4x^2(3x^3 + 2x^2 - 5x + 1)$

$$= (4x^2)(3x^3) + (4x^2)(2x^2)$$

$$- (4x^2)(5x) + (4x^2)(1)$$
$$\qquad Distributive\ property$$

$$= 12x^5 + 8x^4 - 20x^3 + 4x^2$$

35. $(2z - 1)(-z^2 + 3z - 4)$

$$= 2z(-z^2 + 3z - 4) - 1(-z^2 + 3z - 4)$$
$$\qquad Distributive\ property$$

$$= -2z^3 + 6z^2 - 8z + z^2 - 3z + 4$$

$$= -2z^3 + (6z^2 + z^2) + (-8z - 3z)$$

$$+ 4$$

$$= -2z^3 + 7z^2 - 11z + 4$$

We may also multiply vertically.

$$
\begin{array}{r}
-z^2 + 3z - 4 \\
2z - 1 \\
\hline
z^2 - 3z + 4 \\
-2z^3 + 6z^2 - 8z \\
\hline
-2z^3 + 7z^2 - 11z + 4
\end{array}
$$

37. $(m - n + k)(m + 2n - 3k)$

$$= m(m + 2n - 3k) - n(m + 2n - 3k)$$

$$+ k(m + 2n - 3k) \quad Distributive$$
$$\qquad\qquad\qquad\qquad property$$

$$= m^2 + 2mn - 3km - mn - 2n^2 + 3kn$$

$$+ km + 2kn - 3k^2 \quad Distributive$$
$$\qquad\qquad\qquad\qquad property$$

$$= m^2 + (2mn - mn) + (-3km + km)$$

$$- 2n^2 + (3kn + 2kn) - 3k^2$$

$$= m^2 + mn - 2km - 2n^2 + 5kn - 3k^2$$

39. One of the special products for the square of a binomial is

$$(x + y)^2 = x^2 + 2xy + y^2.$$

Since this statement is true for *all* replacements of x and y, the calculator screen will return a 1 no matter what values of x and y are stored in the calculator.

41. **(a)** The area of the largest square is

$$A = s^2$$
$$A = (x + y)^2.$$

(b) The areas of the two squares are x^2 and y^2. The area of each rectangle is xy. Therefore, the area of the largest square can be written as

$$A = x^2 + 2xy + y^2.$$

(c) Since they both represent the area of the largest square, and it can have only one area, the two expressions must be equal.

(d) It reinforces the special product for squaring a binomial:

$$(x + y)^2 = x^2 + 2xy + y^2.$$

43. $(2m + 3)(2m - 3)$

$$= (2m)^2 - 3^2 \quad Difference\ of$$
$$\qquad\qquad\qquad two\ squares$$

$$= 4m^2 - 9$$

45. $(4m + 2n)^2$

$$= (4m)^2 + 2(4m)(2n) + (2n)^2$$
$$\qquad\qquad Square\ of\ a\ binomial$$

$$= 16m^2 + 16mn + 4n^2$$

47. $(5r + 3t^2)^2$

$= (5r)^2 + 2(5r)(3t^2) + (3t^2)^2$
    Square of a binomial

$= 25r^2 + 30rt^2 + 9t^4$

49. $[(2p - 3) + q]^2$

$= (2p - 3)^2 + 2(2p - 3)(q) + q^2$
    Square of a binomial, treating (2p - 3) as one term

$= (2p)^2 - 2(2p)(3) + 3^2$

$+ 2(2p - 3)q + q^2$
    Square the binomial (2p - 3)

$= 4p^2 - 12p + 9 + 4pq - 6q + q^2$

51. $[(3q + 5) - p][(3q + 5) + p]$

$= (3q + 5)^2 - p^2$
    Difference of two squares

$= [(3q)^2 + 2(3q)(5) + 5^2] - p^2$
    Square of a binomial

$= 9q^2 + 30q + 25 - p^2$

53. $[(3a + b) - 1]^2$

$= (3a + b)^2 - 2(3a + b)(1) + 1^2$
    Square of a binomial

$= (9a^2 + 6ab + b^2) - 2(3a + b) + 1$
    Square of a binomial

$= 9a^2 + 6ab + b^2 - 6a - 2b + 1$
    Distributive property

55. $(p^3 - 4p^2 + p) - (3p^2 + 2p + 7)$

$= p^3 - 4p^2 + p - 3p^2 - 2p - 7$

$= p^3 - 7p^2 - p - 7$

57. $(7m + 2n)(7m - 2n)$

$= (7m)^2 - (2n)^2$ *Difference of two squares*

$= 49m^2 - 4n^2$

59. $-3(4q^2 - 3q + 2) + 2(-q^2 + q - 4)$

$= -12q^2 + 9q - 6 - 2q^2 + 2q - 8$
    Distributive property

$= (-12q^2 - 2q^2) + (9q + 2q)$

$+ (-6 - 8)$ *Group like terms*

$= -14q^2 + 11q - 14$
    Combine like terms

61. $p(4p - 6) + 2(3p - 8)$

$= 4p^2 - 6p + 6p - 16$
    Distributive property

$= 4p^2 - 16$ *Combine like terms*

63. $-y(y^2 - 4) + 6y^2(2y - 3)$

$= -y^3 + 4y + 12y^3 - 18y^2$
    Distributive property

$= (-y^3 + 12y^3) - 18y^2 + 4y$
    Group like terms

$= 11y^3 - 18y^2 + 4y$
    Combine like terms

65. $(k^m + 2)(k^m - 2) = (k^m)^2 - 2^2$
    Difference of 2 squares

$= k^{2m} - 4$

67. $(3p^x + 1)(p^x - 2)$

$= 3(p^x)^2 - 6p^x + p^x - 2$

$= 3p^{2x} - 5p^x - 2$

69. $(q^p - 5p^q)^2$

$= (q^p)^2 - 2(q^p)(5p^q) + (5p^q)^2$
    Square of a binomial

$= q^{2p} - 10q^p p^q + 25p^{2q}$

71. The degree of a sum of polynomials is the same as the degree of the higher degree polynomial. Thus, since m > n, the sum has degree m.

73. The product of x^m and x^n is x^{m+n}. Thus, the degree of the product of a polynomial of degree m and one of degree n is m + n.

75. $(x + y)^6$

Use the coefficients from row 6 of Pascal's triangle.

$(x + y)^6$

$= x^6 + 6x^5y + 15x^4y^2 + 20x^3y^3$
$\quad + 15x^2y^4 + 6xy^5 + y^6$

77. $(p - q)^5$

Use the coefficients from row 5 of Pascal's triangle.

$(p - q)^5$

$= p^5 + 5p^4(-q) + 10p^3(-q)^2$
$\quad + 10p^2(-q)^3 + 5p(-q)^4 + (-q)^5$
$= p^5 - 5p^4q + 10p^3q^2 - 10p^2q^3$
$\quad + 5pq^4 - q^5$

79. $(r^2 + s)^5$

Use the coefficients from row 5 of Pascal's triangle.

$(r^2 + s)^5$

$= (r^2)^5 + 5(r^2)^4s + 10(r^2)^3s^2$
$\quad + 10(r^2)^2s^3 + 5(r^2)^1s^4 + s^5$
$= r^{10} + 5r^8s + 10r^6s^2 + 10r^4s^3$
$\quad + 5r^2s^4 + s^5$

81. $(3r - s)^6$

Use the coefficients from row 6 of Pascal's triangle.

$(3r - s)^6$

$= (3r)^6 + 6(3r)^5(-s) + 15(3r)^4(-s)^2$
$\quad + 20(3r)^3(-s)^3 + 15(3r)^2(-s)^4$
$\quad + 6(3r)(-s)^5 + (-s)^6$
$= 729 - 6 \cdot 243r^5s + 15 \cdot 81r^4s^2$
$\quad - 20 \cdot 27r^3s^3 + 15 \cdot 9r^2s^4$
$\quad - 6 \cdot 3rs^5 + s^6$
$= 729r^6 - 1458r^5s + 1215r^4s^2$
$\quad - 540r^3s^3 + 135r^2s^4 - 18rs^5 + s^6$

83. $(4a - 5b)^5$

Use the coefficients from row 5 of Pascal's triangle.

$(4a - 5b)^5$

$= (4a)^5 + 5(4a)^4(-5b)$
$\quad + 10(4a)^3(-5b)^2 + 10(4a)^2(-5b)^3$
$\quad + 5(4a)(-5b)^4 + (-5b)^5$
$= 1024a^5 + 5(256a^4)(-5b)$
$\quad + 10(64a^3)(25b) + 10(16a^2)(-125b^3)$
$\quad + 5(4a)(625b^4) + (-3125b^5)$
$= 1024a^5 - 6400a^4b + 16{,}000a^3b^2$
$\quad - 20{,}000a^2b^3 + 12{,}500ab^4 - 3125b^5$

85. **(a)** The volume is

$V = \frac{1}{3}h(a^2 + ab + b^2)$

$= \frac{1}{3}(200)(314^2 + 314 \times 756 + 756^2)$

$\approx 60{,}501{,}067 \text{ ft}^3$.

(b) The shape becomes a rectangular box with a square base. Its volume is given by length × width × height or b^2h.

(c) If we let a = b, then

$$\frac{1}{3}h(a^2 + ab + b^2)$$

becomes

$$\frac{1}{3}h(b^2 + bb + b^2),$$

which simplifies to hb^2. Yes the Egyptian formula gives the same result.

87. To expand $(x + 1)^0$, use the coefficient in row zero of Pascal's triangle. (See the triangle in the "Connections" section in the textbook. The top row, which contains just a "1" is counted as row zero.)

$$(x + 1)^0 = 1$$
$$(10 + 1)^0 = 1 = 11^0$$

88. To expand $(x + 1)^1$, use the coefficients in row 1 of Pascal's triangle.

$$(x + 1)^1 = x^1 + 1^1$$
$$= x + 1$$
$$(10 + 1)^1 = 10 + 1 = 11 = 11^1$$

89. To expand $(x + 1)^2$, use the coefficients in row 2 of Pascal's triangle.

$$(x + 1)^2 = x^2 + 2(x)(1) + 1^2$$
$$= x^2 + 2x + 1$$
$$(10 + 1)^2 = 10^2 + 2(10) + 1$$
$$\qquad\qquad\qquad Let\ x = 10$$
$$= 100 + 20 + 1$$
$$= 121 = 11^2$$

90. To expand $(x + 1)^3$, use the coefficients in row 3 of Pascal's triangle.

$$(x + 1)^3 = x^3 + 3x^2 \cdot 1 + 3 \cdot x \cdot 1^2 + 1^3$$
$$= x^3 + 3x^2 + 3x + 1$$
$$(10 + 1)^3 = 10^3 + 3 \cdot 10^2 + 3 \cdot 10 + 1$$
$$= 1000 + 300 + 30 + 1$$
$$= 1331$$

91. In each case, the coefficients of the polynomial correspond to the digits in the power of 11.

92. $(x + 1)^4 = x^4 + 4x^3 \cdot 1 + 6x^2 \cdot 1^2$
$$\qquad + 4x \cdot 1^3 + 1^4$$
$$= x^4 + 4x^3 + 6x^2 + 4x + 1$$

A logical prediction for the value of $11^4 = (10 + 1)^4$ is 14,641. A calculator will show that this *is* the value of 11^4.

93. 1900

Substitute 1900 for x in the given polynomial; then evaluate the polynomial.

$-.00102834793874x^2 + 3.9526021723669x$
$- 3791.976211763$
$= -.00102834793874(1900)^2$
$\qquad + 3.9526021723669(1900)$
$\qquad - 3791.976211763\ \ Let\ x = 1900$
≈ 5.6

For the year 1900, the value of the polynomial is approximately 5.6 (representing 5.6 million farms), while the bar graph shows 5.7 (representing 5.7 million farms). The value of the polynomial is .1 off from the bar graph.

95. 1959

$-.00102834793874(1959)^2$

$+ 3.9526021723669(1959)$

$- 3791.976211763$ *Let x = 1959*

≈ 4.7

For the year 1959, the value of the polynomial is approximately 4.7, while the bar graph shows 3.7, so the value of the polynomial is 1.0 off from the bar graph.

97.
$$\frac{-4x^7 - 14x^6 + 10x^4 - 14x^2}{-2x^2}$$

$$= \frac{-4x^7}{-2x^2} - \frac{14x^6}{-2x^2} + \frac{10x^4}{-2x^2} - \frac{14x^2}{-2x^2}$$

$$= 2x^5 + 7x^4 - 5x^2 + 7$$

99.
$$\frac{10x^8 - 16x^6 - 4x^4}{-2x^6}$$

$$= \frac{10x^8}{-2x^6} - \frac{16x^6}{-2x^6} - \frac{4x^4}{-2x^6}$$

$$= -5x^2 + 8 + \frac{2}{x^2}$$

101.
$$\frac{2x^3 + 6x^2 - 8x + 10}{2x - 1}$$

$$
\require{enclose}
\begin{array}{r}
x^2 + \frac{7}{2}x - \frac{9}{4} \\
2x - 1 \enclose{longdiv}{2x^3 + 6x^2 - 8x + 10} \\
\underline{2x^3 - x^2} \\
7x^2 - 8x \\
\underline{7x^2 - \frac{7}{2}x} \\
-\frac{9}{2}x + 10 \\
\underline{-\frac{9}{2}x + \frac{9}{4}} \\
\frac{31}{4}
\end{array}
$$

Thus,

$$\frac{2x^3 + 6x^2 - 8x + 10}{2x - 1}$$

$$= x^2 + \frac{7}{2}x - \frac{9}{4} + \frac{31/4}{2x - 1}.$$

103.
$$\frac{3x^4 + 2x^2 + 6x - 1}{3x^2 - x}$$

$$
\begin{array}{r}
x^2 + \frac{1}{3}x + \frac{7}{9} \\
3x^2 - x \enclose{longdiv}{3x^4 + 0x^3 + 2x^2 + 6x - 1} \\
\underline{3x^4 - x^3} \\
x^3 + 2x^2 \\
\underline{x^3 - \frac{1}{3}x^2} \\
\frac{7}{3}x^2 + 6x \\
\underline{\frac{7}{3}x^2 - \frac{7}{9}x} \\
\frac{61}{9}x - 1
\end{array}
$$

Thus,

$$\frac{3x^4 + 2x^2 + 6x - 1}{3x^2 - x}$$

$$= x^2 + \frac{1}{3}x + \frac{7}{9} + \frac{(61/9)x - 1}{3x^2 - x}.$$

Section 1.4

1. For the polynomial $4x^3y^5 - 8x^2y^4$, the greatest common factor is $4x^2y^4$, so the correct factorization is

$$4x^3y^5 - 8x^2y^4 = 4x^2y^4(xy - 2).$$

In the student's answer, the factor $2xy^2 - 4y$ contains the common factor $2y$. Since the polynomial was not *completely* factored, the teacher was justified in not giving full credit.

3. $4k^2m^3 + 8k^4m^3 - 12k^2m^4$

The greatest common factor is $4k^2m^3$.

$4k^2m^3 + 8k^4m^3 - 12k^2m^4$

$\quad = 4k^2m^3(1) + 4k^2m^3(2k^2)$

$\qquad + 4k^2m^3(-3m)$

$\quad = 4k^2m^3(1 + 2k^2 - 3m)$

5. $2(a + b) + 4m(a + b)$

$\quad = 2(a + b)(1 + 2m)$

$2(a + b)$ is the greatest common factor.

7. $(5r - 6)(r + 3) - (2r - 1)(r + 3)$

$\quad = (r + 3)[(5r - 6) - (2r - 1)]$

$\qquad$ *r + 3 is a common factor*

$\quad = (r + 3)[5r - 6 - 2r + 1]$

$\quad = (r + 3)(3r - 5)$

9. $2(m - 1) - 3(m - 1)^2 + 2(m - 1)^3$

$\quad = (m - 1)[2 - 3(m - 1) + 2(m - 1)^2]$

$\qquad$ *m - 1 is a common factor*

$\quad = (m - 1)[2 - 3m + 3]$

$\qquad + 2(m^2 - 2m + 1)$

$\quad = (m - 1)(2 - 3m + 3 + 2m^2 - 4m + 2)$

$\quad = (m - 1)(2m^2 - 7m + 7)$

11. $6st + 9t - 10s - 15$

$\quad = (6st + 9t) + (-10s - 15)$

$\qquad$ *Group the terms*

$\quad = 3t(2s + 3) - 5(2s + 3)$

$\qquad$ *Factor each group*

$\quad = (2s + 3)(3t - 5)$

$\qquad$ *Factor out 2s + 3*

13. $2m^4 + 6 - am^4 - 3a$

$\quad = (2m^4 + 6) + (-am^4 - 3a)$

$\qquad$ *Group the terms*

$\quad = 2(m^4 + 3) - a(m^4 + 3)$

$\qquad$ *Factor each group*

$\quad = (m^4 + 3)(2 - a)$

$\qquad$ *Factor out $m^4 + 3$*

15. $20z^2 + 18z^2 - 8zx - 45zx$

Rearrange the terms in order to factor by grouping.

$20z^2 - 8zx - 45zx + 18x^2$

$\quad = (20z^2 - 8zx) + (-45zx + 18x^2)$

$\qquad$ *Group the terms*

$\quad = 4z(5z - 2x) - 9x(5z - 2x)$

$\qquad$ *Factor each group*

$\quad = (5z - 2x)(4z - 9x)$

$\qquad$ *Factor out 5z - 2x*

17. $6a^2 - 48a - 120 = 6(a^2 - 8a - 20)$

$\qquad$ *Factor out the greatest common factor, 6*

To factor $a^2 - 8a - 20$, we look for two numbers whose sum is -8 and whose product is -20. 2 and -10 are such numbers.

Thus,

$\quad a^2 - 48a - 120 = 6(a^2 - 8a - 20)$

$\qquad\qquad = 6(a - 10)(a + 2).$

19. $3m^3 + 12m^2 + 9m$

$= 3m(m^2 + 4m + 3)$
*Factor out the greatest
common factor, 3m*

To factor $m^2 + 4m + 3$, look for two numbers whose sum is 4 and whose product is 3. 1 and 3 are such numbers.

Thus,

$3m(m^2 + 4m + 3) = 3m(m^2 + 4m + 3)$
$= 3m(m + 1)(m + 3)$.

21. $6k^2 + 5kp - 6p^2$

The positive factors of 6 could be 2 and 3, or 1 and 6. As factors of -6, we could have -1 and 6, -6 and 1, -2 and 3, or -3 and 2. Try different combinations of these factors until the correct one is found.

$6k^2 + 5kp - 6p^2$
$= (2k + 3p)(3k - 2p)$

23. $5a^2 - 7ab - 6b^2$

The positive factors of 5 can only be 1 and 5. As factors of -6, we could have -1 and 6, -6 and 1, -2 and 3, or -3 and 2. Try different combinations of these factors until the correct one is found.

$5a^2 - 7ab - b^2 = (5a + 3b)(a - 2b)$

25. $9x^2 - 6x^3 + x^4$

$= x^2(9 - 6x + x^2)$
x^2 is greatest common factor
$= x^2(3 - x)^2$
Perfect square trinomial

27. $24a^4 + 10a^3b - 4a^2b^2$

First, factor out the greatest common factor, $2a^2$.

$24a^4 + 10a^3b - 4a^2b^2$

$= 2a^2(12a^2 + 5ab - 2b^2)$

Now factor the trinomial by trial and error.

$12a^2 + 5ab - 2b^2 = (4a - b)(3a + 2b)$

Thus,

$24a^4 + 10a^3b - 4a^2b^2$

$= 2a^2(12a^2 + 5ab - 2b^2)$

$= 2a^2(4a - b)(3a + 2b)$.

29. $9m^2 - 12m + 4$

$= (3m)^2 - 12m + 2^2$

$= (3m)^2 - 2(3m)(2) + 2^2$
Perfect square trinomial

$= (3m - 2)^2$

31. $32a^2 - 48ab + 18b^2$

$= 2(16a^2 - 24ab + 9b^2)$
2 is greatest common factor

$= 2[(4a)^2 - 24ab + (3b)^2]$

$= 2[(4a)^2 - 2(4a)(3b) + (3b)^2]$
Perfect square trinomial

$= 2(4a - 3b)^2$

33. $4x^2y^2 + 28xy + 49$

$= (2xy)^2 + 28xy + 7^2$

$= (2xy)^2 + 2(2xy)(7) + 7^2$
Perfect square trinomial

$= (2xy + 7)^2$

35. $(a - 3b)^2 - 6(a - 3b) + 9$

Let $x = a - 3b$.

Then

$(a - 3b)^2 - 6(a - 3b) + 9$

$= x^2 - 6x + 9$

$= x^2 - 2(x)(3) + 3^2$

         Perfect square trinomial

$= (x - 3)^2$.

Replacing x with a - 3b gives

$(a - 3b)^2 - 6(a - 3b) + 9$

    $= (a - 3b - 3)^2$.

37. $9a^2 - 16 = (3a)^2 - 4^2$

         Difference of two squares

        $= (3a + 4)(3a - 4)$

39. $25s^4 - 9t^2 = (5s^2)^2 - (3t)^2$

         Difference of two squares

        $= (5s^2 + 3t)(5s^2 - 3t)$

41. $(a + b)^2 - 16$

    $= (a + b)^2 - 4^2$

         Difference of two squares

    $= [(a + b) + 4][(a + b) - 4]$

    $= (a + b + 4)(a + b - 4)$

43. $p^4 - 625$

    $= (p^2)^2 - 25^2$

         Difference of two squares

    $= (p^2 + 25)(p^2 - 25)$

    $= (p^2 + 25)(p^2 - 5^2)$

         Difference of two squares

    $= (p^2 + 25)(p + 5)(p - 5)$

45. The correct complete factorization of $x^4 - 1$ is (b): $(x^2 + 1)(x + 1)$ $\cdot (x - 1)$. Choice (a) is not a complete factorization, since $x^2 - 1$ can be factored as $(x + 1)(x - 1)$.

The other choices are not correct factorizations of $x^4 - 1$.

47. $8 - a^3$

    $= 2^3 - a^3$ *Difference of two cubes*

    $= (2 - a)(a^2 + 2 \cdot a + a^2)$

    $= (2 - a)(4 + 2a + a^2)$

49. $125x^3 - 27$

    $= (5x)^3 - 3^3$ *Difference of two cubes*

    $= (5x - 3)[(5x)^2 + 5x \cdot 3 + 3^2]$

    $= (5x - 3)(25x^2 + 15x + 9)$

51. $27y^9 + 125z^6$

    $= (3y^3)^3 + (5z^2)^3$ *Sum of two cubes*

    $= (3y^3 + 5z^2)$

      $\cdot [(3y^3)^2 - (3y^3)(5z^2) + (5z^2)^2]$

    $= (3y^3 + 5z^2)(9y^6 - 15y^3z^2 + 25z^4)$

53. $(r + 6)^3 - 216$

Let $x = r + 6$.   Then

$(r + 6)^3 - 216$

    $= x^3 - 216$

    $= x^3 - 6^3$ *Difference of two cubes*

    $= (x - 6)(x^2 + 6x + 6^2)$

    $= (x - 6)(x^2 + 6x + 36)$.

Replacing x with (r + 6) gives

$(r + 6)^3 - 216$

    $= ((r + 6) - 6)$

      $\cdot [(r + 6)^2 + 6(r + 6) + 36)]$

    $= r(r^2 + 12r + 36 + 6r + 36 + 36)$

    $= r(r^2 + 18r + 108)$.

55. $27 - (m + 2n)^3$

Let $x = m + 2n$.   Then

$27 - (m + 2n)^3$

$= 27 - x^3$

$= 3^3 - x^3$ *Difference of two cubes*

$= (3 - x)(3^2 + 3x + x^2)$

$= (3 - x)(9 + 3x + x^2)$.

Replacing x with $m + 2n$ gives

$27 - (m + 2n)^3$

$= [3 - (m + 2n)]$

$\quad \cdot [9 + 3(m + 2n) + (m + 2n)^2]$

$= (3 - m - 2n)$

$\quad \cdot (9 + 3m + 6n + m^2 + 4mn + 4n^2)$.

57. $x^6 - 1 = (x^3)^2 - 1^2$ *Difference of two squares*

$\qquad = (x^3 + 1)(x^3 - 1)$

$\qquad$ or $(x^3 - 1)(x^3 + 1)$

Use the patterns for the difference of two cubes and sum of two cubes to factor further.   Since

$x^3 - 1 = (x - 1)(x^2 + x + 1)$

and

$x^3 + 1 = (x + 1)(x^2 - x + 1)$,

we obtain the factorization

$x^6 - 1 = (x^3 - 1)(x^3 + 1)$

$\qquad = (x - 1)(x^2 + x + 1)(x + 1)$

$\qquad \quad \cdot (x^2 - x + 1)$.

58. $x^6 - 1 = (x^2)^3 - 1^3$ *Difference of two cubes*

$\qquad = (x^2 - 1)[(x^2)^2 + x \cdot 1 + 1^2]$

$\qquad = (x^2 - 1)(x^4 + x + 1)$

$\qquad = (x - 1)(x + 1)(x^4 + x + 1)$

59. From Exercise 57, we have

$x^6 - 1 = (x - 1)(x^2 + x + 1)$

$\qquad \qquad \cdot (x + 1)(x^2 - x + 1)$.

From Exercise 58, we have

$x^6 - 1 = (x - 1)(x + 1)(x^4 + x^2 + 1)$.

Comparing these answers, we see that

$x^4 + x^2 + 1 = (x^2 - x + 1)(x^2 + x + 1)$.

60. Note: The reason for each step is given following that step.

$x^4 + x^2 + 1 = x^4 + 2x^2 + 1 - x^2$

Additive inverse property

(0 in the form $x^2 - x^2$)

$\qquad = (x^4 + 2x^2 + 1) - x^2$

Associative property of addition

$\qquad = (x^2 + 1)^2 - x^2$

Factoring a perfect square trinomial

$\qquad = (x^2 + 1 - x)(x^2 + 1 + x)$

Factoring the difference of two squares

$\qquad = (x^2 - x + 1)(x^2 + x + 1)$

Commutative property of addition

61. The answer in Exercise 59 and the final line in Exercise 60 are the same.

62. $x^8 + x^4 + 1$

$\qquad = x^8 + 2x^4 + 1 - x^4$

$\qquad = (x^8 + 2x^4 + 1) - x^4$

$\qquad = (x^4 + 1)^2 - (x^2)^2$

$\qquad \qquad$ *Difference of two squares*

$\qquad = (x^4 + 1 - x^2)(x^4 + 1 + x^2)$

$= (x^4 - x^2 + 1)(x^4 + x^2 + 1)$

$= (x^4 - x^2 + 1)(x^2 + x + 1)$

$\quad \cdot (x^2 - x + 1)$
Use result from Exercise 60

63. $m^4 - 3m^2 - 10$

Let $x = m^2$.

Substituting x for m^2, we have

$\quad x^2 - 3x - 10.$

Factor this trinomial as

$\quad x^2 - 3x - 10 = (x - 5)(x + 2).$

Replacing x with m^2 gives

$\quad m^4 - 3m^2 - 10 = (m^2 - 5)(m^2 + 2).$

65. $7(3k - 1)^2 + 26(3k - 1) - 8$

Let $x = 3k - 1$.   This substitution gives

$7(3k - 1)^2 + 26(3k - 1) - 8$

$\quad = 7x^2 + 26x - 8$

$\quad = (7x - 2)(x + 4).$

Replacing x with $3k - 1$ gives

$7(3k - 1)^2 + 26(3k - 1) - 8$

$\quad = [7(3k - 1) - 2][(3k - 1) + 4]$

$\quad = (21k - 7 - 2)(3k - 1 + 4)$

$\quad = (21k - 9)(3k + 3)$

$\quad = 3(7k - 3)(3)(k + 1)$

$\quad = 9(7k - 3)(k + 1).$

67. $9(a - 4)^2 + 30(a - 4) + 25$

Let $x = a - 4$.   With this substitution, we have

$9(a - 4)^2 + 30(a - 4) + 25$

$\quad = 9x^2 + 30x + 25$

$\quad = (3x + 5)^2.$ *Perfect square trinomial*

Replacing x by $a - 4$ gives

$[3(a - 4) + 5]^2$

$\quad = (3a - 12 + 5)^2$

$\quad = (3a - 7)^2.$

69. $4b^2 + 4bc + c^2 - 16$

$\quad = (4b^2 + 4bc + c^2) - 16$

$\quad = (2b + c)^2 - 4^2$ *Difference of two squares*

$\quad = [(2b + c) + 4][(2b + c) - 4]$

$\quad = (2b + c + 4)(2b + c - 4)$

71. $x^2 + xy - 5x - 5y$

$\quad = (x^2 + xy) + (-5x - 5y)$
Group the terms

$\quad = x(x + y) - 5(x + y)$
Factor each group

$\quad = (x + y)(x - 5)$ *Factor out $x + y$*

73. $p^4(m - 2n) + q(m - 2n)$

$\quad = (m - 2n)(p^4 + q)$ *Factor out $m - 2n$*

75. $4z^2 + 28z + 49$

$\quad = (2z)^2 + 2(2z)(7) + 7^2$
Perfect square trinomial

$\quad = (2z + 7)^2$

77. $1000x^3 + 343y^3$

$\quad = (10x)^3 + (7y)^3$ *Sum of two cubes*

$\quad = (10x + 7y)$

$\quad\quad \cdot [(10x^2)^2 - (10x)(7y) + (7y)^2]$

$\quad = (10x + 7y)(100x^2 - 70xy + 49y^2)$

79. $125m^6 - 216$

$\quad = (5m^2)^3 - 6^3$ *Difference of two cubes*

$\quad = (5m^2 - 6)[(5m^2)^2 + 5m^2 \cdot 6 + 6^2]$

$\quad = (5m^2 - 6)(25m^4 + 30m^2 + 36)$

81. $12m^2 + 16mn - 35n^2$

Try different combinations of the factors of 12 and -35 until the correct one is found.

$12m^2 + 16mn - 35n^2$

$= (6m - 7n)(2m + 5n)$

83. $4p^2 + 3p - 1$

The positive factors of 4 could be 2 and 2 or 1 and 4. The factors of -1 can only be 1 and -1. Try different combinations of these factors until the correct one is found.

$4p^2 + 3p - 1 = (4p - 1)(p + 1)$

85. $144z^2 + 121$

The sum of two squares cannot be factored. $144z^2 + 121$ is prime.

87. $(x + y)^2 - (x - y)^2$

Factor this expression as the difference of two squares.

$= [(x + y) - (x - y)]$

$\quad \cdot [(x + y) + (x - y)]$

$= (x + y - x + y)(x + y + x - y)$

$= (2y)(2x)$

$= 4xy$

89. $r^2 + rs^q - 6s^{2q} = (r + 3s^q)(r - 2s^q)$

91. $9a^{4k} - b^{8k}$

$= (3a^{2k})^2 - (b^{4k})^2$ *Difference of two squares*

$= (3a^{2k} + b^{4k})(3a^{2k} - b^{4k})$

93. $4y^{2a} - 12y^a + 9$

$= (2y^a)^2 - 12y^a + 3^2$

$= (2y^a)^2 - 2(2y^a)(3) + 3^2$
Perfect square trinomial

$= (2y^a - 3)^2$

97. $4z^2 + bz + 81 = (2z)^2 + bz + 9^2$

will be a perfect trinomial if

$bz = \pm 2(2z)9$

$bz = \pm 36z$

$b = \pm 36.$

If $b = 36$,

$4z^2 + 36z + 81 = (2z + 9)^2.$

If $b = -36$,

$4z^2 - 36z + 81 = (2z - 9)^2.$

99. $100r^2 - 60r + c$

The perfect square form is

$(10x)^2 - \underbrace{2(10r)(3)}_{-60r} + 3^2.$

Therefore, $c = 9$.

Section 1.5

1. Give restrictions on the variable in the expression

$$\frac{x - 2}{x + 6}.$$

Replacing x with -6 makes the denominator equal 0. Therefore, the restriction for this expression is $x \neq -6$.

3. Give restrictions on the variable in the expression

$$\frac{2x}{5x - 3}.$$

Replacing x with 3/5 makes the denominator equal 0. This can be found by solving the equation $5x - 3 = 0$. Therefore, for this expression the restriction is $x \neq \frac{3}{5}$.

5. Give restrictions on the variable in the expression

$$\frac{-8}{x^2 + 1}.$$

We must exclude any real number x which makes the denominator equal 0. However, the equation $x^2 + 1 = 0$ has no real solution. Therefore, there are no restrictions on the variable for this expression.

7. Write the given expression in lowest terms.

$$\frac{x^2 + 4x + 3}{x + 1} \quad (x \neq -1)$$

$$= \frac{(x + 1)(x + 3)}{x + 1} \quad \textit{Factor the numerator}$$

$$= x + 3 \quad \textit{Fundamental principle}$$

Therefore, the expression which is equivalent to $\frac{x^2 + 4x + 3}{x + 1}$ is

(a): $x + 3$.

9. The calculator-generated table gives an error when $x = -3$. This means that the expression is undefined

when $x = -3$. This will happen if the denominator of the rational expression is 0.

$$x - k = 0$$

$$-3 - k = 0 \quad \textit{Let } x = -3$$

$$-3 = k$$

Thus, $k = -3$, and the rational expression is

$$\frac{x + 5}{x - (-3)} \quad \text{or} \quad \frac{x + 5}{x + 3}.$$

11. $\dfrac{8k + 16}{9k + 18} = \dfrac{8(k + 2)}{9(k + 2)}$ *Factor numerator and denominator*

$$= \frac{8}{9} \quad \textit{Fundamental principle}$$

13. $\dfrac{3(t + 5)}{(t + 5)(t - 3)} = \dfrac{3}{t - 3}$ *Fundamental principle*

15. $\dfrac{8x^2 + 16x}{4x^2}$

$$= \frac{8x(x + 2)}{4x^2} \quad \textit{Factor}$$

$$= \frac{2 \cdot 4x(x + 2)}{x \cdot 4x} \quad \textit{Factor}$$

$$= \frac{2(x + 2)}{x} \quad \textit{Fundamental principle}$$

or $\dfrac{2x + 4}{x}$

17. $\dfrac{m^2 - 4m + 4}{m^2 + m - 6}$

$$= \frac{(m - 2)(m - 2)}{(m - 2)(m + 3)} \quad \textit{Factor}$$

$$= \frac{m - 2}{m + 3} \quad \textit{Use the fundamental principle to write the expression in lowest terms}$$

19. $\dfrac{8m^2 + 6m - 9}{16m^2 - 9}$

$= \dfrac{(2m + 3)(4m - 3)}{(4m + 3)(4m - 3)}$ *Factor*

$= \dfrac{2m + 3}{4m + 3}$ *Use the fundamental principle to write the expression in lowest terms*

21. $\dfrac{15p^3}{9p^2} \div \dfrac{6p}{10p^2}$

$= \dfrac{15p^3}{9p^2} \cdot \dfrac{10p^2}{6p}$ *Definition of division*

$= \dfrac{150p^5}{54p^3}$ *Multiply*

$= \dfrac{25 \cdot 6p^5}{9 \cdot 6p^3}$ *Factor*

$= \dfrac{25p^2}{9}$ *Fundamental principle*

23. $\dfrac{2k + 8}{6} \div \dfrac{3k + 12}{2}$

$= \dfrac{2k + 8}{6} \cdot \dfrac{2}{3k + 12}$ *Definition of division*

$= \dfrac{2(k + 4)^2}{6(3)(k + 4)}$ *Multiply and factor*

$= \dfrac{4}{18} = \dfrac{2}{9}$ *Fundamental principle*

25. $\dfrac{x^2 + x}{5} \cdot \dfrac{25}{xy + y}$

$= \dfrac{x(x + 1)}{5} \cdot \dfrac{25}{y(x + 1)}$ *Factor*

$= \dfrac{25x(x + 1)}{5y(x + 1)}$ *Multiply*

$= \dfrac{5x}{y}$ *Fundamental principle*

27. $\dfrac{4a + 12}{2a - 10} \div \dfrac{a^2 - 9}{a^2 - a - 20}$

$= \dfrac{4a + 12}{2a - 10} \cdot \dfrac{a^2 - a - 20}{a^2 - 9}$ *Definition of division*

$= \dfrac{4(a + 3)(a - 5)(a + 4)}{2(a - 5)(a + 3)(a - 3)}$ *Multiply and factor*

$= \dfrac{2(a + 4)}{(a - 3)}$ *Fundamental principle*

29. $\dfrac{p^2 - p - 12}{p^2 - 2p - 15} \cdot \dfrac{p^2 - 9p + 20}{p^2 - 8p + 16}$

$= \dfrac{(p - 4)(p + 3)(p - 5)(p - 4)}{(p - 5)(p + 3)(p - 4)(p - 4)}$ *Multiply and factor*

$= 1$ *Fundamental principle*

31. $\dfrac{m^2 + 3m + 2}{m^2 + 5m + 4} \div \dfrac{m^2 + 5m + 6}{m^2 + 10m + 24}$

$= \dfrac{m^2 + 3m + 2}{m^2 + 5m + 4} \cdot \dfrac{m^2 + 10m + 24}{m^2 + 5m + 6}$ *Definition of division*

$= \dfrac{(m + 2)(m + 1)(m + 6)(m + 4)}{(m + 4)(m + 1)(m + 3)(m + 2)}$ *Multiply and factor*

$= \dfrac{m + 6}{m + 3}$ *Fundamental principle*

33. $\dfrac{2m^2 - 5m - 12}{m^2 - 10m + 24} \div \dfrac{4m^2 - 9}{m^2 - 9m + 18}$

$= \dfrac{2m^2 - 5m - 12}{m^2 - 10m + 24} \cdot \dfrac{m^2 - 9m + 18}{4m^2 - 9}$ *Definition of division*

$= \dfrac{(2m + 3)(m - 4)(m - 6)(m - 3)}{(m - 6)(m - 4)(2m + 3)(2m - 3)}$ *Multiply and factor*

$= \dfrac{m - 3}{2m - 3}$ *Fundamental principle*

35. $\dfrac{x^3 + y^3}{x^3 - y^3} \cdot \dfrac{x^2 - y^2}{x^2 + 2xy + y^2}$

$= \dfrac{(x + y)(x^2 - xy + y^2)(x - y)(x + y)}{(x - y)(x^2 - xy + y^2)(x + y)(x + y)}$ *Factor, using sum of two cubes and difference of two squares*

$= \dfrac{x^2 - xy + y^2}{x^2 + xy + y^2}$ *Fundamental principle*

37. Expressions (b) and (c) are both equal to −1, since the numerator and denominator are additive inverses.

(b) $\dfrac{-x - 4}{x + 4} = \dfrac{-1(x + 4)}{x + 4} = -1$

(c) $\dfrac{x - 4}{4 - x} = \dfrac{-1(4 - x)}{4 - x} = -1$

39. $\dfrac{3}{2k} + \dfrac{5}{3k} = \dfrac{3 \cdot 3}{2k \cdot 3} + \dfrac{5 \cdot 2}{3k \cdot 2}$ *Fundamental principle*

$= \dfrac{9}{6k} + \dfrac{10}{6k}$ *Common denominator*

$= \dfrac{19}{6k}$ *Add numerators*

41. $\dfrac{a + 1}{2} - \dfrac{a - 1}{2}$

$= \dfrac{(a + 1) - (a - 1)}{2}$ *Subtract numerators*

$= \dfrac{a + 1 - a + 1}{2}$ *Remove parentheses*

$= \dfrac{2}{2} = 1$

43. $\dfrac{3}{p} + \dfrac{1}{2} = \dfrac{3 \cdot 2}{p \cdot 2} + \dfrac{1 \cdot p}{2 \cdot p}$ *Fundamental principle*

$= \dfrac{6}{2p} + \dfrac{p}{2p}$ *Common denominator*

$= \dfrac{6 + p}{2p}$ *Add numerators*

45. $\dfrac{1}{6m} + \dfrac{2}{5m} + \dfrac{4}{m}$

$= \dfrac{1 \cdot 5}{6m \cdot 5} + \dfrac{2 \cdot 6}{5m \cdot 6} + \dfrac{4 \cdot 6 \cdot 5}{m \cdot 6 \cdot 5}$

  Fundamental principle

$= \dfrac{5}{30m} + \dfrac{12}{30m} + \dfrac{120}{30m}$ *Common denominator*

$= \dfrac{137}{30m}$ *Add numerators*

47. $\dfrac{1}{a} - \dfrac{b}{a^2} = \dfrac{1 \cdot a}{a \cdot a} - \dfrac{b}{a^2}$ *Fundamental principle*

$= \dfrac{a}{a^2} - \dfrac{b}{a^2}$ *Common denominators*

$= \dfrac{a - b}{a^2}$ *Subtract denominators*

49. $\dfrac{1}{x + z} + \dfrac{1}{x - z}$

$= \dfrac{1(x - z)}{(x + z)(x - z)} + \dfrac{1(x + z)}{(x - z)(x + z)}$

  Fundamental principle

$= \dfrac{x - z}{(x - z)(x + z)} + \dfrac{x + z}{(x + z)(x - z)}$

  Common denominator

$= \dfrac{2x}{(x + z)(x - z)}$

  Add numerators

51. $\dfrac{3}{a - 2} - \dfrac{1}{2 - a}$

$= \dfrac{3}{a - 2} - \dfrac{1(-1)}{(2 - a)(-1)}$

        $a - 2 = (-1)(2 - a)$

$= \dfrac{3}{a - 2} - \dfrac{-1}{a - 2}$

$= \dfrac{3 + 1}{a - 2} = \dfrac{4}{a - 2}$

We may also use $2 - a$ as the common denominator.

$\dfrac{3(-1)}{(a - 2)(-1)} - \dfrac{1}{2 - a}$

$= \dfrac{-3}{2 - a} - \dfrac{1}{2 - a}$

$= \dfrac{-4}{2 - a}$

The two results, $\dfrac{4}{a - 2}$ and $\dfrac{-4}{2 - a}$, are equivalent rational expressions.

53. $\dfrac{x + y}{2x - y} - \dfrac{2x}{y - 2x}$

$= \dfrac{x + y}{2x - y} - \dfrac{2x(-1)}{(y - 2x)(-1)}$

$\qquad\qquad 2x - y = (-1)(y - 2x)$

$= \dfrac{x + y}{2x - y} - \dfrac{-2x}{2x - y}$

$= \dfrac{x + y + 2x}{2x - y} = \dfrac{3x + y}{2x - y}$

We may also use $y - 2x$ as the common denominator. In this case, our result will be

$$\dfrac{-3x - y}{y - 2x}.$$

The two results are equivalent rational expressions.

55. $\dfrac{1}{x^2 + x - 12} - \dfrac{1}{x^2 - 7x + 12} + \dfrac{1}{x^2 - 16}$

$= \dfrac{1}{(x + 4)(x - 3)} - \dfrac{1}{(x - 4)(x - 3)}$

$\quad + \dfrac{1}{(x - 4)(x + 4)}$ *Factor denominators*

The least common denominator is $(x + 4)(x - 3)(x - 4)$.

$\dfrac{1(x - 4)}{(x + 4)(x - 3)(x - 4)} - \dfrac{1(x + 4)}{(x - 4)(x - 3)(x + 4)}$

$+ \dfrac{1(x - 3)}{(x - 4)(x + 4)(x - 3)}$ *Fundamental principle*

$= \dfrac{(x - 4) - (x + 4) + (x - 3)}{(x - 4)(x + 4)(x - 3)}$

$\qquad\qquad$ *Subtract and add numerators*

$= \dfrac{x - 4 - x - 4 + x - 3}{(x - 4)(x + 4)(x - 3)}$ *Remove parentheses*

$= \dfrac{x - 11}{(x - 4)(x + 4)(x - 3)}$

57. $y = \dfrac{6.7x}{100 - x}$

(a) $x = 75$ (75%)

$y = \dfrac{6.7(75)}{100 - 75}$

$= \dfrac{502.5}{25}$

$= 20.1$

The cost of removing 75% of the pollutant is 20.1 thousand dollars.

(b) $x = 95$ (95%)

$y = \dfrac{6.7(95)}{100 - 95}$

$= \dfrac{636.5}{5}$

$= 127.3$

The cost of removing 95% of the pollutant is 127.3 thousand dollars.

(c) $x = 98.5$ (98.5%)

$y = \dfrac{6.7(98.5)}{100 - 98.5}$

$= \dfrac{659.95}{1.5}$

≈ 439.97

The cost of removing 98.5% of the pollutant is 439.97 thousand dollars.

59. $y = \dfrac{80x - 8000}{x - 110}$

(a) 55% $(x = 55)$

$y = \dfrac{80(55) - 8000}{55 - 110}$

$= \dfrac{-3600}{-55}$

≈ 65.5

For a tax rate of 55%, the revenue is \$65.5 tens of millions, or \$655,000,000.

(b) 60% (x = 60)

$$y = \frac{80(60) - 8000}{60 - 110}$$

$$= \frac{-3200}{-50}$$

$$= 64$$

For a tax rate of 60%, the revenue is \$64 tens of millions, or \$640,000,000.

(c) 70% (x = 70)

$$y = \frac{80(70) - 8000}{70 - 110}$$

$$= \frac{-2400}{-40}$$

$$= 60$$

For a tax rate of 70%, the revenue is \$60 tens of millions, or \$600,000,000.

(d) 90% (x = 90)

$$y = \frac{80(90) - 8000}{90 - 110}$$

$$= \frac{-800}{-20}$$

$$= 40$$

For a tax rate of 90%, the revenue is \$40 tens of millions, or \$400,000,000.

(e) 100% (x = 100)

$$y = \frac{80(100) - 8000}{100 - 110}$$

$$= \frac{0}{-10} = 0$$

For a tax rate of 100%, the revenue is \$0.

61. $\dfrac{1 + \dfrac{1}{x}}{1 - \dfrac{1}{x}}$

Multiply both numerator and denominator by the least common denominator of all the fractions, x.

$$\frac{1 + \frac{1}{x}}{1 - \frac{1}{x}} = \frac{x\left(1 + \frac{1}{x}\right)}{x\left(1 - \frac{1}{x}\right)}$$

$$= \frac{x \cdot 1 + x\left(\frac{1}{x}\right)}{x \cdot 1 - x\left(\frac{1}{x}\right)}$$

Distributive property

$$= \frac{x + 1}{x - 1}$$

63. $\dfrac{\dfrac{1}{x + 1} - \dfrac{1}{x}}{\dfrac{1}{x}}$

Multiply both numerator and denominator by the least common denominator of all the fractions, x(x + 1).

$$\frac{\frac{1}{x + 1} - \frac{1}{x}}{\frac{1}{x}}$$

$$= \frac{x(x + 1)\left(\frac{1}{x + 1} - \frac{1}{x}\right)}{x(x + 1)\left(\frac{1}{x}\right)}$$

$$= \frac{x(x + 1)\left(\frac{1}{x + 1}\right) - x(x + 1)\left(\frac{1}{x}\right)}{x(x + 1)\left(\frac{1}{x}\right)}$$

Distributive property

$$= \frac{x - (x + 1)}{x + 1}$$

$$= \frac{x - x - 1}{x + 1}$$

$$= \frac{-1}{x + 1}$$

65. $\dfrac{1 + \dfrac{1}{1 - b}}{1 - \dfrac{1}{1 + b}}$

Multiply both numerator and denominator by the least common denominator of all the fractions, $(1 - b)(1 + b)$.

$\dfrac{1 + \dfrac{1}{1 - b}}{1 - \dfrac{1}{1 + b}}$

$= \dfrac{(1 - b)(1 + b)\left(1 + \dfrac{1}{1 - b}\right)}{(1 - b)(1 + b)\left(1 - \dfrac{1}{1 + b}\right)}$

$= \dfrac{(1 - b)(1 + b) + (1 + b)}{(1 - b)(1 + b) - (1 - b)}$

$= \dfrac{(1 + b)[(1 - b) + 1]}{(1 - b)[(1 + b) - 1]}$

Factor out common factors in numerator and denominator

$= \dfrac{(1 + b)(2 - b)}{(1 - b)b}$ or $\dfrac{(2 - b)(1 + b)}{b(1 - b)}$

Remove parentheses and simplify

67. $\dfrac{m - \dfrac{1}{m^2 - 4}}{\dfrac{1}{m + 2}} = \dfrac{m - \dfrac{1}{(m + 2)(m - 2)}}{\dfrac{1}{m + 2}}$

Multiply both numerator and denominator by the least common denominator of all the fractions, $(m + 2)(m - 2)$.

$\dfrac{m - \dfrac{1}{m^2 - 4}}{\dfrac{1}{m + 2}}$

$= \dfrac{(m + 1)(m - 2)\left(m - \dfrac{1}{(m + 2)(m - 2)}\right)}{(m + 1)(m - 2)\left(\dfrac{1}{m + 2}\right)}$

$= \dfrac{(m+1)(m-2)(m) - (m+1)(m-2)\left(\dfrac{1}{(m+2)(m-2)}\right)}{(m+2)(m-2)\left(\dfrac{1}{m+2}\right)}$

Distributive property

$= \dfrac{m(m^2 - 4) - 1}{m - 2}$

$= \dfrac{m^3 - 4m - 1}{m - 2}$

69. $\left(\dfrac{3}{p - 1} - \dfrac{2}{p + 1}\right)\left(\dfrac{p - 1}{p}\right)$

$= \left[\dfrac{3(p + 1)}{(p - 1)(p + 1)} - \dfrac{2(p - 1)}{(p + 1)(p - 1)}\right]$

$\cdot \left(\dfrac{p - 1}{p}\right)$

$= \left[\dfrac{3(p + 1) - 2(p - 1)}{(p - 1)(p + 1)}\right]\left(\dfrac{p - 1}{p}\right)$

$= \dfrac{(3p + 3 - 2p + 2)}{(p + 1)(p - 1)}\left(\dfrac{p - 1}{p}\right)$

$= \dfrac{(p - 5)}{(p + 1)(p - 1)} \cdot \dfrac{(p - 1)}{p}$

$= \dfrac{p + 5}{(p + 1)p}$ or $\dfrac{p + 5}{p(p + 1)}$

71. $\dfrac{\dfrac{1}{x + h} - \dfrac{1}{x}}{h}$

To simplify this complex fraction, multiply both numerator and denominator by the least common denominator of all the fractions, $x(x + h)$.

$\dfrac{\dfrac{1}{x + h} - \dfrac{1}{x}}{h}$

$= \dfrac{x(x + h)\left(\dfrac{1}{x + h} - \dfrac{1}{x}\right)}{x(x + h)(h)}$

$= \dfrac{x(x + h)\left(\dfrac{1}{x + h}\right) - x(x + h)\left(\dfrac{1}{x}\right)}{x(x + h)(h)}$

Distributive property

$$= \frac{x - (x + h)}{xh(x + h)}$$

$$= \frac{-h}{xh(x + h)}$$

$$= \frac{-1}{x(x + h)} \quad \textit{Fundamental principle}$$

73. Since one cubic centimeter of aluminum has a mass of 2.7 grams, the sheet of aluminum must contain

$$\frac{47.25}{2.7} = 17.5 \text{ cm}^3.$$

The sheet has area of $30 \times 30 = 900$ cm². Using the hint, the thickness is equal to

$$\frac{17.5}{900} \approx .02 \text{ cm.}$$

Section 1.6

1. $(-4)^{-3} = \dfrac{1}{(-4)^3}$

$$= \frac{1}{(-4)(-4)(-4)}$$

$$= \frac{1}{-64}$$

$$= -\frac{1}{64}$$

3. $\left(\dfrac{1}{2}\right)^{-3} = \dfrac{1}{\left(\frac{1}{2}\right)^3}$

$$= \frac{1}{\frac{1}{8}}$$

$$= \frac{8}{1} = 8$$

5. $-4^{1/2}$

$4^{1/2} = 2$ because $2^2 = 4$.
Thus,

$$-4^{1/2} = -2.$$

7. $8^{2/3} = (8^{1/3})^2 = 2^2 = 4$

This expression can also be evaluated as

$$8^{2/3} = (8^2)^{1/3} = 64^{1/3} = 4.$$

9. $27^{-2/3} = \dfrac{1}{(27)^{2/3}}$

$$= \frac{1}{(27^{1/3})^2}$$

$$= \frac{1}{3^2} = \frac{1}{9}$$

11. $\left(-\dfrac{4}{9}\right)^{-3/2}$ is not a real number (undefined) because the base, $-4/9$, is negative and the exponent, $-3/2$, has an even denominator.

13. $\left(\dfrac{27}{64}\right)^{-4/3} = \left[\left(\dfrac{27}{64}\right)^{1/3}\right]^{-4}$

$$= \left(\frac{27^{1/3}}{64^{1/3}}\right)^{-4}$$

$$= \left(\frac{3}{4}\right)^{-4} = \frac{1}{\left(\frac{3}{4}\right)^4}$$

$$= \frac{1}{\frac{81}{256}} = \frac{256}{81}$$

15. $64^{3/2}$

Enter the following into the calculator.

$$64 \wedge (3/2)$$

$$\blacktriangleright \text{Frac}$$

The calculator will return the value of $64^{3/2}$, which is 512.

17. $8^{-5/3}$

Enter the following into the calculator.

$$8 \wedge (-5/3)$$
$$\blacktriangleright \text{Frac}$$

The calculator will return the value of $8^{-5/3}$, which is $1/32$.

19. $\left(\frac{8}{27}\right)^{-5/3}$

Enter the following into the calculator.

$$(8/27) \wedge (-5/3)$$
$$\blacktriangleright \text{Frac}$$

The calculator will return the value of $(8/27)^{-5/3}$, which is $243/32$.

21. $100^{-2.5}$

Enter the following into the calculator.

$$100 \wedge (-2.5)$$
$$\blacktriangleright \text{Frac}$$

or

$$100 \wedge (-5/2)$$
$$\blacktriangleright \text{Frac}$$

The calculator will return the value of $100^{-2.5}$, which is $1/100{,}000$.

25. $(2x^{-3/2})^2 = 2^2(x^{-3/2})^2$

$\qquad = 2^2 \cdot x^{-3}$

$\qquad = \dfrac{2^2}{x^3}$

Therefore, expression (d) is equivalent to $(2x^{-3/2})^2$.

27. $27^{-2} \cdot 27^{-1} = 27^{-3}$ *Product rule*

$\qquad = \dfrac{1}{27^3}$ *Definition of negative exponent*

29. $\dfrac{4^{-2} \cdot 4^{-1}}{4^{-3}} = \dfrac{4^{-3}}{4^{-3}} = 1$

31. $(m^{2/3})(m^{5/3})$

$\qquad = m^{2/3 + 5/3}$ *Product rule*

$\qquad = m^{7/3}$

33. $(1 + n)^{1/2}(1 + n)^{3/4}$

$\qquad = (1 + n)^{1/2 + 3/4}$

$\qquad = (1 + n)^{5/4}$ *Add exponents*

35. $(2y^{3/4}z)(3y^{-2}z^{-1/3})$

$\qquad = 6y^{3/4 + (-2)}z^{1 + (-1/3)}$

$\qquad = 6y^{3/4 - 8/4}z^{3/3 - 1/3}$

$\qquad = 6y^{-5/4}z^{2/3}$

$\qquad = \dfrac{6z^{2/3}}{y^{5/4}}$

37. $(4a^{-2}b^7)^{1/2} \cdot (2a^{1/4}b^3)^5$

$\qquad = (4^{1/2}s^{-1}b^{7/2})(2^5a^{5/4}b^{15})$

$\qquad = 2 \cdot 2^5 \cdot a^{-1} \cdot a^{5/4} \cdot b^{7/2} \cdot b^{15}$

$\qquad = 2^6a^{-4/4 + 5/4}b^{7/2 + 30/2}$

$\qquad = 2^6a^{1/4}b^{37/2}$

39. $\left(\dfrac{r^{-2}}{s^{-5}}\right)^{-3} = \dfrac{(r^{-2})^{-3}}{(s^{-5})^{-3}}$

$\qquad = \dfrac{r^{(-2)(-3)}}{s^{(-5)(-3)}} = \dfrac{r^6}{s^{15}}$

41. $\left(\dfrac{-a}{b^{-3}}\right)^{-1} = \dfrac{(-a)^{-1}}{(b^{-3})^{-1}}$

$\qquad = \dfrac{1}{-ab^3}$

$\qquad = -\dfrac{1}{ab^3}$

43. $\dfrac{12^{5/4} \, y^{-2}}{12^{-1}y^{-3}} = 12^{5/4 \, - \, (-1/4)}y^{-2-(-3)}$

$\qquad\qquad\qquad$ *Quotient rule*

$\qquad\qquad = 12^{9/4} \, y$

45. $\dfrac{8p^{-3}(4p^2)^{-2}}{p^{-5}} = \dfrac{8p^{-3} \cdot 4^{-2}p^{-4}}{p^{-5}}$

$\qquad\qquad\quad = \dfrac{8 \cdot 4^{-2} \cdot p^{-7}}{p^{-5}}$

$\qquad\qquad\quad = \dfrac{8}{4^2 p^{-5}p^7}$

$\qquad\qquad\quad = \dfrac{8}{16p^2}$

$\qquad\qquad\quad = \dfrac{1}{2p^2}$

47. $\dfrac{m^{7/3} \, n^{-2/5} \, p^{3/8}}{m^{-2/3} \, n^{3/5} \, p^{-5/8}}$

$\qquad = \dfrac{m^{7/3} \, m^{2/3} \, p^{3/8} \, p^{5/8}}{n^{3/5} \, n^{2/5}}$

$\qquad = \dfrac{m^{9/3} \, p^{8/8}}{n^{5/5}} = \dfrac{m^3 p}{n}$

49. $\dfrac{-4a^{-1}a^{2/3}}{a^{-2}} = \dfrac{-4a^{-3/3} \, a^{-2/3}}{a^{-2}}$

$\qquad\qquad = \dfrac{-4a^{-1/3}}{a^{-2}}$

$\qquad\qquad = \dfrac{-4a^{-1/3} \, a^2}{1}$

$\qquad\qquad = -4a^{5/3}$

51. $\dfrac{(k+5)^{1/2} \, (k+5)^{-1/4}}{(k+5)^{3/4}}$

$\qquad = (k+5)^{1/2 \, - 1/4 \, - 3/4}$

$\qquad = (k+5)^{-1/2}$

$\qquad = \dfrac{1}{(k+5)^{1/2}}$

53. $p = 2x^{1/2} + 3x^{2/3}$

$\quad$ If $x = 64$,

$\qquad p = 2(64)^{1/2} + 3(64)^{2/3}$

$\qquad\quad = 2(8) + 3(64^{1/3})^2$

$\qquad\quad = 16 + 3(4)^2$

$\qquad\quad = 16 + 48 = 64.$

When the supply is 64 units, the price is \$64.

55. $p = 1000 - 200x^{-2/3} \quad (x > 0)$

$\quad p = 1000 - 200 \cdot 27^{-2/3} \quad$ *Let* $x = 27$

$\qquad = 1000 - 200 \cdot (27^{1/3})^{-2}$

$\qquad = 1000 - 200 \cdot (3)^{-2}$

$\qquad = 1000 - 200 \cdot \dfrac{1}{3^2}$

$\qquad = 1000 - \dfrac{200}{9}$

$\qquad = 1000 - 22.22$

$\qquad = 977.78$

When the demand is 27 units, the price is \$977.78.

57. $E_{large} = 48; \quad E_{small} = 3$

Amount for small state = \$1,000,000

Amount for large state

$\quad = \left(\dfrac{E_{large}}{E_{small}}\right)^{3/2} \times$ amount for small state

$\quad = \left(\dfrac{48}{3}\right)^{3/2} \times 1,000,000$

$\quad = (16)^{3/2} \times 1,000,000$

$\quad = (16^{1/2})^3 \times 1,000,000$

$\quad = 4^3 \times 1,000,000$

$\quad = 64 \times 1,000,000$

$\quad = 64,000,000$

If \$1,000,000 is spent in the small state, \$64,000,000 should be spent in the large state.

59. $E_{large} = 28$; $E_{small} = 6$

Amount for large state

$$= \left(\frac{28}{6}\right)^{3/2} \times 1,000,000$$

$$\approx 10 \times 1,000,000$$

About $10,000,000 should be spent in the large state.

61. 1988 (x = 0)

$$y = 386(1.18)^x$$
$$= 386(1.18)^0$$
$$= 386 \cdot 1 = 386$$

The amount generated in 1988 was about $386 million.

62. 1989 (x = 1)

$$y = 386(1.18)^1$$
$$= 386 \cdot 1.18$$
$$\approx 455$$

The amount generated in 1989 was about $455 million.

63. 1990 (x = 2)

$$y = 386(1.18)^2$$
$$\approx 537$$

The amount generated in 1990 was about $537 million.

64. 1991 (x = 3)

$$y = 386(1.18)^3$$
$$\approx 634$$

The amount generated in 1991 was about $634 million.

65. 1992 (x = 4)

$$y = 386(1.18)^4$$
$$\approx 748$$

The amount generated in 1992 was about $748 million.

66. (a) For 1988, the bar graph shows $350 million, and the model gives $386 million. (See Exercise 61.)

$$|350 - 386| = |-36| = 36$$

The absolute value of the difference is $36 million.

(b) The amount indicated by the model is greater than the amount indicated on the graph.

67. (a) For 1989, the bar graph shows $500 million, and the model gives $455 million. (See Exercise 62.)

$$|500 - 455| = |45| = 45$$

The absolute value of the difference is $45 million.

(b) The amount indicated by the model is $45 million less than the amount indicated on the graph.

68. (a) For 1990, the bar graph shows $600 million, and the model gives $537 million. (See Exercise 63.)

$$|600 - 537| = |63| = 63$$

The absolute value of the difference is $63 million.

(b) The amount indicated by the model is $63 million less than the amount indicated on the graph.

69. For 1991, the bar graph shows $600 million, and the model gives $634 million. (See Exercise 64.)

$$|600 - 634| = |-34| = 34$$

The absolute value of the difference is $34 million.

(b) The amount indicated by the model is greater than the amount indicated by the graph.

70. For 1992, the bar graph shows $750 million, and the model gives $748 million.

$$|750 - 748| = |2| = 2$$

The absolute value of the difference is $2 million.

(b) The amount indicated by the model is less than the amount indicated by the graph.

71. $S = 28.6 A^{.32}$

If $A = 10$,

$$S = (28.6)(10^{.32})$$
$$\approx 60.$$

If the area of an island is 10 sq mi, the number of land-plant species is approximately 60.

73. $S = 28.6 A^{.32}$

If $A = 300$,

$$S = (28.6)(300)^{.32}$$
$$\approx 177.$$

If the area of an island is 300 sq mi, the number of land-plant species is approximately 177.

75. $(r^{3/p})^{2p}(r^{1/p})p^2$

$$= r^{(3/p)(2p)}r^{(1/p)(p^2)}$$
$$\qquad \textit{Power rule}$$
$$= r^6 r^p \quad \textit{Product rule}$$
$$= r^{6+p}$$

77. $\dfrac{m^{1-a}m^a}{m^{-1/2}}$

$$= \frac{m^{1-a+a}}{m^{-1/2}} \qquad \textit{Product rule}$$

$$= \frac{m^1}{m^{-1/2}}$$

$$= m^{1-(-1/2)} \qquad \textit{Quotient rule}$$

$$= m^{3/2}$$

79. $\dfrac{(x^{n/2})(x^{3n})^{1/2}}{x^{1/n}}$

$$= \frac{x^{n/2}x^{3n/2}}{x^{1/n}}$$

$$= x^{n/2+3n/2-1/n} \qquad \textit{Product and}$$
$$\qquad\qquad\qquad\qquad \textit{quotient rules}$$

$$= x^{2n-1/n}$$

The exponent $2n - \dfrac{1}{n}$ can be written as a single fraction:

$$2n - \frac{1}{n} = \frac{n}{n} \cdot \frac{2n}{1} - \frac{1}{n}$$
$$= \frac{2n^2 - 1}{n}.$$

Therefore, the result can be written

$$x^{(2n^2-1)/n}.$$

81. $\dfrac{(p^{1/n})(p^{1/m})}{p^{-m/n}}$

$$= p^{1/n+1/m-(-m/n)}$$

The exponent in the above expression can be simplified as follows:

$$\frac{1}{n} + \frac{1}{m} + \frac{m}{n} = \frac{m + n + m^2}{mn}.$$

Therefore, the result is

$$p^{(m+n+m^2)/(mn)}.$$

83. $y^{5/8}(y^{3/8} - 10y^{11/8})$

$= y^{5/8}y^{3/8} - 10y^{5/8}y^{11/8}$

$= y^{5/8+3/8} - 10y^{5/8+11/8}$

$= y - 10y^2$

85. $-4k(k^{7/3} - 6k^{1/3})$

$= -4k^1k^{7/3} + 24k^1k^{1/3}$

$= -4k^{10/3} + 24k^{4/3}$

87. $(x + x^{1/2})(x - x^{1/2})$

$= x^2 - (x^{1/2})^2$ *Difference of*
                     two squares

$= x^2 - x$

89. $(r^{1/2} - r^{-1/2})^2$

$= (r^{1/2})^2 - 2(r^{1/2})(r^{-1/2})$

    $+ (r^{-1/2})^2$ *Square of a binomial*

$= r - 2r^0 + r^{-1}$

$= r - 2 + r^{-1}$ or $r - 2 + \dfrac{1}{r}$

91. Factor $4k^{-1} + k^{-2}$, using the common factor k^{-2}.

$$4k^{-1} + k^{-2} = k^{-2}(4k + 1)$$

$$\text{or} \quad \frac{4k + 1}{k^2}$$

93. Factor $9z^{-1/2} + 2z^{1/2}$, using the common factor $z^{-1/2}$.

$$9z^{-1/2} + 2z^{1/2} = z^{-1/2}(9 + z)$$

$$\text{or} \quad \frac{9 + z}{z^{1/2}}$$

95. Factor $p^{-3/4} - 2p^{-7/4}$, using the common factor $p^{-7/4}$.

$$p^{-3/4} - 2p^{-7/4} = p^{-7/4}(p - 2)$$

$$\text{or} \quad \frac{p - 2}{p^{7/4}}$$

97. Factor $(p + 4)^{-3/2} + (p + 4)^{-1/2} + (p + 4)^{1/2}$, using the common factor $(p + 4)^{-3/2}$.

$= (p + 4)^{-3/2}$

    $\cdot [1 + (p + 4) + (p + 4)^2]$

$= (p + 4)^{-3/2}$

    $\cdot (1 + p + 4 + p^2 + 8p + 16)$

$= (p + 4)^{-3/2}(p^2 + 9p + 21)$

or $\dfrac{p^2 + 9p + 21}{(p + 4)^{3/2}}$

99. $\dfrac{a^{-1} + b^{-1}}{(ab)^{-1}}$

$= \dfrac{\dfrac{1}{a} + \dfrac{1}{b}}{\dfrac{1}{ab}}$ *Definition of negative integer exponent*

$= \dfrac{\dfrac{1 \cdot b}{a \cdot b} + \dfrac{1 \cdot a}{b \cdot a}}{\dfrac{1}{ab}}$

$= \dfrac{\dfrac{b + a}{ab}}{\dfrac{1}{ab}}$

$= \dfrac{b + a}{ab} \cdot \dfrac{ab}{1}$ *Definition of division*

$= b + a$

101. $\dfrac{r^{-1} + q^{-1}}{r^{-1} - q^{-1}} \cdot \dfrac{r - q}{r + q}$

$= \dfrac{\dfrac{1}{r} + \dfrac{1}{q}}{\dfrac{1}{r} - \dfrac{1}{q}} \cdot \dfrac{r - q}{r + q}$

$$= \frac{rq\left(\frac{1}{r} + \frac{1}{q}\right)}{rq\left(\frac{1}{r} - \frac{1}{q}\right)} \cdot \frac{r - q}{r + q}$$

Multiply numerator and denominator of first fraction by common denominator, rq

$$= \frac{q + r}{q - r} \cdot \frac{r - q}{r + q}$$

$$= \frac{r - q}{q - r} = \frac{-1(r - q)}{-1(q - r)}$$

$$= \frac{-1(r - q)}{r - q} = -1$$

103. $\dfrac{x - 9y^{-1}}{(x - 3y^{-1})(x + 3y^{-1})}$

$$= \frac{x - \dfrac{9}{y}}{\left(x - \dfrac{3}{y}\right)\left(x + \dfrac{3}{y}\right)}$$

Definition of negative integer exponent

$$= \frac{x - \dfrac{9}{y}}{x^2 - \dfrac{9}{y^2}}$$

Multiply in denominator

$$= \frac{y^2\left(x - \dfrac{9}{y}\right)}{y^2\left(x^2 - \dfrac{9}{y^2}\right)}$$

Multiply numerator and denominator by least common denominator, y

$$= \frac{y^2 x - 9y}{y^2 x^2 - 9}$$

Distributive property

or $\dfrac{y(xy - 9)}{x^2 y^2 - 9}$

Factor numerator

Section 1.7

1. $(-3)^{1/3} = \sqrt[3]{-3x}$ (f)

3. $(-3x)^{-1/3} = \dfrac{1}{(-3x)^{1/3}} = \dfrac{1}{\sqrt[3]{-3x}}$ (h)

5. $(3x)^{1/3} = \sqrt[3]{3x}$ (g)

7. $(3x)^{-1/3} = \dfrac{1}{\sqrt[3]{3x}}$ (c)

9. $(-m)^{2/3} = \sqrt[3]{(-m)^2}$ or $(\sqrt[3]{-m})^2$

11. $(2m + p)^{2/3}$

$= \sqrt[3]{(2m + p)^2}$ or $(\sqrt[3]{2m + p})^2$

13. $\sqrt[5]{k^2} = k^{2/5}$

15. $-3\sqrt{5p^3} = -3(5p^3)^{1/2}$

$= -3 \cdot 5^{1/2} p^{3/2}$

21. $\sqrt[3]{125} = 5$

23. $\sqrt[5]{-3125} = -5$

25. $\sqrt{50} = \sqrt{25 \cdot 2} = \sqrt{25} \cdot \sqrt{2} = 5\sqrt{2}$

27. $\sqrt[3]{81} = \sqrt[3]{27 \cdot 3} = \sqrt[3]{27} \cdot \sqrt[3]{3} = 3\sqrt[3]{3}$

29. $-\sqrt[4]{32} = -\sqrt[4]{16 \cdot 2} = -\sqrt[4]{16} \cdot \sqrt[4]{2}$

$= -2\sqrt[4]{2}$

31. $-\sqrt{\dfrac{9}{5}} = \dfrac{-3}{\sqrt{5}} \cdot \dfrac{\sqrt{5}}{\sqrt{5}} = -\dfrac{3\sqrt{5}}{5}$

33. $-\sqrt[3]{\dfrac{4}{5}} = -\dfrac{\sqrt[3]{4}}{\sqrt[3]{5}} \cdot \dfrac{\sqrt[3]{5^2}}{\sqrt[3]{5^2}}$

$= -\dfrac{\sqrt[3]{4} \cdot \sqrt[3]{5}}{\sqrt[3]{5^3}}$

$= -\dfrac{\sqrt[3]{4 \cdot 25}}{5}$

$= -\dfrac{\sqrt[3]{100}}{5}$

35. $\sqrt[3]{16(-2)^4(2)^8} = \sqrt[3]{2^4 \cdot (-2)^4 2^8}$

$\phantom{\sqrt[3]{16(-2)^4(2)^8}} = \sqrt[3]{2^4 \cdot 2^4 \cdot 2^8}$

$\phantom{\sqrt[3]{16(-2)^4(2)^8}} = \sqrt[3]{2^{16}}$

$\phantom{\sqrt[3]{16(-2)^4(2)^8}} = \sqrt[3]{2^{15} \cdot 2}$

$\phantom{\sqrt[3]{16(-2)^4(2)^8}} = \sqrt[3]{2^{15}} \cdot \sqrt[3]{2}$

$\phantom{\sqrt[3]{16(-2)^4(2)^8}} = 2^5 \cdot \sqrt[3]{2}$

$\phantom{\sqrt[3]{16(-2)^4(2)^8}} = 32\sqrt[3]{2}$

37. $\sqrt{8x^5z^8} = \sqrt{2 \cdot 4 \cdot x^4 \cdot x \cdot z^8}$

$\phantom{\sqrt{8x^5z^8}} = \sqrt{4x^4z^8} \cdot \sqrt{2x}$

$\phantom{\sqrt{8x^5z^8}} = \sqrt{2x} \cdot 2x^2z^4$

$\phantom{\sqrt{8x^5z^8}} = 2x^2z^4\sqrt{2x}$

39. $\sqrt[3]{16z^5x^8y^4} = \sqrt[3]{8 \cdot 2 \cdot z^3z^2x^6x^2y^3y}$

$\phantom{\sqrt[3]{16z^5x^8y^4}} = \sqrt[3]{(8z^3x^6y^3)(2z^2x^2y)}$

$\phantom{\sqrt[3]{16z^5x^8y^4}}$ *Group all perfect cubes*

$\phantom{\sqrt[3]{16z^5x^8y^4}} = \sqrt[3]{8z^3x^6y^3} \cdot \sqrt[3]{2z^2x^2y}$

$\phantom{\sqrt[3]{16z^5x^8y^4}} = 2zx^2y\sqrt[3]{2z^2x^2y}$

41. $\sqrt[4]{m^2n^7p^8} = \sqrt[4]{m^2n^4n^3p^8}$

$\phantom{\sqrt[4]{m^2n^7p^8}} = \sqrt[4]{n^4p^8} \cdot \sqrt[4]{m^2n^3}$

$\phantom{\sqrt[4]{m^2n^7p^8}} = np^2\sqrt[4]{m^2n^3}$

43. $\sqrt[4]{x^4 + y^4}$ cannot be simplified further.

45. $\sqrt{\dfrac{2}{3x}} = \dfrac{\sqrt{2}}{\sqrt{3x}}$

$\phantom{\sqrt{\dfrac{2}{3x}}} = \dfrac{\sqrt{2}}{\sqrt{3x}} \cdot \dfrac{\sqrt{3x}}{\sqrt{3x}} = \dfrac{\sqrt{6x}}{3x}$

47. $\sqrt{\dfrac{x^5y^3}{z^2}} = \dfrac{\sqrt{x^5y^3}}{\sqrt{z^2}} = \dfrac{\sqrt{x^4xy^2y}}{z}$

$\phantom{\sqrt{\dfrac{x^5y^3}{z^2}}} = \dfrac{\sqrt{x^4y^2} \cdot \sqrt{xy}}{z} = \dfrac{x^2y\sqrt{xy}}{z}$

49. $\sqrt[3]{\dfrac{8}{x^2}} = \dfrac{\sqrt[3]{8}}{\sqrt[3]{x^2}}$

$\phantom{\sqrt[3]{\dfrac{8}{x^2}}} = \dfrac{2}{\sqrt[3]{x^2}} \cdot \dfrac{\sqrt[3]{x}}{\sqrt[3]{x}}$

$\phantom{\sqrt[3]{\dfrac{8}{x^2}}} = \dfrac{2\sqrt[3]{x}}{x}$

51. $\sqrt[4]{\dfrac{g^3h^5}{9r^6}} = \dfrac{\sqrt[4]{g^3h^5}}{\sqrt[4]{9r^6}}$

$\phantom{\sqrt[4]{\dfrac{g^3h^5}{9r^6}}} = \dfrac{h\sqrt[4]{g^3h}}{\sqrt[4]{9r^6}} \cdot \dfrac{\sqrt[4]{9r^2}}{\sqrt[4]{9r^2}}$

$\phantom{\sqrt[4]{\dfrac{g^3h^5}{9r^6}}} = \dfrac{h\sqrt[4]{9g^3hr^2}}{\sqrt[4]{81r^8}}$

$\phantom{\sqrt[4]{\dfrac{g^3h^5}{9r^6}}} = \dfrac{h\sqrt[4]{9g^3hr^2}}{3r^2}$

53. $\dfrac{\sqrt[3]{mn} \cdot \sqrt[3]{m^2}}{\sqrt[3]{n^2}} = \sqrt[3]{\dfrac{mnm^2}{n^2}} = \sqrt[3]{\dfrac{m^3}{n}}$

$\phantom{\dfrac{\sqrt[3]{mn} \cdot \sqrt[3]{m^2}}{\sqrt[3]{n^2}}} = \dfrac{\sqrt[3]{m^3}}{\sqrt[3]{n}} \cdot \dfrac{\sqrt[3]{n^2}}{\sqrt[3]{n^2}} = \dfrac{m\sqrt[3]{n^2}}{n}$

55. $\dfrac{\sqrt[4]{32x^5y} \cdot \sqrt[4]{2xy^4}}{\sqrt[4]{4x^3y^2}} = \sqrt[4]{\dfrac{64x^6y^5}{4x^3y^2}}$

$\phantom{\dfrac{\sqrt[4]{32x^5y} \cdot \sqrt[4]{2xy^4}}{\sqrt[4]{4x^3y^2}}} = \sqrt[4]{16x^3y^3}$

$\phantom{\dfrac{\sqrt[4]{32x^5y} \cdot \sqrt[4]{2xy^4}}{\sqrt[4]{4x^3y^2}}} = 2\sqrt[4]{x^3y^3}$

57. $\sqrt[3]{\sqrt{4}} = \sqrt[3]{4^{1/2}} = (4^{1/2})^{1/3} = 4^{1/6}$

$\phantom{\sqrt[3]{\sqrt{4}}} = (2^2)^{1/6} = 2^{2/6} = 2^{1/3} = \sqrt[3]{2}$

59. The display on the screen shows the equation

$$\sqrt[3]{\sqrt{2}} = \sqrt[6]{2}.$$

Since

$$\sqrt[3]{\sqrt{2}} = (2^{1/2})^{1/3} = 2^{1/3} = 2^{1/6} = \sqrt[6]{2},$$

the calculator would indicate that the display is true.

61. $9\sqrt{8k} + 3\sqrt{18k} - \sqrt{32k}$

$\quad = 9\sqrt{4 \cdot 2k} + 3\sqrt{9 \cdot 2k} - \sqrt{16 \cdot 2k}$

$\quad = 9 \cdot 2\sqrt{2k} + 3 \cdot 3\sqrt{2k} - 4\sqrt{2k}$

$\quad = 18\sqrt{2k} + 9\sqrt{2k} - 4\sqrt{2k}$

$\quad = 23\sqrt{2k}$ *Combine like terms*

63. $\sqrt[3]{32} - 5\sqrt[3]{4} + 2\sqrt[3]{108}$

$\quad = \sqrt[3]{8 \cdot 4} - 5\sqrt[3]{4} + 2\sqrt[3]{27 \cdot 4}$

$\quad = 2\sqrt[3]{4} - 5\sqrt[3]{4} + 6\sqrt[3]{4}$

$\quad = 3\sqrt[3]{4}$ *Combine like radicals*

65. $\dfrac{1}{\sqrt{2}} + \dfrac{3}{\sqrt{8}} + \dfrac{1}{\sqrt{32}}$

$\quad = \dfrac{1}{\sqrt{2}} + \dfrac{3}{2\sqrt{2}} + \dfrac{1}{4\sqrt{2}}$

$\quad = \dfrac{4}{4\sqrt{2}} + \dfrac{6}{4\sqrt{2}} + \dfrac{1}{4\sqrt{2}}$ *Least common denominator, $4\sqrt{2}$*

$\quad = \dfrac{11}{4\sqrt{2}}$

$\quad = \dfrac{11}{4\sqrt{2}} \cdot \dfrac{\sqrt{2}}{\sqrt{2}}$ *Rationalize denominator*

$\quad = \dfrac{11\sqrt{2}}{8}$

67. $\dfrac{-4}{\sqrt[3]{3}} + \dfrac{1}{\sqrt[3]{24}} - \dfrac{2}{\sqrt[3]{81}}$

$\quad = \dfrac{-4}{\sqrt[3]{3}} + \dfrac{1}{2\sqrt[3]{3}} - \dfrac{2}{3\sqrt[3]{3}}$

$\quad = \dfrac{-24}{6\sqrt[3]{3}} + \dfrac{3}{6\sqrt[3]{3}} - \dfrac{4}{6\sqrt[3]{3}}$ *Least common denominator, $6\sqrt[3]{3}$*

$\quad = \dfrac{-25}{6\sqrt[3]{3}}$

$\quad = \dfrac{-25}{6\sqrt[3]{3}} \cdot \dfrac{\sqrt[3]{3^2}}{\sqrt[3]{3^2}}$ *Rationalize denominator*

$\quad = \dfrac{-25\sqrt[3]{9}}{18}$

69. $(\sqrt{5} + \sqrt{2})(\sqrt{5} - \sqrt{2}) = (\sqrt{5})^2 - (\sqrt{2})^2$

$\qquad\qquad\qquad\qquad\qquad = 5 - 2 = 3$

71. $(\sqrt[3]{7} + 3)(\sqrt[3]{7^2} - 3\sqrt[3]{7} + 9)$

This product has the pattern $(a + b)(a^2 - ab + b^2) = a^3 + b^3$, the sum of two cubes.

Thus,

$(\sqrt[3]{7} + 3)(\sqrt[3]{7^2} - 3\sqrt[3]{7} + 9)$

$\quad = (\sqrt[3]{7})^3 + 3^3$

$\quad = 7 + 27$

$\quad = 34.$

73. $(\sqrt{2} - 1)^2 = (\sqrt{2})^2 - 2(\sqrt{2}) + 1^2$

$\qquad\qquad\qquad$ *Square of a binomial*

$\qquad\qquad = 2 - 2\sqrt{2} + 1$

$\qquad\qquad = 3 - 2\sqrt{2}$

75. $(4\sqrt{5} - 1)(3\sqrt{5} + 2)$

$\qquad\quad\ \ F \qquad\ O \qquad\ I \quad\ L$

$\quad = 12 \cdot 5 + 8\sqrt{5} - 3\sqrt{5} - 2$

$\quad = 60 + 5\sqrt{5} - 2$

$\quad = 58 + 5\sqrt{5}$

77. $\dfrac{\sqrt{3}}{\sqrt{5} + \sqrt{3}}$

$\quad = \dfrac{\sqrt{3}}{\sqrt{5} + \sqrt{3}} \cdot \dfrac{\sqrt{5} - \sqrt{3}}{\sqrt{5} - \sqrt{3}}$

Multiply numerator and denominator by conjugate of denominator

$\quad = \dfrac{\sqrt{3}(\sqrt{5} - \sqrt{3})}{(\sqrt{5})^2 - (\sqrt{3})^2}$ *Difference of two squares*

$\quad = \dfrac{\sqrt{3}\sqrt{5} - \sqrt{3}\sqrt{3}}{5 - 3}$ *Distributive property*

$\quad = \dfrac{\sqrt{15} - 3}{2}$

79. $\dfrac{1 + \sqrt{3}}{3\sqrt{5} + 2\sqrt{3}}$

$= \dfrac{1 + \sqrt{3}}{3\sqrt{5} + 2\sqrt{3}} \cdot \dfrac{3\sqrt{5} - 2\sqrt{3}}{3\sqrt{5} - 2\sqrt{3}}$

Multiply numerator and denominator by conjugate of denominator

$= \dfrac{(1 + \sqrt{3})(3\sqrt{5} - 2\sqrt{3})}{(3\sqrt{5})^2 - (2\sqrt{3})^2}$

Difference of two squares

$= \dfrac{3\sqrt{5} - 2\sqrt{3} + 3\sqrt{15} - 6}{45 - 12}$

$= \dfrac{3\sqrt{5} - 2\sqrt{3} + 3\sqrt{15} - 6}{33}$

81. $\dfrac{p}{\sqrt{p} + 2} = \dfrac{p}{\sqrt{p} + 2} \cdot \dfrac{\sqrt{p} - 2}{\sqrt{p} - 2}$

Multiply numerator and denominator by conjugate of denominator

$= \dfrac{p(\sqrt{p} - 2)}{(\sqrt{p})^2 - 2^2}$ *Difference of two squares*

$= \dfrac{p(\sqrt{p} - 2)}{p - 4}$

83. $\dfrac{a}{\sqrt{a + b} - 1}$

$= \dfrac{a}{\sqrt{a + b} - 1} \cdot \dfrac{\sqrt{a + b} + 1}{\sqrt{a + b} + 1}$

$= \dfrac{a(\sqrt{a + b} + 1)}{(\sqrt{a + b})^2 - 1^2}$

$= \dfrac{a(\sqrt{a + b} + 1)}{a + b - 1}$

85. **(a)** $T - \left(\dfrac{v}{4} + 7\sqrt{v}\right)\left(1 - \dfrac{T}{90}\right)$

$= -10 - \left(\dfrac{30}{4} + 7\sqrt{30}\right)\left(1 - \dfrac{-10}{90}\right)$

Let T = -10 and v = 30

≈ -60.9

For a temperature of $-10°$F with a wind speed of 30 mph, the wind-chill factor is 60.9°F.

(b) $T - \left(\dfrac{v}{4} + 7\sqrt{v}\right)\left(1 - \dfrac{T}{90}\right)$

$= -40 - \left(\dfrac{5}{4} + 7\sqrt{5}\right)\left(1 - \dfrac{-40}{90}\right)$

Let T = -40 and v = 5

≈ -64.4

For a temperature of $-40°$F with a wind speed of 5 mph the wind-chill factor is $-64.4°$F.

86. **(a)** $91.4 - (91.4 - T)$

$\cdot (.478 + .301\sqrt{v} - .-02v)$

$= 91.4 - [91.4 - (-10)]$

$\cdot [.478 + .301\sqrt{30} - .02(30)]$

Let t = -10 and v = 30

≈ -63.4

For a temperature of $-10°$F and a wind speed of 30 mph, the wind-chill factor is $-63.4°$F.

(b) $91.4 - (91.4 - T)$

$\cdot (.478 + .301\sqrt{v} - .02v)$

$= 91.4 - [91.4 - (-40)]$

$\cdot [.478 + .301\sqrt{5} - .02(5)]$

Let T = -40 and v = 5

≈ -46.7

For a temperature of $-40°$F and a wind speed of 5 mph, the wind-chill factor is $-46.7°$F.

87. **(a)** According to the chart, the wind-chill factor for a temperature of $-10°$F with a wind speed of 30 mph is $-63°$F.

(b) According to the chart, the wind-chill factor for a temperature -40°F with a wind speed of 5 mph is -47°F.

88. The values obtained by using the formula in Exercise 86 are closer to those in the chart than the values obtained by using the formula in Exercise 85, so the formula in Exercise 86 provides a better model.

89. $\sqrt{(m + n)^2} = |m + n|$

since $\sqrt{x^2} = |x|$.

91. $\sqrt{z^2 - 6zx + 9x^2}$

$= \sqrt{(z - 3x)^2}$ *Factor*

$= |z - 3x|$

since

$\sqrt{x} = |x|$.

93. $\dfrac{1 + \sqrt{2}}{2}$

$= \dfrac{1 + \sqrt{2}}{2} \cdot \dfrac{1 - \sqrt{2}}{1 - \sqrt{2}}$

Multiply numerator and denominator by conjugate of numerator

$= \dfrac{1^2 - (\sqrt{2})^2}{2(1 - \sqrt{2})}$ *Difference of two squares*

$= \dfrac{1 - 2}{2(1 - \sqrt{2})}$

$= \dfrac{-1}{2(1 - \sqrt{2})}$

95. $\dfrac{\sqrt{x}}{1 + \sqrt{x}} = \dfrac{\sqrt{x}}{1 + \sqrt{x}} \cdot \dfrac{\sqrt{x}}{\sqrt{x}}$

$= \dfrac{x}{(1 + \sqrt{x})\sqrt{x}}$

$= \dfrac{x}{\sqrt{x} + x}$

97. $\dfrac{\sqrt{x} + \sqrt{x + 1}}{\sqrt{x} - \sqrt{x + 1}}$

$= \dfrac{\sqrt{x} + \sqrt{x + 1}}{\sqrt{x} - \sqrt{x + 1}} \cdot \dfrac{\sqrt{x} - \sqrt{x + 1}}{\sqrt{x} - \sqrt{x + 1}}$

Multiply numerator and denominator by conjugate of numerator

$= \dfrac{(\sqrt{x})^2 - (\sqrt{x + 1})^2}{(\sqrt{x} - \sqrt{x + 1})^2}$ *Difference of two squares*

$= \dfrac{x - (x + 1)}{(\sqrt{x} - \sqrt{x + 1})^2}$

$= \dfrac{-1}{(\sqrt{x} - \sqrt{x + 1})^2}$

$= \dfrac{-1}{x - 2\sqrt{x}\sqrt{x + 1} + x + 1}$

Square of a binomial

$= \dfrac{-1}{2x - 2\sqrt{x(x + 1)} + 1}$

99. **(a)** The radius of the earth is approximately $\dfrac{12,742}{2} = 6371$ km. Its surface area is

$4\pi(6371)^2 \approx 510,064,472$ or

approximately 5.10×10^8 square kilometers.

(b) 71% of $510 \times 10^6 \approx 3.62 \times 10^8$ square kilometers.

(c) The volume of water would be 8% less or 92% of the volume of the ice contained in the Greenland ice cap. This would be equal to 92% of approximately 3 million cubic kilometers or approximately 2.76×10^6 cubic kilometers of water.

(d)

$\dfrac{\text{Volume of Water}}{\text{Area of Oceans}} = \dfrac{2.76 \times 10^6}{362 \times 10^6}$

$\approx .0076$ km $= 7.6$ m

This is approximately 25 ft and very close to the published value of 7.5 m.

(e) An increase in the sea level of 7.6 m would displace millions of people and cause enormous damage to property. Since the elevations of Boston, New Orleans, and San Diego are all less than 7.6 m or about 25 ft, they would be below sea level and under water without a dike system.

(f) If inland seas and fresh water lakes had been accounted for, then the calculated area for the oceans would have been less. Therefore, the increase in sea level would have been greater, but not significantly. The two largest inland bodies of water are the Caspian Sea and Lake Superior which account for only .12% of the total surface area covered by water.

Section 1.8

3. -5 is a real number.

5. $i\sqrt{6}$ is an imaginary number.

7. $2 + 5i$ is an imaginary number.

9. $\sqrt{-100} = i\sqrt{100} = 10i$

11. $-\sqrt{-400} = -i\sqrt{400} = -20i$

13. $-\sqrt{-39} = -i\sqrt{39}$

15. $5 + \sqrt{-4} = 5 + i\sqrt{4} = 5 + 2i$

17. $9 - \sqrt{-50} = 9 - i\sqrt{50}$
$$= 9 - 5i\sqrt{2}$$

19. $\sqrt{-5} \cdot \sqrt{-5} = i\sqrt{5} \cdot i\sqrt{5}$
$$= i^2 \cdot (\sqrt{5})^2$$
$$= (-1)(5) \quad i^2 = -1$$
$$= -5$$

21. $\dfrac{\sqrt{-40}}{\sqrt{-10}} = \dfrac{i\sqrt{40}}{i\sqrt{10}}$
$$= \sqrt{\dfrac{40}{10}}$$
$$= \sqrt{4} = 2$$

23. $(3 + 2i) + (4 - 3i)$
$$= (3 + 4) + [2 + (-3)]i$$
$$= 7 - i$$

25. $(-2 + 3i) - (-4 + 3i)$
$$= [(-2) - (-4)] + (3 - 3)i$$
$$= 2$$

27. $(2 - 5i) - (3 + 4i) - (-2 + i)$
$$= [2 - 3 - (-2)] + [-5 - 4 - 1]i$$
$$= 1 - 10i$$

29. $(2 + 4i)(-1 + 3i)$
$$= 2(-1) + 2(3i) + 4i(-1) + 4i(3i)$$
$$= -2 + 6i - 4i + 12i^2$$
$$= -2 + 2i + 12(-1) \quad i^2 = -1$$
$$= -14 + 2i$$

31. $(-3 + 2i)^2$

$= (-3)^2 + 2(-3)(2i) + (2i)^2$
  Square of a binomial

$= 9 - 12i + 4i^2$

$= 9 - 12i + 4(-1)$

$= 5 - 12i$

33. $(2 + 3i)(2 - 3i)$

$= 2^2 - (3i)^2$ *Difference of two squares*

$= 4 - 9i^2$

$= 4 - 9(-1)$

$= 13$

35. $(\sqrt{6} + i)(\sqrt{6} - i)$

$= (\sqrt{6})^2 - i^2$ *Difference of two squares*

$= 6 - i^2$

$= 6 - (-1)$

$= 7$

37. $i(3 - 4i)(3 + 4i)$

$= i[3^2 - (4i)^2]$ *Difference of two squares*

$= i[9 - (-16)]$

$= 25i$

39. $i^5 = i^4 \cdot i = 1 \cdot i = i$

41. $i^9 = i^8 \cdot i$

$= (i^4)^2 \cdot i$

$= 1^2 \cdot i = i$

43. $i^{12} = (i^4)^3 = 1^3 = 1$

45. $i^{43} = i^{40} \cdot i^3$

$= (i^4)^{10} \cdot i^3$

$= 1^{10} \cdot i^3 = -i$

47. $\dfrac{1}{i^{12}} = \dfrac{1}{(i^4)^3} = \dfrac{1}{1^3} = 1$

49. $i^{-15} = i^{-16} \cdot i$

$= (i^4)^{-4} \cdot i$

$= 1^{-4} \cdot i = i$

53. $\dfrac{1 + i}{1 - i}$

Multiply numerator and denominator by $1 + i$, the conjugate of the denominator.

$\dfrac{1 + i}{1 - i} = \dfrac{(1 + i)(1 + i)}{(1 - i)(1 + i)}$

$= \dfrac{1 + 2i + i^2}{1 - i^2}$ *Multiply*

$= \dfrac{1 + 2i - 1}{1 + 1}$ $i^2 = -1$

$= \dfrac{2i}{2}$

$= i$

55. $\dfrac{4 - 3i}{4 + 3i}$

Multiply numerator and denominator by $4 - 3i$, the conjugate of the denominator.

$\dfrac{4 - 3i}{4 + 3i} = \dfrac{(4 - 3i)(4 - 3i)}{(4 + 3i)(4 - 3i)}$

$= \dfrac{16 - 24i + 9i^2}{16 - 9i^2}$ *Multiply*

$= \dfrac{16 - 24i - 9}{16 + 9}$ $i^2 = -1$

$= \dfrac{7 - 24i}{25}$

$= \dfrac{7}{25} - \dfrac{24}{25}i$ *Standard form*

57. $\dfrac{3 - 4i}{2 - 5i} = \dfrac{(3 - 4i)(2 + 5i)}{(2 - 5i)(2 + 5i)}$

$= \dfrac{6 + 15i - 8i - 20i^2}{4 - 25i^2}$

   Multiply

$= \dfrac{26 + 7i}{29}$

$= \dfrac{26}{29} + \dfrac{7}{29}i$ *Standard form*

59. $\dfrac{-3 + 4i}{2 - i} = \dfrac{(-3 + 4i)(2 + i)}{(2 - i)(2 + i)}$

$= \dfrac{-6 - 3i + 8i + 4i^2}{4 - i^2}$

$= \dfrac{-10 + 5i}{5}$

$= -\dfrac{10}{5} + \dfrac{5}{5}i$

$= -2 + i$ *Lowest terms*

61. $\dfrac{2}{i} = \dfrac{2(-i)}{i(-i)}$ *$-i$ is the conjugate of i*

$= \dfrac{-2i}{-i^2} = \dfrac{-2i}{1} = -2i$

63. Show that $\dfrac{\sqrt{2}}{2} + \dfrac{\sqrt{2}}{2}i$ is a square root of i.

We must show that $\left(\dfrac{\sqrt{2}}{2} + \dfrac{\sqrt{2}}{2}i\right)^2 = i$.

$\left(\dfrac{\sqrt{2}}{2} + \dfrac{\sqrt{2}}{2}i\right)^2$

$= \dfrac{2}{4} + 2\left(\dfrac{2}{4}i\right) + \dfrac{2}{4}i^2$ *Square of a binomial*

$= \dfrac{1}{2} + i - \dfrac{1}{2}$

$= i$

65. Evaluate $3z - z^2$ if $z = 3 - 2i$.

$3z - z^2 = 3(3 - 2i) - (3 - 2i)^2$

$= 9 - 6i - (9 - 12i + 4i^2)$

$= 9 - 6i - (9 - 12i - 4)$

$= 9 - 6i - (5 - 12i)$

$= 9 - 6i - 5 + 12i$

$= 4 + 6i$

67. $(2 + i)^3 = (2 + i)^2(2 + i)$ is a true statement because the product rule for exponents says that

$$a^{m+n} = a^m \cdot a^n,$$

so $a^3 = a^2 \cdot a.$

68. $(2 + i)^2 = (2 + i)(2 + i)$

$= 4 + 2i + 2i + i^2$

$= 4 + 4i - 1$

$= 3 + 4i$

69. $(2 + i)^3 = (2 + i)^2(2 + i)$

$= (3 + 4i)(2 + i)$

$= 6 + 3i + 8i + 4i^2$

$= 6 + 11i - 4$

$= 2 + 11i$

Yes, this product agrees with the one found by expanding a binomial using Pascal's triangle.

70. Use the coefficients from the sixth row of Pascal's triangle.

$(x + y)^6 = x^6 + 6x^5y + 15x^4y^2 + 20x^3y^3$
$+ 15x^2y^4 + 6xy^5 + y^6$

$(1 + i)^6 = 1^6 + 6(1)^5 i + 15(1)^4 i^2$

$\qquad + 20(1)^3 i^3 + 15(1)^2 i^4$

$\qquad + 6(1)i^5 + i^6$

$\qquad\qquad$ *Let x = 1 and y = i*

$\quad = 1 + 6i + 15(-1) + 20(-i)$

$\qquad + 15(1) + 6i - 1$

$\quad = 1 + 6i - 15 - 20i + 15$

$\qquad + 6i - 1$

$\quad = -8i$

Chapter 1 Review Exercises

1. $-12, -6, -\sqrt{4}$ (or -2), 0, and 6 are integers.

3. $-\sqrt{7}, \frac{\pi}{4}$, and $\sqrt{11}$ are irrational numbers.

5. $-\sqrt{36} = -6$ is an integer, a rational number, and a real number.

7. $\frac{4\pi}{5}$ is an irrational number and a real number.

9. $[2^3 - (-5)] - 2^2 = (8 + 5) - 4$

$\qquad\qquad\qquad\quad = 13 - 4$

$\qquad\qquad\qquad\quad = 9$

11. $(6 - 9)(-2 - 7) - (-4)$

$\quad = (-3)(-9) - (-4)$

$\quad = 27 + 4$

$\quad = 31$

13. $\left(-\frac{2^3}{5} - \frac{3}{4}\right) - \left(-\frac{1}{2}\right)$

$\quad = \left(-\frac{32}{20} - \frac{15}{20}\right) - \left(-\frac{1}{2}\right)$

$\quad = \frac{-47}{20} + \frac{1}{2}$

$\quad = \frac{-47}{20} + \frac{10}{20}$

$\quad = -\frac{37}{20}$

15. $\dfrac{(-7)(-3) - (-2^3)(-5)}{(-2^2 - 2)(-1 - 6)} = \dfrac{21 - 40}{(-6)(-7)}$

$\qquad\qquad\qquad\qquad\qquad = -\dfrac{19}{42}$

17. $-4(2a - 5b)$

$\quad = -4[2(-1) - 5(-2)]$ $\quad$ *Let a = -1,*

$\qquad\qquad\qquad\qquad\qquad$ *b = -2*

$\quad = -4(-2 + 10)$

$\quad = -4(8) = -32$

21. $8(5 + 9) = (5 + 9)8$

Commutative property

23. $3 \cdot (4 \cdot 2) = (3 \cdot 4) \cdot 2$

Associative property

25. $(9 + p) + 0 = 9 + p$

Identity property

27. $k(r + s - t) = kr + ks - kt$

29. Simplify each number.

$|6 - 4| = 2, -|-2| = -2,$

$|8 + 1| = 9, -|3 - (-2)| = -5$

The correct order is

$-|3 - (-2)|, -|-2|, |6 - 4|, |8 + 1|.$

31. $-|-6| + |3| = -6 + 3 = -3$

33. $|\sqrt{8} - 3|$

Since $\sqrt{8} < 3$, $\sqrt{8} - 3 < 0$, so

$$|\sqrt{8} - 3| = -(\sqrt{8} - 3)$$
$$= -\sqrt{8} + 3$$
$$= 3 - \sqrt{8}.$$

35. $|m - 3|$ if $m > 3$

If $m > 3$, $m - 3 > 0$, so

$$|m - 3| = m - 3.$$

37. $|\pi - 4|$

Since $\pi < 4$, $\pi - 4 < 0$, so

$$|\pi - 4| = -(\pi - 4)$$
$$= -\pi + 4$$
$$= 4 - \pi.$$

39. $(3q^3 - 9q^2 + 6) + (4q^3 - 8q + 3)$

$$= 3q^3 + 4q^3 - 9q^2 - 8q + 6 + 3$$
$$= 7q^3 - 9q^2 - 8q + 9$$

41. $(8y - 7)(2y + 7)$

$$= 16y^2 + 56y - 14y - 49 \quad \textit{FOIL}$$
$$= 16y^2 + 42y - 49$$

43. $(3k - 5m)^2$

$$= (3k)^2 - 2(3k)(5m) + (5m)^2$$
$$\textit{Square of a binomial}$$
$$= 9k^2 - 30km + 25m^2$$

45. **(a)** For 1992, the bar graph shows 3.0 million users.

(b) 1992 corresponds to x = 0.

$.035x^4 - .266x^3 + 1.005x^2 + .509x + 2.986$

$$= .035(0)^4 - .266(0)^3 + 1.005(0)^2$$
$$+ .509(0) + 2.986 \quad \textit{Let x = 0}$$
$$= 2.986$$

This represents about 2.99 million users.

47. **(a)** For 1994, the graph shows 6.5 million users.

(b) 1994 corresponds to x = 2.

$.035x^4 - .266x^3 + 1.005x^2 + .509x + 2.986$

$$= .035(2)^4 - .266(2)^3 + 1.005(2)^2$$
$$+ .509(2) + 2.986 \quad \textit{Let x = 2}$$
$$= 6.456$$

This represents about 6.46 million users.

49. $(x + 2y)^4$

Use the coefficients from row 4 of Pascal's triangle.

$$(x + 2y)^4 = x^4 + 4x^3(2y)^1 + 6x^2(2y)^2$$
$$+ 4x(2y)^3 + (2y)^4$$
$$= x^4 + 8x^3y + 6x^2(4y^2)$$
$$+ 4x(8y^3) + 16y^4$$
$$= x^4 + 8x^3y + 24x^2y^2$$
$$+ 32xy^3 + 16y^4$$

51. $\dfrac{72r^2 + 59r + 12}{8r + 3}$

$$
\begin{array}{r}
9r + 4 \\
8r + 3 \overline{) 72r^2 + 59r + 12} \\
\underline{72r^2 + 27r } \\
32r + 12 \\
\underline{32r + 12} \\
0
\end{array}
$$

Thus,

$$\frac{72r^2 + 59r + 12}{8r + 3} = 9r + 4.$$

53. $\dfrac{5m^3 - 7m^2 + 14}{m^2 - 2}$

Insert each missing term with a zero coefficient.

$$
\begin{array}{r}
5m \quad\ - 7 \\
m^2 + 0m - 2\overline{\smash{\big)}\,5m^3 + 7m^2 + 0m + 14} \\
\underline{5m^3 + 0m^2 - 10m } \\
7m^2 + 10m + 14 \\
\underline{7m^2 + 0m + 14} \\
10m
\end{array}
$$

Thus,

$$\frac{5m^3 - 7m^2 + 14}{m^2 - 2} = 5m - 7 + \frac{10m}{m^2 - 2}.$$

55. $7z^2 - 9z^3 + z = z(7z - 9z^2 + 1)$
     z is greatest common factor

57. $r^2 + rp - 42p^2$

Find two numbers whose product is -42 and whose sum is 1. They are 7 and -6. Thus,

$$r^2 + rp - 42p^2$$
$$= (r + 7p)(r - 6p).$$

59. $6m^2 - 13m - 5$

The positive factors of 6 could be 2 and 3 or 1 and 6. As factors of -5, we could have -1 and 5 or -5 and 1. Try different combinations of these factors until the correct one is found.

$$6m^2 - 13m - 5 = (3m + 1)(2m - 5)$$

61. $169y^4 - 1$
   $= (13y^2)^2 - 1^2$ *Difference of two squares*
   $= (13y^2 + 1)(13y^2 - 1)$

63. $8y^3 - 1000z^6$
   $= 8(y^3 - 125z^6)$ *Factor out 8*
   $= 8[y^3 - (5z^2)^3]$
        Difference of two cubes
   $= 8(y - 5z^2)$
     $\cdot [y^2 + y(5z^2) + (5z^2)^2]$
   $= 8(y - 5z^2)(y^2 + 5yz^2 + 25z^4)$

65. $ar - 3as + 5rb - 15sb$
   $= (ar - 3as) + (5rb - 15sb)$
        Group the terms
   $= a(r - 3s) + 5b(r - 3s)$
        Factor each group
   $= (r - 3s)(a + 5b)$
        Factor our r - 3s

67. $(16m^2 - 56m + 49) - 25a^2$
   $= (4m - 7)^2 - (5a)^2$
        Difference of two squares
   $= [(4m - 7) + 5a][(4m - 7) - 5a]$
   $= (4m - 7 + 5a)(4m - 7 - 5a)$

69. $\dfrac{2a + b}{4a^2 - b^2}$

   $= \dfrac{2a + b}{(2a + b)(2a - b)}$
        Factor denominator as difference of two squares

   $= \dfrac{1}{2a - b}$ *Use fundamental principle to write expression in lowest terms*

Thus, expression (a) is equal to
$\dfrac{2a + b}{4a^2 - b^2}.$

71. $\dfrac{3r^3 - 9r^2}{r^2 - 9} \div \dfrac{8r^3}{r + 3}$

$= \dfrac{3r^3 - 9r^2}{r^2 - 9} \cdot \dfrac{r + 3}{8r^3}$

       Definition of division

$= \dfrac{3r^2(r - 3)}{(r + 3)(r - 3)} \cdot \dfrac{(r + 3)}{8r^3}$ *Factor*

$= \dfrac{3}{8r}$ *Use fundamental principle to write expression in lowest terms*

73. $\dfrac{27m^3 - n^3}{3m - n} \div \dfrac{9m^2 + 3mn + n^2}{9m^2 - n^2}$

$= \dfrac{27m^3 - n^3}{3m - n} \cdot \dfrac{9m^2 - n^2}{9m^2 + 3mn + n^2}$

       Definition of division

$= \dfrac{(3m)^3 - n^3}{3m - n} \cdot \dfrac{(3m)^2 - n^2}{9m^2 + 3mn + n^2}$

       Difference of two cubes and difference of two squares

$= \dfrac{(3m - n)[(3m)^2 + 3mn + n^2]}{3m - n}$

$\cdot \dfrac{(3m + n)(3m - n)}{9m^2 + 3mn + n^2}$ *Factor numerators*

$= \dfrac{(3m-n)(9m^2+3mn+n^2)(3m+n)(3m-n)}{(3m-n)(9m^2+3mn+n^2)}$

       Multiply

$= (3m + n)(3m - n)$

       Use fundmental principle to write expressions in lowest terms

75. $\dfrac{1}{4y} + \dfrac{8}{5y} = \dfrac{1 \cdot 5}{4y \cdot 5} + \dfrac{8 \cdot 4}{5y \cdot 4}$

$= \dfrac{5}{20y} + \dfrac{32}{20y}$

$= \dfrac{37}{20y}$

77. $\dfrac{3}{x^2 - 4x + 3} - \dfrac{2}{x^2 - 1}$

$= \dfrac{3}{(x - 3)(x - 1)} - \dfrac{2}{(x + 1)(x - 1)}$

The least common denominator is

$(x - 3)(x - 1)(x + 1)$.

$= \dfrac{3(x + 1)}{(x - 3)(x - 1)(x + 1)}$

$- \dfrac{2(x - 3)}{(x + 1)(x - 1)(x - 3)}$

$= \dfrac{3(x + 1) - 2(x - 3)}{(x - 3)(x - 1)(x + 1)}$

$= \dfrac{3x + 3 - 2x + 6}{(x - 3)(x - 1)(x + 1)}$

$= \dfrac{x + 9}{(x - 3)(x - 1)(x + 1)}$

79. $\dfrac{3 + \dfrac{2m}{m^2 - 4}}{\dfrac{2}{m - 2}}$

$= \dfrac{3 + \dfrac{2m}{(m + 2)(m - 2)}}{\dfrac{5}{m - 2}}$

$= \dfrac{(m - 2)(m + 2)\left(3 + \dfrac{2m}{(m - 2)(m + 2)}\right)}{(m - 2)(m + 2)\left(\dfrac{5}{(m - 2)}\right)}$

       Multiply numerator and denominator by common denominator, $(m - 2)(m + 2)$

$= \dfrac{3(m - 2)(m + 2) + 2m}{5(m + 2)}$

$= \dfrac{3m^2 - 12 + 2m}{5m + 10}$

$= \dfrac{3m^2 + 2m - 12}{5(m + 2)}$

81. $2^{-6} = \dfrac{1}{2^6} = \dfrac{1}{64}$

83. $\left(-\dfrac{5}{4}\right)^{-2} = \dfrac{1}{\left(-\dfrac{5}{4}\right)^2}$

$= \dfrac{1}{\dfrac{25}{16}}$

$= \dfrac{16}{25}$

85. $(5z^3)(-2z^5) = -10z^{3+5}$ *Product rule*

$$= -10z^8$$

87. $(-6p^5w^4m^{12})^0 = 1$ *Definition of a^0*

89. $\dfrac{-8y^7p^{-2}}{y^{-4}p^{-3}} = -8y^{7-(-4)}p^{(-2)-(-3)}$ *Quotient rule*

$$= -8y^{11}p$$

91. $\dfrac{(p+q)^4(p+q)^{-3}}{(p+q)^6}$

$= (p+q)^{4+(-3)-6}$ *Product and quotient rules*

$= (p+q)^{-5}$

$= \dfrac{1}{(p+q)^5}$

93. $(7r^{1/2})(2r^{3/4})(-r^{1/6})$

$= -14r^{1/2+3/4+1/6}$ *Product rule*

$= -14r^{17/12}$

95. $\dfrac{y^{5/3} \cdot y^{-2}}{y^{-5/6}} = y^{5/3+(-2)-(-5/6)}$ *Product and quotient rules*

$= y^{10/6-12/6+5/6}$ *Common denominator*

$= y^{3/6} = y^{1/2}$

97. $2z^{1/3}(5z^2 - 2)$

$= 2z^{1/3}(5z^2) - 2z^{1/3}(2)$ *Distributive property*

$= 10z^{7/3} - 4z^{1/3}$

99. $(p + p^{1/2})(3p - 5)$

$= 3p^2 - 5p + 3p^{3/2} - 5p^{1/2}$ *FOIL*

$= 3p^2 + 3p^{3/2} - 5p - 5p^{1/2}$

101. $\sqrt{200} = \sqrt{100 \cdot 2}$

$$= \sqrt{100} \cdot \sqrt{2}$$

$$= 10\sqrt{2}$$

103. $\sqrt[4]{1250} = \sqrt[4]{625 \cdot 2}$

$= \sqrt[4]{625} \cdot \sqrt[4]{2}$

$= 5\sqrt[4]{2}$

105. $-\sqrt[3]{\dfrac{2}{5p^2}} = -\dfrac{\sqrt[3]{2}}{\sqrt[3]{5p^2}}$

$= -\dfrac{\sqrt[3]{2}}{\sqrt[3]{5p^2}} \cdot \dfrac{\sqrt[3]{25p}}{\sqrt[3]{25p}}$ *Rationalize denominator*

$= -\dfrac{\sqrt[3]{50p}}{\sqrt[3]{125p^3}}$

$= -\dfrac{\sqrt[3]{50p}}{5p}$

107. $\sqrt[4]{\sqrt[3]{m}} = (\sqrt[3]{m})^{1/4} = (m^{1/3})^{1/4}$

$= m^{1/3 \cdot 1/4} = m^{1/12} = \sqrt[12]{m}$

109. $(\sqrt[3]{2} + 4)(\sqrt[3]{2^2} - 4\sqrt[3]{2} + 16)$

$= \sqrt[3]{2}(\sqrt[3]{2^2} - 4\sqrt[3]{2} + 16)$

$\quad + 4(\sqrt[3]{2^2} - 4\sqrt[3]{2} + 16)$ *Distributive property*

$= \sqrt[3]{2^3} - 4\sqrt[3]{2^2} + 16\sqrt[3]{2} + 4\sqrt[3]{2^2}$

$\quad - 16\sqrt[3]{2} + 64$ *Distributive property*

$= 2 + 64 = 66$

Alternate solution:

$(\sqrt[3]{2} + 4)(\sqrt[3]{2} - 4\sqrt[3]{2} + 16)$

$= (\sqrt[3]{2} + 4)[(\sqrt[3]{2})^2 - \sqrt[3]{2} \cdot 4 + 4^2]$

$= (\sqrt[3]{2})^3 + 4^3$ *Sum of two cubes*

$= 2 + 64 = 66$

111. $\sqrt{18m^3} - 3m\sqrt{32m} + 5\sqrt{m^3}$

$= \sqrt{9m^2 \cdot 2m} - 3m\sqrt{16 \cdot 2m} + 5\sqrt{m^2 m}$

$= 3m\sqrt{3m} - 12m\sqrt{2m} + 5m\sqrt{m}$

$= -9m\sqrt{2m} + 5m\sqrt{m}$

or $m(-9\sqrt{2m} + 5\sqrt{m})$

113. $\dfrac{6}{3 - \sqrt{2}}$

$= \dfrac{6}{3 - \sqrt{2}} \cdot \dfrac{3 + \sqrt{2}}{3 + \sqrt{2}}$

Multiply numerator and denominator by conjugate of denominator

$= \dfrac{6(3 + \sqrt{2})}{9 - 2}$

$= \dfrac{6(3 + \sqrt{2})}{7}$

115. $\dfrac{\sqrt{x} - \sqrt{x - 2}}{\sqrt{x} + \sqrt{x - 2}}$

$= \dfrac{\sqrt{x} - \sqrt{x - 2}}{\sqrt{x} + \sqrt{x - 2}} \cdot \dfrac{\sqrt{x} - \sqrt{x - 2}}{\sqrt{x} - \sqrt{x - 2}}$

$= \dfrac{(\sqrt{x} - \sqrt{x - 2})^2}{(\sqrt{x})^2 - (\sqrt{x - 2})^2}$

$= \dfrac{(\sqrt{x})^2 - 2(\sqrt{x})(\sqrt{x - 2}) + (\sqrt{x - 2})^2}{x - (x - 2)}$

$= \dfrac{x - 2\sqrt{x}\sqrt{x - 2} + x - 2}{2}$

$= \dfrac{2x - 2\sqrt{x}\sqrt{x - 2} - 2}{2}$

$= \dfrac{2(x - \sqrt{x}\sqrt{x - 2} - 1)}{2}$

$= x - \sqrt{x}\sqrt{x - 2} - 1$

$= x - 1 - \sqrt{x(x - 2)}$

117. $\sqrt{-49} = i\sqrt{49} = 7i$

119. $(6 - i) + (4 - 2i) = 10 - 3i$

121. $15i - (3 + 2i) - 5$

$= 15i - 3 - 2i - 5$

$= -8 + 13i$

123. $(5 - i)(3 + 4i)$

$= 15 + 20i - 3i - 4i^2$

$= 15 + 17i + 4 \quad i^2 = -1$

$= 19 + 17i$

125. $(5 - 11i)(5 + 11i)$

$= 5^2 - (11i)^2$

Difference of two squares

$= 25 - 121i^2$

$= 25 + 121$

$= 146$

127. $(4 - 3i)^2$

$= 4^2 - 2(4)(3i) + (3i)^2$

Square of a binomial

$= 16 - 24i - 9$

$= 7 - 24i$

129. $\dfrac{6 + i}{1 - i}$

Multiply the numerator and denominator by the conjugate of the denominator.

$\dfrac{6 + i}{1 - i} = \dfrac{(6 + i)(1 + i)}{(1 - i)(1 + i)}$

$= \dfrac{6 + 7i + i^2}{1 - i^2}$

$= \dfrac{6 + 7i - 1}{1 + i}$

$= \dfrac{5 + 7i}{2}$

$= \dfrac{5}{2} + \dfrac{7}{2}i \quad$ *Standard form*

131. The product of a complex number and its conjugate is always a *real* number.

$$(a + bi)(a - bi) = a^2 - (bi)^2$$
$$= a^2 - b^2 i^2$$
$$= a^2 - b^2(-1)$$
$$= a^2 + b^2$$

133. $i^7 = i^4 \cdot i^3$
$$= 1(-i)$$
$$= -i$$

135. $i^{-35} = (i^{-4})^{-9} \cdot i$
$$= 1^{-9} \cdot i$$
$$= 1 \cdot i = i$$

Chapter 1 Test

1. (a) -13, $-\dfrac{12}{4}$ (or -3), 0, and $\sqrt{49}$ (or 7) are integers.

(b) -13, $-\dfrac{12}{4}$ (or -3), 0, $\dfrac{3}{5}$, 5.9, and $\sqrt{49}$ (or 7) are rational numbers.

(c) All numbers in the set are real numbers.

2. $\left|\dfrac{x^2 + 2yz}{3(x + z)}\right| = \left|\dfrac{(-2)^2 + 2(-4)(5)}{3(-2 + 5)}\right|$

Let $x = -2$, $y = -4$, $z = 5$

$$= \left|\dfrac{4 + (-40)}{3(3)}\right|$$

$$= \left|\dfrac{-36}{9}\right| = |4| = 4$$

3. (a) $a + (b + c) = (a + b) + c$
Associative property

(b) $a + (c + b) = a + (b + c)$
Commutative property

(c) $a(b + c) = ab + ac$
Distributive property

(d) $a + [b + (-b)] = a + 0$
Inverse property

4. $(x^2 - 3x + 2) - (x - 4x^2)$
$\quad + 3x(2x + 1)$
$\quad = (x^2 - 3x + 2) - (x - 4x^2)$
$\quad\quad + (6x^2 + 3x)$
$\quad = x^2 - 3x + 2 - x + 4x^2 + 6x^2 + 3x$
$\quad = 11x^2 - x + 2$

5. $(6r - 5)^2 = (6r)^2 - 2(6r)(5) + 5^2$
$$= 36r^2 - 60r + 25$$

6. $(t + 2)(3t^2 - t + 4)$

Multiply vertically.

$$
\begin{array}{r}
3t^2 - t + 4 \\
t + 2 \\
\hline
6t^2 - 2t + 8 \\
3t^3 - t^2 + 4t \quad\quad \\
\hline
3t^3 + 5t^2 + 2t + 8
\end{array}
$$

7. $\dfrac{2x^3 - 11x^2 + 28}{x - 5}$

$$
\begin{array}{r}
2x^2 - x - 5 \\
x - 5 \overline{)2x^3 - 11x^2 + 0x + 28} \\
\underline{2x^3 - 10x^2} \quad\quad\quad\quad \\
-x^2 + 0x \quad\quad \\
\underline{-x^2 + 5x} \quad\quad \\
-5x + 28 \\
\underline{-5x + 25} \\
3
\end{array}
$$

Thus,

$$\dfrac{2x^3 - 11x^2 + 28}{x - 5}$$

$$= 2x^2 - x - 5 + \dfrac{3}{x - 5}.$$

8. The year 1987 corresponds to $x = 3$, so substitute 3 for x in the given polynomial.

$18.7x^2 + 105.3x + 4814.1$

$= 18.7(3)^2 + 105.3(3) + 4814.1$

$= 5298.3$

According to this model, the adjusted poverty threshold in 1987 was approximately $5298.

10. $(2x - 3y)^4$

Use the coefficients from row 4 of Pascal's triangle.

$(2x - 3y)^4$

$= (2x)^4 + 4(2x)^3(-3y)^1 + 6(2x)^2(-3y)^2$
$\quad + 4(2x)^1(-3y)^3 + (-3y)^4$

$= 16x^4 + 4(8x^3)(-3y) + 6(4x^2)(9y^2)$
$\quad + 4(2x)(-27y^3) + 81y^4$

$= 16x^4 - 96x^3y + 216x^2y^2 - 216xy^3$
$\quad + 81y^4$

11. $x^4 - 16$

$= (x^2)^2 - 4^2$

$= (x^2 + 4)(x^2 - 4)$

$= (x^2 + 4)(x^2 - 2^2)$

$= (x^2 + 4)(x + 2)(x - 2)$

12. $24m^3 - 14m^2 - 24m$

$= 2m(12m^2 - 7m - 12)$

$= 2m(4m + 3)(3m - 4)$

13. $x^3y^2 - 9x^3 - 8y^2 + 72$

$= (x^3y^2 - 9x^3) + (-8y^2 + 72)$

$= x^3(y^2 - 9) - 8(y^2 - 9)$

$= (x^3 - 8)(y^2 - 9)$

$= (x - 2)(x^2 + 2x + 4)(y + 3)$
$\quad \cdot (y - 3)$

14. $\dfrac{5x^2 - 9x - 2}{30x^3 + 6x^2} \cdot \dfrac{2x^8 + 6x^7 + 4x^6}{x^4 - 3x^2 - 4}$

$= \dfrac{(5x + 1)(x - 2)}{6x^2(5x + 1)}$

$\quad \cdot \dfrac{2x^6(x^2 + 3x + 2)}{(x^2 - 4)(x^2 + 1)}$

$= \dfrac{(5x + 1)(x - 2)(2x^6)(x + 2)(x + 1)}{6x^2(5x + 1)(x + 2)(x - 2)(x^2 + 1)}$

$= \dfrac{2x^6(x + 1)}{6x^2(x^2 + 1)}$

$= \dfrac{x^4(x + 1)}{3(x^2 + 1)}$

15. $\dfrac{x}{x^2 + 3x + 2} + \dfrac{2x}{2x^2 - x - 3}$

$= \dfrac{x}{(x + 2)(x + 1)} + \dfrac{2x}{(2x - 3)(x + 1)}$

The least common denominator is $(x + 2)(x + 1)(2x - 3)$.

$= \dfrac{x(2x - 3)}{(x + 2)(x + 1)(2x - 3)}$

$\quad + \dfrac{2x(x + 2)}{(2x - 3)(x + 1)(x + 2)}$

$= \dfrac{2x^2 - 3x}{(x + 2)(x + 1)(2x - 3)}$

$\quad + \dfrac{2x^2 + 4x}{(x + 2)(x + 1)(2x - 3)}$

$= \dfrac{4x^2 + x}{(x + 2)(x + 1)(2x - 3)}$

$= \dfrac{x(4x + 1)}{(x + 2)(x - 1)(2x - 3)}$

16. $\dfrac{a + b}{2a - 3} - \dfrac{a - b}{3 - 2a}$

$= \dfrac{a + b}{2a - 3} - \dfrac{(a - b)(-1)}{(3 - 2a)(-1)}$

$= \dfrac{a + b}{2a - 3} + \dfrac{a - b}{2a - 3}$

$= \dfrac{2a}{2a - 3}$

If $3 - 2a$ is used as the common denominator, the result will be

$\dfrac{-2a}{3 - 2a}$. The rational expressions

$\dfrac{2a}{2a - 3}$ and $\dfrac{-2a}{3 - 2a}$ are equivalent.

17. $\dfrac{y - 2}{y - \dfrac{4}{y}} = \dfrac{y(y - 2)}{y\left(y - \dfrac{4}{y}\right)}$

$\qquad = \dfrac{y^2 - 2y}{y^2 - 4}$

$\qquad = \dfrac{y(y - 2)}{(y + 2)(y - 2)}$

$\qquad = \dfrac{y}{y + 2}$

18. $\left(\dfrac{x^{-2} y^{-1/3} z}{x^{-5/3} y^{-2/3} z^{2/3}}\right)^3$

$\qquad = \dfrac{x^{-6} y^{-1} z^3}{x^{-5} y^{-2} z^2}$

$\qquad = x^{-6-(-5)} y^{-1-(-2)} z^{3-2}$

$\qquad = x^{-1} y^1 z^1$

$\qquad = \dfrac{yz}{x}$

19. $\sqrt{18x^5 y^8} = \sqrt{(9x^4 y^8)(2x)}$

$\qquad = \sqrt{9x^4 y^8} \cdot \sqrt{2x}$

$\qquad = 3x^2 y^4 \sqrt{2x}$

20. $\sqrt{32x} + \sqrt{2x} - \sqrt{18x}$

$\qquad = \sqrt{16 \cdot 2x} + \sqrt{2x} - \sqrt{9 \cdot 2x}$

$\qquad = 4\sqrt{2x} + \sqrt{2x} - 3\sqrt{2x}$

$\qquad = 2\sqrt{2x}$

21. $(\sqrt{x} - \sqrt{y})(\sqrt{x} + \sqrt{y})$

$\qquad = (\sqrt{x})^2 - (\sqrt{y})^2$

$\qquad = x - y$

22. $t = 2\pi\sqrt{\dfrac{L}{32}} = 2\pi\sqrt{\dfrac{3.5}{32}}$ *Let L = 3.5*

$\qquad \approx 2.1$

The period of a pendulum 3.5 feet long is approximately 2.1 seconds.

23. $(7 - 3i) - (2 + 5i)$

$\qquad = (7 - 2) + (-3 - 5)i$

$\qquad = 5 - 8i$

24. $(4 + 3i)(-5 + 3i)$

$\qquad = -20 + 12i - 15i + 9i^2$

$\qquad = -20 + 12i - 15i - 9$

$\qquad = -29 - 3i$

25. $\dfrac{5 - 5i}{1 - 3i} = \dfrac{(5 - 5i)(1 + 3i)}{(1 - 3i)(1 + 3i)}$

$\qquad = \dfrac{5 + 15i - 5i - 15i^2}{1 - 9i^2}$

$\qquad = \dfrac{20 + 10i}{10}$

$\qquad = 2 + i$

26. $i^{301} = (i^4)^{75} \cdot i = 1^{75} \cdot i = 1 \cdot i = i$

i^{301} is equal to i.

CHAPTER 2 EQUATIONS AND INEQUALITIES

Section 2.1

1. The solution set of $2x + 3 = x - 5$ is $\{-8\}$.

   Replacing x with -8 gives

   $$2(-8) + 3 = -8 - 5$$
   $$-13 = -13,$$

   so the given statement is true.

3. The equations $x^2 = 9$ and $x = 3$ are equivalent equations.

   The solution set for $x = 3$ is $\{3\}$, while the solution set for $x^2 = 9$ is $\{3, -3\}$. Since the equations do not have the same solution set, they are not equivalent. The given statement is false.

7. $x^2 + 6x = x(x + 6)$

   Since the product of x and $x + 6$ is $x^2 + 6x$, the given equation is true for every value of x, and is an identity. The solution set is $\{$all real numbers$\}$.

9. $3t + 4 = 5(t - 2)$

   Replacing t with 7 gives

   $$3 \cdot 7 + 4 = 5(7 - 2)$$
   $$25 = 25,$$

   a true statement. However, using $t = 1$ gives

   $$3 \cdot 1 + 4 = 5(1 - 2)$$
   $$7 = -5,$$

   which is false. The equation is true for some values of t only, and is a conditional equation. The solution set is $\{7\}$.

11. $2x - 4 = 2(x + 2)$

    Applying the distributive property on the right side gives $2x + 4$, so the equation becomes

    $$2x - 4 = 2x + 4.$$

    Since this equation is false for all values of x, the equation is a contradiction.

    The solution set is $\emptyset$.

13. $\dfrac{5x}{x - 2} = \dfrac{20}{x - 2}$

    Solution set: $\{4\}$

    $$5x = 20$$

    Solution set: $\{4\}$

    Since the solution sets are equal, the equations are equivalent.

15. $\dfrac{x + 3}{x + 1} = \dfrac{2}{x + 1}$

    Solution set: $\emptyset$

    $$x = -1$$

    Solution set: $\{-1\}$

    Since the solution sets are not equal, the equations are not equivalent.

17. (b) $8x^2 - 4x + 3 = 0$

Because of the x^2 that appears, this equation cannot be written in the form $ax + b = 0$. It is not a linear equation. All of the other choices are equations that can be written in the form $ax + b = 0$ and therefore are linear equations.

19. $2m - 5 = m + 7$

$2m - 5 - m = m + 7 - m$
         Subtract m from both sides

      $m - 5 = 7$

$m - 5 + 5 = 7 + 5$ *Add 5 to both sides*

         $m = 12$

Solution set: $\{12\}$

21. $\frac{5}{6}k - 2k + \frac{1}{3} = \frac{2}{3}$

Multiply both sides of the equation by the least common denominator, 6.

$6\left(\frac{5}{6}k - 2k + \frac{1}{3}\right) = 6\left(\frac{2}{3}\right)$

$5k - 12k + 2 = 4$ *Distributive property*

   $-7k + 2 = 4$

      $-7k = 2$ *Subtract 2*

         $k = -\frac{2}{7}$ *Divide by 7*

Solution set: $\left\{-\frac{2}{7}\right\}$

23. $3r + 2 - 5(r + 1) = 6r + 4$

   $3r + 2 - 5r - 5 = 6r + 4$
            Distributive property

   $-2r - 3 = 6r + 4$
            Combine terms

$-2r - 3 + 2r = 6r + 4 + 2r$
            Add 2r to both sides

      $-3 = 8r + 4$
            Combine terms

   $-3 - 4 = 8r + 4 - 4$
            Subtract 4 from both sides

      $-7 = 8r$
            Combine terms

      $-\frac{7}{8} = r$
            Divide both sides by 8

Solution set: $\left\{-\frac{7}{8}\right\}$

25. $2[m - (4 + 2m) + 3] = 2m + 2$

   $2[m - 4 - 2m + 3] = 2m + 2$

      $2[-m - 1] = 2m + 2$

      $-2m - 2 = 2m + 2$

$-2m - 2 + 2m = 2m + 2 + 2m$

         $-2 = 4m + 2$

      $-2 - 2 = 4m + 2 - 2$

         $-4 = 4m$

         $-1 = m$

Solution set: $\{-1\}$

27. $\frac{3x - 2}{7} = \frac{x + 2}{5}$

Multiply both sides of the equation by the least common denominator, 35.

$35\left(\frac{3x - 2}{7}\right) = 35\left(\frac{x + 2}{5}\right)$

   $5(3x - 2) = 7(x + 2)$

   $15x - 10 = 7x + 14$

$15x - 10 - 7x = 7x + 14 - 7x$

      $8x - 10 = 14$

   $8x - 10 + 10 = 14 + 10$

      $8x = 24$

      $x = 3$

Solution set: $\{3\}$

29.
$$\frac{1}{4p} + \frac{2}{p} = 3$$

Multiply both sides of the equation by the least common denominator, $4p$, assuming $p \neq 0$.

$$4p\left(\frac{1}{4p}\right) + 4p\left(\frac{2}{p}\right) = 4p \cdot 3$$
$$1 + 8 = 12p$$
$$9 = 12p$$
$$\frac{1}{12} \cdot 9 = \frac{1}{12} \cdot 12p$$
$$\frac{3}{4} = p$$

Solution set: $\left\{\frac{3}{4}\right\}$

31.
$$\frac{m}{2} - \frac{1}{m} = \frac{6m + 5}{12}$$

Multiply both sides of the equation by the least common denominator, $12m$, assuming $m \neq 0$.

$$12m\left(\frac{m}{2} - \frac{1}{m}\right) = 12m\left(\frac{6m + 5}{12}\right)$$
$$6m^2 - 12 = 6m^2 + 5m$$
$$6m^2 - 12 - 6m^2 = 6m^2 + 5m - 6m^2$$
$$-12 = 5m$$
$$\frac{1}{5}(-12) = \frac{1}{5} \cdot 5m$$
$$-\frac{12}{5} = m$$

Solution set: $\left\{-\frac{12}{5}\right\}$

33.
$$\frac{2r}{r - 1} = 5 + \frac{2}{r - 1}$$

Multiply both sides of the equation by the least common denominator, $r - 1$, assuming $r \neq 1$.

$$(r - 1)\left(\frac{2r}{r - 1}\right) = (r - 1)\left(5 + \frac{2}{r - 1}\right)$$
$$2r = 5(r - 1) + 2$$
$$2r = 5r - 5 + 2$$
$$2r = 5r - 3$$
$$2r + 3 = 5r$$
$$3 = 3r$$
$$1 = r$$

Substituting 1 for r in the original equation would result in a denominator of 0, so 1 is not a solution. Solution set: $\emptyset$

35.
$$\frac{5}{2a + 3} + \frac{1}{a - 6} = 0$$

Multiply both sides by the least common denominator, $(2a + 3)(a - 6)$, assuming $a \neq -3/2$ and $a \neq 6$.

$$(2a + 3)(a - 6)\left(\frac{5}{2a + 3} + \frac{1}{a - 6}\right)$$
$$= (2a + 3)(a - 6) \cdot 0$$
$$5(a - 6) + (2a + 3) = 0$$
$$5a - 30 + 2a + 3 = 0$$
$$7a - 27 = 0$$
$$7a = 27$$
$$a = \frac{27}{7}$$

Solution set: $\left\{\frac{27}{7}\right\}$

37.
$$\frac{4}{x - 3} - \frac{8}{2x + 5} + \frac{3}{x - 3} = 0$$
$$\frac{7}{x - 3} - \frac{8}{2x + 5} = 0$$

Multiply both sides by the least common denominator, $(x - 3)(2x + 5)$, assuming $x \neq 3$ and $x \neq -5/2$.

$$(x - 3)(2x + 5)\left(\frac{7}{x - 3} - \frac{8}{2x + 5}\right)$$

$$= (x - 3)(2x + 5) \cdot 0$$

$$7(2x + 5) - 8(x - 3) = 0$$

$$14x + 35 - 8x + 24 = 0$$

$$6x + 59 = 0$$

$$6x = -59$$

$$x = -\frac{59}{6}$$

Solution set: $\left\{-\frac{59}{6}\right\}$

39.
$$\frac{2p}{p - 2} = 3 + \frac{4}{p - 2}$$

Multiply both sides by the least common denominator, $p - 2$, assuming $p \neq 2$.

$$(p - 2)\left(\frac{2p}{p - 2}\right) = (p - 2)\left(3 + \frac{4}{p - 2}\right)$$

$$2p = (p - 2) \cdot 3 + 4$$

$$2p = 3p - 6 + 4$$

$$2p = 3p - 2$$

$$0 = p - 2$$

$$2 = p$$

However, substituting 2 for p in the original equation would result in a denominator at 0, so 2 is not a solution.

Solution set: ∅

41.
$$\frac{3}{y - 2} + \frac{1}{y + 1} = \frac{1}{y^2 - y - 2}$$

$$\frac{3}{y - 2} + \frac{1}{y + 1} = \frac{1}{(y - 2)(y + 1)}$$

Multiply both sides by the least common denominator, $(y - 2)(y + 1)$, assuming $y \neq 2$ and $y \neq -1$.

$$(y - 2)(y + 1)\left(\frac{3}{y - 2}\right)$$

$$+(y - 2)(y + 1)\left(\frac{1}{y + 1}\right)$$

$$= (y - 2)(y + 1)\left(\frac{1}{(y - 2)(y + 1)}\right)$$

$$3(y + 1) + (y - 2) = 1$$

$$3y + 3 + y - 2 = 1$$

$$4y + 1 = 1$$

$$4y = 0$$

$$y = 0$$

Solution set: $\{0\}$

43. $.08w + .06(w + 12) = 7.72$

$$.08w + .06w + .72 = 7.72$$
Distributive property

$$.14w + .72 = 7.72$$

$$.14w + .72 - .72 = 7.72 - .72$$
Subtract .72

$$.14w = 7$$

$$\frac{.14w}{.14} = \frac{7}{.14}$$ *Divide by .14*

$$w = 50$$

Solution set: $\{50\}$

45. $2x + 6 = x + 2$

The solution set is $\{-4\}$, so the value -4 must be stored in X to make the equation true.

47. $3(2x - 5) + 4x = $ _____

Replacing x with 5 on the left side gives

$$3(2 \cdot 5 - 5) + 4 \cdot 5 = 3(5) + 20$$

$$= 35.$$

The blank on the right side must be replaced by 35 to make the solution set $\{5\}$.

48. $3(2x - 5) + 4x =$ _____

Replacing x with -4 on the left side gives

$$3[2(-4) - 5] + 4(-4) = 3(-13) - 16$$
$$= -55.$$

The blank on the right side must be replaced by -55 to make the solution set $\{-4\}$.

49. $-5x + 2(4 - 2x) = x +$ _____

Replacing x by 1 on the left side gives

$$-5(1) + 2[4 - 2(1)] = -5 + 2 \cdot 2$$
$$= -1.$$

Replacing x by 1 on the right side gives the value 1. The value -2 must be substituted in the blank so that the right side becomes 1 - 2 or -1, which equals the left side.

50. $-5x + 2(4 - 2x) = x +$ _____

Replacing x by 3.5 on the left side gives
$-5(3.5) + 2[4 - 2(3.5)]$
$= -17.5 + 2(-3)$
$= -23.5$

Replacing x by 3.5 on the right side gives the value 3.5. The value -27 must be substituted in the blank so that the right side becomes 3.5 - 27 or -23.5, which equals the left side.

51. 20°C

$$F = \frac{9}{5}C + 32$$
$$= \frac{9}{5}(20) + 32$$
$$= 36 + 32 = 68$$

Therefore, 20°C = 68°F.

53. 59°F

$$C = \frac{5(F - 32)}{9}$$
$$= \frac{5(59 - 32)}{9}$$
$$= \frac{5(27)}{9} = 15$$

Therefore, 59°F = 15°C.

55. 100°F

$$C = \frac{5(F - 32)}{9}$$
$$= \frac{5(100 - 32)}{9}$$
$$= \frac{5}{9}(68) = \frac{340}{9}$$
$$= 37.8$$

Therefore, 100°F = 37.8°C.

57. $p = 12$, $f = \$800$, $b = \$4000$, $q = 36$; find A

$$A = \frac{2pf}{b(q + 1)}$$

$$= \frac{2(12)(800)}{4000(36 + 1)}$$

$$= \frac{19,200}{148,000}$$

$$\approx .13$$

The annual interest rate, to the nearest percent, is 13%.

59. $A = 14\%$ (or $.14$), $p = 12$, $b = \$2000$, $q = 36$, find f

$$A = \frac{2pf}{b(q + 1)}$$

$$.14 = \frac{2(12)f}{2000(36 + 1)}$$

$$.14(2000)(37) = 24f$$

$$\frac{.14(2000)(37)}{24} = f$$

$$431.67 = f$$

The finance charge, to the nearest dollar, is $432.

61. $f = \$800$, $q = 36$, $n = 18$

$$u = f \cdot \frac{n(n + 1)}{q(q + 1)}$$

$$= 800 \cdot \frac{18(18 + 1)}{36(36 + 1)}$$

$$= \frac{273,600}{1332}$$

$$= 205.41$$

The amount of unearned interest is $205.41.

63. $f = \$950$, $q = 24$, $n = 6$

$$u = f \cdot \frac{n(n + 1)}{q(q + 1)}$$

$$u = 950 \cdot \frac{6(6 + 1)}{24(24 + 1)}$$

$$= 66.50$$

The amount of unearned interest is $66.50.

65. We must determine F when P equals 50. Thus

$$P = 1.06F + 7.18$$

$$50 = 1.06F + 7.18$$

$$F = \frac{42.82}{1.06} \approx 40$$

The required flow rate is approximately 40 liters per second.

67. **(a)** Let $y = 50$ in the equation.

$$y = -1.18x + 57.03$$

$$50 = -1.18x + 57.03$$

$$-7.03 = -1.18x$$

$$6 \approx x$$

In year 6, or 1988, the percent is approximately 50%.

(b) The line graph crosses the 50% line around 1988, which corresponds very closely with the answer from the model equation.

(c) In 1997, $x = 15$.

$$y = -1.18(15) + 57.03$$

$$y = 39.33$$

Using the model, the percent of alcohol-related deaths in 1997 would be 39.33%.

69. $2(x - 5) + 3x - x - 6 = 0$

Enter

$$y_1 = 2(x - 5) + 3x - x - 6$$

into the graphing calculator. Using the "root" option under the CALC menu yields the value x = 4.
The solution set is $\{4\}$.

71. $4x - 3(4 - 3x) - 2(x - 3) - 6x - 2 = 0$

Enter

$$y_1 = 4x - 3(4 - 3x) - 2(x - 3) - 6x - 2$$

into the graphing calculator. Using the "root" option under the CALC menu yields the value x = 1.6.
The solution set is $\{1.6\}$.

73. $\frac{x}{2} + \frac{x}{3} = 5$

Rewrite the equation as

$$\frac{x}{2} + \frac{x}{3} - 5 = 0.$$

Enter

$$y_1 = x/2 + x/3 - 5$$

into the graphing calculator. Using the "root" option under the CALC menu yields the value x = 6.
The solution set is $\{6\}$.

75. $3(2x + 1) - 2(x - 2) = 5$

Enter

$$y_1 = 3(2x + 1) - 2(x - 2)$$

and $\quad y_2 = 5$

into the graphing calculator. Using the "intersect" option under the CALC menu yields the value x = −.5.
The solution set is $\{-.5\}$.

77. $-(8 + 3x) + 5 = 2x + 3$

Enter:

$$y_1 = -(8 + 3x) + 5$$
$$y_2 = 2x + 3$$

Intersect: x = −1.2
Solution set: $\{-1.2\}$

79. $\frac{x - 2}{4} + \frac{x + 1}{2} = 1$

Enter:

$$y_1 = (x - 2)/4 + (x + 1)/2$$
$$y_2 = 1$$

Intersect: x = 1.3333...
Solution set: $\{1.\overline{3}\}$

81. $2x + 7 = 3(x + 3) - 8$

Enter

solve $(2x + 7 - 3(x + 3) + 8, x, 3)$

on the graphing calculator. The value returned is 6.
The solution set is $\{6\}$.

83. $5x - 2(x + 4) = 6x + 3$

Enter:

solve $(5x - 2(x + 4) - 6x - 3, x, 3)$

Return: −3.6666...
Solution set: $\{-3.\overline{6}\}$

85. $-2(x - 1) - 10 = 2(2 + x)$

Enter:

solve $(-2(x - 1) - 10 - 2(2 + x), x, 2)$

Return: −3
Solution set: $\{-3\}$

87. $4(.23x + \sqrt{5}) = \sqrt{2} + 1$

Using the SOLVE feature, the solution set is $\{16.07\}$.

89. $2\pi x + \sqrt[3]{4} = .5\pi x - \sqrt{28}$

Using the SOLVE feature, the solution set is $\{-1.46\}$.

91. $.23(\sqrt{3} + 4x) - .82(\pi x + 2.3) = 5$

Using the SOLVE feature, the solution set is $\{-3.92\}$.

Section 2.2

1. 15 minutes is 1/4 of an hour, so multiply 80 miles per hour by 1.4 to get a distance of 20 miles.

3. Multiply $100 by .04(4%) and by 2 years to get interest of $8.

5. $V = \ell wh$ for ℓ (volume of a rectangular box)

$$\frac{1}{wh} \cdot V = \frac{1}{wh} \cdot \ell wh$$

$$\frac{V}{wh} = \ell$$

7. $P = a + b + c$ for c (perimeter of a triangle)

$$P - a = b + c \quad \textit{Subtract } a$$
$$P - a - b = c \quad \textit{Subtract } b$$

9. $A = \frac{1}{2}(B + b)h$ for B (area of a trapezoid)

$$2A = (B + b)h \quad \textit{Multiply by 2}$$
$$2A = Bh + bh \quad \textit{Distributive property}$$
$$2A - bh = Bh \quad \textit{Subtract } bh$$
$$\frac{2A - bh}{h} = B \quad \textit{Divide by } h$$

or $\dfrac{2A}{h} - b = B$

11. $S = 2\pi rh + 2\pi r^2$ for h (surface area of a right circular cylinder)

$$S - 2\pi r^2 = 2\pi rh \quad \textit{Subtract } 2\pi r^2$$
$$\frac{S - 2\pi r^2}{2\pi r} = h \quad \textit{Divide by } 2\pi r$$

or $\dfrac{S}{2\pi r} - r = h$

13. $C = \frac{5}{9}(F - 32)$ for F (Fahrenheit to Celsius)

$$\frac{9}{5}C = F - 32 \quad \textit{Multiply by 9/5}$$
$$\frac{9}{5}C + 32 = F \quad \textit{Add 32}$$

or $F = \dfrac{9C}{5} + 32$

15. $u = f \cdot \dfrac{k(k + 1)}{n(n + 1)}$ for f (unearned interest)

$$un(n + 1) = f \cdot k(k + 1) \quad \textit{Multiply by } n(n + 1)$$
$$\frac{un(n + 1)}{k(k + 1)} = f \quad \textit{Divide by } k(k + 1)$$

17. It is not solved for x, since x appears on the left side of the equation as well.

19. Expression (d) does not represent the sales price. $x - .30$ represents x dollars discounted by 30 cents, not x dollars discounted by 30%. All of the other choices are equivalent and represent the sales price.

21. Let w = the width of the rectangle.

Then $2w - 3$ = the length.

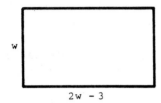

Use the formula for the perimeter of a rectangle.

$$P = 2L + 2W$$
$$54 = 2(2w - 3) + 2w$$
$$54 = 4w - 6 + 2w$$
$$60 = 6w$$
$$10 = w$$

The width is 10 cm.

23. Let w = the width of the tablecloth. Then $w + 11,757.6$

= the length.

Use the formula for the perimeter of a rectangle.

$$P = 2L + 2W$$
$$23,803.2 = 2(w + 11,757.6) + 2w$$
$$23,803.2 = 2w + 23,515.2 + 2w$$
$$288 = 4w$$
$$72 = w$$
$$w + 11,757.6 = 11,829.6$$

The length is 11,829.6 inches or 328.6 yd.

25. Let h = the height of the cylinder. Use the formula for the volume of a right circular cylinder.

$$V = \pi r^2 h$$
$$144\pi = \pi \cdot 6^2 \cdot h \quad \textit{Let } V = 144\pi, \; r = 6$$
$$144\pi = 36\pi h$$
$$4 = h$$

The height of the cylinder is 4 in.

27. (b) and (c) cannot be correct equations.

In (b), $-2x + 7(5 - x) = 62$
$$-2x + 35 - 7x = 62$$
$$35 - 9x = 62$$
$$-9x = 27$$
$$x = -3,$$

but the length of a rectangle cannot be negative.

In (c), $4(x + 2) + 4x = 8$
$$4x + 8 + 4x = 8$$
$$8 + 8x = 8$$
$$8x = 0$$
$$x = 0,$$

but the length of a rectangle cannot be zero.

29. Let x = Doug's score on the final exam.

$$\frac{78 + 94 + 60 + 2x}{5} = 80$$
$$\frac{232 + 2x}{5} = 80$$
$$232 + 2x = 400$$
$$2x = 168$$
$$x = 84$$

To earn an average of 80, Doug must score 84 on his final exam.

31. Let x = the amount of 92-octane gasoline.

Then 12 - x
= the amount of 98-octane gasoline.

Percent of isooctane	Amount of gasoline	Amount of isooctane
.92	x	.92x
.98	12 - x	.98(12 - x)
.96	12	.96(12)

Amount of isooctane in 92-octane	and	Amount of isooctane in 98-octane	is	Amount of isooctane in 96-octane
↓	↓	↓	↓	↓
.92x	+	.98(12 - x)	=	.96(12)

$$92x + 98(12 - x) = 96(12)$$
Multiply by 100
$$92x + 1176 - 98x = 1152$$
$$-6x = -24$$
$$x = 4$$
$$12 - x = 8$$

The mix requires 4 liters of 92-octane gasoline and 8 liters of 98-octane gasoline.

33. Let x = the person's IQ.

$$x = \frac{100 \cdot 20}{16}$$

$$= \frac{2000}{16}$$

$$= 125$$

The IQ is 125.

35. Let x = biking speed;

x + 4.5 = driving speed.

Set up a chart, using d = rt.

	d	r	t
Car	$\frac{1}{3}(x + 4.5)$	x + 4.5	$\frac{1}{3}$
Bike	$\frac{3}{4}x$	x	$\frac{3}{4}$

Since the speeds are given in miles per hour, the times must be changed from minutes to hours.

$$\frac{\text{Distance}}{\text{driving}} = \frac{\text{Distance}}{\text{biking}}$$

$$\frac{1}{3}(x + 4.5) = \frac{3}{4}x$$

$$12 \cdot \frac{1}{3}(x + 4.5) = 12 \cdot \frac{3}{4}x$$

$$4(x + 4.5) = 9x$$

$$4x + 18 = 9x$$

$$18 = 5x$$

$$\frac{18}{5} = x$$

To find the distance, use

$$d = \frac{3}{4}x = \frac{3}{4}\left(\frac{18}{5}\right) = \frac{27}{10} = 2.7.$$

Johnny travels 2.7 mi to work.

37. Let x = time on trip from Denver to Minneapolis.

Set up a chart, using the relationship d = rt.

	d	r	t
Denver to Minneapolis	50x	50	x
Minneapolis to Denver	55(32 - x)	55	32 - x

Distance to Minneapolis = Return distance

$$50x = 55(32 - x)$$
$$50x = 1760 - 55x$$
$$105x = 1760$$
$$x = 16.76$$
$$d = rt$$
$$= 50(16.76)$$
$$\approx 840$$

The distance between the two cities is about 840 mi.

39.   Let d = distance Janet runs.

Then $d + \frac{1}{2}$ = distance Russ runs.

Set up a chart, using d = rt.
This is equivalent to t = d/r.

	d	r	t
Russ	$d + \frac{1}{2}$	7	$\dfrac{d + \frac{1}{2}}{7}$
Janet	d	5	$\dfrac{d}{5}$

Since they both traveled for the same time, we have the equation

$$\frac{d + \frac{1}{2}}{7} = \frac{d}{5}.$$

Multiply both sides by the least common denominator, 35.

$$35\left(\frac{d + \frac{1}{2}}{7}\right) = 35\left(\frac{d}{5}\right)$$

$$5\left(d + \frac{1}{2}\right) = 35\left(\frac{d}{5}\right)$$

$$5d + \frac{5}{2} = 7d$$

$$\frac{5}{2} = 2d$$

$$\frac{5}{4} = d$$

To find t, use either

$$t = \frac{d + \frac{1}{2}}{7}$$

$$\text{or}\quad t = \frac{d}{5}.$$

Then $t = \dfrac{d}{5} = \dfrac{\frac{5}{4}}{5} = \dfrac{1}{4}.$

It will take 1/4 hr or 15 min until they are 1/2 mi apart.

41.   Let   x = the number of hours it takes Plant A to produce the maximum amount of pollutant.

Then 2x = the number of hours it takes Plant B to produce the maximum amount of pollutant.

	Rate	Time	Part of the job accomplished
Plant B	$\frac{1}{2x}$	26	$\frac{1}{2x}(26) = \frac{13}{x}$
Plant A	$\frac{1}{x}$	26	$\frac{1}{x}(26) = \frac{26}{x}$

$$\begin{array}{ccccc}\text{Part done by} & + & \text{Part done by} & = & \text{1 whole} \\ \text{Plant B} & & \text{Plant A} & & \text{job} \\ \downarrow & \downarrow & \downarrow & \downarrow & \downarrow \\ \frac{13}{x} & + & \frac{26}{x} & = & 1\end{array}$$

Multiply both sides by the least common denominator, x.

$$x\left(\frac{13}{x} + \frac{26}{x}\right) = x \cdot 1$$

$$13 + 26 = x$$

$$39 = x$$

$$78 = 2x$$

It will take plant B 78 hr to produce the maximum pollutant alone.

43. Let x = the number of hours to fill
the pool with both pipes
open.

	Rate	Time	Part of the job accomplished
Inlet pipe	$\frac{1}{5}$	x	$\frac{1}{5}x$
Outlet pipe	$\frac{1}{8}$	x	$\frac{1}{8}x$

Part done by inlet pipe	–	Part done by outlet pipe	=	Full pool
↓	↓	↓	↓	↓
$\frac{1}{5}x$	–	$\frac{1}{8}x$	=	1

Multiply both sides by the least
common denominator, 40.

$$40\left(\frac{1}{5}x - \frac{1}{8}x\right) = 40 \cdot 1$$

$$8x - 5x = 40$$

$$3x = 40$$

$$x = \frac{40}{3}$$

It took 40/3 hr to fill the pool.

45. Let x = number of liters of pure
alcohol to be added.

Strength	Liters of solution	Liters of pure alcohol
10%	7	.10(7)
100%	x	1(x)
30%	7 + x	.30(7 + x)

Liters of alcohol in 10% solution	+	Liters of alcohol in 100% solution	=	Liters of alcohol in 30% solution
↓	↓	↓	↓	↓
.10(7)	+	1(x)	=	.30(7 + x)

$$.10(7) + 1(x) = .30(7 + x)$$

$$.7 + x = 2.1 + .3x$$

$$.7x = 1.4$$

$$x = 2$$

He should add 2 liters of pure
alcohol.

47. Let x = the number of liters of pure
acid to be added.

Strength	Liters of solution	Liters of pure acid
30%	6	.30(6)
100%	x	1(x)
50%	6 + x	.50(6 + x)

Liters of alcohol in 30% solution	+	Liters of acid in 100% solution	=	Liters of acid in 30% solution
↓	↓	↓	↓	↓
.30(6)	+	1(x)	=	.50(6 + x)

$$.30(6) + 1(x) = .50(6 + x)$$
$$1.8 + x = 3 + .5x$$
$$.5x = 1.2$$
$$x = 2.4$$

2.4 liters of pure acid should be added.

49. Let x = amount of short-term note. Then 125,000 - x

= amount of long-term note.

$$.12x + .10(125,000 - x) = 13,700$$
$$12x + 10(125,000 - x) = 1,370,000$$
Multiply by 1000
$$12x + 1,250,000 - 10x = 1,370,000$$
$$2x = 120,000$$
$$x = 60,000$$

The amount of the short-term note is $60,000 and the amount of the long-term note is $125,000 - $60,000 = $65,000.

51. Let x = amount invested at 7%. Then 4x = amount invested at 11%.

$$.07x + .11(4x) = 7650$$

Multiply by 100.

$$7x + 11(4x) = 765,000$$
$$7x + 44x = 765,000$$
$$51x = 765,000$$
$$x = 15,000$$
$$4x = 60,000$$

The church invested $15,000 at 7% and $60,000 at 11%.

53. 28% of $48,000 is $13,440, so after paying her income tax, Majorie had $34,560 left to invest.

Let x = amount invested at 6.5%. Then 34,560 - x

= amount invested at 6.25%.

$$.065x + .0625(34,560 - x) = 2210$$

Multiply by 10,000 to clear decimals.

$$650x + 625(34,560 - x) = 22,100,000$$
$$650x + 21,600,000 - 625x = 22,100,000$$
$$25x = 500,000$$
$$x = 20,000$$

Marjorie invested $20,000 at 6.5% and $34,560 - $20,000 = $14,560 at 6.25%.

55. **(a)** Since each student needs 15 cu ft each minute and there are 60 minutes in an hour, the ventilation required by x students per hour would be

$$V = 60(15x) = 900x.$$

(b) The number of air exchanges per hour would be

$$A = \frac{900x}{15,000} = \frac{3}{50}x.$$

(c) If x = 40, then

$$A = \frac{3}{50}(40) = 2.4 \text{ ach.}$$

(d) It should be increased by $\frac{50}{15} = 3\frac{1}{3}$ times. Smoking areas require more than triple the ventilation.

57. **(a)** The risk for one year would be

$$\frac{R}{72} = \frac{1.5 \times 10^{-3}}{72} \approx .000021$$

for each individual.

(b) $C = .000021x$

(c) $C = .000021(100,000)$

$C = 2.1$

There are approximately 2.1 cancer cases for every 100,000 passive smokers.

(d) $C = \frac{.44(260,000,000)(.26)}{72}$

$C \approx 413,111$

There are approximately 413,000 excess deaths caused by smoking each year.

59. **(a)** In 1991, $x = 1$.

$y = 1.082(1) + 16.882$
$y = 17.964$

The average monthly rate in 1991 was about $17.96.

(b) In 1993, $x = 3$.

$y = 1.082(3) + 16.882$
$y = 20.128$

The average monthly rate in 1993 was about $20.13.

(c) In 1994, $x = 4$.

$y = 1.082(4) + 16.882$
$y = 21.21$

In 1994, the rate would be $21.21.

(d) Congress regulated rates for cable television in 1992 and called for more competition. We cannot use the model reliably because of influences like this.

Section 2.3

1. **(d)** is the only one set up for direct use of the zero-factor property.

$$(3x + 1)(x - 7) = 0$$
$$3x + 1 = 0 \quad \text{or} \quad x - 7 = 0$$
$$x = -\frac{1}{3} \quad \text{or} \qquad x = 7$$

Solution set: $\left\{-\frac{1}{3}, 7\right\}$

3. **(c)** is the only one that does not require Step 1 of the method of completing the square.

$$x^2 + x = 12$$
$$x^2 + x + \frac{1}{4} = 12 + \frac{1}{4}$$
$$\left(x + \frac{1}{2}\right)^2 = \frac{49}{4}$$
$$x + \frac{1}{2} = \pm\sqrt{\frac{49}{4}}$$
$$x = -\frac{1}{2} \pm \frac{7}{2}$$
$$x = -\frac{1}{2} + \frac{7}{2} \quad \text{or} \quad x = -\frac{1}{2} - \frac{7}{2}$$
$$x = 3 \qquad \text{or} \quad x = -4$$

Solution set: $\{3, -4\}$

5. $p^2 = 16$

$p = \sqrt{16} \quad \text{or} \quad p = -\sqrt{16}$
  Square root property

$p = 4 \quad \text{or} \quad p = -4$

Solution set: $\{\pm 4\}$

7. $x^2 = 27$

$x = \sqrt{27}$ or $x = -\sqrt{27}$
$\qquad\qquad\qquad$ *Square root property*

$x = 3\sqrt{3}$ or $x = -3\sqrt{3}$

Solution set: $\left\{\pm 3\sqrt{3}\right\}$

9. $t^2 = -16$

$t = \sqrt{-16}$ or $t = -\sqrt{-16}$
$\qquad\qquad\qquad$ *Square root property*

$t = 4i$ or $t = -4i$

Solution set: $\left\{\pm 4i\right\}$

11. $x^2 = -18$

$x = \sqrt{-18}$ or $x = -\sqrt{-18}$

$x = 3i\sqrt{2}$ or $x = -3i\sqrt{2}$

Solution set: $\left\{\pm 3i\sqrt{2}\right\}$

13. $(3k - 1)^2 = 12$

$3k - 1 = \pm\sqrt{12}$

$3k - 1 = \pm 2\sqrt{3}$

$3k = 1 \pm 2\sqrt{3}$

$k = \dfrac{1 \pm 2\sqrt{3}}{3}$

Solution set: $\left\{\dfrac{1 \pm 2\sqrt{3}}{3}\right\}$

15. $p^2 - 5p + 6 = 0$

$(p - 2)(p - 3) = 0$ *Factor*

$p - 2 = 0$ or $p - 3 = 0$
$\qquad\qquad\qquad$ *Zero-factor property*

$p = 2$ or $\quad p = 3$

Solution set: $\left\{2, 3\right\}$

17. $(5r - 3)^2 = -3$

$5r - 3 = \pm\sqrt{-3}$

$5r - 3 = \pm i\sqrt{3}$

$5r = 3 \pm i\sqrt{3}$

$r = \dfrac{3 \pm i\sqrt{3}}{5}$

$= \dfrac{3}{5} \pm \dfrac{\sqrt{3}}{5}i$ *Standard form*

Solution set: $\left\{\dfrac{3}{5} \pm \dfrac{\sqrt{3}}{5}i\right\}$

19. $p^2 - 8p + 15 = 0$

$p^2 - 8p = -15$

Half the coefficient of p is -4, and $(-4)^2 = 16$. Add 16 to both sides.

$p^2 - 8p + 16 = -15 + 16$

Factor on the left and combine terms on the right.

$(p - 4)^2 = 1$

Use the square root property to complete the solution.

$p - 4 = \pm 1$

$p = 4 + 1 = 5$ or $p = 4 - 1 = 3$

Solution set: $\left\{5, 3\right\}$

21. $x^2 - 2x - 4 = 0$

$x^2 - 2x = 4$

$x^2 - 2x + 1 = 4 + 1$

$(x - 1)^2 = 5$

$x - 1 = \pm\sqrt{5}$

$x = 1 \pm \sqrt{5}$

Solution set: $\left\{1 \pm \sqrt{5}\right\}$

23. $2p^2 + 2p + 1 = 0$

$p^2 + p + \dfrac{1}{2} = 0$ *Multiply by 1/2*

$p^2 + p = -\dfrac{1}{2}$ *Subtract 1/2*

Half the coefficient of p is 1/2, and $(1/2)^2 = 1/4$. Add 1/4 to both sides.

$p^2 + p + \dfrac{1}{4} = -\dfrac{1}{2} + \dfrac{1}{4}$

$\left(p + \dfrac{1}{2}\right)^2 = -\dfrac{1}{4}$

$p + \dfrac{1}{2} = \pm\sqrt{-\dfrac{1}{4}} = \dfrac{\pm i}{2}$

$p = -\dfrac{1}{2} \pm \dfrac{1}{2}i$

Solution set: $\left\{-\dfrac{1}{2} \pm \dfrac{1}{2}i\right\}$

25. He is incorrect. If the constant term is missing from the equation, c = 0.

27. $m^2 - m - 1 = 0$

Here a = 1, b = -1, and c = 1. Substitute these values into the quadratic formula.

$x = \dfrac{-b \pm \sqrt{b^2 - 4ac}}{2a}$

$= \dfrac{-(-1) \pm \sqrt{(-1)^2 - 4(1)(-1)}}{2(1)}$

$= \dfrac{1 \pm \sqrt{1 + 4}}{2}$

$x = \dfrac{1 \pm \sqrt{5}}{2}$

Solution set: $\left\{\dfrac{1 \pm \sqrt{5}}{2}\right\}$

29. $x^2 - 6x + 7 = 0$

Here a = 1, b = -6, and c = 7. Substitute these values into the quadratic formula.

$x = \dfrac{-b \pm \sqrt{b^2 - 4ac}}{2a}$

$= \dfrac{-(-6) \pm \sqrt{(-6)^2 - 4(1)(7)}}{2(1)}$

$= \dfrac{6 \pm \sqrt{36 - 28}}{2} = \dfrac{6 \pm \sqrt{8}}{2}$

$= \dfrac{6 \pm 2\sqrt{2}}{2}$

$= \dfrac{2(3 \pm \sqrt{2})}{2}$ *Factor numerator*

$x = 3 \pm \sqrt{2}$ *Lowest terms*

Solution set: $\left\{3 \pm \sqrt{2}\right\}$

31. $4z^2 - 12z + 11 = 0$

Here a = 4, b = -12, and c = 11. Substitute these values into the quadratic formula.

$z = \dfrac{-(-12) \pm \sqrt{(-12)^2 - 4(4)(11)}}{2(4)}$

$= \dfrac{12 \pm \sqrt{144 - 176}}{8}$

$= \dfrac{12 \pm \sqrt{-32}}{8} = \dfrac{12 \pm i\sqrt{32}}{8}$

$= \dfrac{12 \pm 4i\sqrt{2}}{8} = \dfrac{4(3 \pm i\sqrt{2})}{8}$

$z = \dfrac{3}{2} \pm \dfrac{\sqrt{2}}{2}i$

Solution set: $\left\{\dfrac{3}{2} \pm \dfrac{\sqrt{2}}{2}i\right\}$

33. $\dfrac{1}{2}t^2 + \dfrac{1}{4}t - 3 = 0$

Multiply both sides by the least common denominator, 4.

$4\left(\dfrac{1}{2}t^2 + \dfrac{1}{4}t - 3\right) = 4(0)$

$2t^2 + t - 12 = 0$

Substitute $a = 2$, $b = 1$, and $c = -12$ into the quadratic formula.

$$t = \frac{-1 \pm \sqrt{1^2 - 4(2)(-12)}}{2(2)}$$

$$= \frac{-1 \pm \sqrt{1 + 96}}{4}$$

$$t = \frac{-1 \pm \sqrt{97}}{4}$$

Solution set: $\left\{\dfrac{-1 \pm \sqrt{97}}{4}\right\}$

35. $4 + \dfrac{3}{x} - \dfrac{2}{x^2} = 0$

Multiply both sides of the equation by the least common denominator, x^2.

$$4x^2 + 3x - 2 = 0$$

Here $a = 4$, $b = 3$, and $c = -2$. Substitute these values into the quadratic formula.

$$x = \frac{-(3) \pm \sqrt{3^2 - 4(4)(-2)}}{2(4)}$$

$$= \frac{-3 \pm \sqrt{9 + 32}}{8}$$

$$x = \frac{-3 \pm \sqrt{41}}{8}$$

Solution set: $\left\{\dfrac{-3 \pm \sqrt{41}}{8}\right\}$

37. $x^3 - 8 = 0$

Factor the left side as the difference of two cubes.

$$(x - 2)(x^2 + 2x + 4) = 0$$

Set each factor to zero.

$x - 2 = 0$ or $x^2 + 2x + 4 = 0$

$x = 2$

Use the quadratic formula to solve $x^2 + 2x + 4 = 0$.

$x^2 + 2x + 4 = 0$

$$x = \frac{-2 \pm \sqrt{4 - 16}}{2}$$

$$= \frac{-2 \pm \sqrt{-12}}{2}$$

$$= \frac{-2 \pm 2i\sqrt{3}}{2}$$

$x = -1 \pm i\sqrt{3}$

Solution set: $\{2, -1 \pm i\sqrt{3}\}$

39. $x^3 - 27 = 0$

$(x - 3)(x^2 + 3x + 9) = 0$

$x - 3 = 0$

$x = 3$

or

$x^2 + 3x + 9 = 0$

$$x = \frac{-3 \pm \sqrt{9 - 36}}{2}$$

$$= \frac{-3 \pm \sqrt{-27}}{2}$$

$$= \frac{-3 \pm 3i\sqrt{3}}{2}$$

$x = 3$ or $x = -\dfrac{3}{2} \pm \dfrac{3\sqrt{3}}{2}i$ *Standard form*

Solution set: $\left\{3, -\dfrac{3}{2} \pm \dfrac{3\sqrt{3}}{2}i\right\}$

41. $x^2 + 64 = 0$

$(x + 4)(x^2 - 4x + 16) = 0$

$x + 4 = 0$

$x = -4$

or

$x^2 - 4x + 16 = 0$

$$x = \frac{4 \pm \sqrt{16 - 64}}{2}$$

$$= \frac{4 \pm \sqrt{-48}}{2}$$

$$= \frac{4 \pm 4i\sqrt{3}}{2}$$

$x = -4$ or $x = 2 \pm 2i\sqrt{3}$

Solution set: $\{-4, \ 2 \pm 2i\sqrt{3}\}$

43.
$$8p^3 + 125 = 0$$

Factor the left side as the sum of two cubes.

$(2p + 5)(4p^2 - 10p + 25) = 0$

$2p + 5 = 0$ or $4p^2 - 10p + 25 = 0$
                               Square root property

$p = -\dfrac{5}{2}$ or $p = \dfrac{10 \pm \sqrt{100 - 400}}{8}$
                                      Quadratic formula

$\qquad\qquad\qquad = \dfrac{10 \pm \sqrt{-300}}{8}$

$\qquad\qquad\qquad = \dfrac{10 \pm 10i\sqrt{3}}{8}$

$\qquad\qquad\qquad = \dfrac{2(5 \pm i\sqrt{3})}{2 \cdot 4}$

$\qquad\qquad\qquad = \dfrac{5 \pm 5i\sqrt{3}}{4}$

$p = -\dfrac{5}{2}$ or $p = \dfrac{5}{4} \pm \dfrac{5\sqrt{3}}{4}i$
                               Standard form

Solution set: $\left\{-\dfrac{5}{2}, \ \dfrac{5}{4} \pm \dfrac{5\sqrt{3}}{4}i\right\}$

45. $(m - 3)^2 = 5$

$m - 3 = \pm\sqrt{5}$ *Square root property*

$m = 3 \pm \sqrt{5}$

Solution set: $\{3 \pm \sqrt{5}\}$

47.
$$x^2 + x = -1$$
$$x^2 + x + 1 = 0$$

Substitute $a = 1$, $b = 1$, and $c = 1$ into the quadratic formula.

$x = \dfrac{-1 \pm \sqrt{1^2 - 4 \cdot 1 \cdot 1}}{2 \cdot 1}$

$\quad = \dfrac{-1 \pm \sqrt{1 - 4}}{2}$

$\quad = \dfrac{-1 \pm \sqrt{-3}}{2}$

$\quad = \dfrac{-1 \pm i\sqrt{3}}{2}$

$\quad = -\dfrac{1}{2} \pm \dfrac{\sqrt{3}}{2}i$ *Standard form*

Solution set: $\left\{-\dfrac{1}{2} \pm \dfrac{\sqrt{3}}{2}i\right\}$

49. $(3y + 1)^2 = -7$

Use the square root property.

$3y + 1 = \pm i\sqrt{7}$

$3y = -1 \pm i\sqrt{7}$

$y = \dfrac{-1 \pm i\sqrt{7}}{3}$

$\quad = -\dfrac{1}{3} \pm \dfrac{\sqrt{7}}{3}i$ *Standard form*

Solution set: $\left\{-\dfrac{1}{3} \pm \dfrac{\sqrt{7}}{3}i\right\}$

Note for Exercises 51–55: The answers given here show all decimal places which are displayed at the bottom of the screen when the graphing method is used. However, the internal memory of the calculator holds more decimal places. To see these extra decimal places, call up X from the home screen after the graphing routine is completed.

51. $x^2 - \sqrt{2}x - 1 = 0$

Enter $y_1 = x^2 - \sqrt{2}x - 1$ on the graphing calcultor. Using the "root" option under the CALC menu yields $x = -.5176381$ or $x = -1.9318517$.

Solution set:

$\qquad \{-.5176381, \ 1.9318517\}$

53. $\sqrt{2}x^2 - 3x = -\sqrt{2}$

Rewrite as $\sqrt{2}x^2 - 3x + \sqrt{2} = 0$.

Enter: $y_1 = \sqrt{2}x^2 - 3x + \sqrt{2}$

Roots: $x = .70710678, 1.4142136$

Solution set:

$\{.70710678, 1.4142136\}$

55. $x(x + \sqrt{5}) = -1$

Simplify and rewrite as

$x^2 + \sqrt{5}x = -1$

$x^2 + \sqrt{5}x + 1 = 0$

Enter: $y_1 = x^2 + \sqrt{5}x + 1$

Roots: $x = -1.618034, x = -.618034$

Solution set: $\{-1.618034, -.618034\}$

57. $2x^2 + \sqrt{6}x + 7 = 0$

Using a quadratic formula program, enter $a = 2$, $b = \sqrt{6}$, $c = 7$.

Solution set:

$\{-.6123724357 \pm 1.7677669953i\}$

59. $\sqrt{6}x^2 + 5x = -\sqrt{10}$

Rewrite as $\sqrt{6}x^2 + 5x + \sqrt{10} = 0$.

Enter: $a = \sqrt{6}$, $b = 5$, $c = \sqrt{10}$

Solution set:

$\{-1.020620726 \pm .49932732296i\}$

61. $8.4x(x - 1) = -8$

Simplify and rewrite as

$8.4x^2 - 8.4x = -8$

$8.4x^2 - 8.4x + 8 = 0$.

Enter: $a = 8.4$, $b = -8.4$, $c = 8$

Solution set: $\{.5 \pm .83808170981i\}$

63. $2x^2 - 13x - 7 = 0$

Using the SOLVE feature of the graphing calculator, the results are

Guess: -5 Root: $-.5$

Guess: 5 Root: $7.$

Solution set: $\{-.5, 7\}$

65. $\sqrt{6}x^2 - 2x - 1.4 = 0$

Guess: -1 Root: $-.4509456768$

Guess: 2 Root: 1.267442258

Solution set:

$\{-.4509456768, 1.267442258\}$

67. $s = \frac{1}{2}gt^2$ for t

First, multiply both sides by 2.

$2s = gt^2$

Now divide by g.

$\frac{2s}{g} = t^2$

Use the square root property and rationalize the denominator on the right.

$t = \pm\sqrt{\dfrac{2s}{g}} = \pm\dfrac{\sqrt{2s}}{\sqrt{g}} \cdot \dfrac{\sqrt{g}}{\sqrt{g}}$

$= \pm\dfrac{\sqrt{2sg}}{g}$

69. $F = \dfrac{kMv^4}{r}$ for r

First, multiply both sides by r.

$Fr = kMv^4$

Next, divide both sides by kM.

$\dfrac{Fr}{kM} = v^4$

Now take the fourth root of both sides.

$$v = \pm \sqrt[4]{\frac{Fr}{kM}}$$

Finally, rationalize the denominator on the right.

$$v = \pm\frac{\sqrt[4]{Fr}}{\sqrt[4]{kM}} \cdot \frac{\sqrt[4]{k^3M^3}}{\sqrt[4]{k^3M^3}}$$

$$= \frac{\pm\sqrt[4]{Frk^3M^3}}{kM}$$

71. $P = \dfrac{E^2R}{(r + R)^2}$ for R

$$P(r + R)^2 = E^2R$$
$$P(r^2 + 2rR + R^2) = E^2R$$
$$Pr^2 + 2PrR + PR^2 = E^2R$$
$$PR^2 - E^2R + 2PrR + Pr^2 = 0$$
$$PR^2 + (2Pr - E^2)R + Pr^2 = 0$$

To find R, use the quadratic formula with $a = P$, $b = 2Pr - E^2$, and $c = Pr^2$.

$$R = \frac{-(2Pr - E^2) \pm \sqrt{(2Pr - E^2) - 4(P)(Pr^2)}}{2(P)}$$

$$= \frac{-2Pr + E^2 \pm \sqrt{4P^2r^2 - 4PrE^2 + E^4 - 4P^2r^2}}{2P}$$

$$= \frac{-2Pr + E^2 \pm \sqrt{E^4 - 4PrE^2}}{2P}$$

$$= \frac{-2Pr + E^2 \pm \sqrt{E^2(E^2 - 4Pr)}}{2P}$$

$$R = \frac{E^2 - 2Pr \pm E\sqrt{E^2 - 4Pr}}{2P}$$

73. $4x^2 - 2xy + 3y^2 = 2$

(a) Solve for x in terms of y.

$$4x^2 - 2yx + 3y^2 - 2 = 0$$
$$a = 4, \ b = -2y, \ c = 3y^2 - 2$$

$$x = \frac{2y \pm \sqrt{4y^2 - 16(3y^2 - 2)}}{8}$$

$$= \frac{2y \pm \sqrt{4y^2 - 48y^2 + 32}}{8}$$

$$= \frac{2y \pm \sqrt{32 - 44y^2}}{8}$$

$$= \frac{2y \pm \sqrt{4(8 - 11y^2)}}{8}$$

$$= \frac{2y \pm 2\sqrt{8 - 11y^2}}{4}$$

$$x = \frac{y \pm \sqrt{8 - 11y^2}}{4}$$

(b) Solve for y in terms of x.

$$3y^2 - 2xy + 4x^2 - 2 = 0$$
$$a = 3, \ b = -2x, \ c = 4x^2 - 2$$

$$y = \frac{2x \pm \sqrt{4x^2 - 12(4x^2 - 2)}}{6}$$

$$= \frac{2x \pm \sqrt{4x^2 - 48x^2 + 24}}{6}$$

$$= \frac{2x \pm \sqrt{24 - 44x^2}}{6}$$

$$= \frac{2x \pm \sqrt{4(6 - 11x^2)}}{6}$$

$$= \frac{2x \pm 2\sqrt{6 - 11x^2}}{6}$$

$$y = \frac{x \pm \sqrt{6 - 11x^2}}{3}$$

75. $x^2 + 8x + 16 = 0$

$a = 1, \ b = 8, \ c = 16$

$$b^2 - 4ac = 8^2 - 4(1)(16)$$
$$= 64 - 64 = 0$$

The equation has one rational solution since the discriminant is 0.

77. $3m^2 - 5m + 2 = 0$

$a = 3, \ b = -5, \ c = 2$

$$b^2 - 4ac = (-5)^2 - 4(3)(2)$$
$$= 25 - 24 = 1$$

The equation has two different rational solutions since the discriminant is a positive perfect square.

79. $4p^2 = 6p + 3$

$a = 4$, $b = -6$, $c = -3$

$$b^2 - 4ac = (-6)^2 - 4(4)(-3)$$
$$= 36 + 48$$
$$= 84$$

The equation has two different irrational solutions since the discriminant is positive but not a perfect square.

81. $9k^2 + 11k + 4 = 0$

$a = 9$, $b = 11$, $c = 4$

$$b^2 - 4ac = 11^2 - 4(9)(4)$$
$$= 121 - 144$$
$$= -23$$

The equation has two different imaginary solutions since the discriminant is negative.

83. $8x^2 - 72 = 0$

$a = 8$, $b = 0$, $c = -72$

$$b^2 - 4ac = 0^2 - 4(8)(-72)$$
$$= 0 + 2304$$
$$= 2304 = 48^2$$

The equation has two different rational solutions since the discriminant is a positive perfect square.

85. For a quadratic equation to have only one solution, the value of the discriminant must be 0.

86. $x^2 + 11x + k = 0$

The discriminant is

$$b^2 - 4ac = 11^2 - 4(1)(k)$$
$$= 121 - 4k.$$

87. The discriminant must equal zero, so the equation that must be solved is

$$121 - 4k = 0.$$

88. Solve $121 - 4k = 0$.

$$121 - 4k = 0$$
$$121 = 4k$$
$$\frac{121}{4} = k$$

Solution set: $\left\{\dfrac{121}{4}\right\}$

89. With $k = \dfrac{121}{4}$, the original equation is

$$x^2 + 11x + \frac{121}{4} = 0.$$

90. Solve $x^2 + 11x + \dfrac{121}{4} = 0$.

$$x^2 + 11x + \frac{121}{4} = 0$$
$$\left(x + \frac{11}{2}\right)^2 = 0$$
$$x + \frac{11}{2} = 0$$
$$x = -\frac{11}{2}$$

Solution set: $\left\{-\dfrac{11}{2}\right\}$

91. $25x^2 - 10x + k = 0$

Calculate the discriminant.

$$b^2 - 4ac = (-10)^2 - 4(25)(k)$$
$$= 100 - 100k$$

Set the discriminant equal to zero and solve for k.

$$100 - 100k = 0$$
$$100 = 100k$$
$$1 = k$$

Substitute $k = 1$ into the original equation and solve.

$$25x^2 - 10x + 1 = 0$$
$$(5x - 1)^2 = 0$$
$$5x - 1 = 0$$
$$5x = 1$$
$$x = \frac{1}{5}$$

Solution set: $\left\{\dfrac{1}{5}\right\}$

92. $16x^2 + kx + 25 = 0$

Calculate the discriminant.

$$b^2 - 4ac = k^2 - 4(16)(25)$$
$$= k^2 - 1600$$

Set the discriminant equal to zero and solve for k.

$$k^2 - 1600 = 0$$
$$k^2 = 1600$$
$$k = \pm 40$$

The two equations are

$$16x^2 + 40x + 25 = 0$$

and

$$16x^2 - 40x + 25 = 0.$$

Solve the first equation.

$$16x^2 + 40x + 25 = 0$$
$$(4x + 5)^2 = 0$$
$$4x + 5 = 0$$
$$4x = -5$$
$$x = -\frac{5}{4}$$

Solution set: $\left\{-\dfrac{5}{4}\right\}$

Solve the second equation.

$$16x^2 - 40x + 25 = 0$$
$$(4x - 5)^2 = 0$$
$$4x - 5 = 0$$
$$4x = 5$$
$$x = \frac{5}{4}$$

Solution set: $\left\{\dfrac{5}{4}\right\}$

93. $\dfrac{3 + 2x}{3x^2 - 19x - 14}$

To find the restrictions on the variable, set the denominator equal to zero and solve.

$$3x^2 - 19x - 14 = 0$$
$$(3x + 2)(x - 7) = 0$$
$$3x + 2 = 0 \quad \text{or} \quad x - 7 = 0$$
$$x = -\frac{2}{3} \quad \text{or} \quad x = 7$$

The restrictions on the variable are $x \neq -\dfrac{2}{3}$ and $x \neq 7$.

95. $\dfrac{y - 3}{16y^2 + 8y + 1}$

To find the restrictions on the variable, set the denominator equal to zero and solve.

$$16y^2 + 8y + 1 = 0$$
$$(4y + 1)^2 = 0$$
$$4y + 1 = 0$$
$$y = -\frac{1}{4}$$

The restriction on the variable is $y \neq -\dfrac{1}{4}$.

97. $\dfrac{-8 + 7x}{x^2 + x + 1}$

To find the restrictions on the variable, set the denominator equal to zero and solve by the quadratic formula with $a = 1$, $b = 1$, and $c = 1$.

$$x^2 + x + 1 = 0$$
$$x = \frac{-1 \pm \sqrt{1 - 4(1)(1)}}{2(1)}$$
$$= \frac{-1 \pm \sqrt{-3}}{2}$$
$$= \frac{-1 \pm i\sqrt{3}}{2}$$

We can also see that the equation has no real solutions by finding the discriminant:

$$b^2 - 4ac = 1 - 4 = -3.$$

Since there are no real solutions to this equation, there are no real numbers that make the denominator equal to zero. Thus, there are no real number restrictions on x.

Section 2.4

1. The length of the parking area is $2x + 200$, while the width is x, so the area is

$$(2x + 200)x.$$

Set the area equal to 40,000.

$$(2x + 200)x = 400,000.$$

The correct choice is (a).

3. Use the Pythagorean theorem with $a = x$, $b = 2x - 2$, and $c = x + 4$.

$$x^2 + (2x - 2)^2 = (x + 4)^2$$

The correct choice is (d).

5. Use the figure and equation (a) from Exercise 1.

$$x(2x + 200) = 40,000$$
$$2x^2 + 200x = 40,000$$
$$2x^2 + 200x - 40,000 = 0$$
$$x^2 + 100x - 20,000 = 0$$
$$(x - 100)(x + 200) = 0$$
$$x = 100 \quad \text{or} \quad x = -200$$

A rectangle cannot have a negative width, so reject -200 as a solution. If $x = 100$, then $2x + 200 = 400$. The dimensions of the lot are 100 yd by 400 yd.

7. Let $\quad x$ = outside width of frame. Then $x + 3$ = outside length of frame.

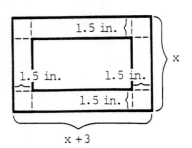

Since 70 sq inches is the area of the unframed picture, we must express the dimensions of the unframed picture in terms of x. Since the frame extends 1.5 inches beyond the picture on each side, the width of the picture is $x - 2(1.5) = x - 3$, and the length of the picture is $x + 3 - 2(1.5) = x + 3 - 3 = x$. Apply the formula $A = LW$ to the rectangular picture.

$$A = LW$$
$$70 = x(x - 3)$$
$$70 = x^2 - 3x$$
$$0 = x^2 - 3x - 70$$
$$0 = (x + 7)(x - 10)$$
$$x = -7 \quad \text{or} \quad x = 10$$

The width of a rectangle cannot be negative, so reject −7 as a solution.

If $x = 10$, $x + 3 = 13$.

The outside dimensions of the frame are 10 inches by 13 inches.

9. Let x = width of margins.
Then $23 + 2x$ = length of page;
$18 + 2x$ = width of page.

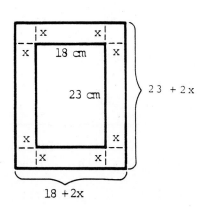

Write an equation for the area of the page using the formula $A = LW$.

$$594 = (23 + 2x)(18 + 2x)$$
$$594 = 414 + 36x + 46x + 4x^2$$
$$594 = 414 + 82x + 4x^2$$
$$0 = 4x^2 + 82x - 180$$
$$0 = 2x^2 + 41x - 90$$
$$0 = (2x + 45)(x - 2)$$
$$x = -\frac{45}{2} \quad \text{or} \quad x = 2$$

The width of the margin cannot be negative. The width of the margins should be 2 cm.

11. Let x = the number of hours for Felix to clean garage working alone.

Then $x - 9$ = the number of hours for Felipe to clean garage working alone.

	Rate	Time	Part of the job accomplished
Felix	$\frac{1}{x}$	20	$20\left(\frac{1}{x}\right) = \frac{20}{x}$
Felipe	$\frac{1}{x - 9}$	20	$20\left(\frac{1}{x - 9}\right) = \frac{20}{x - 9}$

The sum of the two parts of the job accomplished is 1, since one whole job is done.

$$\frac{20}{x} + \frac{20}{x - 9} = 1$$

To clear fractions, multiply both sides by the least common denominator $x(x - 9)$.

$$x(x - 9)\left(\frac{20}{x} + \frac{20}{x - 9}\right) = x(x - 9) \cdot 1$$

$$20(x - 9) + 20x = x^2 - 9x$$

$$20x - 180 + 20x = x^2 - 9x$$

$$40x - 180 = x^2 - 9x$$

$$0 = x^2 - 49x + 180$$

$$0 = (x - 45)(x - 4)$$

$$x = 45 \quad \text{or} \quad x = 4$$

$$x - 9 = 36 \quad \text{or} \quad x - 9 = -5$$

Since the solution $x = 4$ leads to $x - 9 = -5$, this solution must be rejected because the number of hours for Felix working alone cannot be negative.

It would take Felix 45 hr and Felipe 36 hr to clean the garage alone.

13. Let x = number of hours for new data processor to complete project.

Then $x - 2$ = number of hours for experienced data processor to complete project.

	Rate	Time	Part of the job accomplished
New data processor	$\frac{1}{x}$	2.4	$2.4\left(\frac{1}{x}\right) = \frac{2.4}{x}$
Experienced data processor	$\frac{1}{x - 2}$	2.4	$2.4\left(\frac{1}{x - 2}\right) = \frac{2.4}{x - 2}$

Together, the two data processors do one whole job, so

$$\frac{2.4}{x} + \frac{2.4}{x - 2} = 1.$$

To clear fractions, multiply both sides of the equation by the least common denominator, $x(x - 2)$.

$$x(x - 2)\left(\frac{2.4}{x} + \frac{2.4}{x - 2}\right) = x(x - 2) \cdot 1$$

$$2.4(x - 2) + 2.4x = x^2 - 2x$$

$$2.4x - 4.8 + 2.4x = x^2 - 2x$$

$$4.8x - 4.8 = x^2 - 2x$$

$$0 = x^2 - 6.8x + 4.8$$

To clear decimals, multiply both sides of the equation by 10.

$$0 = 10x^2 - 68x + 48$$

$$0 = 5x^2 - 34x + 24$$

$$0 = (5x - 4)(x - 6)$$

$$5x - 4 = 0 \quad \text{or} \quad x - 6 = 0$$

$$x = \frac{4}{5} \quad \text{or} \quad x = 6$$

$$x - 2 = -\frac{6}{5} \qquad x - 2 = 4$$

Since the solution $x = \frac{4}{5}$ leads to $x - 2 = -\frac{6}{5}$, this solution must be rejected because the number of hours for the experienced data processor working alone cannot be negative.

It would take the experienced data processor 4 hr to complete the project alone.

15. Let r = Steve's speed.

	d	r	t
Paula	100	$r + 10$	$\frac{100}{r + 10}$
Steve	100	r	$\frac{100}{r}$

Steve's time	is	1/3 hour longer than Paula's time.
↓	↓	↓
$\frac{100}{r}$	$=$	$\frac{100}{r + 10} + \frac{1}{3}$

To clear fractions, multiply both sides of the equation by the least common denominator, $3r(r + 10)$.

$$3r(r + 10) \cdot \frac{100}{r} = 3r(r + 10)\left(\frac{100}{r + 10} + \frac{1}{3}\right)$$

$$300(r + 10) = 300r + r(r + 10)$$

$$300r + 3000 = 300r + r^2 + 10r$$

$$0 = r^2 + 10r - 3000$$

$$0 = (r - 50)(r + 60)$$

$$r = 50 \quad \text{or} \quad r = -60$$

Since speed cannot be negative, reject −60 as a solution.

Steve's average speed is 50 mph.

17. Let   x = the speed of the plane for the first trip.

    Then x + 50 = the speed of the plane for the second trip.

    Since d = rt, t = d/r.

	d	r	t
1st trip	1000	x	$\frac{1000}{x}$
2nd trip	2025	x + 50	$\frac{2025}{x + 50}$

Time for 2nd trip	=	Time for 1st trip	+	2 hr
↓	↓	↓	↓	↓
$\frac{2025}{x + 50}$	=	$\frac{1000}{x}$	+	2

To clear fractions, multiply both sides by the least common denominator, $x(x + 50)$.

$$x(x+50)\left(\frac{2025}{x + 50}\right) = x(x+50)\left(\frac{1000}{x}\right) + x(x+50)(2)$$

$$2025x = 1000x + 50,000 + 2x^2 + 100x$$

$$0 = 2x^2 - 925x + 50,000$$

$$0 = (2x - 125)(x - 400)$$

$$2x - 125 = 0 \quad \text{or} \quad x - 400 = 0$$

$$x = \frac{125}{2} \quad \text{or} \quad x = 400$$

The first solution, $x = \frac{125}{2}$, means that the speed of the first airplane would be 62.5 mph.

This solution satisfies the equation but is not a realistic airplane speed. The speed of the airplane for the first trip is 400 mph.

To find the time for the first trip, use $t = \frac{d}{r} = \frac{1000}{x} = \frac{1000}{400} = \frac{2}{5}$. The time for the first trip was 2.5 hr.

19. Let h = the height of the kite.

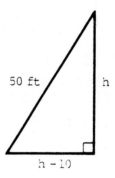

Apply the Pythagorean theorem to the right triangle.

$$a^2 + b^2 = c^2$$

$$h^2 + (h - 10)^2 = 50^2$$

$$h^2 + h^2 - 20h + 100 = 2500$$

$$2h^2 - 20h - 2400 = 0$$

$$h^2 - 10h - 1200 = 0$$

$$(h - 40)(h + 30) = 0$$

$$h = 40 \quad \text{or} \quad h = -30$$

Reject the negative solution.

The kite is 40 ft above the ground.

21. Let x = the number of hours they can talk to each other on the walkie-talkies.

Use d = rt to determine how far each boy walks in x hours.

Then 2.5x = the number of miles Chris walks north

and   3x = the number of miles Josh walks east.

This forms a right triangle with legs of length 2.5x and 3x, and length of the hypotenuse is the distance between the boys. We want to find x when the length of the hypotenuse is 4 mi.

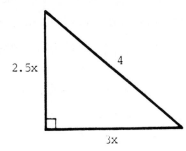

$$(2.5x)^2 + (3x)^2 = 4^2$$
$$6.25x^2 + 9x^2 = 16$$
$$15.25x^2 = 16$$
$$x^2 = 1.049$$
$$x = \pm 1.02$$

Reject the negative solution.

1.02 hr = 1.02 (60 min)
$$\approx 61 \text{ min}$$
They will be able to talk for about 61 min.

23. The height of the rocket is given by

$$h = -16t^2 + 128t.$$

Set h = 80, and solve for t.

$$80 = -16t^2 + 128t$$
$$16t^2 - 128t + 80 = 0$$

Divide by 16.

$$t^2 - 8t + 5 = 0$$

Using the quadratic formula.

$$t = \frac{8 \pm \sqrt{64 - 20}}{2}$$
$$= \frac{8 \pm \sqrt{44}}{2}$$
$$= \frac{8 \pm 2\sqrt{11}}{2}$$
$$t = 4 \pm \sqrt{11}$$
$$4 + \sqrt{11} \approx 7.32 \quad 4 - \sqrt{11} \approx .68$$

The rocket will reach a height of 80 ft after .68 sec (on the way up) and after 7.32 sec (on the way down).

25. The height of the ball is given by

$$h = -2.7t^2 + 30t + 6.5.$$

When the ball is 12 ft above the moon's surface, h = 12.
Set h = 12 and solve for t.

$$12 = -2.7t^2 + 30t + 6.5$$
$$2.7t^2 - 30t + 5.5 = 0$$

Use the quadratic formula with a = 2.7, b = -30, and c = 5.5

$$t = \frac{30 \pm \sqrt{900 - 4(2.7)(5.5)}}{2(2.7)}$$
$$= \frac{30 \pm \sqrt{840.6}}{5.4}$$

$$\frac{30 + \sqrt{840.6}}{5.4} \approx 10.92$$

$$\frac{30 - \sqrt{840.6}}{5.4} \approx .19$$

Therefore, the ball reaches 12 ft first after .19 sec (on the way up), then again after 10.92 sec (on the way down.

When the ball returns to the surface, h = 0.

$$0 = -2.7t^2 + 30t + 6.5$$

Use the quadratic formula with a = -2.7, b = 30, and c = 6.5.

$$t = \frac{-30 \pm \sqrt{900 - 4(-2.7)(6.5)}}{2(-2.7)}$$

$$= \frac{-30 \pm \sqrt{970.2}}{-5.4}$$

$$\frac{-30 + \sqrt{970.2}}{-5.4} \approx -.21$$

$$\frac{-30 - \sqrt{970.2}}{-5.4} \approx 11.32$$

Reject the first solution because time cannot be negative. Therefore, the ball returns to the surface after 11.32 sec.

27. Let x = 50.

$$T = .00787(50)^2 - 1.528(50) + 75.89$$
$$\approx 19.2 \text{ hr.}$$

29. Let x = 600 and solve for T.

$$T = .0002x^2 - .316x + 127.9$$
$$= .0002(600)^2 - .316(600) + 127.9$$
$$= 10.3$$

The exposure time when x = 600 ppm is 10.3 hr.

31. Let y = 7500 and solve for x.

$$7500 = -21.99x^2 + 353.44x + 6507.5$$
$$0 = -21.99x^2 + 353.44x - 992.5$$

Using the quadratic formula, a quadratic formula program, or the SOLVE feature on a graphing calculator, we obtain the solutions x ≈ 3.6 and x ≈ 12.5

We reject the second potential solution because the model only applies to a 7-year period.

Since 3.6 occurs during year 3, which is 1988, and year 4 corresponds to 1989, we would expect that in 1989, the number was about 7500 million.

33. For each $20 increase in rent over $300, one unit will remain vacant. Therefore, for x $20 increases, x units will remain vacant. Therefore, the number of rented units will be 80 - x.

34. x is the number of $20 increases in rent. Therefore, the rent will be 300 + 20x dollars.

35. 300 + 20x is the rent for each apartment, and 80 - x is the number of apartments that will be rented at that cost. The revenue generated will then be the product of 80 - x and 300 + 2x, so the correct expression is

$$(80 - x)(300 + 20x)$$
or $24,000 + 1300x - 20x^2$.

36. Set the revenue equal to $35,000.
This gives the equation

$$35,000 = 24,000 + 1300x - 20x^2.$$

Rewrite this equation in standard form.

$$20x^2 - 1300x + 11,000 = 0$$

37.
$$20x^2 - 1300x + 11,000 = 0$$
$$x^2 - 65x + 550 = 0$$
$$(x - 10)(x - 55) = 0$$
$$x - 10 = 0 \quad \text{or} \quad x - 55 = 0$$
$$x = 10 \quad \text{or} \quad x = 55$$

Solution set: $\{10, 55\}$

If $x = 10$, $80 - x = 70$.
If $x = 55$, $80 - x = 25$.
Because of the restriction that at least 30 units must be rented, only $x = 10$ is valid here, and the number of units rented is 70.

38. Let x = number of passengers in excess of 75.

Then $225 - 5x$

$\quad$ = the cost per passenger (in dollars);

$75 + x$ = the number of passengers.

$$\begin{array}{ccccc} \text{Cost per} & & \text{Number of} & & \\ \text{passenger} & \cdot & \text{passengers} & = & \text{revenue} \\ \downarrow & \downarrow & \downarrow & \downarrow & \downarrow \\ (225 - 5x) & \cdot & (75 + x) & = & 16,000 \end{array}$$

$$16,875 - 150x - 5x^2 = 16,000$$
$$0 = 5x^2 + 150x - 875$$
$$0 = x^2 + 30x - 175$$
$$0 = (x + 35)(x - 5)$$
$$x = -35 \quad \text{or} \quad x = 5$$

Reject the negative solution. Since there are 5 passengers in excess of 75, the total number of passengers is 80.

Section 2.5

1. For an equation to be solved by the method of substitution, the highest power of the variable must be twice the next highest power. The correct choice is (d).

3. Since the original equation is a radical equation, each proposed solution must be checked in the original equation. A check will show that the number -1 is not a solution of the equation

$$x = \sqrt{3x + 4}.$$

5. $m^4 + 2m^2 - 15 = 0$

Let $x = m^2$; then $x^2 = m^4$. With this substitution, the equation becomes

$$x^2 + 2x - 15 = 0.$$

Solve this equation by factoring.

$$(x - 3)(x + 5) = 0$$
$$x = 3 \quad \text{or} \quad x = -5$$

To find m, replace x with m^2.

$$m^2 = 3 \quad \text{or} \quad m^2 = -5$$
$$m = \pm\sqrt{3} \quad \text{or} \quad m = \pm i\sqrt{5}$$

Solution set: $\left\{ \pm\sqrt{3}, \ \pm i\sqrt{5} \right\}$

7. $2r^4 - 7r^2 + 5 = 0$

Let $x = r^2$; then $x^2 = r^4$. With this substitution, the equation becomes

$$2x^2 - 7x + 5 = 0.$$

Solve this equation by factoring.

$$(x - 1)(2x - 5) = 0$$
$$x = 1 \quad \text{or} \quad x = \frac{5}{2}$$

To find r, replace x with r^2.

$$r^2 = 1 \quad \text{or} \quad r^2 = \frac{5}{2}$$
$$r = \pm\sqrt{\frac{5}{2}}$$
$$r = \pm 1 \quad \text{or} \quad r = \pm\frac{\sqrt{10}}{2}$$

Solution set: $\left\{ \pm 1, \ \pm\frac{\sqrt{10}}{2} \right\}$

9. $(g - 2)^2 - 6(g - 2) + 8 = 0$

Let $x = g - 2$. Solve the resulting equation by factoring.

$$x^2 - 6x + 8 = 0$$
$$(x - 2)(x - 4) = 0$$
$$x = 2 \quad \text{or} \quad x = 4$$

To find g, replace x with $g - 2$.

$$g - 2 = 2 \quad \text{or} \quad g - 2 = 4$$
$$g = 4 \quad \text{or} \quad g = 6$$

Solution set: $\{4, 6\}$

11. $-(r + 1)^2 - 3(r + 1) + 3 = 0$

Let $x = r + 1$. With this substitution, the equation becomes

$$-x^2 - 3x + 3 = 0$$
$$\text{or} \quad x^2 + 3x - 3 = 0.$$

This equation cannot be solved by factoring, so use the quadratic formula.

$$x = \frac{-3 \pm \sqrt{9 + 12}}{2}$$
$$= \frac{-3 \pm \sqrt{21}}{2}$$

To find r, replace x with $r + 1$.

$$r + 1 = \frac{-3 \pm \sqrt{21}}{2}$$
$$r = \frac{-3 \pm \sqrt{21}}{2} - \frac{2}{2}$$
$$= \frac{-5 \pm \sqrt{21}}{2}$$

Solution set: $\left\{ \frac{-5 \pm \sqrt{21}}{2} \right\}$

13. $6(k + 2)^4 - 11(k + 2)^2 + 4 = 0$

Let $x = (k + 2)^2$. Solve the resulting equation by factoring.

$$6x^2 - 11x + 4 = 0$$
$$(3x - 4)(2x - 1) = 0$$
$$x = \frac{4}{3} \quad \text{or} \quad x = \frac{1}{2}$$

To find k, replace x with $(k + 2)^2$.

$$(k + 2)^2 = \frac{4}{3}$$
$$k + 2 = \pm\sqrt{\frac{4}{3}}$$
$$= \pm\frac{2\sqrt{3}}{3}$$

$$k = -2 \pm \frac{2\sqrt{3}}{3}$$

$$= -\frac{6}{3} \pm \frac{2\sqrt{3}}{3}$$

$$= \frac{-6 \pm 2\sqrt{3}}{3}$$

or

$$(k + 2)^2 = \frac{1}{2}$$

$$k + 2 = \pm\sqrt{\frac{1}{2}}$$

$$= \pm\frac{\sqrt{2}}{2}$$

$$k = -2 \pm \frac{\sqrt{2}}{2}$$

$$= -\frac{4}{2} \pm \frac{\sqrt{2}}{2}$$

$$= \frac{-4 \pm \sqrt{2}}{2}$$

Solution set: $\left\{\dfrac{-6 \pm 2\sqrt{3}}{3}, \dfrac{-4 \pm \sqrt{2}}{2}\right\}$

15. $7p^{-2} + 19p^{-1} = 6$

Let $x = p^{-1}$. Solve the resulting equation by factoring.

$$7x^2 + 19x - 6 = 0$$

$$(7x - 2)(x + 3) = 0$$

$$x = \frac{2}{7} \quad\text{or}\quad x = -3$$

To find p, replace x with p^{-1}.

$$p^{-1} = \frac{2}{7} \quad\text{or}\quad p^{-1} = -3$$

$$p = \frac{7}{2} \quad\text{or}\quad p = -\frac{1}{3}$$

Solution set: $\left\{\dfrac{7}{2}, -\dfrac{1}{3}\right\}$

17. $(r - 1)^{2/3} + (r - 1)^{1/3} = 12$

Let $u = (r - 1)^{1/3}$.

Then $u^2 = [(r - 1)^{1/3}]^2 = (r - 1)^{2/3}$.

Solve the resulting equation by factoring.

$$u^2 + u = 12$$

$$u^2 + u - 12 = 0$$

$$(u + 4)(u - 3) = 0$$

$$u = -4 \quad\text{or}\quad u = 3$$

To find r, replace u with $(r - 1)^{1/3}$.

$$(r - 1)^{1/3} = -4 \quad\text{or}\quad (r - 1)^{1/3} = 3$$

Cube both sides in each equation.

$$[(r - 1)^{1/3}]^3 = (-4)^3$$

$$r - 1 = -64$$

$$r = -63$$

or

$$[(r - 1)^{1/3}]^3 = 3^3$$

$$r - 1 = 27$$

$$r = 28$$

Because the original equation contained rational exponents, both solutions must be checked.

Solution set: $\{-63, 28\}$

19.
$$\sqrt{3z + 7} = 3z + 5$$

$$(\sqrt{3z + 7})^2 = (3z + 5)^2$$

Square both sides

$$3z + 7 = 9z^2 + 30z + 25$$

Square of a binomial

$$0 = 9z^2 + 27z + 18$$

$$0 = z^2 + 3z + 2$$

Divide by 9

$$0 = (z + 2)(z + 1) \quad\text{Factor}$$

$$z = -2 \quad\text{or}\quad z = -1$$

Check each proposed solution in the original equation.

Let $z = -2$.

$$\sqrt{3z + 7} = 3z + 5$$
$$\sqrt{3(-2) + 7} = 3(-2) + 5 \quad ?$$
$$\sqrt{-6 + 7} = -6 + 5 \quad ?$$
$$\sqrt{1} = -1 \quad ?$$
$$1 = -1 \qquad \textit{False}$$

Let $z = -1$.

$$\sqrt{3z + 7} = 3z + 5$$
$$\sqrt{3(-1) + 7} = 3(-1) + 5 \quad ?$$
$$\sqrt{-3 + 7} = -3 + 5 \quad ?$$
$$\sqrt{4} = 2 \quad ?$$
$$2 = 2 \qquad \textit{True}$$

These checks show that only -1 is a solution.

Solution set: $\{-1\}$

21.
$$\sqrt{4k + 5} - 2 = 2k - 7$$
$$\sqrt{4k + 5} = 2k - 5$$
$$(\sqrt{4k + 5})^2 = (2k - 5)^2$$
$$\textit{Square both sides}$$
$$4k + 5 = 4k^2 - 20k + 25$$
$$\textit{Square of a binomial}$$
$$0 = 4k^2 - 24k + 20$$
$$0 = k^2 - 6k + 5$$
$$\textit{Divide by 4}$$
$$k = 1 \quad \text{or} \quad k = 5$$

Check each proposed solution in the original equation.

Let $k = 1$.

$$\sqrt{4k + 5} - 2 = 2k - 7$$
$$\sqrt{4 + 5} - 2 = 2 - 7 \quad ?$$
$$\sqrt{9} - 2 = -5 \quad ?$$
$$3 - 2 = -5 \quad ?$$
$$1 = -5 \qquad \textit{False}$$

Let $k = 5$.

$$\sqrt{4k + 5} - 2 = 2k - 7$$
$$\sqrt{20 + 5} - 2 = 10 - 7 \quad ?$$
$$\sqrt{25} - 2 = 3 \quad ?$$
$$5 - 2 = 3 \quad ?$$
$$3 = 3 \qquad \textit{True}$$

Solution set: $\{5\}$

23.
$$\sqrt{4x} - x + 3 = 0$$
$$\sqrt{4x} = x - 3$$
$$(\sqrt{4x})^2 = (x - 3)^2$$
$$\textit{Square both sides}$$
$$4x = x^2 - 6x + 9$$
$$\textit{Square of a binomial}$$
$$0 = x^2 - 10x + 9$$
$$0 = (x - 1)(x - 9)$$
$$x = 1 \quad \text{or} \quad x = 9$$

Check each proposed solution in the original equation.

Let $x = 1$.

$$\sqrt{4 \cdot 1} - 1 + 3 = 0 \quad ?$$
$$\sqrt{4} - 1 + 3 = 0 \quad ?$$
$$2 - 1 + 3 = 0 \quad ?$$
$$4 = 0 \qquad \textit{False}$$

Let $x = 9$.

$$\sqrt{4 \cdot 9} - 9 + 3 = 0 \quad ?$$
$$\sqrt{36} - 9 + 3 = 0 \quad ?$$
$$6 - 9 + 3 = 0 \quad ?$$
$$0 = 0 \qquad \textit{True}$$

Solution set: $\{9\}$

25. When squaring both sides of

$$\sqrt{3z + 7} = 3z + 5,$$

the right side, the binomial $3x + 5$, must be squared to get

$$(3z + 5)^2 = 9z^2 + 30z + 25.$$

The student did not square 3z + 5 correctly.

27.
$$\sqrt{y} = \sqrt{y - 5} + 1$$
$$(\sqrt{y})^2 = (\sqrt{y - 5} + 1)^2$$
Square both sides
$$y = (\sqrt{y - 5})^2 + \sqrt{y - 5} + 1$$
Square of a binomial
$$y = y - 5 + 2\sqrt{y - 5} + 1$$
$$0 = -4 + 2\sqrt{y - 5}$$
$$4 = 2\sqrt{y - 5} \quad \text{reduce}$$
$$2 = \sqrt{y - 5}$$
$$2^2 = (\sqrt{y - 5})^2 \quad \text{Square both sides}$$
$$4 = y - 5$$
$$9 = y$$

Check this proposed solution in the original equation.

$$\sqrt{y} = \sqrt{y - 5} + 1$$
$$\sqrt{9} = \sqrt{9 - 5} + 1 \quad ?$$
$$3 = \sqrt{4} + 1 \quad\quad ?$$
$$3 = 3 \quad\quad\quad \text{True}$$

Solution set: $\{9\}$

29.
$$\sqrt{m + 7} + 3 = \sqrt{m - 4}$$
$$(\sqrt{m + 7} + 3)^2 = (\sqrt{m - 4})^2$$
Square both sides
$$m + 7 + 6\sqrt{m + 7} + 9 = m - 4$$
Square of a binomial
$$6\sqrt{m + 7} = -20 \quad \text{Simplify}$$
$$3\sqrt{m + 7} = -10$$
$$(3\sqrt{m + 7})^2 = (-10)^2$$
Square both sides
$$9(m + 7) = 100$$
$$9m + 63 = 100$$
$$9m = 37$$
$$m = \frac{37}{9}$$

Check this proposed solution in the original equation.

$$\sqrt{m + 7} + 3 = \sqrt{m - 4}$$
$$\sqrt{\frac{37}{9} + 7} + 3 = \sqrt{\frac{37}{9} - 4} \quad ?$$
$$\sqrt{\frac{37}{9} + \frac{63}{9}} + 3 = \sqrt{\frac{37}{9} + \frac{36}{9}} \quad ?$$
$$\sqrt{\frac{100}{9}} + 3 = \sqrt{\frac{1}{9}} \quad\quad ?$$
$$\frac{10}{3} + \frac{9}{3} = \frac{1}{3} \quad\quad\quad ?$$
$$\frac{19}{3} = \frac{1}{3} \quad\quad\quad\quad \text{False}$$

Since the only proposed solution is not a solution of the original equation, the equation has no solution.
Solution set: ∅

31.
$$\sqrt{2z} = \sqrt{3z + 12} - 2$$
$$(\sqrt{2z})^2 = (\sqrt{3z + 12} - 2)^2$$
Square both sides
$$2z = 3z + 12 - 4\sqrt{3z + 12} + 4$$
Square of a binomial
$$4\sqrt{3z + 12} = z + 16$$
$$(4\sqrt{3z + 12})^2 = (z + 16)^2$$
Square both sides
$$16(3z + 12) = z^2 + 32z + 256$$
Square of a binomial
$$48z + 192 = z^2 + 32z + 256$$
Distributive property
$$0 = z^2 - 16z + 64$$
$$0 = (z - 8)^2 \quad \text{Factor}$$
$$z - 8 = 0$$
$$z = 8$$

Check this proposed solution in the original equation.

$$\sqrt{2z} = \sqrt{3z + 12} - 2$$
$$\sqrt{2(8)} = \sqrt{3(8) + 12} - 2 \quad ?$$
$$\sqrt{16} = \sqrt{36} - 2 \qquad ?$$
$$4 = 6 - 2 \qquad ?$$
$$4 = 4 \qquad\qquad True$$

Solution set: $\{8\}$

33.
$$\sqrt{r + 2} = 1 - \sqrt{3r + 7}$$
$$(\sqrt{r + 2})^2 = (1 - \sqrt{3r + 7})^2$$
$$r + 2 = 1 - 2\sqrt{3r + 7} + 3r + 7$$
$$2\sqrt{3r + 7} = 2r + 6$$
$$2\sqrt{3r + 7} = 2(r + 3)$$
$$\sqrt{3r + 7} = r + 3$$
$$(\sqrt{3r + 7})^2 = (r + 3)^2$$
$$3r + 7 = r^2 + 6r + 9$$
$$0 = r^2 + 3r + 2$$
$$0 = (r + 1)(r + 2)$$
$$r = -1 \quad or \quad r = -2$$

Check $r = -1$.
$$\sqrt{-1 + 2} = 1 - \sqrt{-3 + 7} \quad ?$$
$$\sqrt{1} = 1 - \sqrt{4} \qquad ?$$
$$1 = 1 - 2 \qquad\qquad ?$$
$$1 = -1 \qquad\qquad False$$

Check $r = -2$.
$$\sqrt{-2 + 2} = 1 - \sqrt{-6 + 7} \quad ?$$
$$0 = 1 - \sqrt{1} \qquad ?$$
$$0 = 1 - 1 \qquad\qquad ?$$
$$0 = 0 \qquad\qquad True$$

Solution set: $\{-2\}$

35.
$$\sqrt[3]{4n + 3} = \sqrt[3]{2n - 1}$$
$$(\sqrt[3]{4n + 3})^3 = (\sqrt[3]{2n - 1})^3$$
$$\qquad\qquad Cube \ both \ sides$$
$$4n + 3 = 2n - 1$$
$$2n = -4$$
$$n = -2$$

Check this proposed solution in the original equation.

$$\sqrt[3]{4n + 3} = \sqrt[3]{2n - 1}$$
$$\sqrt[3]{4(-2) + 3} = \sqrt[3]{2(-2) - 1} \quad ?$$
$$\sqrt[3]{-5} = \sqrt[3]{-5} \qquad\qquad True$$

Solution set: $\{-2\}$

37.
$$\sqrt[3]{t^2 + 2t - 1} = \sqrt[3]{t^2 + 3}$$
$$(\sqrt[3]{t^2 + 2t - 1})^3 = (\sqrt[3]{t^2 + 3})^3$$
$$\qquad\qquad Cube \ both \ sides$$
$$t^2 + 2t - 1 = t^2 + 3$$
$$2t = 4$$
$$t = 2$$

Check this proposed solution in the original equation.

$$\sqrt{t^2 + 2t - 1} = \sqrt{t^2 = 3}$$
$$\sqrt[3]{2^2 + 2(2) - 1} = \sqrt[3]{2^2 + 3} \quad ?$$
$$\sqrt[3]{4 + 4 - 1} = \sqrt[3]{4 + 3} \quad ?$$
$$\sqrt[3]{7} = \sqrt[3]{7} \qquad\qquad True$$

Solution set: $\{2\}$

39.
$$(2r + 5)^{1/3} = (6r - 1)^{1/3}$$
$$[(2r + 5)^{1/3}]^3 = [(6r - 1)^{1/3}]^3$$
$$\qquad\qquad Cube \ both \ sides$$
$$2r + 5 = 6r - 1$$
$$6 = 4r$$
$$r = \frac{3}{2}$$

Check this proposed solution in the original equation.

$$(2r + 5)^{1/3} = (6r - 1)^{1/3}$$
$$\left[2\left(\frac{3}{2}\right) + 5\right]^{1/3} = \left[6\left(\frac{3}{2}\right) - 1\right]^{1/3}$$
$$(3 + 5)^{1/3} = (9 - 1)^{1/3} \qquad ?$$
$$8^{1/3} = 8^{1/3} \qquad\qquad True$$

Solution set: $\left\{\frac{3}{2}\right\}$

41.

$$(z^2 + 24z)^{1/4} = 3$$

$$[(z^2 + 24z)^{1/4}]^4 = 3^4$$

Raise both sides to 4th power

$$z^2 + 24z = 81$$

$$z^2 + 24z - 81 = 0$$

$$(z + 27)(z - 3) = 0$$

$$z = -27 \quad \text{or} \quad z = 3$$

Checking will show that both of these proposed solutions are solutions of the original equation.

Solution set: $\{-27, 3\}$

43.

$$(2r - 1)^{2/3} = r^{1/3}$$

$$[(2r - 1)^{2/3}]^3 = (r^{1/3})^3$$

Cube both sides

$$(2r - 1)^2 = r$$

$$4r^2 - 4r + 1 = r$$

$$4r^2 - 5r + 1 = 0$$

$$(4r - 1)(r - 1) = 0$$

$$r = \frac{1}{4} \quad \text{or} \quad r = 1$$

Checking will show that both of these proposed solutions are solutions of the original equation.

Solution set: $\left\{\frac{1}{4}, 1\right\}$

45. $x - \sqrt{x} - 12 = 0$

Let $u = \sqrt{x}$. Solve the resulting equation by factoring.

$$u^2 - u - 12 = 0$$

$$(u - 4)(u + 3) = 0$$

$$u = 4 \quad \text{or} \quad u = -3$$

To find x, replace u with $\sqrt{x}$.

$$\sqrt{x} = 4$$

$$x = 16$$

or

$$\sqrt{x} = -3$$

When $u = -3$, there is no solution for x.

Solution set: $\{16\}$

46. $x - \sqrt{x} - 12 = 0$

Solve by isolating $\sqrt{x}$, then squaring both sides.

$$x - 12 = \sqrt{x}$$

$$(x - 12)^2 = (\sqrt{x})^2$$

$$x^2 - 24x + 144 = x$$

$$x^2 - 25x + 144 = 0$$

$$(x - 16)(x - 9) = 0$$

$$x = 16 \quad \text{or} \quad x = 9$$

Check $x = 16$.

$$16 - \sqrt{16} - 12 = 0 \ ?$$

$$16 - 4 - 12 = 0 \ ?$$

$$0 = 0 \quad \textit{True}$$

Check $x = 9$.

$$9 - \sqrt{9} - 12 = 0 \ ?$$

$$9 - 3 - 12 = 0 \ ?$$

$$-6 = 0 \quad \textit{False}$$

The checks show that 16 satisfies the equation but 9 does not.

Solution set: $\{16\}$

47. Answer will vary depending on your preference.

48. $3x - 2\sqrt{x} - 8 = 0$

Solve by substitution. Let $u = \sqrt{x}$ and solve the resulting equation by factoring.

$$3u^2 - 2u - 8 = 0$$
$$(3u + 4)(u - 2) = 0$$
$$u = -\frac{4}{3} \quad \text{or} \quad u = 2$$

To find x, replace u with $\sqrt{x}$.

$$\sqrt{x} = -\frac{4}{3} \quad \text{or} \quad \sqrt{x} = 2$$

No solution or x = 2

Solution set: $\{4\}$

49. $2\sqrt{x} - \sqrt{3x + 4} = 0$

Using the SOLVE feature of the graphing calculator yields the value 4.

Solution set: $\{4\}$

51. $\sqrt{x^2 - \sqrt{5}x + 6} = x + \pi$

Rewrite as

$$\sqrt{x^2 - \sqrt{5}x + 6} - x - \pi = 0.$$

Solution set: $\{-.4542187292\}$

53. When solving

$$x^4 - x^2 = 0,$$

dividing by x^2 is incorrect because if x = 0, you will be dividing by zero. Dividing both sides by a variable expression causes us to "lose" the solution 0. The correct method is to factor out x^2.

$$x^4 - x^2 = 0$$
$$x^2(x^2 - 1) = 0$$
$$x^2(x + 1)(x - 1) = 0$$
$$x = 0 \quad \text{or} \quad x = -1 \quad \text{or} \quad x = 1$$

Solution set: $\{0, -1, 1\}$

55. $x^{2/3} + y^{2/3} = a^{2/3}$ for y

$$y^{2/3} = a^{2/3} - x^{2/3}$$
$$(y^{2/3})^3 = (a^{2/3} - x^{2/3})^3 \quad \textit{Cube both sides}$$
$$y^2 = (a^{2/3} - x^{2/3})^3$$
$$y = \sqrt{(a^{2/3} - x^{2/3})^3}$$
$$y = (a^{2/3} - x^{2/3})^{3/2}$$

Section 2.6

1. A = 5w; Area of a rectangle with length 5

The area of this rectangle varies directly as its *width*. The constant of the variation is *5*.

3. $A = \pi r^2$; Area of a circle with radius r

The area of a circle varies *directly* as the *square* of its *radius*. The constant of the variation is π.

5. $C = 2\pi r$; Circumference of a circle with radius r

The circumference of a circle varies *directly* as its *radius*. The constant of the variation is 2π.

7. $b = \frac{24}{h}$; Base of a parallelogram with area 24

The base of this parallelogram varies *inversely* as its *height*. The constant of the variation is *24*.

9. $m = kxy$

Substitute $m = 10$, $x = 4$, and $y = 7$ to find k.

$$10 = k \cdot 4 \cdot 7$$
$$10 = 28k$$
$$\frac{5}{14} = k$$

Thus, the relationship between m, x, and y is given by

$$m = \frac{5}{14}xy.$$

Now find m when $x = 11$ and $y = 8$.

$$m = \frac{5}{14} \cdot 11 \cdot 8$$
$$= \frac{440}{14}$$
$$= \frac{220}{7}.$$

11. $r = \dfrac{km^2}{s}$

Substitute $r = 12$, $m = 6$, and $s = 4$ to find k.

$$12 = \frac{k \cdot 6^2}{4}$$
$$12 = 9k$$
$$k = \frac{4}{3}$$

Thus, the relationship between r, m, and s is given by

$$r = \frac{4}{3} \cdot \frac{m^2}{s}.$$

Now find r when $m = 4$ and $s = 10$.

$$r = \frac{4}{3} \cdot \frac{4^2}{10}$$
$$= \frac{32}{15}$$

13. $a = \dfrac{kmn^2}{y^3}$

Substitute $a = 9$, $m = 4$, $n = 9$, and $y = 3$ to find k.

$$9 = \frac{k \cdot 4 \cdot 9^2}{3^3}$$
$$9 = 12k$$
$$k = \frac{3}{4}$$

Thus, $a = \dfrac{3}{4} \cdot \dfrac{mn^2}{y^3}.$

If $m = 6$, $n = 2$, and $y = 5$, then

$$a = \frac{3}{4} \cdot \frac{6 \cdot 2^2}{5^3}$$
$$= \frac{18}{125}.$$

15. Let x = the total fish population. On May 1,

$$\frac{\text{number tagged}}{\text{fish population}} = \frac{300}{x}.$$

On June 1,

$$\frac{\text{number tagged}}{\text{fish population}} = \frac{5}{400}.$$

Set the fractions equal to each other.

$$\frac{300}{x} = \frac{5}{400}$$

16. The least common denominator is $400x$.

17. $\dfrac{300}{x} = \dfrac{5}{400}$

Multiply both sides by $400x$.

$$400x\left(\frac{300}{x}\right) = 400x\left(\frac{5}{400}\right)$$
$$120,000 = 5x$$

18. $120,000 = 5x$

$24,000 = x$

Solution set: $\{24,000\}$

There are about 24,000 fish in False River.

19. For $k > 0$, if y varies directly as x, when x increases, y *increases*, and when x decreases, y *decreases*.

21. $y = \dfrac{k}{x}$

If x is doubled, the right side is divided by 2, or equivalently, multiplied by 1/2. Thus, y is half as large as it was before.

23. $y = kx$

If x is replaced by $(1/3)x$, the right side is multiplied by 1/3, so y is one-third as large as it was before.

25. $p = \dfrac{kr^3}{t^2}$

If r is halved, the right side will be multiplied by $(1/2)^3 = 1/8$.
If t is doubled, by the right side will be divided by $2^2 = 4$, or, equivalently, multiplied by 1/4. Thus, the right side is multiplied by $(1/8)(1/4) = 1/32$, so p is 1/32 as large as it was before.

27. Let d = distance the spring stretches;

f = force applied.

$$d = kf$$

Substitute d = 8 and f = 15 to find k.

$$8 = k \cdot 15$$

$$\frac{8}{15} = k$$

Thus, $d = \dfrac{8}{15} \cdot f$.

If f = 30,

$$d = \frac{8}{15} \cdot 30$$

$$d = 16.$$

The spring will stretch 16 in.

29. Let I = illumination;

d = distance from source.

$$I = \frac{k}{d^2}$$

Substitute I = 70 and d = 5 to find k.

$$70 = \frac{k}{5^2}$$

$$25 \cdot 70 = k$$

$$1750 = k$$

Thus, $I = \dfrac{1750}{d^2}$.

If d = 12,

$$I = \frac{1750}{12^2}$$

$$= \frac{1750}{144}$$

$$= \frac{875}{72}.$$

The illumination is $\dfrac{875}{72}$ candela.

31. Let d = distance a person can see to horizon;

h = height from the surface of the earth.

$$d = k\sqrt{h}$$

Substitute d = 15 and h = 121 to find k.

$$15 = k\sqrt{121}$$
$$15 = k \cdot 11$$
$$\frac{15}{11} = k$$

Thus, $k = \frac{15}{11}\sqrt{h}$.

If h = 900,

$$d = \frac{15}{11}\sqrt{900}$$
$$= \frac{15}{11} \cdot 30$$
$$= \frac{450}{11}.$$

The distance from the hill to the horizon is $\frac{450}{11}$ km.

33. Let V = volume of right circular cylinder;

r = radius of the base;

h = height of the cylinder.

$$V = kr^2h$$

Substitute V = 300, r = 3, and h = 10.62 to find k.

$$300 = k \cdot 3^2 \cdot 10.62$$
$$300 = 95.58k$$
$$3.1387 \approx k$$

Thus, $V = 3.1387r^2h$.

If h = 15.92,

$$V = 3.1387 \cdot 4^2 \cdot 15.92$$
$$= 3.1387 \cdot 16 \cdot 15.92$$
$$\approx 799.5.$$

The volume is 799.5 cu cm.

35. Let L = load;

w = width;

h = height;

ℓ = length between supports.

$$L = \frac{kwh^2}{\ell}$$

Substitute L = 400, w = 12, h = 15, and ℓ = 8 to find k.

$$400 = \frac{k \cdot 12 \cdot 15^2}{8}$$
$$400 = \frac{675}{2} \cdot k$$
$$\frac{2}{675} \cdot 400 = k$$
$$\frac{32}{27} = k$$

Thus,

$$L = \frac{\frac{32}{27}wh^2}{\ell}.$$

If w = 24, h = 8, and ℓ = 16,

$$L = \frac{\frac{32}{27} \cdot 24 \cdot 8^2}{16}$$
$$= \frac{32 \cdot 24 \cdot 64}{27 \cdot 16}$$
$$= \frac{1024}{9}.$$

The maximum load is $\frac{1024}{9}$ kg.

37. Let s = maximum speed;

A = amount of money spent.

$$s = k\sqrt[3]{A}$$

Substitute s = 25 and A = 450,000 to find k.

$$25 = \sqrt[3]{450,000}$$

$$\frac{25}{\sqrt[3]{450,000}} = k$$

$$\frac{25}{10\sqrt[3]{450}} = k$$

$$\frac{5}{2\sqrt[3]{450}} = k$$

Thus,

$$s = \frac{5}{2\sqrt[3]{450}} \cdot \sqrt[3]{A}.$$

If A = 1,750,000,

$$s = \frac{5}{2\sqrt[3]{450}} \cdot \sqrt[3]{1,750,000}$$

$$= \frac{5}{2\sqrt[3]{450}} \cdot 50\sqrt[3]{14}$$

$$= \frac{125\sqrt[3]{14}}{\sqrt[3]{450}}$$

$$= \frac{125 \cdot 2.41}{7.66}$$

$$\approx 39.3.$$

The maximum speed would be 39.3 km per hr.

39. $R = \dfrac{k\ell}{r^4}$

Substitute R = 25, ℓ = 12, and r = .2 to find k.

$$25 = \frac{k(12)}{.0016}$$

$$k = \frac{1}{300}$$

Thus, $R = \dfrac{1}{300} \cdot \dfrac{\ell}{r^4}.$

If r = .3 and ℓ = 12, then

$$R = \frac{1}{300} \cdot \frac{12}{(.3)^4}$$

$$\approx 4.94.$$

41. Let D = distance;

Y = yield.

$$D = k\sqrt[3]{y}$$

Substitute Y = 100 and D = 3 to find k.

$$3 = k\sqrt[3]{100}$$

$$k = \frac{3}{\sqrt[3]{100}}.$$

Thus, $D = \dfrac{3}{\sqrt[3]{100}} \cdot \sqrt[3]{Y}.$

If Y = 1500,

$$D = \frac{3}{\sqrt[3]{100}} \cdot \sqrt[3]{1500}$$

$$= 3\sqrt[3]{15}$$

$$\approx 7.4$$

The distance is 7.4 km.

43. $L = \dfrac{25E^2}{st}$

Here, L = 500, s = 200, and t = 1/250. Substitute these values into the formula and solve for F.

$$500 = \frac{25F^2}{200 \cdot \frac{1}{250}}$$

$$400 = 25F^2$$

$$16 = F^2$$

$$\pm 4 = F$$

A negative value of F is not meaningful in this problem. The appropriate F-stop is 4.

45. Let x = the total fish population. On October 5,

$$\frac{\text{number tagged}}{\text{fish population}} = \frac{250}{x}.$$

At a later date,

$$\frac{\text{number tagged}}{\text{fish population}} = \frac{7}{350}.$$

Set the fractions equal to each other and solve for x.

$$\frac{250}{x} = \frac{7}{350}$$
$$87,500 = 7x$$
$$12,500 = x$$

There are approximately 12,500 fish in Willow Lake.

Section 2.7

1. x < -4

The interval includes all real numbers less than -4, not including -4. The correct interval notation is (-∞, -4), so the correct choice is F.

3. -2 < x ≤ 6

The interval includes all real numbers from -2 to 6, not including -2, but including 6. The correct interval notation is (-2, 6], so the correct choice is A.

5. x ≥ -3

The interval includes all real numbers greater than or equal to -3, so it includes -3. The correct interval notation is [-3, ∞), so the correct choice is I.

7. The interval shown on the number line includes all real numbers between -2 and 6, including -2, but not including 6. The correct interval notation is [-2, 6), so the correct choice is B.

9. The interval shown on the number line includes all real numbers greater than 3, not including 3. The correct interval notation is (3, ∞), so the correct choice is E.

For Exercises 13–25, see the answer graphs in the back of the textbook.

13.   -3p - 2 ≤ 1

Add 2 to both sides of the inequality.

$$-3p - 2 + 2 \le 1 + 2$$
$$-3p \le 3$$

Multiply both sides of the inequality by -1/3 and reverse the direction of the inequality symbol.

$$\left(-\frac{1}{3}\right)(-3p) \ge \left(-\frac{1}{3}\right)(3)$$
$$p \ge -1$$

Solution set: [-1, ∞)

15. $2(m + 5) - 3m + 1 \geq 5$

$2m + 10 - 3m + 1 \geq 5$
Distributive property

$-m + 11 \geq 5$ *Combine terms*

$-m \geq -6$ *Subtract 11*

$-1(-m) \leq -1(-6)$
Multiply by -1; reverse inequality symbol

$m \leq 6$

Solution set: $(-\infty, 6]$

17. $8k - 3k + 2 < 2(k + 7)$

$5k + 2 < 2k + 14$
Distributive property

$5k + 2 - 2k < 2k + 14 - 2k$
Subtract 2k

$3k + 2 < 14$

$3k + 2 - 2 < 14 - 2$ *Subtract 2*

$3k < 12$

$\frac{1}{3} \cdot 3k < \frac{1}{3} \cdot 12$ *Multiply by 1/3*

$k < 4$

Solution set: $(-\infty, 4)$

19. $\frac{4x + 7}{-3} \leq 2x + 5$

$(-3)\left(\frac{4x + 7}{-3}\right) \geq (-3)(2x + 5)$
Multiply by -3; reverse inequality symbol

$4x + 7 \geq -6x - 15$

$4x + 7 + 6x \geq -6x - 15 + 6x$
Add 6x

$10x + 7 \geq -15$

$10x + 7 - 7 \geq -15 - 7$ *Subtract 7*

$10x \geq -22$

$\frac{1}{10}(10x) \geq \frac{1}{10}(-22)$ *Multiply by 1/10*

$x \geq -\frac{11}{5}$

Solution set: $\left[-\frac{11}{5}, \infty\right)$

21. $2 \leq y + 1 \leq 5$

$1 \leq y \leq 4$ *Subtract 1*

Solution set: $[1, 4]$

23. $-10 > 3r + 2 > -16$

$-10 - 2 > 3r + 2 - 2 > -16 - 2$
Subtract 2

$-12 > 3r > -18$

$\frac{1}{3}(-12) > \frac{1}{3}(3r) > \frac{1}{3}(-18)$ *Multiply by 1/3*

$-4 > r > -6$

or $-6 < r < -4$

Solution set: $(-6, -4)$

25. $-3 \leq \frac{x - 4}{-5} < 4$

$(-5)(-3) \geq (-5)\left(\frac{x - 4}{-5}\right) > (-5)(4)$
Multiply by -5; reverse inequality symbol

$15 \geq x - 4 > -20$

$15 + 4 \geq x - 4 + 4 > -20 + 4$ *Add 4*

$19 \geq x > -16$

or $-16 < x \leq 19$

Solution set: $(-16, 19]$

27. $y = 2480x + 12{,}726$

Set y to 17,600 and solve for x.

$17{,}600 = 2480x + 12{,}726$

$4874 = 2480x$

$2 \approx x$

The capital outlay first exceeded $17,600 million in 1987.

29. $C = 50x + 5000; \quad R = 60x$

The product will at least break even when $R \geq C$.

Set $R \geq C$ and solve for x.

$$60x \geq 50x + 5000$$
$$10x \geq 5000$$
$$x \geq 500$$

The break-even point is at x = 500. This product will at least break even if the number of units produced is in the interval [500, ∞).

31. $C = 85x + 900$; $R = 105x$

$$R \geq C$$
$$105x \geq 85x + 900$$
$$20x \geq 900$$
$$x \geq 45$$

The break-even point is at x = 45. This product will at least break even if the number of units produced is in the interval [45, ∞).

For Exercises 33-45, see the answer graphs in the back of the textbook.

33. $x^2 \leq 9$
$$x^2 - 9 \leq 0$$

Solve the corresponding quadratic equation by factoring.

$$x^2 - 9 = 0$$
$$(x + 3)(x - 3) = 0$$
$$x = -3 \quad \text{or} \quad x = 3$$

These two points, -3 and 3, divide a number line into the three regions shown on the following sign graph.

```
x + 3 - - o + + + + + + : + + +
x - 3 - - : - - - - - - o + + +
      ──┼┼┼┼┼┼┼┼┼┼┼┼──►
          -3      0     3
        +        -        +
       Sign of
      (x + 3) (x - 3)
```

The factor x + 3 is positive when x > -3, and x - 3 is positive when x > 3.

We want the product to be negative, which happens if the two factors have opposite signs. As the sign graph shows, this happens when x is between -3 and 3. The endpoints of the interval are included in the solution because the product is zero (and therefore "less than or equal to zero") when x = -3 or x = 3. Solution set: [-3, 3]

35. $r^2 + 4r + 6 \geq 3$
$$r^2 + 4r + 3 \geq 0$$

Solve the corresponding quadratic equation.

$$r^2 + 4r + 3 = 0$$
$$(r + 3)(r + 1) = 0$$
$$r = -3 \quad \text{or} \quad r = -1$$

These two points, -3 and -1, divide a number line into the three regions shown on the following sign graph.

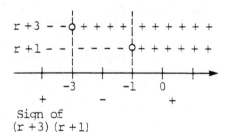

```
r + 3 - - o + + + + + : + + + + + +
r + 1 - - : - - - - - o + + + + + +
      ──┼──┼──┼──┼──►
         -3     -1    0
        +        -        +
       Sign of
      (r + 3) (r + 1)
```

The factor r + 3 is positive when r > -3, and r + 1 is positive when r > -1.

We want the product to be positive, which happens if both factors have the same sign. As the sign graph shows, this happens when r < -3 or when r > -1.

The inequality is also satisfied when r = -3 and when r = -1 because of the ≤ sign, so square brackets should be used at these endpoints.
Solution set: $(-\infty, -3] \cup [-1, \infty)$

37.
$$x^2 - x \le 6$$
$$x^2 - x - 6 \le 0$$

Solve the corresponding quadratic equation.

$$x^2 - x - 6 = 0$$
$$(x + 2)(x - 3) = 0$$
$$x + 2 = 0 \quad \text{or} \quad x - 3 = 0$$
$$x = -2 \quad \text{or} \quad x = 3$$

These two points divide a number line into three regions.

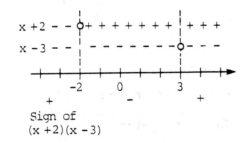

Sign of
$(x + 2)(x - 3)$

The factor x - 3 is positive when x > 3 and x + 2 is positive when x > -2.
The product is negative if the two factors have opposite signs, which happens when x is between -2 and 3. Use square brackets since the original inequality is ≤, which means that both -2 and 3 are part of the solution set.
Solution set: $[-2, 3]$

39.
$$2k^2 - 9k > -4$$
$$2k^2 - 9k + 4 > 0$$

Solve the corresponding quadratic equation.

$$2k^2 - 9k + 4 = 0$$
$$(2k - 1)(k - 4) = 0$$
$$k = \frac{1}{2} \quad \text{or} \quad k = 4$$

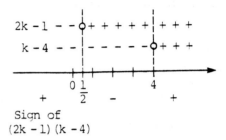

Sign of
$(2k - 1)(k - 4)$

The product is positive when $k < \frac{1}{2}$ or when k > 4. No endpoints are included since the inequality symbol is >.

Solution set: $\left(-\infty, \frac{1}{2}\right) \cup (4, \infty)$

41. $x^2 > 0$ is true for all values of x except 0, so that the solution is $(-\infty, 0) \cup (0, \infty)$.

43. $x^2 + 5x - 2 < 0$

Solve the corresponding quadratic equation.

$$x^2 + 5x - 2 = 0$$

The expression $x^2 + 5x - 2$ cannot be factored, so use the quadratic formula.

$$x = \frac{-5 \pm \sqrt{25 + 8}}{2}$$

$$x = \frac{-5 \pm \sqrt{33}}{2}$$

$$x \approx .4 \quad \text{or} \quad x \approx -5.4$$

Use these approximations and $x - .4$ and $x + 5.4$ to make a sign graph.

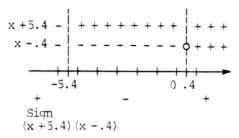

Solution set: $\left(\dfrac{-5 - \sqrt{33}}{2}, \dfrac{-5 + \sqrt{33}}{2}\right)$

45. $m^2 - 2m \le 1$

$m^2 - 2m - 1 \le 0$

Solve the corresponding quadratic equation.

$m^2 - 2m - 1 = 0$

The expression $m^2 - 2m - 1$ cannot be factored, so use the quadratic formula.

$$m = \frac{2 \pm \sqrt{4 + 4}}{2}$$

$$= \frac{2 \pm \sqrt{8}}{2}$$

$$= \frac{2 \pm 2\sqrt{2}}{2}$$

$$m = 1 \pm \sqrt{2}$$

$m \approx 2.4$ or $m \approx -.4$

Use a sign graph.

Sign
$(m + .4)(m - 2.4)$

The sign graph shows that the product is negative when x is between $1 - \sqrt{2}$ and $1 + \sqrt{2}$. The endpoints

are included because the inequality symbol is $\le$.

Solution set: $[1 - \sqrt{2},\ 1 + \sqrt{2}]$

47. $\dfrac{m - 3}{m + 5} \le 0$

To draw a sign graph, first solve the equations

$m - 3 = 0$ and $m + 5 = 0$,

getting the solutions

$m = 3$ and $m = -5$.

Use the values -5 and 3 to divide the number line into three regions.

Sign of
$\dfrac{m - 3}{m + 5}$

The quotient is negative when the factors have different signs, or when m is between -5 and 3. Since the inequality symbol is $\le$, we must check each endpoint separately. Here, -5 gives a 0 denominator, but 3 satisfies the inequality.

Solution set: $(-5,\ 3]$

49. $\dfrac{k - 1}{k + 2} > 1$

$\dfrac{k - 1}{k + 2} - 1 > 0$ *Subtract 1*

$\dfrac{k - 1}{k + 2} - \dfrac{k + 2}{k + 2} > 0$ *Common denominator is k + 2*

$\dfrac{k - 1 - (k + 2)}{k + 2} > 0$

$\dfrac{k - 1 - k - 2}{k + 2} > 0$

$\dfrac{-3}{k + 2} > 0$ *Combine terms*

Since −3 is negative, this inequality is true when k + 2 is negative, or when k < −2.

Solution set: (−∞, −2)

51.

$$\frac{3}{x - 6} \le 2$$

$$\frac{3}{x - 6} - 2 \le 0$$

$$\frac{3}{x - 6} - \frac{2(x - 6)}{x - 6} \le 0$$

$$\frac{3 - 2(x - 6)}{x - 6} \le 0$$

$$\frac{3 - 2x + 12}{x - 6} \le 0$$

$$\frac{-2x + 15}{x - 6} \le 0$$

Solve the equations

−2x + 15 = 0 and x − 6 = 0,

getting the solutions

$$x = -\frac{15}{2} \quad \text{and} \quad x = 6.$$

Draw a sign graph.

-2x +15 + + + + + + + | + +○ − −
x − 6 − − − − − − − ○+ +| + +

0   6   15
                2
        −   + −

Sign of
$$\frac{-2x + 15}{x - 6}$$

The quotient is negative when the factors have different signs, that is, when x < 6 or x > 15/2. Since we have a quotient with "less than or equal to," we must check each endpoint separately. x = 6 doesn't work since it makes a denominator 0. x = 15/2 is acceptable, since it doesn't make any denominator 0.

Thus, the solution set is

$$(-\infty, \ 6) \cup \left[\frac{15}{2}, \ \infty\right).$$

53.

$$\frac{1}{m - 1} < \frac{5}{4}$$

$$\frac{1}{m - 1} - \frac{5}{4} < 0$$

$$\frac{4 - 5(m - 1)}{4(m - 1)} < 0$$

$$\frac{4 - 5m + 5}{4(m - 1)} < 0$$

$$\frac{9 - 5m}{4(m - 1)} < 0$$

Solve the equations

9 − 5m = 0 and 4(m − 1) = 0,

getting the solutions

$$m = \frac{9}{5} \quad \text{and} \quad m = 1.$$

Draw a sign graph.

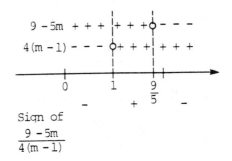

Sign of
$$\frac{9 - 5m}{4(m - 1)}$$

The quotient is negative when m < 1 or m > 9/5. There is no need to check the endpoints, since the original inequality symbol is <. The solution set is

$$(-\infty, \ 1) \cup \left(\frac{9}{5}, \ \infty\right).$$

55.

$$\frac{10}{3 + 2x} \le 5$$

$$\frac{10}{3 + 2x} - 5 \le 0$$

$$\frac{10 - 5(3 + 2x)}{3 + 2x} \le 0$$

$$\frac{10 - 15 - 10x}{3 + 2x} \le 0$$

$$\frac{-5 - 10x}{3 + 2x} \le 0$$

$$\frac{-5(1 + 2x)}{3 + 2x} \le 0$$

Make a sign graph.

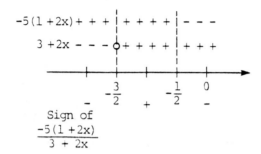

The quotient is negative when x < -3/2 or x > -1/2. Since the inequality symbol is ≤, we must check each endpoint separately. Here, -3/2 gives a 0 denominator but -1/2 satisfies the inequality. Thus, the solution set is

$$\left(-\infty, -\frac{3}{2}\right) \cup \left[-\frac{1}{2}, \infty\right).$$

57. $\dfrac{7}{k + 2} \ge \dfrac{1}{k + 2}$

Subtract $\dfrac{1}{k + 2}$ from both sides.

$$\frac{7}{k + 2} - \frac{1}{k + 2} \ge 0$$

$$\frac{6}{k + 2} \ge 0$$

Since 6 is positive, this inequality is true whenever k + 2 is positive, or when k > -2.

-2 does not satisfy the inequality because it makes both denominators 0 in the original inequality. The solution set is (-2, ∞).

59. $\dfrac{3}{2r - 1} > -\dfrac{4}{r}$

Add $\dfrac{4}{r}$ to both sides.

$$\frac{3}{2r - 1} + \frac{4}{r} > 0 \quad \text{Add } 4/r$$

$$\frac{3r + 4(2r - 1)}{r(2r - 1)} > 0 \quad \begin{array}{l}\text{Common denomina-} \\ \text{tor is } r(2r - 1)\end{array}$$

$$\frac{3r + 8r - 4}{r(2r - 1)} > 0$$

$$\frac{11r - 4}{r(2r - 1)} > 0$$

Make a sign graph showing the three factors 11r - 4, 2r - 1, and r.

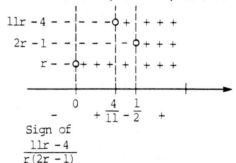

The quotient is positive when 0 < r < 4/11 or r > 1/2. Neither endpoint satisfies the inequality since the inequality symbol is >. The solution set is

$$\left(0, \frac{4}{11}\right) \cup \left(\frac{1}{2}, \infty\right).$$

61. $\dfrac{4}{y - 2} \le \dfrac{3}{y - 1}$

Subtract $\dfrac{3}{y - 1}$ from both sides.

$$\frac{4}{y-2} - \frac{3}{y-1} \leq 0$$

$$\frac{4(y-1) - 3(y-2)}{(y-1)(y-2)} \leq 0$$

Common denominator is $(y-1)(y-2)$

$$\frac{4y-4-3y+6}{(y-1)(y-2)} \leq 0$$

$$\frac{y+2}{(y-1)(y-2)} \leq 0$$

Make a sign graph showing the three factors $y+2$, $y-1$, and $y-2$.

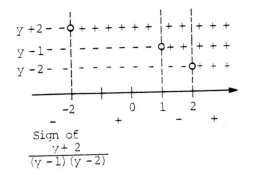

The quotient is negative when $y < -2$ or $1 < y < 2$. Checking the endpoints gives $y = -2$ as an acceptable solution. However, $y = 1$ and $y = 2$ are not acceptable since the denominator is zero at these points. Thus the solution set is

$$(-\infty, -2] \cup (1, 2).$$

63.
$$\frac{y+3}{y-5} \leq 1$$

$$\frac{y+3}{y-5} - 1 \leq 0$$

$$\frac{y+3 - (y-5)}{y-5} \leq 0$$

$$\frac{y+3-y+5}{y-5} \leq 0$$

$$\frac{8}{y-5} \leq 0$$

Since 8 is positive, this inequality is true whenever $y - 5$ is negative, that is, when $y < 5$. $y = 5$ cannot be used since it makes the denominator 0. Thus, the solution set is $(-\infty, 5)$.

65. $(3x - 4)(x + 2)(x + 6) = 0$

Set each factor to zero and solve.

$3x - 4 = 0$ or $x + 2 = 0$ or $x + 6 = 0$

$x = \dfrac{4}{3}$ or $\quad x = -2$ or $\quad x = -6$

Solution set: $\left\{ \dfrac{4}{3}, -2, -6 \right\}$

66. Plot the solutions -6, -2, and $\dfrac{4}{3}$ on a number line. See the answer graph in the back of the textbook.

67. Choose $x = -7$, from the region farthest to the left, which is the interval $(-\infty, -6)$, and substitute in the original inequality.

$$[3(-7) - 4](-7 + 2)(-7 + 6) \leq 0 \ ?$$
$$(-25)(-5)(-1) \leq 0 \ ?$$
$$-125 \leq 0 \quad True$$

Since -7 satisfies the original inequality, graph this region on the number line. See the answer graph in the back of the textbook.

68. Choose x = -3 from the second region from the left, which is the interval (-6, -2), and substitute in the original inequality.

$$[3(-3) - 4](-3 + 2)(-3 + 6) \le 0 \ ?$$
$$(-13)(-1)(3) \le 0 \ ?$$
$$39 \le 0$$
False

Since -3 does not satisfy the original inequality, do not graph this region on the number line.

69. Choose x = 0 from the third region from the left, which is the interval (-2, 4/3), and substitute in the original inequality.

$$(3 \cdot 0 - 4)(0 + 2)(0 + 6) \le 0 \ ?$$
$$(-4)(2)(6) \le 0 \ ?$$
$$-48 \le 0 \quad \textit{True}$$

Since 0 satisfies the original inequality, graph this region on the number line. See the answer graph in the back of the textbook.

70. Choose x = 2 from the region farthest to the right, which is the interval (4/3, ∞) and substitute in the original inequality.

$$[3(2) - 4](2 + 2)(2 + 6) \le 0 \ ?$$
$$(2)(4)(8) \le 0 \ ?$$
$$64 \le 0 \quad \textit{False}$$

Since 2 does not satisfy the original inequality, do not graph this region on the number line.

71. The solution includes all real numbers to the left of -6, including -6, and all real numbers between -2 and 4/3, including -2 and 4/3. Plot the regions of the solution on a number line. See the answer graph in the back of the textbook.

72. $x^3 + 4x^2 - 9x - 36 > 0$

Solve the associated equation.

$$x^3 + 4x^2 - 9x - 36 = 0$$
$$x^2(x + 4) - 9(x + 4) = 0$$
$$(x^2 - 9)(x + 4) = 0$$
$$(x + 3)(x - 3)(x + 4) = 0$$
$$x = -3 \quad \text{or} \quad x = 3 \quad \text{or} \quad x = -4$$

Plot these points on a number line.

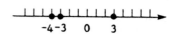

Choose points from each of the four regions, and substitute in the original inequality.

x = -5:

$$(-5)^3 + 4(-5)^2 - 9(-5) - 36 > 0 \ ?$$
$$-16 > 0$$
False

Do not graph the region farthest to the left.

x = -3.5:

$$(-3.5)^3 + 4(-3.5)^2 - 9(-3.5) - 36 > 0 \ ?$$
$$1.625 > 0$$
True

Graph the second region from the left.

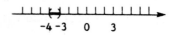

x = 0:

$(0)^3 + 4(0)^2 - 9(0) - 36 > 0$?

$\qquad\qquad -36 > 0$ *False*

Do not graph the third region from the left.

x = 4:

$(4)^3 + 4(4)^2 - 9(4) - 36 > 0$?

$\qquad\qquad 56 > 0$ *True*

Graph the region farthest to the right.

Plot the regions of the solution on a number line. See the answer graph in the back of the textbook.

73. (a) $(x + 3)^2$ is never negative, so $(x + 3)^2 \geq 0$ has solution set $(-\infty, \infty)$.

(e) $\dfrac{x^2 + 7}{2x^2 + 4}$ is never smaller than $\dfrac{7}{4}$.

This certainly implies that $\dfrac{x^2 + 7}{2x^2 + 4}$ is never negative, so $\dfrac{x^2 + 7}{2x^2 + 4} \geq 0$ has solution set $(-\infty, \infty)$.

(a) and (e) are the only inequalities that have solution set $(-\infty, \infty)$.

75. Rewrite the inequality with 0 on one side.

$$3(x + 2) - 6x < 0$$

Graph $y_1 = 3(x + 2) - 6x$. The graph shows that $y_1 < 0$ (the graph is below the x-axis) when x > 2. Thus,

$3(x + 2) - 6x < 0$ when x > 2, so the solution set of the original inequality is $(2, \infty)$.

77. $x^2 - 4x - 5 \leq 0$

Graph $y_1 = x^2 - 4x - 5$. The graph intersects the x-axis at x = -1 and x = 5, and y_1 is less than 0 for values of x between -1 and 5. The solution set is $[-1, 5]$.

79. $\dfrac{x - 2}{x + 2} \leq 2$

We will use the intersection-of-graphs method.

Graph $y_1 = \dfrac{x - 2}{x + 2}$

and $y_2 = 2$.

The graphs intersect at x = -6, and y_1 has a vertical asymptote at x = -2. y_1 is less than y_2 for values of x less than -6 and for values of x greater than -2. The solution set is $(-\infty, -6] \cup (-2, \infty)$.

83. (a) $1.5 \times 10^{-3} \leq R \leq 6.0 \times 10^{-3}$

$2.08 \times 10^{-5} \leq \dfrac{R}{72} \leq 8.33 \times 10^{-5}$

(b) Let N be the number of additional lung cancer deaths each year. Then N would be determined by taking the annual individual risk times the total number of people. Thus,

$$N = (260 \times 10^6)\left(\frac{R}{72}\right).$$

The range for N would be approximately

$(260 \times 10^6)(2.08 \times 10^{-5}) \leq N$

$\leq (260 \times 10^6)(8.33 \times 10^{-5})$

$5417 \leq N \leq 21{,}667.$

Thus, radon gas exposure is expected by the EPA to cause between 5400 and 21,700 cases of lung cancer each year in the United States.

Section 2.8

1. $|x| = 4$

The solution set includes any value of x whose absolute value is 4; thus x = 4 or x = -4 are both solutions. The correct graph is F.

3. $|x| > -4$

The solution set is all real numbers, since the absolute value of any real number is always greater than -4. The correct graph is D, which shows the entire number line.

5. $|x| < 4$

The solution set includes any value of x whose absolute value is less than 4; thus x must be between -4 and 4, not including -4 or 4. The correct graph is G.

7. $|x| \leq 4$

The solution set includes any value of x whose absolute value is less than or equal to 4; thus x must be between -4 and 4, including -4 and 4. The correct graph is C.

9. $|3m - 1| = 2$

$3m - 1 = 2 \quad \text{or} \quad 3m - 1 = -2$

$3m = 3 \qquad\qquad 3m = -1$

$m = 1 \quad \text{or} \qquad m = -\dfrac{1}{3}$

Solution set: $\left\{ 1, -\dfrac{1}{3} \right\}$

11. $|5 - 3x| = 3$

$5 - 3x = 3 \quad \text{or} \quad 5 - 3x = -3$

$2 = 3x \qquad\qquad 8 = 3x$

$\dfrac{2}{3} = x \quad \text{or} \qquad \dfrac{8}{3} = x$

Solution set: $\left\{ \dfrac{2}{3}, \dfrac{8}{3} \right\}$

13. $\left| \dfrac{z - 4}{2} \right| = 5$

$\dfrac{z - 4}{2} = 5 \quad \text{or} \quad \dfrac{z - 4}{2} = -5$

$z - 4 = 10 \qquad z - 4 = -10$

$z = 14 \quad \text{or} \qquad z = -6$

Solution set: $\left\{ 14, -6 \right\}$

15. $\left| \dfrac{5}{r - 3} \right| = 10$

$\dfrac{5}{r - 3} = 10$

$5 = 10(r - 3)$

$r = 10r - 30$

$35 = 10r$

$\dfrac{7}{2} = r$

or

$\dfrac{5}{r - 3} = -10$

$5 = -10(r - 3)$

$5 = -10r + 30$

$$-25 = -10r$$

$$\frac{5}{2} = r$$

Solution set: $\left\{\dfrac{7}{2}, \dfrac{5}{2}\right\}$

17. $|4w + 3| - 2 = 7$

$$|4w + 3| = 9$$

$4w + 3 = 9$ or $4w + 3 = 9$

$$4w = 6 \qquad\qquad 4w = -12$$

$w = \dfrac{3}{2}$ or $\qquad w = -3$

Solution set: $\left\{\dfrac{3}{2}, -3\right\}$

19. $|6x + 9| = 0$

This absolute value equation has only one case.

$$6x + 9 = 0$$

$$6x = -9$$

$$x = -\frac{9}{6} = -\frac{3}{2}$$

Solution set: $\left\{-\dfrac{3}{2}\right\}$

21. $\left|\dfrac{6y + 1}{y - 1}\right| = 3$

$$\frac{6y + 1}{y - 1} = 3$$

$$6y + 1 = 3(y - 1)$$

$$6y + 1 = 3y - 3$$

$$3y = -4$$

$$y = -\frac{4}{3}$$

or

$$\frac{6y + 1}{y - 1} = -3$$

$$6y + 1 = -3(y - 1)$$

$$6y + 1 = -3y + 3$$

$$9y = 2$$

$$y = \frac{2}{9}$$

Solution set: $\left\{-\dfrac{4}{3}, \dfrac{2}{9}\right\}$

23. $|2k - 3| = |5k + 4|$

$$2k - 3 = 5k + 4$$

$$-7 = 3k$$

$$-\frac{7}{3} = k$$

or

$$2k - 3 = -(5k + 4)$$

$$2k - 3 = -5k - 4$$

$$7k = -1$$

$$k = -\frac{1}{7}$$

Solution set: $\left\{-\dfrac{7}{3}, -\dfrac{1}{7}\right\}$

25. $|4 - 3y| = |2 - 3y|$

$$4 - 3y = 2 - 3y$$

$$0 = -2 \; \textit{False}$$

or

$$4 - 3y = -(2 - 3y)$$

$$4 - 3y = -2 + 3y$$

$$-6y = -6$$

$$y = 1$$

Solution set: $\{1\}$

29. $|x - 2| = 1$

Graph $y_1 = |x - 2| - 1$.
The graph intersects the x-axis at 1 and 3. The solution set is $\{1, 3\}$.

31. $|2x + 7| = 5$

Graph $y_1 = |2x + 7| - 5$.
The graph intersects the x-axis at
-6 and -1. The solution set is
$\{-6, -1\}$.

33. $|2 + 5x| = |4 - 6x|$

Graph $y_1 = |2 + 5x| - |4 - 6x|$.
The graph intersects the x-axis
at $\frac{2}{11}$ and 6. The solution set is

$\left\{\frac{2}{11}, 6\right\}$.

35. $|m| > 1$

$m < -1$ or $m > 1$

Solution set: $(-\infty, -1) \cup (1, \infty)$

37. $|2x + 5| < 3$

$-3 < 2x + 5 < 3$
$-8 < 2x < -2$
$-4 < x < -1$

Solution set: $(-4, -1)$

39. $4|x - 3| > 12$

$4(x - 3) > 12$ or $4(x - 3) < -12$
$4x - 12 > 12$ $4x - 12 < -12$
$\quad\quad 4x > 24$ $4x < 0$
$\quad\quad x > 6$ or $x < 0$

Solution set: $(-\infty, 0) \cup (6, \infty)$

41. $|3z + 1| \geq 7$

$3z + 1 \leq -7$ or $3z + 1 \geq 7$
$\quad 3z \leq -8$ $3z \geq 6$
$\quad z \leq -\frac{8}{3}$ or $z \geq 2$

Solution set: $\left(-\infty, -\frac{8}{3}\right] \cup [2, \infty)$

43. $\left|\frac{2}{3}t + \frac{1}{2}\right| \leq \frac{1}{6}$

$-\frac{1}{6} \leq \frac{2}{3}t + \frac{1}{2} \leq \frac{1}{6}$

Multiply by the least common denominator, 6, to clear fractions.

$-1 \leq 4t + 3 \leq 1$
$-4 \leq 4t \leq -2$
$-1 \leq t \leq -\frac{1}{2}$

Solution set: $\left[-1, -\frac{1}{2}\right]$

45. $\left|5x + \frac{1}{2}\right| - 2 < 5$

$\left|5x + \frac{1}{2}\right| < 7$ Add 2

$-7 < 5x + \frac{1}{2} < 7$

$-\frac{15}{2} < 5x < \frac{13}{2}$ Subtract 1/2

$-\frac{3}{2} < x < \frac{13}{10}$ Divide by 5

Solution set: $\left(-\frac{3}{2}, \frac{13}{10}\right)$

47. $|6x + 3| \geq -2$

Since the absolute value of a number
is always nonnegative, $|6x + 3| \geq -2$
is always true.

Solution set: $(-\infty, \infty)$

49. $\left|\frac{1}{2}x + 6\right| > 0$

The absolute value of a number will
be positive so long as the number is
negative or positive (but not zero).

$\frac{1}{2}x + 6 < 0 \qquad$ or $\quad \frac{1}{2}x + 6 > 0$

$\qquad \frac{1}{2}x < -6 \qquad\qquad \frac{1}{2}x > -6$

$\qquad x < -12 \quad$ or $\qquad x > -12$

Solution set: $(-\infty, -12) \cup (-12, \infty)$

51. $|p - q| = 5$, which is equivalent to $|q - p| = 5$, indicates that the distance between p and q is 5 units.

55. "m is no more than 8 units from 9" is written

$$|m - 9| \le 8.$$

57. "p is at least 5 units from 9" is written

$$|p - 9| \ge 5.$$

59. "r is 5 units from 3" is written

$$|r - 3| = 5.$$

61. For $x^2 - x$ to have an absolute value equal to 6, the expression must equal -6 or 6.

62. For $x^2 - x$ to equal 6, the equation is

$$x^2 - x = 6.$$

63. $\qquad x^2 - x = 6$

$\qquad x^2 - x - 6 = 0$

$\qquad (x + 2)(x - 3) = 0$

$\qquad x + 2 = 0 \quad$ or $\quad x - 3 = 0$

$\qquad\qquad x = -2 \quad$ or $\qquad x = 3$

Solution set: $\{-2, 3\}$

64. For $x^2 - x$ to equal -6, the equation is

$$x^2 - x = -6.$$

65. $\qquad x^2 - x = -6$

$x^2 - x + 6 = 0$

Use the quadratic formula with $a = 1$, $b = -1$, and $c = 6$.

$$x = \frac{1 \pm \sqrt{1 - (1)(6)}}{2}$$

$$x = \frac{1 \pm \sqrt{-23}}{2}$$

$$x = \frac{1 \pm i\sqrt{23}}{2}$$

$$= \frac{1}{2} \pm \frac{\sqrt{23}}{2}i$$

Solution set: $\left\{ \frac{1}{2} \pm \frac{\sqrt{23}}{2}i \right\}$

66. The complete solution set of

$$|x^2 - x| = 6$$

is $\left\{ -2, 3, \frac{1}{2} \pm \frac{\sqrt{23}}{2}i \right\}$.

67. $\qquad |R_L - 26.75| \le 1.42$

$\qquad -1.42 \le R_L - 26.75 \le 1.42$

$\qquad\qquad 25.33 \le R_L \le 28.17$

$\qquad |R_E - 38.75| \le 2.17$

$\qquad -2.17 \le R_E - 38.75 \le 2.17$

$\qquad\qquad 36.58 \le R_E \le 40.92$

69. There are many possible explanations. Students may work harder during an exam or there may be more stress, so students may breathe more frequently.

71. $|y - 8.0| \leq 1.5$

$-1.5 \leq y - 8.0 \leq 1.5$

$6.5 \leq y \leq 9.5$

The range of weights, in pounds, is $[6.5, 9.5]$.

73.   Let x = the speed of the kite.

148 is 25 more than 123, and 98 is 25 less than 123, so all the speeds are within 25 ft per sec of 123 ft per sec, that is, $|x - 123| \leq 25$.

Let x = the speed of the wind.

26 is 5 more than 21, and 16 is 5 less than 21, so all the speeds are within 5 ft per sec of 21 ft per sec, that is, $|x - 21| \leq 5$.

Chapter 2   Review Exercises

1. $2m + 7 = 3m + 1$

$7 = m + 1$ *Subtract 2m*

$6 = m$ *Subtract 1*

Solution set: $\{6\}$

3. $5y - 2(y + 4) = 3(2y + 1)$

$5y - 2y - 8 = 6y + 3$
             Distributive property

$3y - 8 = 6y + 3$
       Combine like terms

$-8 = 3y + 3$ *Subtract 3y*

$-11 = 3y$ *Subtract 3*

$\frac{1}{3}(-11) = \frac{1}{3}(3y)$ *Multiply by 1/3*

$-\frac{11}{3} = y$

Solution set: $\left\{-\frac{11}{3}\right\}$

5. $\dfrac{10}{4z - 4} = \dfrac{1}{1 - z}$

Multiply both sides by the common denominator $(4z - 4)(1 - z)$, assuming $z = 1$.

$10(1 - z) = 1(4z - 4)$

$10 - 10z = 4z - 4$

$14 = 14z$

$1 = z$

However, substituting 1 for z in the original equation would result in a denominator of 0, so 1 is not a solution.

Solution set: $\emptyset$

7. $\dfrac{5}{3r} - 10 = \dfrac{3}{2r}$

Multiply both sides by the least common denominator, $6r$, assuming $r \neq 0$.

$6r\left(\dfrac{5}{3r}\right) - 6r(10) = 6r\left(\dfrac{3}{2r}\right)$

$10 - 60r = 9$

$-60r = -1$

$r = \dfrac{1}{60}$

Solution set: $\left\{\dfrac{1}{60}\right\}$

9. $3(x + 2b) + a = 2x - 6$ for x

$3x + 6b + a = 2x - 6$

$x + 6b + a = -6$

$x = -6b - a - 6$

11. $A = \dfrac{24f}{B(p + 1)}$ for f   (approximate

annual interest rate)

$$AB(p + 1) = \dfrac{24f}{B(p + 1)} \cdot B(p + 1)$$
$$\qquad\qquad \textit{Multiply by } B(p + 1)$$

$$AB(p + 1) = 24f$$

$$\dfrac{AB(p + 1)}{24} = f \quad \textit{Divide by 24}$$

13. $A = P\left(1 + \dfrac{i}{m}\right)$ for m   (compound

interest)

$$A = P + \dfrac{Pi}{m} \quad \begin{array}{l}\textit{Distributive}\\\textit{property}\end{array}$$

$$A - P = \dfrac{Pi}{m}$$

$$m(A - P) = Pi \qquad \textit{Multiply by } m$$

$$m = \dfrac{Pi}{A - P} \qquad \textit{Divide by } A - P$$

15. Let x = the number of earned runs
        Clemens allowed.

$$3.13 = \dfrac{9x}{253\frac{1}{3}}$$

Multiply numerator and denominator
of the fraction by 3.

$$3.13 = \dfrac{27x}{760}$$

Multiply both sides of the equation
by 760.

$$760(3.13) = 760 \cdot \dfrac{27x}{760}$$

$$2378.8 = 27x$$

Divide both sides by 27.

$$\dfrac{2378.8}{27} = x$$

$$88.10 \approx x$$

Roger Clemens allowed 88 earned
runs.

17. Let x = Swindell's E.R.A.

$$x = \dfrac{9(69)}{184\frac{1}{3}}$$

$$= \dfrac{621}{184\frac{1}{3}}$$

Multiply numerator and denominator
of the fraction by 3.

$$x = \dfrac{1863}{553}$$

$$\approx 3.37$$

Greg Swindell had an E.R.A. of 3.37.

19. Let x = the number of innings
        pitched by Blyleven.

$$2.73 = \dfrac{9(73)}{x}$$

$$2.73 = \dfrac{657}{x}$$

$$2.73x = x \cdot \dfrac{657}{x}$$

$$2.73x = 657$$

$$x = \dfrac{657}{2.73}$$

$$x \approx 240.66$$

Bert Blyleven pitched 241 innings.

21. Let   x = the number of pounds of
            chocolate hearts at $5
            per pound.

Then 30 − x = the number of pounds of
            candy kisses at $3.50
            per pound.

$$5x + 3.5(30 - x) = 4.5(30)$$

$$5x + 105 - 3.5x = 135$$

$$1.5x = 30$$

$$x = 20$$

20 lb of hearts and 10 lb of kisses should be used.

23. Let x = the number of liters of pure alcohol to be added.

Strength	Liters of solution	Liters of pure alcohol
10%	12	.10(12)
100%	x	1(x)
30%	12 + x	.30(12 + x)

$$\begin{matrix}\text{Liters of}\\\text{alcohol}\\\text{in 10\%}\\\text{solution}\end{matrix} + \begin{matrix}\text{Liters of}\\\text{alcohol}\\\text{in 100\%}\\\text{solution}\end{matrix} = \begin{matrix}\text{Liters of}\\\text{alcohol}\\\text{in 30\%}\\\text{solution}\end{matrix}$$

$$.10(12) \quad + \quad 1(x) \quad = .30(12 + x)$$
$$.10(12) + 1(x) = .30(12 + x)$$
$$1.2 + x = 3.6 + .3x$$
$$.7x = 2.4$$
$$x = \frac{2.4}{.7}$$
$$= \frac{24}{7} = 3\frac{3}{7}$$

Add $3\frac{3}{7}$ liters of pure alcohol.

25. Let   x = Lynn's gross weekly pay.

Then .26x = Lynn's weekly deductions.

$$x - .26x = 592$$
$$.74x = 592$$
$$x = 800$$

Lynn's weekly pay is $800 before deductions.

27. Let   x = average speed upriver.

Then x + 5 = average speed on return trip.

	d	r	t
Upstream	1.2x	x	1.2
Downstream	.9(x + 5)	x + 5	.9

$$1.2x = .9(x + 5)$$
$$1.2x = .9x + 4.5$$
$$.3x = 4.5$$
$$x = 15$$

The average speed of the boat upriver is 15 mph.

29. **(a)** In one year, the maximum amount of lead ingested would be

.05 mg/liter × 2 liters/day

  × 365.25 days/year

   = 36.525 mg/year.

The maximum amount A of lead (in milligrams) ingested in x years would be A = 36.525x.

(b) If x = 72,

   A = 36.525(72) = 2629.8 mg.

The EPA maximum lead intake from water over a lifetime is 2629.8 mg.

31. $(b + 7)^2 = 5$

Use the square root property.

$$b + 7 = \pm\sqrt{5}$$
$$b = -7 \pm \sqrt{5}$$

Solution set: $\{-7 \pm \sqrt{5}\}$

33. $2a^2 + a - 15 = 0$

Solve the equation by factoring.

$(2a - 5)(a + 3) = 0$

$2a - 5 = 0$ or $a + 3 = 0$
    Zero-factor property

$a = \dfrac{5}{2}$ or $\qquad a = -3$

Solution set: $\left\{\dfrac{5}{2}, -3\right\}$

35. $\qquad 2q^2 - 11q = 21$

$2q^2 - 11q - 21 = 0$

$(2q + 3)(q - 7) = 0$

$2q + 3 = 0$ or $q - 7 = 0$

$q = -\dfrac{3}{2}$ or $\qquad q = 7$

Solution set: $\left\{-\dfrac{3}{2}, 7\right\}$

37. $\qquad 2 - \dfrac{5}{p} = \dfrac{3}{p^2}$

Multiply both sides by the least common denominator, p^2, assuming $p \neq 0$.

$$p^2\left(2 - \dfrac{5}{p}\right) = p^2\left(\dfrac{3}{p^2}\right)$$

$$p^2(2) - p^2\left(\dfrac{5}{p}\right) = p^2\left(\dfrac{3}{p^2}\right)$$

$$2p^2 - 5p = 3$$

$$2p^2 - 5p - 3 = 0$$

$$(2p + 1)(p - 3) = 0$$

$$p = -\dfrac{1}{2} \text{ or } p = 3$$

Solution set: $\left\{-\dfrac{1}{2}, 3\right\}$

39. $\sqrt{2}x^2 - 4x + \sqrt{2} = 0$

Use the quadratic formula with
$a = \sqrt{2}$, $b = -4$, and $c = \sqrt{2}$.

$x = \dfrac{4 \pm \sqrt{(-4)^2 - 4 \cdot \sqrt{2} \cdot \sqrt{2}}}{2 \cdot \sqrt{2}}$

$= \dfrac{4 \pm \sqrt{16 - 8}}{2\sqrt{2}} = \dfrac{4 \pm \sqrt{8}}{2\sqrt{2}}$

$= \dfrac{(4 \pm \sqrt{8})\sqrt{2}}{(2\sqrt{2})\sqrt{2}}$

$= \dfrac{4\sqrt{2} \pm \sqrt{16}}{2 \cdot 2} = \dfrac{4\sqrt{2} \pm 4}{4}$

$= \sqrt{2} \pm 1$ *Lowest terms*

Solution set: $\left\{\sqrt{2} \pm 1\right\}$

41. (b) and (c) are the equations that have exactly one real solution.

43. $\qquad 8y^2 = 2y - 6$

$8y^2 - 2y + 6 = 0$

$a = 8$, $b = -2$, $c = 6$

$b^2 - 4ac = (-2)^2 - 4(8)(6)$

$= 4 - 192$

$= -188$

The equation has two different imaginary solutions since the discriminant is negative.

45. $\qquad 16r^2 + 3 = 26r$

$16r^2 - 26r + 3 = 0$

$a = 16$, $b = -26$, $c = 3$

$b^2 - 4ac = (-26)^2 - 4(16)(3)$

$= 676 - 192$

$= 484 = 22^2$

The equation has two different rational solutions since the discriminant is a positive perfect square.

47. The projectile will be 624 ft above the ground whenever

$$220t - 16t^2 = 624.$$

Solve this equation for t.

$$220t - 16t^2 = 624$$
$$-16t^2 + 220t - 624 = 0$$

Simplify by dividing by −4.

$$4t^2 - 55t + 156 = 0$$
$$(t - 4)(4t - 39) = 0$$
$$t - 4 = 0 \quad \text{or} \quad 4t - 39 = 0$$
$$t = 0 \quad \text{or} \quad t = \frac{39}{4} = 9.75$$

The projectile will be 624 ft high at 4 sec and at 9.75 sec.

49. Let x = the length of one side of the building.

Building

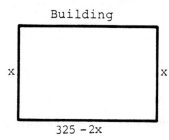

325 − 2x

Apply the formula A = LW to this rectangle.

$$11,250 = x(325 - 2x)$$
$$11,250 = 325x - 2x^2$$
$$2x^2 - 325x - 11,250 = 0$$
$$(x - 50)(2x - 225) = 0$$
$$x = 50 \quad \text{or} \quad x = 112.5$$

If x = 50, then

$$325 - 2(50) = 225.$$

If x = 112.5, then

$$325 - 2(112.5) = 100.$$

The building is either 50 m by 225 m or 112.5 m by 100 m.

51. Let   x = the length of the middle side.

Then x − 7 = the length of the shorter side

and x + 1 = the length of the hypotenuse.

Use the Pythagorean theorem.

$$x^2 + (x - 7)^2 = (x + 1)^2$$
$$x^2 + x^2 - 14x + 49 = x^2 + 2x + 1$$
$$x^2 - 16x + 48 = 0$$
$$(x - 12)(x - 4) = 0$$
$$x = 12 \quad \text{or} \quad x = 4$$

If x = 12, then x − 7 = 5
and x + 1 = 13.

If x = 4, then x − 7 = −3,
which is not possible.

The sides are 5 inches, 12 inches, and 13 inches long.

53. $r = \dfrac{kx}{y^2}$

Substitute r = 10, x = 5, and y = 3 to find k.

$$10 = \frac{k \cdot 5}{3^2}$$
$$k = 18$$

Thus, $r = \dfrac{18x}{y^2}.$

If x = 12 and y = 4,

$$r = \frac{18 \cdot 12}{4^2}$$
$$= \frac{27}{2}.$$

55. Let P = pressure;

D = distance.

$$P = kD$$

Substitute D = 4 and P = 60 to find k.

$$60 = k(4)$$
$$k = 15$$

Thus

$$P = 15D.$$

If D = 10, then

$$P = 15 \cdot 10$$
$$= 150.$$

The pressure is 150 kg per sq m.

57. Let p = power;

v = wind velocity.

$$p = kv^3$$

Substitute p = 10,000 and v = 10 to find k.

$$10,000 = k \cdot 10^3$$
$$\frac{10,000}{1000} = k$$
$$10 = k$$

Thus, $p = 10v^3$.

If v = 15,

$$p = 10 \cdot 15^3$$
$$= 33,750$$

33,750 units of power are produced.

59. $4a^4 + 3a^2 - 1 = 0$

Let $u = a^2$; then $u^2 = a^4$.

With this substitution, the equation becomes

$$4u^2 + 3u - 1 = 0.$$

Solve this equation by factoring.

$$(u + 1)(4u - 1) = 0$$
$$u + 1 = 0 \quad \text{or} \quad 4u - 1 = 0$$
$$u = -1 \quad \text{or} \quad u = \frac{1}{4}$$

To find a, replace u with a^2.

$$a^2 = -1 \quad \text{or} \quad u = \frac{1}{4}$$
$$a = \pm\sqrt{-1} \quad \text{or} \quad u = \pm\sqrt{\frac{1}{4}}$$
$$a = \pm i \quad \text{or} \quad u = \pm\frac{1}{2}$$

Solution set: $\left\{\pm i, \ \pm\frac{1}{2}\right\}$

61. $(2z + 3)^{2/3} + (2z + 3)^{1/3} = 6$
$(2z + 3)^{2/3} + (2z + 3)^{1/3} - 6 = 0$

Let $x = (2z + 3)^{1/3}$; then

$x^2 = [(2z + 3)^{1/3}]^2 = (2z + 3)^{2/3}$.

With this substitution, the equation becomes

$$x^2 + x - 6 = 0.$$

Solve this equation by factoring.

$$(x + 3)(x - 2) = 0$$
$$x = -3 \quad \text{or} \quad x = 2$$

To find z, replace x with $(2z + 3)^{1/3}$.

If x = -3, then

$$(2z + 3)^{1/3} = -3$$
$$2z + 3 = -27 \quad \textit{Cube both sides}$$
$$2z = -30$$
$$z = -15.$$

If x = 2, then

$$(2z + 3)^{1/3} = 2$$

$$2z + 3 = 8 \quad \textit{Cube both sides}$$

$$2z = 5$$

$$z = \frac{5}{2}.$$

Since the original equation involves rational exponents, both solutions should be checked. $z = -15$ and $z = \frac{5}{2}$ both satisfy the original equation.

Solution set: $\left\{-15, \frac{5}{2}\right\}$

63. $$\sqrt{4y - 2} = \sqrt{3y + 1}$$

$$(\sqrt{4y - 2})^2 = (\sqrt{3y + 1})^2$$

$$\textit{Square both sides}$$

$$4y - 2 = 3y + 1$$

$$y = 3$$

Check this proposed solution in the original equation.

$$\sqrt{4y - 2} = \sqrt{3y + 1}$$

$$\sqrt{4(3) - 2} = \sqrt{3(3) + 1} \quad ?$$

$$\sqrt{10} = \sqrt{10} \qquad \textit{True}$$

Solution set: $\{3\}$

65. $$\sqrt{p + 2} = 2 + p$$

$$(\sqrt{p + 2})^2 = (2 + p)^2$$

$$\textit{Square both sides}$$

$$p + 2 = p^2 + 4p + 4$$

$$\textit{Square of a binomial}$$

$$0 = p^2 + 3p + 2$$

$$0 = (p + 2)(p + 1)$$

$$p = -2 \quad \text{or} \quad p = -1$$

Checking these proposed solutions will show that both satisfy the original equation.

Solution set: $\{-2, -1\}$

67. $$\sqrt{x + 3} - \sqrt{3x + 10} = 1$$

$$\sqrt{x + 3} = 1 + \sqrt{3x + 10}$$

$$(\sqrt{x + 3})^2 = (1 + \sqrt{3x + 10})^2$$

$$\textit{Square both sides}$$

$$x + 3 = 1 + 2\sqrt{3x + 10} + 3x + 10$$

$$\textit{Square of a binomial}$$

$$-2x - 8 = 2\sqrt{3x + 10}$$

$$2(-x - 4) = 2\sqrt{3x + 10}$$

$$-x - 4 = \sqrt{3x + 10}$$

$$\textit{Divide by 2}$$

$$(-x - 4)^2 = (\sqrt{3x + 10})^2$$

$$\textit{Square both sides}$$

$$x^2 + 8x + 16 = 3x + 10$$

$$\textit{Square of a binomial}$$

$$x^2 + 5x + 6 = 0$$

$$(x + 2)(x + 3) = 0$$

$$x = -2 \quad \text{or} \quad x = -3$$

Check x = -2.

$$\sqrt{-2 + 3} - \sqrt{-6 + 10} = 1 \quad ?$$

$$\sqrt{1} - \sqrt{4} = 1 \quad ?$$

$$1 - 2 = 1 \quad ?$$

$$-1 = 1 \qquad \textit{False}$$

Check x = -3.

$$\sqrt{-3 + 3} - \sqrt{3(-3) + 10} = 1 \quad ?$$

$$\sqrt{0} - \sqrt{1} = 1 \quad ?$$

$$-1 = 1 \qquad \textit{False}$$

Since neither of the proposed solutions satisfies the original equation, the equation has no solution.

Solution set: ∅

69. $$\sqrt[3]{6y + 2} = \sqrt[3]{4y}$$

$$(\sqrt[3]{6y + 2})^3 = (\sqrt[3]{4y})^3 \quad \textit{Cube both sides}$$

$$6y + 2 = 4y$$

$$2y = -2$$

$$y = -1$$

A check will show that this proposed solution satisfies the original equation.

Solution set: $\{-1\}$

71. $-9x < 4x + 7$

$0 < 13x + 7$

$-7 < 13x$

$-\dfrac{7}{13} < x$ or $x > -\dfrac{7}{13}$

Solution set: $\left(-\dfrac{7}{13},\ \infty\right)$

73. $-5z - 4 \geq 3(2z - 5)$

$-5z - 4 \geq 6z - 15$

$-5z - 4 + 5z \geq 6z - 15 + 5z$

$-4 \geq 11z - 15$

$-4 + 15 \geq 11z - 15 + 15$

$11 \geq 11z$

$1 \geq z$ or $z \leq 1$

Solution set: $(-\infty,\ 1]$

75. $3r - 4 + r > 2(r - 1)$

$4r - 4 > 2r - 2$

$2r - 4 > -2$

$2r > 2$

$r > 1$

Solution set: $(1,\ \infty)$

77. $5 \leq 2x - 3 \leq 7$

$8 \leq 2x \leq 10$

$4 \leq x \leq 5$

Solution set: $[4,\ 5]$

79. $x^2 + 3x - 4 \leq 0$

Solve the corresponding quadratic equation by factoring.

$x^2 + 3x - 4 = 0$

$(x + 4)(x - 1) = 0$

$x = -4$ or $x = 1$

These two points, -4 and 1, divide a number line into three regions shown on the following sign graph.

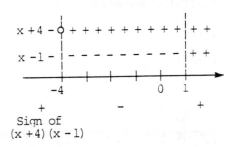

The product is negative when x is between -4 and 1. The endpoints satisfy the inequality because the inequality symbol is $\leq$.

Solution set: $[-4,\ 1]$

81. $6m^2 - 11m - 10 < 0$

Solve the corresponding quadratic equation by factoring.

$6m^2 - 11m - 10 = 0$

$(3m + 2)(2m - 5) = 0$

$3m + 2 = 0$ or $2m - 5 = 0$

$m = -\dfrac{2}{3}$ or $m = \dfrac{5}{2}$

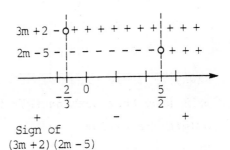

The product is negative when the two factors have opposite signs, which occurs when x is between $-2/3$ and $5/2$. The endpoints are not included because the inequality symbol is <.

Solution set: $\left(-\frac{2}{3}, \frac{5}{2}\right)$

83. $x^2 - 6x + 9 \leq 0$

First, solve the corresponding quadratic equation.

$$x^2 - 6x + 9 = 0$$
$$(x - 3)^2 = 0$$
$$x - 3 = 0$$
$$x = 3$$

This point, 3, divides a number line into the region to the left of 3 and the region to the right of 3. Testing any number from either region will result in $(x - 3)^2$ being positive and therefore not less than or equal to zero. The point 3 causes $(x - 3)^2$ to equal 0, so x = 3 is the only solution of $x^2 - 6x + 9 \leq 0$.
Solution set: $\{3\}$

85. $\dfrac{5p + 2}{p} < -1$

$$\dfrac{5p + 2}{p} + 1 < 0$$

$$\dfrac{5p + 2}{p} + \dfrac{p}{p} < 0$$

$$\dfrac{6p + 2}{p} < 0$$

Sign of
$\dfrac{6p + 2}{p}$

The quotient is negative in the interval $\left(-\frac{1}{3}, 0\right)$.

Solution set: $\left(-\frac{1}{3}, 0\right)$

87. $\dfrac{3}{x + 2} > \dfrac{2}{x - 4}$

$$\dfrac{3}{x + 2} - \dfrac{2}{x - 4} > 0$$

$$\dfrac{3(x - 4) - 2(x + 2)}{(x + 2)(x - 4)} > 0$$

Common denominator is $(x + 2)(x + 4)$

$$\dfrac{3x - 12 - 2x - 4}{(x + 2)(x - 4)} > 0$$

$$\dfrac{x - 16}{(x + 2)(x - 4)} > 0$$

Sign of
$\dfrac{x - 16}{(x + 2)(x - 4)}$

The quotient is positive in the intervals $(-2, 4)$ and $(16, \infty)$. None of the endpoints are included.
Solution set: $(-2, 4) \cup (16, \infty)$

91.   The amount of ozone remaining after filtration is $140 - .43(140) = 79.8$ ppb. Since $79.8 > 50$, the filter did not remove enough of the ozone.

93. $C = 3x + 1500$, $R = 8x$

The company will at least break even when $R \geq C$.

$$8x \geq 3x + 1500$$
$$5x \geq 1500$$
$$x \geq 300$$

The break—even point is at x = 300. The company will at least break even if the number of units produced is in the interval $[300, \infty)$.

95. $|a + 4| = 7$

$a + 4 = 7$ or $a + 4 = -7$

$a = 3$ or $\quad\quad a = -11$

Solution set: $\{3, -11\}$

97. $\left|\dfrac{7}{2 - 3a}\right| = 9$

$\dfrac{7}{2 - 3a} = 9$

$7 = 9(2 - 3a)$

$7 = 18 - 27a$

$27a = 11$

$a = \dfrac{11}{27}$

or

$\dfrac{7}{2 - 3a} = -9$

$7 = -9(2 - 3a)$

$7 = -18 + 27a$

$25 = 27a$

$a = \dfrac{25}{27}$

Solution set: $\left\{\dfrac{11}{27}, \dfrac{25}{27}\right\}$

99. $|5r - 1| = |2r + 3|$

$5r - 1 = 2r + 3$

$3r = 4$

$r = \dfrac{4}{3}$

or

$5r - 1 = -(2r + 3)$

$5r - 1 = -2r - 3$

$7r = -2$

$r = -\dfrac{2}{7}$

Solution set: $\left\{\dfrac{4}{3}, -\dfrac{2}{7}\right\}$

101. $|m| \le 7$

$-7 \le m \le 7$

Solution set: $[-7, 7]$

103. $|p| > 3$

$p < -3$ or $p > 3$

Solution set: $(-\infty, -3) \cup (3, \infty)$

105. $|2z + 9| \le 3$

$-3 \le 2z + 9 \le 3$

$-12 \le 2z \le -6$

$-6 \le z \le -3$

Solution set: $[-6, -3]$

107. $|7k - 3| < 5$

$-5 < 7k - 3 < 5$

$-2 < 7k < 8$

$-\dfrac{2}{7} < k < \dfrac{8}{7}$

Solution set: $\left(-\dfrac{2}{7}, \dfrac{8}{7}\right)$

109. $|3r + 7| - 5 > 0$

$|3r + 7| > 5$

$3r + 7 < -5$ or $3r + 7 > 5$

$3r < -12 \quad\quad\quad 3r > -2$

$r < -4$ or $\quad\quad r > -\dfrac{2}{3}$

Solution set: $(-\infty, -4) \cup \left(-\dfrac{2}{3}, \infty\right)$

Chapter 2   Test

1. $3(x - 4) - 5(x + 2) = 2 - (x + 24)$

$3x - 12 - 5x - 10 = 2 - x - 24$

$-2x = 0$

$x = 0$

Solution set: $\{0\}$

2. $\dfrac{2}{t - 3} - \dfrac{3}{t + 3} = \dfrac{12}{t^2 - 9}$

$\dfrac{2}{t - 3} - \dfrac{3}{t + 3} = \dfrac{12}{(t + 3)(t - 3)}$

Multiply both sides by the common denominator, $(t + 3)(t - 3)$, assuming $t \neq -3$ and $t \neq 3$.

$2(t + 3) - 3(t - 3) = 12$

$2t + 6 - 3t + 9 = 12$

$-t = -3$

$t = 3$

However, substituting 3 for t in the original equation would result in a denominator of 0, so 3 is not a solution.
Solution set: $\emptyset$

3. $S = 2HW + 2LW + 2LH$ for W (width)

$S - 2LH = 2HW + 2LW$

$S - 2LH = W(2H + 2L)$

$\dfrac{S - 2LH}{2H + 2L} = W$

$W = \dfrac{S - 2LH}{2H + 2L}$

4. Substituting $q = 3.1 \times 10^{-13}$ and $T = 20$ into the equation we have

$$C = \frac{5.48 \times 10^3 (3.1 \times 10^{-13})^{.571}(20 + 273)^{-1}}{[2.10 \times 10^{-11} - 6.58 \times 10^{-14}(20 + 273)]^{.571}} \approx 7.029 \text{ pCi.}$$

Since the level is above 4 pCi/L, it is unsafe.

5. On possible action would be to seal any cracks in the basement walls and increase ventilation.

6. Let x = number of quarts of 60% alcohol solution.

Strength	Quarts of solution	Quarts of pure alcohol
60%	x	.60x
20%	40	.20(40)
30%	x + 40	.30(x + 40)

$$.60x + .20(40) = .30(x + 40)$$
$$.60x + 8 = .30x + 12$$
$$.30x = 4$$
$$x = \frac{4}{.30} = 13\frac{1}{3}$$

$13\frac{1}{3}$ quarts of 60% alcohol must be added.

7. Let x = time Fred travels.
   Then x − 3 = time Wilma travels.

	d	r	t
Fred	30x	30	x
Wilma	50(x − 3)	50	x − 3

They both travel the same distance, so

$$30x = 50(x - 3)$$
$$30x = 50x - 150$$
$$-20x = -150$$
$$x = 7.5$$

and $d = 30(7.5) = 225.$

They travel 225 mi before meeting.

8. $3x^2 - 5x = -2$
   $3x^2 - 5x + 2 = 0$

   Solve by factoring.

   $(3x - 2)(x - 1) = 0$
   $3x - 2 = 0$ or $x - 1 = 0$
   $x = \frac{2}{3}$ or $x = 1$

   Solution set: $\left\{\frac{2}{3}, 1\right\}$

9. $(5t - 3)^2 = 17$

   Use the square root property.

   $5t - 3 = \pm\sqrt{17}$
   $5t = 3 \pm \sqrt{17}$
   $t = \frac{3 \pm \sqrt{17}}{5}$

   Solution set: $\left\{\frac{3 \pm \sqrt{17}}{5}\right\}$

10. $6s(2 - s) = 7$
    $12s - 6s^2 = 7$
    $0 = 6s^2 - 12s + 7$

    Use the quadratic formula with
    $a = 6$, $b = -12$, and $c = 7$.

    $$s = \frac{12 \pm \sqrt{144 - 4(6)(7)}}{12}$$
    $$= \frac{12 \pm \sqrt{-24}}{12}$$
    $$= \frac{12 \pm 2i\sqrt{6}}{12}$$
    $$s = 1 \pm \frac{\sqrt{6}}{6}i$$

    Solution set: $\left\{1 \pm \frac{\sqrt{6}}{6}i\right\}$

11. In order for $4x^2 - 5x - k = 0$ to have a single real solution, the value of the discriminant must be 0.

$$b^2 - 4ac = (-5)^2 - 4(4)(-k)$$
$$= 25 + 16k$$

Set the discriminant to 0 and solve for k.

$$25 + 16k = 0$$
$$16k = -25$$
$$k = -\frac{25}{16}$$

The value of k must be $-\frac{25}{16}$.

Note: The equation will be

$$4x^2 - 5x - \left(-\frac{25}{16}\right) = 0$$

or $4x^2 - 5x + \frac{25}{16} = 0$.

12. The table shows each equation evaluated at the years 1975, 1994, and 2006. Equation (b) best models the data.

13. **(a)** In the year 2000, x = 25.

$$y = .24x + .6$$
$$y = .24(25) + .6 = 6.6$$

Using this equation, we estimate that there will be 6.6 million commuter fliers in the year 2000.

(b) $y = .0138x^2 - .172x + 1.4$

$$y = .0138(25)^2 - .172(25) + 1.4$$
$$= 5.725$$

Using this equation, we estimate that there will be 5.725 million commuter fliers in the year 2000.

(c) $y = .0125x^2 - .193x + 1.4$

$$y = .0125(25)^2 - .193(25) + 1.4$$
$$\approx 4.388$$

Using this equation, we estimate that there will be 4.388 commuter fliers in the year 2000.

14. $h = -16t^2 + 96t$

(a) Let h = 80 and solve for t.

$$80 = -16t^2 + 96t$$
$$16t^2 - 96t + 80 = 0$$

Divide by 16.

$$t^2 - 6t + 5 = 0$$
$$(t - 1)(t - 5) = 0$$
$$t = 1 \quad \text{or} \quad t = 5$$

The projectile will reach a height of 80 ft at 1 sec and 5 sec.

(b) Let h = 0 and solve for t.

$$0 = -16t^2 + 96t$$
$$0 = -16t(t - 6)$$
$$t = 0 \quad \text{or} \quad t = 6$$

The projectile will return to the ground at 6 sec.

15. $\sqrt{5 + 2x} - x = 1$

$$\sqrt{5 + 2x} = x + 1$$
$$(\sqrt{5 + 2x})^2 = (x + 1)^2$$
$$5 + 2x = x^2 + 2x + 1$$
$$0 = x^2 - 4$$
$$0 = (x + 2)(x - 2)$$
$$x = -2 \quad \text{or} \quad x = 2$$

Check: x = -2.

$$\sqrt{5 + 2(-2)} - (-2) = 1 \ ?$$
$$\sqrt{1} + 2 = 1 \ ?$$
$$3 = 1 \quad \textit{False}$$

Check: $x = 2$.

$$\sqrt{5 + 2(2)} - 2 = 1 \; ?$$
$$\sqrt{9} - 2 = 1 \; ?$$
$$1 = 1 \quad \textit{True}$$

The checks show that 2 satisfies the equation but -2 does not.

Solution set: $\{2\}$

16. $x^4 + 6x^2 - 40 = 0$

Let $u = x^2$, and solve the resulting equation for u.

$$u^2 + 6u - 40 = 0$$
$$(u + 10)(u - 4) = 0$$
$$u = -10 \quad \text{or} \quad u = 4$$

Replace u with x^2 and solve for x.

If $y = -10$, then

$$x^2 = -10$$
$$x = \pm\sqrt{-10}$$
$$x = \pm i\sqrt{10}.$$

If $u = 4$, then

$$x^2 = 4$$
$$x = \pm 2.$$

Solution set: $\{\pm 2, \; \pm i\sqrt{10}\}$

17.
$$\sqrt[3]{3x - 8} = \sqrt[3]{9x + 4}$$
$$(\sqrt[3]{3x - 8})^3 = (\sqrt[3]{9x + 4})^3$$
$$3x - 8 = 9x + 4$$
$$-6x = 12$$
$$x = -2$$

Solution set: $\{-2\}$

18. $6 = \dfrac{7}{2y - 3} + \dfrac{3}{(2y - 3)^3}$

Multiply by the common denominator, $(2y - 3)^2$, assuming $y \neq 3/2$.

$$6(2y - 3)^2 = 7(2y - 3) + 3$$
$$6(4y^2 - 12y + 9) = 14y - 21 + 3$$
$$24y^2 - 72y + 54 = 14y - 18$$
$$24y^2 - 86y + 72 = 0$$

Divide by 2.

$$12y^2 - 43y + 36 = 0$$
$$(3y - 4)(4y - 9) = 0$$
$$3y - 4 = 0 \quad \text{or} \quad 4y - 9 = 0$$
$$y = \frac{4}{3} \quad \text{or} \quad y = \frac{9}{4}$$

Solution set: $\left\{\dfrac{4}{3}, \dfrac{9}{4}\right\}$

19. Since Dobson units are linear, we can use proportions. Let x be the thickness of the Antarctic ozone layer. Then,

$$\frac{x}{110} = \frac{3}{300}$$
$$300x = 330$$
$$x = 1.1.$$

The thickness of the ozone layer in 1991 at the Antartic ozone hole was 1.1 mm.

20. Let D = number of days to ripen; t = average maximum temperature.

$$D = \frac{k}{t}$$

Substitute $D = 25$ and $t = 80$ to find k.

$$25 = \frac{k}{80}$$

$$80 \cdot 25 = k$$

$$2000 = k$$

Thus,

$$D = \frac{2000}{k}.$$

If $t = 75$,

$$D = \frac{2000}{75} = \frac{80}{3} \quad \text{or} \quad 26\frac{2}{3}.$$

It would take $26\frac{2}{3}$ days for the fruit to ripen.

21. $-2(x - 1) - 10 < 2(2 + x)$

$$-2x + 2 - 10 < 4 + 2x$$

$$-2x - 8 < 4 + 2x$$

$$-4x < 12$$

$$\left(-\frac{1}{4}\right)(-4x) > \left(-\frac{1}{4}\right)(12)$$
Multiply by $-1/4$; reverse inequality symbol

$$x > -3$$

Solution set: $(-3, \infty)$

22. $-2 \le \frac{1}{2}x + 3 \le 4$

Multiply all parts by 2.

$$-4 \le x + 6 \le 8$$

$$-10 \le x \le 2$$

Solution set: $[-10, 2]$

23. $2x^2 - x - 3 \ge 0$

Solve the corresponding quadratic equation by factoring.

$$2x^2 - x - 3 = 0$$

$$(x + 1)(2x - 3) = 0$$

$$x = -1 \quad \text{or} \quad x = \frac{3}{2}$$

These two points divide the number line into three regions.

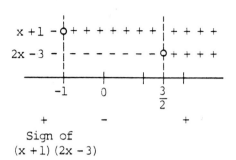

The sign graph shows that the product is positive in the intervals $(-\infty, -1)$ and $\left(\frac{3}{2}, \infty\right)$. The endpoints satisfy the inequality since the inequality symbol is $\ge$.

Solution set: $(-\infty, -1] \cup \left[\frac{3}{2}, \infty\right)$

24.

$$\frac{x + 1}{x - 3} < 5$$

$$\frac{x + 1}{x - 3} - 5 < 0$$

$$\frac{x + 1 - 5(x - 3)}{x - 3} < 0$$

$$\frac{x + 1 - 5x + 15}{x - 3} < 0$$

$$\frac{-4x + 16}{x - 3} < 0$$

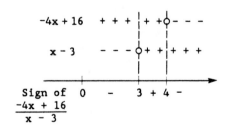

The quotient is negative in the intervals $(-\infty, 3)$ and $(4, \infty)$. None of the endpoints are included.

Solution set: $(-\infty, 3) \cup (4, \infty)$

25. $|3x + 5| = 4$

$3x + 5 = 4$ or $3x + 5 = -4$

$\quad\quad 3x = -1 \quad\quad\quad\quad 3x = -9$

$\quad\quad x = -\dfrac{1}{3}$ or $x = -3$

Solution set: $\left\{-3, \ -\dfrac{1}{3}\right\}$

26. $|2x - 5| < 9$

$\quad\quad -9 < 2x - 5 < 9$

$\quad\quad -4 < 2x < 14$

$\quad\quad -2 < x < 7$

Solution set: $(-2, \ 7)$

27. $|2x + 1| \geq 11$

$2x + 1 \leq -11$ or $2x + 1 \geq 11$

$\quad\quad 2x \leq -12 \quad\quad\quad\quad 2x \geq 10$

$\quad\quad x \leq -6$ or $x \geq 5$

Solution set: $(-\infty, \ -6] \cup [5, \ \infty)$

Cumulative Review Exercises (Chapters 1-2)

1. Which of the following real numbers are not rational numbers? $\frac{3}{4}$, $\sqrt{2}$, π, 1.2

2. Evaluate the expression $-\frac{1}{2}x^2 + 3y - 5z$ where $x = -4$, $y = \frac{2}{3}$, $z = 1$.

3. What property is illustrated by the statement $6x\left(x + \frac{1}{6}\right) = 6x^2 + x$?

4. Write $\left|\sqrt{3} - 2\right|$ without absolute value bars.

5. Name the inequality property that justifies the following statement:
   If $m < -4$, then $-m > 4$.

6. For what values of x does $\left|x\right| = -x$?

7. Find the product $(x - 4)\left(2x + \frac{1}{2}\right)$.

8. Use the properties of exponents to simplify $(-r^2 s^3)^4$.

9. Write out the binomial expansion of $(2a - b)^3$.

10. Factor out the greatest common factor from the polynomial $6x^3 y^2 - 9x^2 y^3$.

11. Factor the trinomial $6x^2 + 7x - 3$.

12. Factor $27x^3 - 8y^6$.

13. Write the rational expression $\dfrac{6x - 2x^2}{x^2 - 9}$ in lowest terms.

14. Find the quotient $\dfrac{3x^2 - x - 2}{9x^2 - 4} \div \dfrac{2x^2 + x - 3}{2x^3 + 3x^2}$.

15. Find $\dfrac{3x}{x^2 + 2x} - \dfrac{4}{x^2 - 2x}$.

16. Simplify $\dfrac{(3x)^{-2}(2x^{-1})^4}{(4x^{-3})^2}$.

17. Simplify $\dfrac{-16^{3/4} \cdot 27^{10/3}}{4^{3/2} \cdot 27^2}$.

18. Rewrite $\dfrac{a^{-1} - b^{-1}}{(a - b)^{-1}}$ with positive integer exponents.

19. Simplify $\sqrt{25x^6y^{12}}$. Use absolute value if necessary.

20. Add $3\sqrt[3]{2} + 5\sqrt[3]{54} - 2\sqrt[3]{16}$.

21. Simplify $\dfrac{\sqrt[3]{a^3b^7c^7} \cdot \sqrt[3]{a^6b^8c^9}}{\sqrt[3]{a^7b^3c^5}}$. Assume that all variables represent positive numbers.

22. Rationalize the denominator of $\dfrac{1 - \sqrt{3}}{\sqrt{3} + \sqrt{5}}$.

23. Multiply $\sqrt{-15} \cdot \sqrt{-6}$.

24. Multiply $(2 - 3i)(4 + 5i)$. Write the answer in standard form.

25. Find the quotient $\dfrac{4 - 5i}{1 + 2i}$. Write the answer in standard form.

26. Solve the equation $5(k + 2) - 4(k - 1) = 18 - 3k$.

27. Solve the equation $\dfrac{3}{p - 1} + \dfrac{4}{9} = \dfrac{3p + 4}{p^2 - p}$.

28. Solve $y = \dfrac{2 - x}{1 + x}$ for x.

29. How many pounds of coffee worth $6 per lb should be mixed with 20 lb of coffee selling for $4.50 per pound to get a mixture that can be sold for $5 per pound?

30. After winning a lottery, Mr. Rodriquez has $90,000 to invest. He puts part of the money in a certificate of deposit at 8%, and the rest into a real estate scheme paying 12%. The total annual income from the investments is $8800. How much does he have invested at each rate?

31. A boat can go 15 km upstream in the same time that it takes to go 27 km downstream. The speed of the current is 2 km per hr. Find the speed of the boat in still water.

32. Solve $3m^2 - 12m - 63 = 0$ by completing the square.

33. Solve $9r^2 + 6r + 1 = 0$ by the quadratic formula.

34. Solve $27k^3 - 8 = 0$ by first factoring and then using the quadratic formula.

35. One leg of a right triangle is 3 cm longer than three times the length of the shorter leg. The hypotenuse is 1 cm longer than the longer leg. Find the length of the sides of the triangle.

36. Amanda can do a job in 5 hr. When Amanda is working with Brittany, the job takes 3 hr. How long would it take Brittany working alone to do the job?

37. A poster is to have an area of 125 square inches. The printed material is to be surrounded by a margin of 3 inches at the top and margins of 2 inches at the bottom and sides. Find the dimensions of the poster for which the area of the printed material is 45 square inches.

38. Solve $\sqrt{r + 3} - \sqrt{2r - 1} = -1$.

39. Solve $(z^2 - 18z)^{1/4} = 0$.

40. Solve $6z^{-2} + 7z^{-1} + 2 = 0$.

41. Write the statement "h varies directly as the fifth power of t and inversely as the cube of v" as an equation.

42. Suppose u varies directly as x and y and inversely as the cube of z. If u is 15 when x is 6, y is 4, and z is 2, find u when x is 10, y is 20, and z is 5.

43. The kinetic energy of a moving object is proportional to the mass and the square of the velocity. If a 2000-kg car traveling at 30 meters per second has a kinetic energy of 900,000 units, how much kinetic energy is produced when the velocity is increased to 40 meters per second?

Solve each equation or inequality.

44. $12m - 17 \geq 8m + 8$

45. $x^2 + 6x + 16 < 8$

46. $\dfrac{a - 6}{a + 2} < -1$

47. $|3m - 1| = |2m + 5|$

48. $|5 - 2r| - 3 < 1$

49. Write the statement "x is at most 3 units from 5" as an absolute value inequality.

Solutions to Cumulative Review Exercises (Chapters 1-2)

1. $\sqrt{2}$ and π cannot be expressed as a ratio of integers. Therefore, $\sqrt{2}$ and π are not rational numbers.

2. $-\frac{1}{2}x^2 + 3y - 5z$

   $= -\frac{1}{2}(-4)^2 + 3\left(\frac{2}{3}\right) - 5(1)$

   Let $x = -4$, $y = 2/3$, $z = 1$

   $= -\frac{1}{2}(16) + 2 - 5$

   $= -8 + 2 - 5 = -11$

3. $6x\left(x + \frac{1}{6}\right) = 6x^2 + x$ illustrates the distributive property, $a(b + c) = ab + ac$.

4. Since $\sqrt{3} < 2$, $\sqrt{3} - 2$ is negative. Therefore,

   $$\left|\sqrt{3} - 2\right| = -(\sqrt{3} - 2) = -\sqrt{3} + 2$$
   $$= 2 - \sqrt{3}.$$

5. "If $m < -4$, then $-m > 4$" is justified by the multiplication property.

   $$m < -4$$
   $$(-1)m > (-1)(-4)$$
   $$-m > 4$$

6. If x is a negative number, $|x| = -x$. For example, $|-3| = -(-3) = 3$. Also, If $x = 0$, $|x| = -x$ since $|0| = -0$. Therefore, $|x| = -x$ if $x \leq 0$.

7. $(x - 4)\left(2x + \frac{1}{2}\right) = 2x^2 + \frac{1}{2}x - 8x - 2$

   $$= 2x^2 - \frac{15}{2}x - 2$$

8. $(-r^2s^3)^4 = (-r^2)^4(s^3)^4$

   $$= r^8s^{12}$$

9. $(2a - b)^3$

   To expand $(2a - b)^3$, use the coefficients in row 3 of Pascal's triangle.

   $(2a - b)^3$

   $= (2a)^3 + 3(2a)^2(-b)$

   $\quad + 3(2a)(-b)^2 + (-b)^3$

   $= 8a^3 - 12a^2b + 6ab^2 - b^3$

10. $6x^3y^2 - 9x^2y^3$

    $= (3x^2y^2)(2x) - (3x^2y^2)(3y)$

    $\qquad\qquad\qquad GCF = 3x^2y^2$

    $= 3x^2y^2(2x - 3y)$

11. $6x^2 + 7x + 3$

    To factor $6x^2 + 7x - 3$, there are eight possibilities that will give us a first term of $6x^2$ and a third term of -3. These eight possibilities are

    1. $(6x + 1)(x - 3)$ 2. $(6x - 1)(x + 3)$
    3. $(6x + 3)(x - 1)$ 4. $(6x - 3)(x + 1)$
    5. $(3x + 3)(2x - 1)$ 6. $(3x - 3)(2x + 1)$
    7. $(3x + 1)(2x - 3)$ 8. $(3x - 1)(2x + 3)$.

    Using FOIL, we must find which of these products would give a middle term of $+7x$. We see that the last one will. Therefore,

    $$6x^2 + 7x - 3 = (3x - 1)(2x + 3).$$

12. $27x^3 - 8y^6$

$= (3x)^3 - (2y^2)^3$ *Difference of two cubes*

$= (3x - 2y^2)$

$\quad \cdot [(3x)^2 + (3x)(2y^2) + (2y^2)^2]$

$= (3x - 2y^2)(9x^2 + 6xy^2 + 4y^4)$

13. $\dfrac{6x - 2x^2}{x^2 - 9} = \dfrac{-2x(-3 + x)}{(x + 3)(x - 3)}$

$\qquad = \dfrac{-2x(x - 3)}{(x + 3)(x - 3)}$

$\qquad = \dfrac{-2x}{x + 3}$

14. $\dfrac{3x^2 - x - 2}{9x^2 - 4} \div \dfrac{2x^2 + x - 3}{2x^3 + 3x^2}$

$= \dfrac{3x^2 - x - 2}{9x^2 - 4} \cdot \dfrac{2x^3 + 3x^2}{2x^2 + x - 3}$

$= \dfrac{(3x + 2)(x - 1)}{(3x + 2)(3x - 2)} \cdot \dfrac{x^2(2x + 3)}{(2x + 3)(x - 1)}$

$= \dfrac{x^2(3x + 2)(x - 1)(2x + 3)}{(3x + 2)(3x - 2)(2x + 3)(x - 1)}$

$= \dfrac{x^2}{3x - 2}$

15. $\dfrac{3x}{x^2 + 2x} - \dfrac{4}{x^2 - 2x}$

$= \dfrac{3x}{x(x + 2)} - \dfrac{4}{x(x - 2)}$

$= \dfrac{3x(x - 2)}{x(x + 2)(x - 2)} - \dfrac{4(x + 2)}{x(x - 2)(x + 2)}$

$\qquad\qquad LCD = x(x + 2)(x - 2)$

$= \dfrac{3x^2 - 6x - 4x - 8}{x(x + 2)(x - 2)}$

$= \dfrac{3x^2 - 10x - 8}{x(x^2 - 4)}$

16. $\dfrac{(3x)^{-2}(2x^{-1})^4}{(4x^{-3})^2} = \dfrac{3^{-2} \cdot x^{-2} \cdot 2^4 \cdot x^{-4}}{4^2 x^{-6}}$

$\qquad = \dfrac{16x^{-6}}{9 \cdot 16x^{-6}}$

$\qquad = \dfrac{1}{9}$

17. $\dfrac{-16^{3/4} \cdot 27^{10/3}}{4^{3/2} \, 27^2} = -\dfrac{(\sqrt[4]{16})^3 27^{4/3}}{4^{3/2}}$

$= -\dfrac{2^3(\sqrt[3]{27})^4}{(\sqrt{4})^3}$

$= -\dfrac{8(3^4)}{2^3}$

$= -\dfrac{8 \cdot 81}{8}$

$= -81$

18. $\dfrac{a^{-1} - b^{-1}}{(a - b)^{-1}}$

$= \dfrac{\dfrac{1}{a} - \dfrac{1}{b}}{\dfrac{1}{a - b}}$

$= \dfrac{ab(a - b)\left(\dfrac{1}{a}\right) - ab(a - b)\left(\dfrac{1}{b}\right)}{ab(a - b)\left(\dfrac{1}{a - b}\right)}$

$\qquad\qquad LCD = ab(a - b)$

$= \dfrac{b(a - b) - a(a - b)}{ab}$

$= \dfrac{ab - b^2 - a^2 + ab}{ab}$

$= \dfrac{-a^2 + 2ab - b^2}{ab}$

$= -\dfrac{a^2 - 2ab + b^2}{ab}$

$= -\dfrac{(a - b)^2}{ab}$

19. $\sqrt{25x^6y^{12}} = \sqrt{5^2(x^3)^2(y^6)^2}$

$= 5\,|x^3| \cdot |y^6|$

$= 5\,|x^3|\,y^6$

20. $3\sqrt[3]{2} + 5\sqrt[3]{54} - 2\sqrt[3]{16}$

$= 3\sqrt[3]{2} + 5\sqrt[3]{3^3 \cdot 2} - 2\sqrt[3]{2^3 \cdot 2}$

$= 3\sqrt[3]{2} + 5 \cdot 3\sqrt[3]{2} - 2 \cdot 2\sqrt[3]{2}$

$= 3\sqrt[3]{2} + 15\sqrt[3]{2} - 4\sqrt[3]{2}$

$= 14\sqrt[3]{2}$

21. $\dfrac{\sqrt[3]{a^3b^7c^7} \cdot \sqrt[3]{a^6b^8c^9}}{\sqrt[3]{a^7b^3c^5}}$

$= \sqrt[3]{\dfrac{(a^3b^7c^7)(a^6b^8c^9)}{a^7b^3c^5}}$

$= \sqrt[3]{\dfrac{a^9b^{15}c^{16}}{a^7b^3c^5}}$

$= \sqrt[3]{a^2b^{12}c^{11}}$

$= \sqrt[3]{a^2(b^4)^3(c^3)^3c^2}$

$= b^4c^3\sqrt[3]{a^2c^2}$

22. $\dfrac{1 - \sqrt{3}}{\sqrt{3} + \sqrt{5}} = \dfrac{(1 - \sqrt{3})(\sqrt{3} - \sqrt{5})}{(\sqrt{3} + \sqrt{5})(\sqrt{3} - \sqrt{5})}$

$= \dfrac{\sqrt{3} - \sqrt{5} - 3 + \sqrt{15}}{3 - 5}$

$= \dfrac{\sqrt{3} - \sqrt{5} - 3 + \sqrt{15}}{-2}$

$= \dfrac{-\sqrt{3} + \sqrt{5} + 3 - \sqrt{15}}{2}$

23. $\sqrt{-15} \cdot \sqrt{-6} = (i\sqrt{15})(i\sqrt{6})$

$= i^2\sqrt{90}$

$= 3\sqrt{10}(-1)$

$= -3\sqrt{10}$

24. $(2 - 3i)(4 + 5i)$

$= 8 + 10i - 12i - 15i^2$

$= 8 - 2i - 15(-1)$

$= 8 - 2i + 15$

$= 23 - 2i$

25. $\dfrac{4 - 5i}{1 + 2i} = \dfrac{(4 - 5i)(1 - 2i)}{(1 + 2i)(1 - 2i)}$

$= \dfrac{4 - 8i - 5i + 10i^2}{1 - 4i^2}$

$= \dfrac{4 - 13i + 10(-1)}{1 - 4(-1)}$

$= \dfrac{4 - 13i - 10}{1 + 4}$

$= \dfrac{-6 - 13i}{5}$

$= -\dfrac{6}{5} - \dfrac{13}{5}i$ *Standard form*

26. $5(k + 2) - 4(k - 1) = 18 - 3k$

$5k + 10 - 4k + 4 = 18 - 3k$

$k + 14 = 18 - 3k$

$4k + 14 = 18$

$4k = 4$

$k = 1$

Solution set: $\{1\}$

27. $\dfrac{3}{p - 1} + \dfrac{4}{p} = \dfrac{3p + 4}{p^2 - p}$

$\dfrac{3}{p - 1} + \dfrac{4}{p} = \dfrac{3p + 4}{p(p - 1)}$

Notice the restrictions, $p \neq 1$ and $p \neq 0$. Multiply both sides of the equation by the common denominator $p(p - 1)$.

$p(p - 1)\dfrac{3}{p - 1} + p(p - 1)\dfrac{4}{p}$

$\qquad\qquad = p(p - 1)\dfrac{3p + 4}{p(p - 1)}$

$3p + 4(p - 1) = 3p + 4$

$3p + 4p - 4 = 3p + 4$

$7p - 4 = 3p + 4$

$4p - 4 = 4$

$4p = 8$

$p = 2$

Solution set: $\{2\}$

28. $\qquad\qquad y = \dfrac{2 - x}{1 + x}$

$(1 + x)y = (1 + x)\dfrac{2 - x}{1 + x}$

$y + xy = 2 - x$

$y + xy + x = 2$

$xy + x = 2 - y$

$(y + 1)x = 2 - y$

$x = \dfrac{2 - y}{y + 1}$

29. Let x = the number of pounds of
     coffee worth $6 per pound.

$$6x + 4.50(20) = 5(x + 20)$$
$$6x + 90 = 5x + 100$$
$$x + 90 = 100$$
$$x = 10$$

Use 10 pounds of coffee worth $6 per
pound.

30. Let x = amount of money invested
     at 8%.

Then 90,000 - x

     = amount of money invested
       at 12%.

$$.08x + .12(90,000 - x) = 8800$$

Multiply both sides by 100.

$$8x + 12(90,000 - x) = 880,000$$
$$8x + 1,080,000 - 12x = 880,000$$
$$-4x + 1,080,000 = 880,000$$
$$-4x = -200,000$$
$$x = 50,000$$

He has $50,000 invested at 8% and
$40,000 invested at 12%.

31. Let x = the speed of the boat in
     still water.

We use t = d/r.

$$\frac{15}{x - 2} = \frac{27}{x + 2}$$
$$15(x + 2) = 27(x - 2)$$
$$15x + 30 = 27x - 54$$
$$30 = 12x - 54$$
$$84 = 12x$$
$$7 = x$$

The speed of the boat in still water
is 7 km per hr.

32. $3m^2 - 12m - 63 = 0$

Divide both sides by 3.

$$m^2 - 4m - 21 = 0$$
$$m^2 - 4m = 21$$
$$m^2 - 4m + 4 = 21 + 4$$
$$(m - 2)^2 = 25$$
$$m - 2 = \pm\sqrt{25}$$
$$m = 2 \pm 5$$
$$2 + 5 = 7 \quad \text{and} \quad 2 - 5 = -3$$
$$m = 7 \quad \text{or} \quad m = -3$$

Solution set: $\{-3, 7\}$

33. $9r^2 + 6r + 1 = 0$

Substitute a = 9, b = 6, and c = 1
in the quadratic formula.

$$r = \frac{-(6) \pm \sqrt{(6)^2 - 4(9)(1)}}{2(9)}$$
$$= \frac{-6 \pm \sqrt{36 - 36}}{18}$$
$$= \frac{-6 \pm 0}{18}$$
$$r = -\frac{6}{18} = -\frac{1}{3}$$

Solution set: $\left\{-\frac{1}{3}\right\}$

34. $$27k^3 - 8 = 0$$
     $$(3k)^3 - (2)^3 = 0$$
     $$(3k - 2)(9k^2 + 6k + 4) = 0$$
     $$3k - 2 = 0$$
     $$3k = 2$$
     $$k = \frac{2}{3}$$

or

$9k^2 + 6k + 4 = 0$

$a = 9, b = 6, c = 4$

$$k = \frac{-(6) \pm \sqrt{(6)^2 - 4(9)(4)}}{2(9)}$$

$$= \frac{-6 \pm \sqrt{36 - 144}}{18}$$

$$= \frac{-6 \pm \sqrt{-108}}{18}$$

$$= \frac{-6 \pm i\sqrt{108}}{18}$$

$$= \frac{-6 \pm 6i\sqrt{3}}{18}$$

$$= \frac{6(-1 \pm i\sqrt{3})}{6(3)}$$

$$= \frac{-1 \pm i\sqrt{3}}{3}$$

$$k = -\frac{1}{3} \pm \frac{\sqrt{3}}{3}i$$

Solution set:

$$\left\{ \frac{2}{3}, -\frac{1}{3} + \frac{\sqrt{3}}{3}i, -\frac{1}{3} - \frac{\sqrt{3}}{3}i \right\}$$

35. Let x = length of the shorter leg;

3x + 3 = length of the longer leg;

3x + 4 = length of the hypotenuse.

Use the Pythagorean theorem with
a = x, b = 3x + 3, and c = 3x + 4.

$$x^2 + (3x + 3)^2 = (3x + 4)^2$$

$$x^2 + 9x^2 + 18x + 9 = 9x^2 + 24x + 16$$

$$10x^2 + 18x + 9 = 9x^2 + 24x + 16$$

$$x^2 - 6x - 7 = 0$$

$$(x - 7)(x + 1) = 0$$

$$x = 7 \quad \text{or} \quad x = -1$$

The solution x = −1 makes no sense
in the problem.

If x = 7, 3x + 3 = 3(7) + 3 = 24

and   3x + 4 = 3(7) + 4 = 25.

The lengths of the sides are 7 cm,
24 cm, and 25 cm.

36. Let x = the number of hours required for Brittany to
complete the job working
alone.

$$3\left(\frac{1}{5}\right) + 3\left(\frac{1}{x}\right) = 1$$

$$\frac{3}{5} + \frac{3}{x} = 1$$

$$5x\left(\frac{3}{5}\right) + 5x\left(\frac{3}{x}\right) = 5x(1)$$

$$3x + 15 = 5x$$

$$15 = 2x$$

$$7.5 = x$$

Working alone, Brittany would take
7.5 hr to complete the job.

37. Let x = length of the poster.
     y = width of the poster.

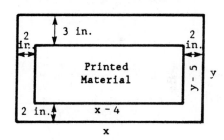

Since the area of the poster is 125
square inches,

$$xy = 125.$$

Solve this equation for y.

$$y = \frac{125}{x}$$

Since the area of the printed material is 45 square inches

$$(x - 4)(y - 5) = 45.$$

Substitute 125/x for y in this equation.

$$(x - 4)\left(\frac{125}{x} - 5\right) = 45$$

$$125 - 5x - \frac{500}{x} + 20 = 45$$

$$-5x + 100 - \frac{500}{x} = 0$$

Multiply by $-x$.

$$5x^2 - 100x + 500 = 0$$

Divide by 5.

$$x^2 - 20x + 100 = 0$$
$$(x - 10)^2 = 0$$
$$x = 10$$

If $x = 10$,

$$y = \frac{125}{x} = \frac{125}{10} = 12.5.$$

The dimensions of the poster are 10 in. by 12.5 in.

38.
$$\sqrt{r + 3} - \sqrt{2r - 1} = -1$$
$$\sqrt{r + 3} = \sqrt{2r - 1} - 1$$
$$(\sqrt{r + 3})^2 = (\sqrt{2r - 1} - 1)^2$$
$$r + 3 = (2r - 1)$$
$$- 2\sqrt{2r - 1} + 1$$
$$-r + 3 = -2\sqrt{2r - 1}$$
$$(-r + 3)^2 = (-2\sqrt{2r - 1})^2$$
$$r^2 - 6r + 9 = 4(2r - 1)$$
$$r^2 - 6r + 9 = 8r - 4$$
$$r^2 - 14r + 13 = 0$$
$$(r - 1)(r - 13) = 0$$
$$r = 1 \quad \text{or} \quad r = 13$$

Checking these proposed solutions in the original equation will show that 13 satisfies the equation but 1 does not.

Solution set: $\{13\}$

39.
$$(z^2 - 18z)^{1/4} = 0$$
$$[(z^2 - 18z)^{1/4}]^4 = 0^4$$
$$z^2 - 18z = 0$$
$$z(z - 18) = 0$$
$$z = 0 \quad \text{or} \quad z = 18$$

Both proposed solutions check in the original equation.

Solution set: $\{0, 18\}$

40. $6z^{-2} + 7z^{-1} + 2 = 0$

Let $y = z^{-1}$.

With this subtitution, the equation becomes

$$6u^2 + 7u + 2 = 0.$$

Solve this equation by factoring.

$$(3u + 2)(2u + 1) = 0$$
$$3u + 2 = 0 \quad \text{or} \quad 2u + 1 = 0$$
$$u = -\frac{2}{3} \qquad\qquad u = -\frac{1}{2}$$

Replace u with z^{-1} and solve for z.

$$z^{-1} = -\frac{2}{3} \qquad\qquad z^{-1} = -\frac{1}{2}$$
$$z = -\frac{3}{2} \quad \text{or} \qquad z = -2$$

Solution set: $\left\{-\frac{3}{2}, -2\right\}$

41. The given statement translates into the equation

$$h = \frac{kt^5}{v^3}.$$

42. $u = \dfrac{kxy}{z^3}$

Since u is 15 when x is 6, y is 4, and z = 2,

$$15 = \frac{k(6)(4)}{(2)^3}$$

$$15 = \frac{24k}{8}$$

$$15 = 3k$$

$$k = 5.$$

Therefore,

$$u = \frac{5xy}{z^3}.$$

Now find u when x is 10, y is 20, and z is 5.

$$u = \frac{5(10)(20)}{5^3}$$

$$u = \frac{1000}{125}$$

$$u = 8$$

43. $E = kmv^2$

Since E = 900,000 when m = 2000 and v = 30,

$$900,000 = k(2000)(30)^2$$

$$900,000 = 1,800,000k$$

$$k = \frac{1}{2}.$$

Therefore,

$$E = \frac{1}{2}mv^2.$$

Now find E when v = 40.

$$E = \frac{1}{2}(2000)(40)^2$$

$$E = 1,600,000$$

When the velocity is 40 m per sec, the kinetic energy is 1,600,000 units.

44.

$$12m - 17 \geq 8m + 8$$

$$4m - 17 \geq 8$$

$$4m \geq 25$$

$$m \geq \frac{25}{4}$$

Solution set: $\left[\dfrac{25}{4},\ \infty\right)$

45.

$$x^2 + 6x + 16 < 8$$

$$x^2 + 6x + 8 < 0$$

Find the values of x that satisfy

$$x^2 + 6x + 8 = 0.$$

Solve this equation by factoring.

$$(x + 2)(x + 4) = 0$$

$$x = -2 \quad \text{or} \quad x = -4$$

Use these two points in a sign graph.

The product of (x + 2) and (x + 4) is less than zero when the two factors have different signs. This occurs in the middle interval. The solution set is the open interval (−4, −2).

46.

$$\frac{a - 6}{a + 2} < -1$$

$$\frac{a - 6}{a + 2} + 1 < 0$$

$$\frac{a - 6}{a + 2} + \frac{a + 2}{a + 2} < 0$$

$$\frac{2a - 4}{a + 2} < 0$$

$$\frac{2(a - 2)}{a + 2} < 0$$

The numerator is zero when a = 2.
The denominator is zero when a = −2.
Use these two points in a sign
graph.

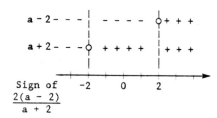

The quotient $\dfrac{2(a - 2)}{a + 2}$ will be less

than zero when the two factors have
different signs.   This occurs in the
middle interval.   The solution set
is the open interval (−2, 2).

47. $|3m - 1| = |2m + 5|$

$3m - 1 = 2m + 5$ or $3m - 1 = -(2m + 5)$

$m - 1 = 5$ or $3m - 1 = -2m - 5$

$m = 6$ or $5m - 1 = -5$

$5m = -4$

$m = -\dfrac{4}{5}$

Solution set: $\left\{6, -\dfrac{4}{5}\right\}$

48. $|5 - 2r| - 3 < 1$

$|5 - 2r| < 4$

$-4 < 5 - 2r < 4$

$-9 < -2r < -1$

$\dfrac{-9}{-2} > \dfrac{-2r}{-2} > \dfrac{-1}{-2}$

$\dfrac{9}{2} > r > \dfrac{1}{2}$

or $\dfrac{1}{2} < r < \dfrac{9}{2}$

Solution set: $\left(\dfrac{1}{2}, \dfrac{9}{2}\right)$

49. If x is at most 3 units from 5, then
the absolute value of the difference
between x and 5 is at most 3.   The
phrase "at most" means "less than or
equal to," so we have $|x - 5| \le 3$.

CHAPTER 3   RELATIONS AND FUNCTIONS

Section 3.1

1. For a particular point (x, y), if
    xy > 0, then the point will lie in
    either quadrant I or quadrant II.

    If xy > 0, then x and y have the
    same sign.   Since x and y have the
    same sign in quadrants I and III,
    the statement is true.

3. The point (a, 0) lies on the x-axis
    and its distance from the origin is
    $|a|$.

    If a ≥ 0, then the distance from the
    point (a, 0) to the origin is a.
    If a < 0, then the distance from the
    point (a, 0) to the origin is -a,
    which will be a positive number.   If
    a = 0, the point is (0, 0), so its
    distance to the origin is 0.   In all
    three of these cases, the distance
    from the point (a, 0) to the origin
    is $|a|$.
    The statement is true.

5. $\{(-4, 6), (3, 2), (5, 7)\}$

    Three ordered pairs which belong to
    the relation are (-4, 6), (3, 2),
    and (5, 7).
    The domain is the set of first
    elements, $\{-4, 3, 5\}$.
    The range is the set of second
    elements, $\{6, 2, 7\}$.

In Exercises 7–13, there are other possible answers for the three points.

7. y = 9x - 3

    If x = -1, y = 9(-1) - 3 = -12.
    If x = 1, y = 9(1) - 3 = 6.
    If x = 2, y = 9(2) - 3 = 15.

    Three ordered pairs which belong to
    the relation are (-1, -12), (1, 6),
    and (2, 15).
    The domain is (-∞, ∞).
    The range is (-∞, ∞).

9. $y = -\sqrt{x}$

    If x = 0, $y = -\sqrt{0} = 0$.
    If x = 1, $y = -\sqrt{1} = -1$.
    If x = 4, $y = -\sqrt{4} = -2$.

    Three ordered pairs which belong to
    the relation are (0, 0), (1, -1),
    and (4, -2).
    In order for $\sqrt{x}$ to be a real number,
    x ≥ 0.   Therefore, the domain is
    [0, ∞).
    Since $y = -\sqrt{x}$, y will never be posi-
    tive, so the range is (-∞, 0].

11. $y = |x + 2|$

    If x = 0, $y = |0 + 2| = 2$.
    If x = 1, $y = |1 + 2| = 3$.
    If x = -1, $y = |-1 + 2| = 1$.

    Three ordered pairs which belong to
    the relation are (0, 2), (1, 3), and
    (-1, 1).
    Since we can substitute any real
    number for x in $|x + 2|$, the domain
    is (-∞, ∞).
    Since absolute value is never nega-
    tive, the range is [0, ∞).

13. Three ordered pairs which belong to the relation are (0, 3), (1, 5.1), and (2, 7.2).
The domain is $\{0, 1, 2, 3, 4, 5, 6\}$.
The range is
$\{3, 5.1, 7.2, 9.3, 11.4, 13.5, 15.6\}$.

15. P(-5, -7), Q(-13, 1)

(a) d(P, Q)
$$= \sqrt{[1 - (-7)]^2 + [-13 - (-5)]^2}$$
$$= \sqrt{8^2 + (-8)^2}$$
$$= \sqrt{128}$$
$$= 8\sqrt{2}$$

(b) The midpoint M of the segment joining points P and Q has coordinates $\left(\dfrac{-5 + (-13)}{2}, \dfrac{-7 + 1}{2}\right)$ or
(-9, -3).

17. $P(3\sqrt{2}, 4\sqrt{5})$, $Q(\sqrt{2}, -\sqrt{5})$

(a) d(P, Q)
$$= \sqrt{(5 - 2)^2 + (3 - 8)^2}$$
$$= \sqrt{3^2 + (-5)^2}$$
$$= \sqrt{34}$$

(b) The midpoint M of the segment joining points P and Q has coordinates $\left(\dfrac{8 + 3}{2}, \dfrac{2 + 5}{2}\right)$ or (11/2, 7/2).

19. $P(3\sqrt{2}, 4\sqrt{5})$, $Q(\sqrt{2}, -\sqrt{5})$

(a) d(P, Q)
$$= \sqrt{(\sqrt{2} - 3\sqrt{2})^2 + (-\sqrt{5} - 4\sqrt{5})^2}$$
$$= \sqrt{(-2\sqrt{2})^2 + (-5\sqrt{5})^2}$$
$$= \sqrt{8 + 125}$$
$$= \sqrt{133}$$

(b) The midpoint M of the segment joining points P and Q has coordinates $\left(\dfrac{3\sqrt{2} + \sqrt{2}}{2}, \dfrac{4\sqrt{5} + (-\sqrt{5})}{2}\right)$ or
$(2\sqrt{2}, 3\sqrt{5}/2)$.

21. P(-4, 3), Q(2, 5)

(a) d(P, Q)
$$= \sqrt{[2 - (-4)]^2 + (5 - 3)^2}$$
$$= \sqrt{6^2 + 2^2}$$
$$= \sqrt{40}$$
$$= 2\sqrt{10}$$

(b) The midpoint M of the segment joining points P and Q has coordinates $\left(\dfrac{-4 + 2}{2}, \dfrac{3 + 5}{2}\right)$ or (-1, 4).

23. (a) The points to use would be (1960, 3022) and (1970, 3968). Their midpoint is
$$\left(\dfrac{1960 + 1970}{2}, \dfrac{3022 + 3968}{2}\right)$$
$$= (1965, 3495).$$

In 1965 it was approximately $3495.

(b) If the data are related linearly, the midpoint formula will be exact.

(c) No. The midpoint of (1970, 3968) and (1980, 8414) is (1975, 6191). The actual cutoff for 1975 is $5500, not $6191.

25. (-6, -4), (0, -2), (-10, 8)

Label the points A(-6, -4), B(0, -2), and C(-10, 8).
Use the distance formula to find the length of each side of the triangle.

d(A, B)

$= \sqrt{[0 - (-6)]^2 + [-2 - (-4)]^2}$

$= \sqrt{6^2 + 2^2} = \sqrt{40}$

d(B, C)

$= \sqrt{(-10 - 0)^2 + [8 - (-2)]^2}$

$= \sqrt{10^2 + 10^2} = \sqrt{200}$

d(A, C)

$= \sqrt{[-10 - (-6)]^2 + [8 - (-4)]^2}$

$= \sqrt{(-4)^2 + 12^2} = \sqrt{160}$

Since

$$(\sqrt{40})^2 + (\sqrt{160})^2 = (\sqrt{200})^2,$$

triangle ABC is a right triangle.

27.   (-4, 1), (1, 4), (-6, -1)

Label the points A(-4, 1), B(1, 4), and C(-6, -1).

d(A, B)

$= \sqrt{[1 - (-4)]^2 + (4 - 1)^2}$

$= \sqrt{5^2 + 3^2} = \sqrt{34}$

d(B, C)

$= \sqrt{(-6 - 1)^2 + (-1 - 4)^2}$

$= \sqrt{(-7)^2 + (-5)^2} = \sqrt{74}$

d(A, C)

$= \sqrt{[-6 - (-4)]^2 + (-1 - 1)^2}$

$= \sqrt{(-2)^2 + (-2)^2} = \sqrt{8}$

Since

$$(\sqrt{8})^2 + (\sqrt{34})^2 \neq (\sqrt{74})^2,$$

triangle ABC is not a right triangle.

29.   (0, 7), (3, -5), (-2, 15)

Label the given points M(0, 7), N(3, -5), and P(-2, 15). Find the distance between each pair of points.

d(M, N) $= \sqrt{(0 - 3)^2 + [7 - (-5)]^2}$

$= \sqrt{(-3)^2 + 12^2}$

$= \sqrt{153} = 3\sqrt{17}$

d(N, P) $= \sqrt{[3 - (-2)]^2 + [-5 - 15]^2}$

$= \sqrt{5^2 + (-20)^2}$

$= \sqrt{425} = 5\sqrt{17}$

d(M, P) $= \sqrt{[0 - (-2)]^2 + (7 - 15)^2}$

$= \sqrt{2^2 + (-8)^2}$

$= \sqrt{68} = 2\sqrt{17}$

Since

d(M, N) + d(M, P) = d(N, P)

or $3\sqrt{17} + 2\sqrt{17} = 5\sqrt{17}$,

the given points lie on a straight line.

31.   Label the points A(0, 9), B(-3, -7), and C(2, 19).

d(A, B)

$= \sqrt{(-3 - 0)^2 + (-7 - 9)^2}$

$= \sqrt{(-3)^2 + (-16)^2} = \sqrt{265} \approx 16.279$

d(B, C)

$= \sqrt{[2 - (-3)]^2 + [19 - (-7)]^2}$

$= \sqrt{5^2 + 26^2} = \sqrt{701} \approx 26.476$

d(A, C)

$= \sqrt{(2 - 0)^2 + (19 - 9)^2}$

$= \sqrt{2^2 + 10^2} = \sqrt{104} \approx 10.198$

Since

d(A, B) + d(A, C) $\neq$ d(B, C)

or $\sqrt{265} + \sqrt{104} \neq \sqrt{701}$

$16.279 + 10.198 \neq 26.479$,

the three given points are not collinear. (Note, however, that these points are very close to lying on a straight line and may appear to lie on a straight line when graphed.)

33. Let the unknown endpoint have coordinates (x_2, y_2). Then,

$$\frac{13 + x_2}{2} = 5 \quad \text{and} \quad \frac{10 + y_2}{2} = 8$$

$$13 + x_2 = 10 \quad \text{and} \quad 10 + y_2 = 16$$

$$x_2 = -3 \quad \text{and} \quad y_2 = 6.$$

The other endpoint has coordinates $(-3, 6)$.

35. Let the unknown endpoint have coordinates (x_2, y_2). Then,

$$\frac{19 + x_2}{2} = 12 \quad \text{and} \quad \frac{16 + y_2}{2} = 6$$

$$19 + x_2 = 24 \quad \text{and} \quad 16 + y_2 = 12$$

$$x_2 = 5 \quad \text{and} \quad y_2 = -4.$$

The other endpoint has coordinates $(5, -4)$.

37. We use $P(x_1, y_1)$, $Q(x_2, y_2)$, and $M\left(\dfrac{x_1 + x_2}{2}, \dfrac{y_1 + y_2}{2}\right)$.

$d(P, M)$

$$= \sqrt{\left(\frac{x_1 + x_2}{2} - x_1\right)^2 + \left(\frac{y_1 + y_2}{2} - y_1\right)^2}$$

$$= \sqrt{\left(\frac{x_1 + x_2}{2} - \frac{2x_1}{2}\right)^2 + \left(\frac{y_1 + y_2}{2} - \frac{2y_1}{2}\right)^2}$$

$$= \sqrt{\left(\frac{x_2 - x_1}{2}\right)^2 + \left(\frac{y_2 - y_1}{2}\right)^2}$$

$$= \frac{\sqrt{(x_2 - x_1)^2 + (y_2 - y_1)^2}}{4}$$

$$= \frac{1}{2}\sqrt{(x_2 - x_1)^2 + (y_2 - y_1)^2}$$

$d(Q, M)$

$$= \sqrt{\left(\frac{x_1 + x_2}{2} - x_2\right)^2 + \left(\frac{y_1 + y_2}{2} - y_2\right)^2}$$

$$= \sqrt{\left(\frac{x_1 + x_2}{2} - \frac{2x_2}{2}\right)^2 + \left(\frac{y_1 + y_2}{2} - \frac{2y_2}{2}\right)^2}$$

$$= \sqrt{\left(\frac{x_1 - x_2}{2}\right)^2 + \left(\frac{y_1 - y_2}{2}\right)^2}$$

$$= \frac{1}{2}\sqrt{(x_1 - x_2)^2 + (y_1 - y_2)^2}$$

Since $(x_2 - x_1)^2 = (x_1 - x_2)^2$ and $(y_2 - y_1)^2 = (y_1 - y_2)^2$, we see that $d(P, M) = d(Q, M)$.

$d(P, M) + d(Q, M)$

$$= \frac{1}{2}\sqrt{(x_2 - x_1)^2 + (y_2 - y_1)^2}$$

$$+ \frac{1}{2}\sqrt{(x_1 - x_2)^2 + (y_1 - y_2)^2}$$

$$= \frac{1}{2}\sqrt{(x_2 - x_1)^2 + (y_2 - y_1)^2}$$

$$+ \frac{1}{2}\sqrt{(x_2 - x_1)^2 + (y_2 - y_1)^2}$$

$$= \sqrt{(x_2 - x_1)^2 + (y_2 - y_1)^2}$$

$$= d(P, Q)$$

For Exercises 39–45, see the answer graphs in the back of the textbook.

39. Center $(0, 0)$, radius 6

$$x^2 + y^2 = r^2$$

$$x^2 + y^2 = 6^2 \quad \textit{Let } r = 6$$

$$x^2 + y^2 = 36$$

From the graph, we see that the domain is $[-6, 6]$ and the range is $[-6, 6]$.

41. Center $(2, 0)$, radius 6

$$(x - h)^2 + (y - k)^2 = r^2$$

$$(x - 2)^2 + (y - 0)^2 = 6^2$$

$$\textit{Let } h = 2, k = 0,$$
$$r = 6$$

$$(x - 2)^2 + y^2 = 36$$

From the graph, we see that the domain is $[-4, 8]$ and the range is $[-6, 6]$.

43. Center (-2, 5), radius 4

$$[x - (-2)]^2 + (y - 5)^2 = 4^2$$
$$(x + 2)^2 + (y - 5)^2 = 16$$

From the graph, we see that the domain is [-6, 2] and the range is [1, 9].

45. Center (5, -4), radius 7

$$(x - 5)^2 + [y - (-4)]^2 = 7^2$$
$$(x - 5)^2 + (y + 4)^2 = 49$$

From the graph, we see that the domain is [-2, 12] and the range is [-11, 3].

47. The radius of the circle is the distance from the center C(3, 2) to the x-axis. This distance is 2, so r = 2.

$$(x - 3)^2 + (y - 2)^2 = 2^2$$
$$(x - 3)^2 + (y - 2)^2 = 4$$

49.
$$(x + 4)^2 + (y - 2)^2 = 25$$
$$(y - 2)^2 = 25 - (x + 4)^2$$
$$y - 2 = \pm \sqrt{25 - (x + 4)^2}$$
$$y = 2 \pm \sqrt{25 - (x + 4)^2}$$
$$Y_1 = 2 + \sqrt{25 - (x + 4)^2}$$
$$Y_2 = 2 - \sqrt{25 - (x + 4)^2}$$

See the answer graph in the back of the textbook.

51.
$$x^2 + 6x + y^2 + 8y + 9 = 0$$

Complete the square on x and y separately.

$$(x^2 + 6x \quad) + (y^2 + 8y \quad) = -9$$
$$(x^2 + 6x + 9) + (y^2 + 8y + 16) = -9 + 9 + 16$$
$$(x + 3)^2 + (y + 4)^2 = 16$$

The circle has its center at (-3, -4) and radius 4.

53.
$$x^2 - 4x + y^2 + 12y = -4$$
$$(x^2 - 4x \quad) + (y^2 + 12y \quad) = -4$$
$$(x^2 - 4x + 4) + (y^2 + 12y + 36) = -4 + 4 + 36$$
$$(x - 2)^2 + (y + 6)^2 = 36$$

The circle has its center at (2, -6) and radius 6.

55.
$$x^2 + y^2 - 2y - 48 = 0$$
$$x^2 + (y^2 - 2y \quad) = 48$$
$$x^2 + (y^2 - 2y + 1) = 48 + 1$$
$$x^2 + (y - 1)^2 = 49$$

The circle has its center at (0, 1) and radius 7.

57.
$$x^2 - 6x + y^2 - 6y + 18 = 0$$
$$(x^2 - 6x \quad) + (y^2 - 6y \quad) = -18$$
$$(x^2 - 6x + 9) + (y^2 - 6y + 9) = -18 + 9 + 9$$
$$(x - 3)^2 + (y - 3)^2 = 0$$

The only point whose coordinates satisfy the equation is the point with coordinates (3, 3).

61. The midpoint M has coordinates

$$\left(\frac{-1 + 5}{2}, \frac{3 + (-9)}{2} \right) \text{ or } (2, -3).$$

62. Use points C(2, -3) and P(-1, 3).

$$d(C, P) = \sqrt{[3 - (-3)]^2 + (-1 - 2)^2}$$
$$= \sqrt{6^2 + (-3)^2}$$
$$= \sqrt{45}$$
$$= 3\sqrt{5}$$

The radius is $3\sqrt{5}$.

63. Use points C(2, -3) and Q(5, -9).

$$d(C, Q)$$
$$= \sqrt{(5 - 2)^2 + [-9 - (-3)]^2}$$
$$= \sqrt{3^2 + (-6)^2}$$
$$= \sqrt{45}$$
$$= 3\sqrt{5}$$

The radius is $3\sqrt{5}$.

64. Use the points P(-1, 3) and Q(5, -9).

$$d(P, Q) = \sqrt{[5 - (-1)]^2 + (-9 - 3)^2}$$
$$= \sqrt{6^2 + (-12)^2}$$
$$= \sqrt{180}$$
$$= 6\sqrt{5}$$

The radius is $\frac{1}{2}d(P, Q)$.

$$r = \frac{1}{2}(6\sqrt{5}) = 3\sqrt{5}$$

65. The center-radius form for this circle is

$$(x - 2)^2 + (y + 3)^2 = (3\sqrt{5})^2$$
$$(x - 2)^2 + (y + 3)^2 = 45.$$

66. Label the endpoints of the diameter P(3, -5) and Q(-7, 3).
The midpoint M of the segment joining P and Q has coordinates $\left(\frac{3 + (-7)}{2}, \frac{-5 + 3}{2}\right)$ or (-2, -1). The center is C(-2, -1).

$$d(C, P)$$
$$= \sqrt{[3 - (-2)]^2 + [-5 - (-1)]^2}$$
$$= \sqrt{5^2 + (-4)^2} = \sqrt{41}$$

The radius is $r = \sqrt{41}$.

The center-radius form of the equation of the circle is

$$[x - (-2)]^2 + [y - (-1)]^2 = (\sqrt{41})^2$$
$$(x + 2)^2 + (y + 1)^2 = 41.$$

67. To show algebraically that the epicenter lies at (-3, 4), determine the equation for each circle and substitute x = -3 and y = 4.

Station 1:
Center (1, 4), r = 4

$$(x - 1)^2 + (y - 4)^2 = 16$$
$$(-3 - 1)^2 + (4 - 4)^2 = 16$$
$$(-4)^2 + 0^2 = 16$$
$$16 + 0 = 16$$
$$16 = 16$$

Station 2:
Center (-6, 0), r = 5

$$(x + 6)^2 + y^2 = 25$$
$$(-3 + 6)^2 + 4^2 = 25$$
$$3^2 + 4^2 = 25$$
$$9 + 16 = 25$$
$$25 = 25$$

Station 3:
Center (5, -2), r = 10

$$(x - 5)^2 + (y + 2)^2 = 100$$
$$(-3 - 5)^2 + (4 + 2)^2 = 100$$
$$(-8)^2 + 6^2 = 100$$
$$64 + 36 = 100$$
$$100 = 100$$

We have shown that the point (-3, 4) lies on all three circles, so the epicenter lies at (-3, 4).

69. Label the points A(-2, 2), B(13, 10), C(21, -5), and D(6, -13).

d(A, B)

$$= \sqrt{[13 - (-2)]^2 + (10 - 2)^2}$$

$$= \sqrt{15^2 + 8^2} = \sqrt{289} = 17$$

d(B, C)

$$= \sqrt{(21 - 13)^2 + (-5 - 10)^2}$$

$$= \sqrt{8^2 + (-15)^2} = \sqrt{289} = 17$$

d(C, D)

$$= \sqrt{(6 - 21)^2 + [-13 - (-5)]^2}$$

$$= \sqrt{(-15)^2 + (-8)^2} = \sqrt{289} = 17$$

d(D, A)

$$= \sqrt{[6 - (-2)]^2 + (-13 - 2)^2}$$

$$= \sqrt{8^2 + (-15)^2} = \sqrt{289} = 17$$

Since

d(A, B) = d(B, C) = d(C, D) = d(D, A),

the points are the vertices of a rhombus.

71. Since the center is in the third quadrant, the radius is $\sqrt{2}$, and the circle is tangent to both axes, the center must be at $(-\sqrt{2}, -\sqrt{2})$.

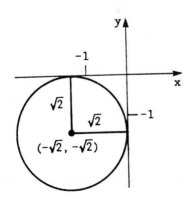

Using the center-radius of the equation of a circle, we have

$$[x - (-\sqrt{2})]^2 + [y - (-\sqrt{2})]^2 = (\sqrt{2})^2$$

$$(x + \sqrt{2})^2 + (y + \sqrt{2})^2 = 2.$$

73. Label the points P(x, y) and Q(1, 3).

If d(P, Q) = 4,

$$\sqrt{(1 - x)^2 + (3 - y)^2} = 4$$

$$(1 - x)^2 + (3 - y)^2 = 16.$$

If x = y, then

$$(1 - y)^2 + (3 - y)^2 = 16$$

$$1 - 2y + y^2 + 9 - 6y + y^2 = 16$$

$$2y^2 - 8y - 6 = 0$$

$$y^2 - 4y - 3 = 0.$$

To solve this equation, use the quad- ratic formula with a = 1, b = -4, and c = -3.

$$y = \frac{4 \pm \sqrt{(-4)^2 - 4(1)(-3)}}{2}$$

$$= \frac{4 \pm \sqrt{28}}{2}$$

$$= \frac{4 \pm 2\sqrt{7}}{2}$$

$$y = 2 \pm \sqrt{7}$$

Since x = y, the points are

$$(2 + \sqrt{7}, 2 + \sqrt{7}) \text{ and}$$

$$(2 - \sqrt{7}, 2 - \sqrt{7}).$$

75. Let P(x, y) be a point whose distance from A(1, 0) is $\sqrt{10}$ and whose distance from B(5, 4) is $\sqrt{10}$.

d(P, A) = $\sqrt{10}$, so

$$\sqrt{(1 - x)^2 + (0 - y)^2} = \sqrt{10}$$

$$(1 - x)^2 + y^2 = 10.$$

d(p, B) = $\sqrt{10}$, so

$$\sqrt{(5 - x)^2 + (4 - y)^2} = \sqrt{10}$$

$$(5 - x)^2 + (4 - y)^2 = 10.$$

Thus,

$$(1 - x)^2 + y^2 = (5 - x)^2 + (4 - y)^2$$
$$1 - 2x + x^2 + y^2 = 25 - 10x + x^2 + 16$$
$$- 8y + y^2$$
$$8x = -8y + 40$$
$$x = -y + 5$$

or
$$y = 5 - x.$$

Substitute $x - 5$ for y in the equation $(1 - x)^2 + y^2 = 10$ and solve for x.

$$(1 - x)^2 + (5 - x)^2 = 10$$
$$1 - 2x + x^2 + 25 - 10x + x^2 = 10$$
$$2x^2 - 12x + 26 = 10$$
$$2x^2 - 12x + 16 = 0$$
$$x^2 - 6x + 8 = 0$$
$$(x - 2)(x - 4) = 0$$
$$x - 2 = 0 \quad \text{or} \quad x - 4 = 0$$
$$x = 2 \quad \text{or} \quad x = 4$$

To find the corresponding values of x, use the equation $y = 5 - x$.
If $x = 2$, $y = 5 - 2 = 3$.
If $x = 4$, $y = 5 - 4 = 1$.
The points are $(2, 3)$ and $(4, 1)$.

77. Label the points $A(3, y)$ and $B(-2, 9)$.
If $d(A, B) = 12$,

$$\sqrt{(-2 - 3)^2 + (9 - y)^2} = 12$$
$$(-5)^2 + (9 - y)^2 = 12^2$$
$$25 + 81 - 18y + y^2 = 144$$
$$y^2 - 18y - 38 = 0.$$

Solve this equation by using the quadratic formula with $a = 1$, $b = -18$, and $c = -38$.

$$y = \frac{-(-18) \pm \sqrt{(-18)^2 - 4(1)(-38)}}{2(1)}$$
$$y = \frac{18 \pm \sqrt{476}}{2}$$
$$y = \frac{18 \pm 2\sqrt{119}}{2}$$
$$y = 9 \pm \sqrt{119}$$

The values of y are $9 + \sqrt{119}$ and $9 - \sqrt{119}$.

Section 3.2

1. $\{(1, 3), (2, 4), (3, 5)\}$

Since each element in the domain corresponds to exactly one element in the range, the set represents y as a function of x.

3. $\{(x, y) \mid x = y^2\}$

Since the domain element $x = 4$ corresponds to the two range elements $y = 2$ and $y = -2$, the set does not represent y as a function of x.

5. $\{(x, y) \mid y = 3x - 7\}$

Since every domain element x corresponds to exactly one range element y, the set represents y as a function of x.

7. Since each vertical line intersects the graph at no more than one point, the graph is the graph of a function.

9. Since some vertical lines intersect the graph at two points, the graph is not the graph of a function.

11. Since some vertical lines intersect the graph at two points, the graph is not the graph of a function.

13. Since each vertical line intersects the graph at no more than one point, the graph is the graph of a function.

15. Since some vertical lines intersect the graph at two points, the graph is not the graph of a function.

For Exercises 17–25, $f(x) = 4x - 2$ and $g(x) = x^2 + 3$.

17. $f(-3) = 4(-3) - 2 = -12 - 2 = -14$

19. $g(-8) = (-8)^2 + 3 = 64 + 3 = 67$

21. $f(k) = 4(k) - 2 = 4k - 2$

23. $f(-1) + g(9)$
    $= [4(-1) - 2] + [(9)^2 + 3]$
    $= (-4 - 2) + (81 + 3)$
    $= -6 + 84$
    $= 78$

25. $g(x + k) = (x + k)^2 + 3$
    $\qquad\quad = x^2 + 2xk + k^2 + 3$

27. Since $x = 3$ and $y = -4$, $f(3) = -4$.

29. **(a)** $f(-2) = 0$ since, when $x = -2$, $y = 0$.

    (b) $f(0) = 4$ since, when $x = 0$, $y = 4$.

    (c) $f(1) = 2$ since, when $x = 1$, $y = 2$.

    (d) $f(4) = 4$ since, when $x = 4$, $y = 4$.

31. **(a)** $f(-2) = -3$ since, when $x = -2$, $y = -3$.

    (b) $f(0) = -2$ since, when $x = -2$, $y = -2$.

    (c) $f(1) = 0$ since, when $x = 1$, $y = 0$.

    (d) $f(4) = 2$ since, when $x = 4$, $y = 2$.

33. When $x = 0$, $y_1 = -15$, so $f(0) = -15$.
    When $x = 2$, $y_2 = 1$, so $g(2) = 1$.
    Therefore,
    $$f(0) + g(2) = -15 + 1 = -14.$$

35. When $x = 2$, $y_1 = 1$, so $f(2) = 1$.
    Therefore, $g[f(2)] = g(1)$.
    When $x = 1$, $y_2 = 1.5$, so $g(1) = 1.5$.
    Therefore,
    $$g[f(2)] = 1.5.$$

37. If $f(x) = g(x)$, $Y_1 = Y_2$.
    We see that $Y_1 = Y_2$ when $x = 2$.
    Thus, $f(x) = g(x)$ when $x = 2$.

39. (a) $P(x) = \dfrac{x - 1}{x}$

$P(9) = \dfrac{9 - 1}{9} = \dfrac{8}{9}$

$P(9) \approx .89$

Approximately 89% of the lung cancer deaths can be attributed to smoking.

(b) $P(x) = \dfrac{x - 1}{x}$

$P(1.9) = \dfrac{1.9 - 1}{1.9} = \dfrac{.9}{1.9}$

$P(1.9) \approx .47$

Approximately 47% of the coronary heart disease deaths can be attributed to smoking.

41. From the graph, we see that when the initial speed is 40 miles per hour, the coast-down time is approximately 12 seconds.

43. The domain is $[-5, 4]$, since x takes all values from -5 to 4 inclusive. The range is $[-2, 6]$, since y takes all values from -2 to 6 inclusive.

45. The domain is $(-\infty, \infty)$, since x can be any real number. The range is $(-\infty, 12)$ since y never takes values greater than 12.

47. The domain is $[-3, 4]$, since x takes all values from -3 to 4 inclusive. The range is $[-6, 8]$, since y takes all values from -6 to 8 inclusive.

49. The function graphed in Exercise 27

The domain is $(-\infty, \infty)$, since x can be any real number.
The range is $[-4, \infty)$, since y takes on all values greater than or equal to -4.

51. $f(x) = 3x - 9$

Since x can be any real-number, the domain is $(-\infty, \infty)$.
As x takes on all real-number values, $3x - 9$ will also take on all real number values, so the range is also $(-\infty, \infty)$.

53. $f(x) = x^6$

Since x can be any real number, the domain is $(-\infty, \infty)$.
Since x^6 is never negative (it is zero or positive), the range is $[0, \infty)$.

55. $h(x) = \sqrt{9 + x}$

Since $9 + x \geq 0$, $x \geq -9$, and the domain is $[-9, \infty)$.
The expression $\sqrt{9 + x}$ means the positive square root, so $\sqrt{9 + x}$ is never negative (it is zero or positive). The range is $[0, \infty)$.

57. $f(x) = -\sqrt{4 - x^2}$

Since $4 - x^2$ must be nonnegative,

$$4 - x^2 \geq 0$$

and

$$(2 + x)(2 - x) \geq 0.$$

Using a sign graph, we see that $-2 \leq x \leq 2$. The domain is $[-2, 2]$. $f(-2) = 0$ and $f(2) = 0$. When we let $x = 0$, $f(0) = -2$. The largest function value is 0, and the smallest function value is -2. Thus, the range is $[-2, 0]$.

59. $g(x) = \dfrac{3}{7 + x}$

Since denominators can never be zero, x cannot be -7. The domain is $(-\infty, -7) \cup (-7, \infty)$. As x takes on all real–number values except -7, $\dfrac{3}{7 + x}$ will take on all real–number values except zero. The range is $(-\infty, 0) \cup (0, \infty)$.

61. $f(x) = \sqrt[3]{x + 2}$

Unlike square roots, cube roots of negative real numbers are real numbers, and the cube root of a negative real number is also a negative real number.

Therefore, the domain of the function is $(-\infty, \infty)$, and the range is also $(-\infty, \infty)$.

63. **(a)** As x is getting larger on the interval $[4, \infty)$, the value of y is increasing.

(b) As x is getting larger on the interval $(-\infty, -1]$, the value of y is decreasing.

(c) As x is getting larger on the interval $[-1, 4]$, the value of y is constant.

65. **(a)** As x is getting larger on the interval $(-\infty, 4]$, the value of y is increasing.

(b) As x is getting larger on the interval $[4, \infty)$, the value of y is decreasing.

(c) There is no interval where the value of y is constant.

67. **(a)** There is no interval where the value of y is increasing.

(b) As x is getting larger on the intervals $(-\infty, 2]$ and $[3, \infty)$, the value of y is decreasing.

(c) As x is getting larger on the interval $(-2, 3)$, the value of y is constant.

69. **(a)** The function is increasing over the interval $[0, 25]$.

(b) The function is decreasing over the interval $[50, 75]$.

(c) The function is constant over the intervals $[25, 50]$ and $[75, 100]$.

71. **(a)** Since the cost function is linear, it will have the form $C(x) = mx + b$, with $m = 10$ and $b = 500$. That is,

$$C(x) = 10x + 500.$$

(b) Since each item sells for $35, the revenue function is

$$R(x) = px = 35x.$$

(c) The profit function is given by

$$P(x) = R(x) - C(x)$$
$$= 35x - (10x + 500)$$
$$= 35x - 10x - 500$$
$$= 25x - 500.$$

(d)
$$C(x) = R(x)$$
$$10x + 500 = 35x$$
$$500 = 25x$$
$$20 = x$$

The break–even point is 20 units. Do not produce the product, since it is possible to sell only 18 units and no profit is made until after the 20th unit is sold.

73. (a) $C(x) = mx + b$, with $m = 150$ and $b = 2700$.

$$C(x) = 150x + 2700$$

(b) $R(x) = 280x$

(c)
$$P(x) = R(x) - C(x)$$
$$= 280x - (150x + 2700)$$
$$= 280x - 150x - 2700$$
$$= 130x - 2700$$

(d)
$$C(x) = R(x)$$
$$150x + 2700 = 280x$$
$$2700 = 130x$$
$$x \approx 20.77$$

The break–even point is 21 units. Produce the product, since it is possible to sell up to 25 units and a profit will be realized starting with the 21st unit.

75. $C(x) = 200x + 1000$

$R(x) = 240x$

(a)
$$C(x) = R(x)$$
$$200x + 1000 = 240x$$
$$1000 = 40x$$
$$25 = x$$

The break–even point is 25 units.

(b) See the answer graph in the textbook.

(c)
$$C(25) = 200(25) + 1000$$
$$= 5000 + 1000$$
$$= 6000$$
$$R(25) = 240(25)$$
$$= 6000$$

At the break–even point of 25 units, the cost and the revenue are each $6000.

Section 3.3

1. To find the x–intercept of the graph of a linear function $f(x) = ax + b$, we solve $f(x) = 0$, and to find the y–intercept, we evaluate $f(0)$.

This statement is true.

3. The graph of $f(x) = ax$ is a straight line that passes through the origin.

This statement is true.

5. The slope of the line in Figure 17 is 3/2.

The line in Figure 17 goes down from left to right, so its slope must be negative. (In fact, the slope is −3/2.)
The statement is false.

For Exercises 7–15, see the answer graphs in the back of the textbook.

7. −x + y = −4

Use the intercepts.

$$0 + y = -4$$
$$y = -4 \quad y\text{-}intercept$$
$$-x + 0 = -4$$
$$x = 4 \quad x\text{-}intercept$$

Graph the line through (0, −4) and (4, 0). The domain and range are both (−∞, ∞).

9. y = 3x − 6

Use the intercepts.

$$y = 3(0) - 6$$
$$y = -6 \quad y\text{-}intercept$$
$$0 = 3x - 6$$
$$x = 2 \quad x\text{-}intercept$$

Graph the line through (0, −6) and (2, 0).
The domain and range are both (−∞, ∞).

11. 2x + 5y = 10

The y-intercept is 2, and the x-intercept is 5. Graph the line through (0, 2) and (5, 0). The domain and range are both (−∞, ∞).

13. y = −4

The graph of y = −4 is a horizontal line with a y-intercept of −4. The domain is (−∞, ∞), and the range is {−4}.

15. y = 3x

The x- and y-intercepts are both 0. (1, 3) is one other point on the line. Graph the line through (0, 0) and (1, 3). The domain and range are both (−∞, ∞).

17. y = 2 is a horizontal line with y-intercept 2.
Choice A resembles this.

19. x = 2 is a vertical line with x-intercept 2.
Choice D resembles this.

For Exercises 21–25, see the answer graphs in the back of the textbook.

21. y = 3x + 4

Use $Y_1 = 3x + 4$.

23. 3x + 4y = 6

Solve for y.

$$4y = -3x + 6$$
$$y = -\frac{3}{4}x + \frac{3}{2}$$

Use $Y_1 = (-3/4)x + (3/2)$.

25. y = −3x

Use $Y_1 = -3x$.

27. Through $(2, -1)$ and $(-3, -3)$

Let $x_1 = 2$, $y_1 = -1$, $x_2 = -3$, and $y_2 = -3$. Then

$$\Delta y = -3 - (-1) = -2$$

and $\Delta x = -3 - 2 = -5.$

The slope is

$$m = \frac{\Delta y}{\Delta x} = \frac{-2}{-5} = \frac{2}{5}.$$

29. Through $(5, 9)$ and $(-2, 9)$

$$m = \frac{\Delta y}{\Delta x} = \frac{9 - 9}{-2 - 5} = \frac{0}{-7} = 0$$

31. Horizontal, through $(3, -7)$

The slope of every horizontal line is zero, so $m = 0$.

33. Vertical, through $(3, -7)$

The slope of every vertical line is undefined; m is undefined.

35. Two points on the line are $(2, -2)$ and $(4, -9)$. Use these two points in the slope formula.

$$m = \frac{\Delta y}{\Delta x} = \frac{-9 - (-2)}{4 - 2} = \frac{-7}{2} = -3.5$$

For Exercises 39–43, see the answer graphs in the back of the textbook.

39. Through $(-1, 3)$, $m = \frac{3}{2}$

First locate the point $(-1, 3)$. Since the slope is $\frac{3}{2}$, a change of 2 units horizontally (2 units to the right) produces a change of 3 units vertically (3 units up). This gives a second point, $(1, 6)$, which can be used to complete the graph.

41. Through $(3, -4)$, $m = -\frac{1}{3}$

First locate the point $(3, -4)$. Since the slope is $-\frac{1}{3}$, a change of 3 units horizontally (3 units to the right) produces a change of -2 units vertically (2 units down). This gives a second point, $(6, -5)$, which can be used to complete the graph.

43. Through $(-1, 4)$, $m = 0$

The graph is the horizontal line through $(-1, 4)$.

45. **(a)** The slope of $-.0221$ indicates that, on the average, from 1912 to 1992 the 5000-meter run is being run .0221 second faster every year. It is negative because the times are generally decreasing as time progresses.

(b) World War II (1939–1945) included the years 1940 and 1944.

(c) Yes. If it contains a linear pattern, eventually the winning time is 0, which is unrealistic.

47. The first two points are $(0, -6)$ and $(1, -3)$.

$$m = \frac{-3 - (-6)}{1 - 0} = \frac{3}{1} = 3$$

48. The second and third points are (1, −3) and (2, 0).

$$m = \frac{0 - (-3)}{2 - 1} = \frac{3}{1} = 3$$

49. If we use any two points on a line to find its slope, we find that the slope is *equal* in all cases.

50. The first two points are A(0, −6) and B(1, −3).

$$d(A, B) = \sqrt{[-3 - (-6)]^2 + (1 - 0)^2}$$
$$= \sqrt{3^2 + 1^2} = \sqrt{10}$$

51. The second and fourth points are B(1, −3) and D(3, 3).

$$d(B, D) = \sqrt{[3 - (-3)]^2 + (3 - 1)^2}$$
$$= \sqrt{6^2 + 2^2} = \sqrt{40} = 2\sqrt{10}$$

52. The first and fourth points are A(0, −6) and D(3, 3).

$$d(A, D) = \sqrt{[3 - (-6)]^2 + (3 - 0)^2}$$
$$= \sqrt{9^2 + 3^2} = \sqrt{90} = 3\sqrt{10}$$

53. $\sqrt{10} + 2\sqrt{10} = 3\sqrt{10}$

The sum is $3\sqrt{10}$, which is equal to the answer in Exercise 52.

54. If points A, B, and C lie on a line in that order, then the distance between A and B added to the distance between *B* and *C* is equal to the distance between *A* and *C*. (The order of the last two may be reversed.)

55. The midpoint of the segment joining A(0, −6) and G(6, 12) has coordinates $M\left(\frac{0 + 6}{2}, \frac{-6 + 12}{2}\right)$ or M(3, 3). The midpoint is M(3, 3), which is the same as the middle entry in the table.

56. The midpoint of the segment joining E(4, 6) and F(5, 9) has coordinates $M\left(\frac{4 + 5}{2}, \frac{6 + 9}{2}\right)$ or M(4.5, 7.5). If the x−value 4.5 were in the table, the corresponding y−value would be 7.5.

57. $f(x) = 6x + 2$

(a) $f(x + h) = 6(x + h) + 2$
$$= 6x + 6h + 2$$

(b) $f(x + h) - f(x)$
$$= (6x + 6h + 2) - (6x + 2)$$
$$= 6x + 6h + 2 - 6x - 2$$
$$= 6h$$

(c) $\dfrac{f(x + h) - f(x)}{h} = \dfrac{6h}{h}$
$$= 6$$

59. $f(x) = -2x + 5$

(a) $f(x + h) = -2(x + h) + 5$
$$= -2x - 2h + 5$$

(b) $f(x + h) - f(x)$
$$= (-2x - 2h + 5) - (-2x + 5)$$
$$= -2x - 2h + 5 + 2x - 5$$
$$= -2h$$

(c) $\dfrac{f(x + h) - f(x)}{h}$
$$= \frac{-2h}{h}$$
$$= -2$$

61. $f(x) = x^2 - 4$

(a) $f(x + h)$

$= (x + h)^2 - 4$

$= x^2 + 2xh + h^2 - 4$

(b) $f(x + h) - f(x)$

$= (x^2 + 2xh + h^2 - 4)$

$\quad - (x^2 - 4)$

$= x^2 + 2xh + h^2 - 4 - x^2 + 4$

$= 2xh + h^2$

(c) $\dfrac{f(x + h) - f(x)}{h}$

$= \dfrac{2hx + h^2}{h}$

$= 2x + h$

Section 3.4

1. An equation of the line through $(3, 4)$ with slope 6 is $y - 4 = 6(x - 3)$.

The point-slope form of the equation of this line is

$$y - 4 = 6(x - 3),$$

so the statement is true.

3. The graph of $y = x + 4$ is parallel to the graph of $y = -x + 4$.

The two slopes are 1 and -1. Since the slopes are unequal, the lines are not parallel, so the given statement is false.

5. Through $(1, 3)$, $m = -2$

Write the equation in point-slope form.

$$y - y_1 = m(x - x_1)$$
$$y - 3 = -2(x - 1)$$

Then, change to standard form.

$$y - 3 = -2x + 2$$
$$2x + y = 5$$

7. Through $(-5, 4)$, $m = -\dfrac{3}{2}$

Write the equation in point-slope form.

$$y - 4 = -\dfrac{3}{2}(x + 5)$$

Change to standard form.

$$2(y - 4) = -3(x + 5)$$
$$2y - 8 = -3x - 15$$
$$3x + 2y = -7$$

9. Through $(-8, 4)$, undefined slope

Since undefined slope indicates a vertical line, the equation will have the form $x = k$. The equation of the line is $x = -8$.

11. Through $(-1, 3)$ and $(3, 4)$

First find m.

$$m = \dfrac{4 - 3}{3 - (-1)} = \dfrac{1}{4}$$

Use either point and the point-slope form.

$$y - 4 = \dfrac{1}{4}(x - 3)$$

PSF

$y - y_1 = m(x - y_1)$

$y = mx + b$

Change to slope-intercept form.

$$y - 4 = \frac{1}{4}x - \frac{3}{4}$$

$$y = \frac{1}{4}x - \frac{3}{4} + \frac{16}{4}$$

$$y = \frac{1}{4}x + \frac{13}{4}$$

13. x-intercept 3, y-intercept -2

The line passes through $(3, 0)$ and $(0, -2)$. Use these points to find m.

$$m = \frac{-2 - 0}{0 - 3} = \frac{2}{3}$$

Then use point-slope form.

$$y - 0 = \frac{2}{3}(x - 3)$$

Change to slope-intercept form.

$$y = \frac{2}{3}x - 2$$

15. Vertical, through $(-6, 4)$

The equation of a vertical line has an equation of the form $x = k$. Since the line passes through $(-6, 4)$, the equation is $x = -6$. (Since this slope of a vertical line is undefined, this equation cannot be written in slope-intercept form.)

17. The line $x + 2 = 0$ has x-intercept -2. It *does not* have a y-intercept. The slope of this line is *undefined*. The line $4y = 2$ has y-intercept *1/2*. It *does not* have an x-intercept. The slope of this line is *zero*.

19. **(a)** The graph of $y = 2x + 3$ has a positive slope and a positive y-intercept. These conditions match graph B.

(b) The graph of $y = -2x + 3$ has a negative slope and a positive y-intercept. These conditions match graph D.

(c) The graph of $y = 2x - 3$ has a positive slope and a negative y-intercept. These conditions match graph A.

(d) The graph of $y = -2x - 3$ has a negative slope and a negative y-intercept. These conditions match graph C.

21. $y = 3x - 1$

This equation is in the slope-intercept form, $y = mx + b$. The slope is $m = 3$ and the y-intercept is $b = -1$.

23. $4x - y = 7$

Solve for y to write the equation in slope-intercept form.

$$-y = -4x + 7$$
$$y = 4x - 7$$

The slope is 4 and the y-intercept is -7.

25. $4y = -3x$

$$y = -\frac{3}{4}x$$

The slope is $-\frac{3}{4}$ and the y-intercept is 0.

27. Through (-1, 4), parallel to
x + 3y = 5

First, find the slope of the line
x + 3y = 5 by writing this equation
in slope-intercept form.

$$x + 3y = 5$$
$$3y = -x + 5$$
$$y = -\frac{1}{3}x + \frac{5}{3}$$

The slope is -1/3. Since the lines
are parallel, -1/3 is also the slope
of the line whose equation is to be
found. Substitute m = -1/3, x_1 =
-1, and y_1 = 4 into the point-slope
form.

$$y - y_1 = m(x - x_1)$$
$$y - 4 = -\frac{1}{3}[x - (-1)]$$
$$y - 4 = -\frac{1}{3}(x + 1)$$
$$3(y - 4) = -1(x + 1)$$
$$3y - 12 = -x - 1$$
$$x + 3y = 11$$

29. Through (1, 6), perpendicular to
3x + 5y = 1

First, find the slope of the line
3x + 5y = 1 by writing this equation
in slope-intercept form. $y - y_1 = m(x$

$$3x + 5y = 1$$
$$5y = -3x + 1$$
$$y = -\frac{3}{5}x + \frac{1}{5}$$

This line has a slope of -3/5.
Call the line whose equation is to
be found L. Since line L is perpendicular to the line 3x + 5y = 1,
the product of their slopes is -1.

If line L has slope m, then

$$-\frac{3}{5}m = -1$$
$$m = \frac{5}{3}.$$

To find the equation of the line L,
substitute m = 5/3, x_1 = 1, and
y_1 = 6 into the point-slope form.

$$y - 6 = \frac{5}{3}(x - 1)$$
$$3(y - 6) = 5(x - 1)$$
$$3y - 18 = 5x - 5$$
$$-13 = 5x - 3y$$
or $5x - 3y = -13$

31. Through (-5, 6), perpendicular to
y = -2. $y - 6 = 0(x - (-3))$ $hor = 0$ slope

Since y = -2 is a horizontal line,
any line perpendicular to this line
will be vertical and have an equation of the form x = k. Since the
line passes through (-5, 6), the
equation is x = -5.

33. (a) Find the slope of the line
3y + 2x = 6.

$$3y + 2x = 6$$
$$3y = -2x + 6$$
$$y = -\frac{2}{3}x + 2$$
$$m = -\frac{2}{3}$$

A line parallel to 3y + 2x = 6 also
has slope $-\frac{2}{3}$.

$$-\frac{2}{3} = \frac{-1 - 2}{4 - k}$$

Solve for k using the slope formula.

$$\frac{-2}{3} = \frac{-3}{4 - k}$$

$$-9 = -8 + 2k$$

$$-\frac{1}{2} = k$$

(b) Find the slope of the line $2x - 5y = 1$.

$$2y - 5x = 1$$

$$2y = 5x + 1$$

$$y = \frac{5}{2}x + \frac{1}{2}$$

$$m = \frac{5}{2}$$

A line perpendicular to $2y - 5x = 1$ has slope $-\frac{2}{5}$, since $\frac{5}{2}(-\frac{2}{5}) = -1$.

Solve for k using the slope formula.

$$-\frac{2}{5} = \frac{-1 - 2}{4 - k}$$

$$-\frac{2}{5} = \frac{-3}{4 - k}$$

$$-15 = -8 + 2k$$

$$k = -\frac{7}{2}$$

35. The Pythagorean Theorem and its converse assure us that in triangle OPQ, angle POQ is a right angle if and only if

$$[d(O, P)]^2 + [d(O, Q]^2 = [d(P, Q)]^2.$$

36. $d(O, P) = \sqrt{(x_1 - 0)^2 + (m_1 x_1 - 0)^2}$

$$= \sqrt{x_1^2 + m_1^2 x_1^2}$$

37. $d(O, Q) = \sqrt{(x_2 - 0)^2 + (m_2 x_2 - 0)^2}$

$$= \sqrt{x_2^2 + m_2^2 x_2^2}$$

38. $d(P, Q)$

$$= \sqrt{(x_2 - x_1)^2 + (m_2 x_2 - m_1 x_1)^2}$$

39. $[d(O, P]^2 + [d(O, Q)]^2 = [(P, Q)]^2$

$$(\sqrt{x_1^2 + m_1^2 x_1^2})^2 + (\sqrt{x_2^2 + m_2^2 x_2^2})^2$$
$$= (\sqrt{(x_2 - x_1)^2 + (m_2 x_2 - m_1 x_1)^2})^2$$

$$x_1^2 + m_1^2 x_1^2 + x_2^2 + m_2^2 x_2^2$$
$$= (x_2 - x_1)^2 + (m_2 x_2 - m_1 x_1)^2$$

Expand the squares of binomials on the right.

$$x_1^2 + m_1^2 x_1^2 + x_2^2 + m_2^2 x_2^2$$
$$= x_2^2 - 2x_2 x_1 + x_1^2 + m_2^2 x_2^2$$
$$- 2m_2 x_2 m_1 x_1 + m_1^2 x_1^2$$

Simplify this equation

$$0 = -2x_2 x_1 - 2m_1 m_2 x_1 x_2$$
$$\text{or} \quad -2m_1 m_2 x_1 x_2 - 2x_1 x_2 = 0$$

40. $-2m_1 m_2 x_1 x_2 - 2x_1 x_2 = 0$

$$-2x_1 x_2 (m_1 m_2 + 1) = 0$$

41. $-2x_1 x_2 (m_1 m_2 + 1) = 0$

Since $x_1 \neq 0$ and $x_2 \neq 0$, $-2x_1 x_2 \neq 0$. By the zero-factor property,

$$m_1 m_2 + 1 = 0$$
$$m_1 m_2 = -1.$$

42. The product of the slopes is -1.

43. **(a)** See the answer art in the back of the textbook.

There appears to be a linear relationship between the data. The farther the galaxy is from Earth, the faster it is receding.

(b) Using the points (520, 40,000) and (0, 0), we obtain

$$m = \frac{40,000 - 0}{520 - 0} = \frac{40,000}{520} \approx 76.9.$$

The equation of the line through these two points is

$$y = 76.9x.$$

See the answer art in the back of the textbook.

(c) $76.9x = 60,000$

$$x = \frac{60,000}{76.9}$$

$$\approx 780$$

The galaxy Hydra is approximately 780 megaparsecs away. This is about 1.5×10^{22} miles.

(d) $A(m) = \dfrac{9.5 \times 10^{11}}{m}$

$$A(76.9) = \frac{9.5 \times 10^{11}}{76.9}$$

$$\approx 1.24 \times 10^{10}$$

$$\text{or } 12.4 \times 10^{9}$$

Using $m = 76.9$, we estimate that the age of the universe is approximately 12.4 billion years.

(e) $A(50) = \dfrac{9.5 \times 10^{11}}{50}$

$$= 1.9 \times 10^{10} \text{ or } 19 \times 10^{9}$$

$$A(100) = \frac{9.5 \times 10^{11}}{100} = 9.5 \times 10^{9}$$

The range for the age of the universe is between 9.5 billion and 19 billion years.

45. **(a)** See the answer art in the back of the textbook.

The debt is increasing and the data appears to be linear.

(b) First, find the slope of the line through the points (0, 1828) and (4, 2881).

$$m = \frac{2881 - 1828}{4 - 0}$$

$$= \frac{1053}{4} = 263.25$$

The slope of the line is 263.25, and, from the point (0, 1828), the y-intercept is 1828, so the linear function is

$$f(x) = 263.25x + 1828.$$

The slope of 263.25 indicates that the federal debt is increasing at a rate of $263.25 billion per year.

See the answer art in the back of the textbook.

(c) In 1984, x = -1.

f(-1) = 263.25(-1) + 1828 = 1564.75

In 1990, x = 5.

f(5) = 263.25(5) + 1828 = 3144.25

Using f, we predict that the national debt was $1564.75 billion in 1984 and $3144.25 billion in 1990. In both cases, the results are less than the true values.

(d) In 1980, x = -5.

f(-5) = 263.25(-5) + 1828 = 511.75

In 1994, x = 9.

f(9) = 263.25(9) + 1828 = 4197.25

Using f, we predict that the national debt was $511.75 billion in 1980 and $4197.25 billion in 1994. In both cases, the results are less than the true values.

(e) A linear approximation is more accurate over a smaller time interval. As the time interval increases, the approximation becomes less accurate.

47. **(a)** Using the points A(4, .17) and B(7, .33), we obtain

$$m = \frac{.33 - .17}{7 - 4}$$

$$= \frac{.16}{3} = \frac{16}{300} = \frac{4}{75}.$$

Use the point-slope form with m = 4/75, $x_1 = 4$, and $y_1 = .17 = 17/100$.

$$y - \frac{17}{100} = \frac{4}{75}(x - 4)$$

$$y - \frac{17}{100} = \frac{4}{75}x - \frac{16}{75}$$

$$y = \frac{4}{75}x - \frac{64}{300} + \frac{51}{300}$$

$$y = \frac{4}{75}x - \frac{13}{300}$$

$$f(x) = \frac{4}{75}x - \frac{13}{300}$$

or f(x) = .05$\overline{3}$x - .04$\overline{3}$

(b) The slope of .05$\overline{3}$ tells us that the percent of HIV patients who get AIDS will increase by 5.3̄% a year.

49. If x = 0 corresponds to 1988, then x = 6 corresponds to 1994. The numbers of employees are given in thousands, so we use the points (0, 717) and (6, 589). Find the slope of the line through these two points.

$$m = \frac{589 - 717}{6 - 0}$$

$$= \frac{-128}{6} = -\frac{64}{3}$$

The slope is -64/3 and, from the point (0, 717), the y-intercept is 717, so the linear model is

$$f(x) = -\frac{64}{3}x + 717.$$

51. Use the data points (50, 320) and (80, 440) to find the equation of the line.

$$m = \frac{440 - 320}{80 - 50}$$

$$= \frac{120}{30} = 4$$

$$y - 320 = 4(x - 50)$$
$$y - 320 = 4x - 200$$
$$y = 4x + 120$$

Since the slope of the line is 4, the ball will travel 4 ft further for each one-mile-per-hour increase in the speed of the bat.

55. It is not true if one line is vertical and the other is horizontal.

Section 3.5

1. The equation $y = x$ matches graph H.

3. The equation $x = |y|$ matches graph F.

5. The equation $y = x^3$ matches graph A.

7. The equation $y = \sqrt[3]{x}$ matches graph J.

9. The equation $y = [x]$ matches graph B.

11. If $y = \sqrt[3]{x}$, then

$$(y)^3 = (\sqrt[3]{x})^3$$
$$y^3 = x$$
$$x = y^3.$$

Therefore, the graph of $x = y^3$ is the same as the graph of $y = \sqrt[3]{x}$. This is graph J.

For Exercises 13–31, see the answer graphs in the back of the textbook.

13. $y = 3x - 2$

Since x and y can take any real number values, both the domain and the range are the set of all real numbers, $(-\infty, \infty)$.
To graph the relation, find several ordered pairs by selecting values for x and finding the corresponding values for y.

x	−3	−2	0	1	4
y	−11	−8	−2	1	10

Use these points to draw the graph, which is a straight line.

15. $3x = y^2$

Since y^2 cannot be negative, $x = (1/3)y^2$ cannot be negative, so the domain is $[0, \infty)$. Since y can have any value, the range is $(-\infty, \infty)$.
Find several ordered pairs by selecting values for y and finding the corresponding x-values.

x	12	3	0	3	12
y	−6	−3	0	3	6

Use these points to draw the graph.

17. $16x^2 = -y$

Rewrite this equation as

$$y = -16x^2.$$

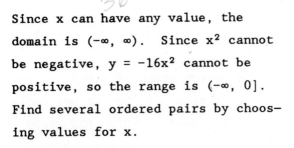

Since x can have any value, the domain is $(-\infty, \infty)$. Since x^2 cannot be negative, $y = -16x^2$ cannot be positive, so the range is $(-\infty, 0]$. Find several ordered pairs by choosing values for x.

x	-2	-1	0	1	2
y	-64	-16	0	-16	-64

Use these points to draw the graph.

19. $y = |x| + 4$

Find several ordered pairs by selecting values for x.

x	-3	-1	0	1	3
y	7	5	4	5	7

Use these points to draw the graph. From the graph, we see that the domain is $(-\infty, \infty)$ and the range is $[4, \infty)$.

21. $x = |y| + 1$

Find several ordered pairs by selecting values for y.

x	5	3	1	3	5
y	-4	-2	0	2	4

Use these points to draw the graph. From the graph, we see that the domain is $[1, \infty)$ and the range is $(-\infty, \infty)$.

23. $y = -|x + 1|$

Find several ordered pairs by selecting values for x.

x	-5	-3	-1	1	3
y	-4	-2	0	-2	-4

Use these points to draw the graph. From the graph, we see that the domain is $(-\infty, \infty)$ and the range is $(-\infty, 0]$.

25. $x = \sqrt{y} - 2$

Find several ordered pairs by selecting values for y.

x	-2	-1	0	1	2
y	0	1	4	9	16

Use these points to draw the graph. From the graph, we see that the domain is $[-2, \infty)$, and the range is $[0, \infty)$.

27. $x = -\sqrt{y} - 2$

Find several ordered pairs by selecting values for y.

x	0	-1	-2	-3	-4
y	2	3	6	11	18

Use these points to draw the graph. From the graph, we see that the domain is $(-\infty, 0]$, and the range is $[2, \infty)$.

29. $y = \sqrt{2x + 4}$

Find several ordered pairs by selecting convenient values for x.

x	-2	-3/2	0	5/2	6
y	0	1	2	3	4

Use these points to draw the graph. From the graph, we see that the domain is $[-2, \infty)$, and the range is $[0, \infty)$.

31. $y = -2\sqrt{x}$

Find several ordered pairs by selecting convenient values for x. Note that x must be nonnegative.

x	0	1	4	9	16
y	0	-2	-4	-6	-8

Use these points to draw the graph. From the graph, we see that the domain is $[0, \infty)$ and the range is $(-\infty, 0]$.

33. The relations in Exercises 13, 14, 17, 18, 19, 20, 23, 24, 29, 30, 31, and 32 express y as a function of x.

35. $f(x) = \begin{cases} 2x & \text{if } x \le -1 \\ x - 1 & \text{if } x > -1 \end{cases}$

(a) $f(-5) = 2(-5) = -10$

(b) $f(-1) = 2(-1) = -2$

(c) $f(0) = 0 - 1 = -1$

(d) $f(3) = 3 - 1 = 2$

For Exercises 37–41, see the answer graphs in the textbook.

37. $f(x) = \begin{cases} x - 1 & \text{if } x \le 3 \\ 2 & \text{if } x > 3 \end{cases}$

Draw the graph of $y = x - 1$ to the left of $x = 3$, including the endpoint at $x = 3$. Draw the graph of $y = 2$ to the right of $x = 3$, but do not include the endpoint at $x = 3$.

39. $f(x) = \begin{cases} 4 - x & \text{if } x < 2 \\ 1 + 2x & \text{if } x \ge 2 \end{cases}$

Draw the graph of $y = 4 - x$ to the left of $x = 2$, but do not include the endpoint. Draw the graph of $y = 1 + 2x$ to the right of $x = 2$, including the endpoint.

41. $f(x) = \begin{cases} 2 + x & \text{if } x < -4 \\ -x & \text{if } -4 \le x \le 5 \\ 3x & \text{if } x > 5 \end{cases}$

Draw the graph of $y = 2 + x$ to the left of -4, but do not include the endpoint at $x = 4$. Draw the graph of $y = -x$ between -4 and 5, including both endpoints. Draw the graph of $y = 3x$ to the right of 5, but do not include the endpoint at $x = 5$.

43. The solid circle on the graph shows that the endpoint $(0, -1)$ is part of the graph, while the open circle shows that the endpoint $(0, 1)$ is not part of the graph. The graph is made up of parts of two horizontal lines. The function which fits this graph is

$$f(x) = \begin{cases} -1 & \text{if } x \le 0 \\ 1 & \text{if } x > 0. \end{cases}$$

The domain of this function is $(-\infty, \infty)$, and the range is $\{-1, 1\}$.

45. The graph is made up of parts of two horizontal lines. The solid circle shows that the endpoint $(0, 2)$ of the one on the left belongs to the graph while the open circle shows that the endpoint $(0, -1)$ of the one on the right does not belong to the graph. The function that fits this graph is

$$f(x) = \begin{cases} 2 & \text{if } x \le 0 \\ -1 & \text{if } x > 1. \end{cases}$$

The domain of this function is $(-\infty, 0] \cup (1, \infty)$, and the range is $\{-1, 2\}$.

For Exercises 47–59, see the answer graphs in the textbook.

47. $f(x) = \left[-x\right]$

Plot points.

x	-x	$f(x) = [-x]$
-2	2	2
-1.5	1.5	1
-1	1	1
-.5	.5	0
0	0	0
.5	-.5	-1
1	-1	-1
1.5	-1.5	-2
2	-2	2

More generally, to get $y = 0$, we need

$$0 \le -x < 1$$
$$0 \ge x > -1 \quad \text{or}$$
$$-1 < x \le 0.$$

To get $y = 1$, we need

$$-1 \le -x < 2$$
$$-1 \ge x > -2 \quad \text{or}$$
$$-2 < x \le -1.$$

Follow this pattern to graph the step function.
The domain of this function is $(-\infty, \infty)$ and the range is $[\ldots, -2, -1, 0, 1, 2, \ldots]$.

49. $f(x) = \left[2x - 1\right]$

To get $y = 0$, we need

$$0 \le 2x - 1 < 1$$
$$1 \le 2x < 2$$
$$\frac{1}{2} \le x < 1.$$

To get $y = 1$, we need

$$1 \le 2x - 1 < 2$$
$$2 \le 2x < 3$$
$$1 \le x < \frac{3}{2}.$$

Follow this pattern to graph the step function.
The domain of this function is $(-\infty, \infty)$, and the range is $\{\ldots, 2, -1, 0, 1, 2, \ldots\}$.

51. To graph $f(x) = x$, enter $Y_1 = x$.

53. To graph $f(x) = x^3$, enter $Y_1 = x \wedge 3$.

55. To graph $f(x) = \sqrt[3]{x}$, enter $Y_1 = x \wedge (1/3)$.

57. If $x = y^2$, then $y = \pm\sqrt{x}$.
To graph $x = y^2$, enter $Y_1 = \sqrt{x}$ and $Y_2 = -\sqrt{x}$.

59. We see that

$y = 2.65$ when $1978 \leq x < 1979$,

$y = 2.90$ when $1979 \leq x < 1980$,

$y = 3.10$ when $1980 \leq x < 1981$,

$y = 3.35$ when $1981 \leq x < 1991$,

$y = 4.50$ when $1991 \leq x < 1996$.

We graph these to obtain the graph of the piecewise-defined function.

61. $i(t) = \begin{cases} 40t + 100 & \text{if } 0 \leq t \leq 3 \\ 220 & \text{if } 3 < t \leq 8 \\ -80t + 860 & \text{if } 8 < t \leq 10 \\ 60 & \text{if } 10 < t \leq 24 \end{cases}$

(a) 6 A.M. is the starting time, so 7 A.M. corresponds to $t = 1$.

Use $i(t) = 40(t) + 100$.

Since $t = 1$,

$$i(1) = 40 + 100 = 140.$$

(b) 9 A.M. corresponds to $t = 3$.

$$i(3) = 40 \cdot 3 + 100 = 220$$

(c) 10 A.M. corresponds to $t = 4$.

Use $i(t) = 220$.

$$i(4) = 220$$

(d) Noon corresponds to $t = 6$.

Use $i(t) = 220$.

$$i(6) = 220$$

(e) 2 P.M. corresponds to $t = 8$.

Use $i(t) = 220$.

$$i(8) = 220$$

(f) 5 P.M. corresponds to $t = 11$.

Use $i(t) = 60$.

$$i(11) = 60$$

(g) Midnight corresponds to $t = 18$.

Use $i(t) = 60$.

$$i(11) = 60$$

(h) See the answer graph in the back of the textbook.

63. $f(x) = \begin{cases} x^2 - 4 & \text{if } x \geq 0 \\ -x + 5 & \text{if } x < 0 \end{cases}$

The graph is a line with negative slope for $x < 0$ and a parabola opening upward for $x \geq 0$. This matches graph B.

65. $f(x) = \begin{cases} 6 & \text{if } x \geq 0 \\ -6 & \text{if } x < 0 \end{cases}$

The graph is the horizontal line $y = -6$ for $x < 0$ and the horizontal line $y = 6$ for $x \geq 0$. This matches graph D.

Section 3.6

1. The graph of a nonzero function cannot be symmetric with respect to the x-axis.

Such a graph would fail the vertical line test, so the statement is true.

3. The graph of an odd function is symmetric with respect to the origin.

This statement is true.

5. If (a, b) is on the graph for an odd function, so is (-a, b).

This statement is false. The correct statement would be "If (a, b) is on the graph of an odd function, so is (-a, -b)."

For Exercises 7–11, see the answer graphs in the back of the textbook.

7. (a) The point that is symmetric to (5, -3) with respect to the x-axis is (5, 3).

(b) The point that is symmetric to (5, -3) with respect to the y-axis is (-5, -3).

(c) The point that is symmetric to (5, -3) with respect to the origin is (-5, 3).

9. (a) The point that is symmetric to (-4, -2) with respect to the x-axis is (-4, 2).

(b) The point that is symmetric to (-4, -2) with respect to the y-axis is (4, -2).

(c) The point that is symmetric to (-4, -2) with respect to the origin is (4, 2).

11. (a) $y = g(-x) + 1$

The graph of $g(x)$ is reflected about the y-axis and translated 1 unit up to obtain the graph of $y = g(-x) + 1$.

(b) $y = g(x - 2)$

The graph of $g(x)$ is translated to the right 2 units to obtain the graph of $y = g(-x) + 1$.

(c) $y = g(x + 1) - 2$

The graph of $g(x)$ is translated to the left 1 unit and down 2 units to obtain the graph of $y = g(x + 1) - 2$.

(d) $y = -g(x) + 2$

The graph of $g(x)$ is reflected about the x-axis and translated up 2 units to obtain the graph of $y = -g(x) + 2$.

13. $y = x^2 + 2$

Replace x with -x to obtain

$$y = (-x)^2 + 2 = x^2 + 2.$$

The result is the same as the original equation, so the graph is symmetric with respect to the y-axis. Since y is a function of x, the graph cannot be symmetric with respect to the x-axis.

Replace x with -x and y with -y to obtain

$$-y = (-x)^2 + 2$$
$$-y = x^2 + 2$$
$$y = -x^2 - 2.$$

The result is not the same as the original equation, so the graph is not symmetric with respect to the origin.

Therefore, the graph is symmetric with respect to the y-axis only.

15. $x^2 + y^2 = 10$

Replace x with −x to obtain

$$(-x)^2 + y^2 = 10$$
$$x^2 + y^2 = 10.$$

The result is the same as the original equation, so the graph is symmetric with respect to the y-axis.

Replace y with −y to obtain

$$x^2 + (-y)^2 = 10$$
$$x^2 + y^2 = 10.$$

The result is the same as the original equation, so the graph is symmetric with respect to the x-axis.

Since the graph is symmetric with respect to the x-axis and y-axis, it is also symmetric with respect to the origin.

17. $y = -3x^3$

Replace x with −x to obtain

$$y = -3(-x)^3$$
$$y = -3(-x^3)$$
$$y = 3x^3.$$

The result is not the same as the original equation, so the graph is not symmetric with respect to the y-axis.

Replace y with −y to obtain

$$-y = -3x^3$$
$$y = 3x^3.$$

The result is not the same as the original equation, so the graph is not symmetric with respect to the x-axis.

Replace x with −x and y with −y to obtain

$$-y = -3(-x)^3$$
$$-y = -3(-x^3)$$
$$-y = 3x^3$$
$$y = -3x^3.$$

The result is the same as the original equation, so the graph is symmetric with respect to the origin. Therefore, the graph is symmetric with respect to the origin only.

19. $y = x^2 - x + 7$

Replace x with −x to obtain

$$y = (-x)^2 - (-x) + 7$$
$$y = x^2 + x + 7.$$

The result is not the same as the original equation, so the graph is not symmetric with respect to the y-axis.

Since y is a function of x, the graph cannot be symmetric with respect to the x-axis.

Replace x with −x and y with −y to obtain

$$-y = (-x)^2 - (-x) + 7$$
$$-y = x^2 + x + 7$$
$$y = -x^2 - x - 7.$$

The result is not the same as the original equation, so the graph is not symmetric with respect to the origin.

Therefore, the graph has none of the listed symmetries.

For Exercises 21–27, see the answer graphs in the back of the textbook.

21. $y = |x| - 1$

The graph is obtained by translating the graph of $y = |x|$ 1 unit down.

23. $y = \dfrac{1}{x}$

Since x and y must have the same sign and since neither x nor y can be 0, the graph is made up of two separate portions in Quadrants I and III. Find some points in Quadrant I.

x	4	2	1	1/2	1/4
y	1/4	1/2	1	2	4

Connect the points with a smooth curve. Using the tests of symmetry, we determine that the graph is symmetric with respect to the origin. Using this symmetry, plot corresponding points, such as $(-4, -1/4)$ in Quadrant III. Connect these points with a smooth curve to complete the graph.

25. $y = -(x + 1)^3$

This graph may be obtained by translating it 1 unit to the left and then reflecting it about the x-axis.

27. $y = 2x^2 - 1$

If $f(x) = x^2$, $y = 2f(x) - 1$.
We start with the familiar graph of $f(x) = x^2$. The graph of $y = 2f(x)$ stretches the graph of $f(x) = x^2$ vertically.
The graph of $y = 2f(x) - 1$ translates the graph of $y = 2f(x)$ 1 unit down.

29. Since $f(3) = 6$, the point $(3, 6)$ is on the graph.
Since the graph is symmetric with respect to the origin, the point $(-3, -6)$ is on the graph.
Therefore, $f(-3) = -6$.

31. Since $f(3) = 6$, the point $(3, 6)$ is on the graph.
Since the graph is symmetric with respect to the line $x = 6$ and since the point $(3, 6)$ is 3 units to the left of the line $x = 6$, the image point of $(3, 6)$, 3 units to the right of the line $x = 6$, is $(9, 6)$.
Therefore, $f(9) = 6$.

33. An odd function is a function whose graph is symmetric with respect to the origin. Since $(3, 6)$ is on the graph, $(-3, -6)$ must also be on the graph.
Therefore, $f(-3) = -6$.

35. If the graph of $f(x) = 2x + 5$ is translated up 2 units, the new graph will correspond to the function

$$t(x) = (2x + 5) + 2$$
$$= 2x + 7.$$

Now translate the graph of $t(x) = 2x + 7$ to the left 3 units. The final graph will correspond to the function

$$g(x) = 2(x + 3) + 7$$
$$= 2x + 13.$$

(Note that if the original graph is first translated to the left 3 units and then up 2 units, the final result will be the same.)

37. **(a)** Since $f(-x) = f(x)$, the graph is symmetric with respect to the y-axis. See the answer graph in the back of the textbook.

(b) Since $f(-x) = -f(x)$, the graph is symmetric with respect to the origin. See the answer graph in the back of the textbook.

41. Many answers are possible. Three possible graphs are shown below.

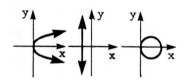

43. The graph of $Y_2 = -Y_1$ is the graph of Y_1 reflected about the x-axis. This is graph F.

45. The graph of $Y_4 = 2Y_1$ is the graph of Y_1 stretched vertically by a factor of 2. This is graph D.

47. The graph of $Y_6 = Y_1 - 2$ is the graph of Y_1 translated 2 units down. This is graph B.

49. See the answer graph in the back of the textbook.

50. $m = \dfrac{2 - (-2)}{3 - 1} = \dfrac{4}{2} = 2$

51. Use the point-slope form; then rewrite the equation in the form $y_1 = mx + b$.

$$y - 2 = 2(x - 3)$$
$$y - 2 = 2x - 6$$
$$y = 2x - 4$$
$$y_1 = 2x - 4$$

52. $(1, -2)$ becomes $(1, 4)$.
$(3, 2)$ becomes $(3, 8)$.

53. $m = \dfrac{8 - 4}{3 - 1} = \dfrac{4}{2} = 2$

54. Use the point-slope form; then rewrite the equation in the form $y_2 = mx + b$.

$$y - 8 = 2(x - 3)$$
$$y - 8 = 2x - 6$$
$$y = 2x + 2$$
$$y_2 = 2x + 2$$

55. See the answer graph in the back of the textbook.

The graph of y_2 is obtained by shifting the graph of y_1 up 6 units. The constant, 6, comes from the 6 added in Exercise 52.

56. If the points (x_1, y_1) and (x_2, y_2) lie on a line, then when we add the positive constant c to each y-value, we obtain the points $(x_1, y_1 + c)$ and $(x_2, y_2 + c)$. The slope of the new line is *the same as* the slope of the original line.

The graph of the new line can be obtained by shifting the graph of the original line c units in the *upward* direction.

57. $f(x) = -.012053x^2 - .046607x + 9.125$

For 1982, $x = 0$.

$f(0) = -.012053(0)^2 - .046607(0)$
$\qquad\quad + 9.125$
$\qquad = 9.125$

For 1992, $x = 10$.

$f(10) = -.012053(10)^2 - .046607(10)$
$\qquad\qquad + 9.125$
$\qquad \approx 7.454$

The general trend in the carbon monoxide levels has been to gradually decrease from 9.125 to 7.454 parts per million.

Section 3.7

1. If $f(x) = x$ and $g(x) = x^2$, then $(f + g)(2) = 6$.

$$(f + g)(2) = f(2) + g(2)$$
$$= 2 + 4 = 6$$

The statement is true.

3. If $f(x) = x$ and $g(x) = \dfrac{1}{x}$, then $(fg)(x) = 1 \ (x \neq 0)$.

$$(fg)(x) = f(x) \cdot g(x)$$
$$= (x)\left(\frac{1}{x}\right) = 1$$

(Note that $g(x)$ is undefined when $x = 0$.) The statement is true.

5. $f(x) = 3x + 4, \ g(x) = 2x - 5$

$$(f + g)(x) = f(x) + g(x)$$
$$= (3x + 4) + (2x - 5)$$
$$= 5x - 1$$

The domain of $f + g$ is $(-\infty, \infty)$.

$$(f - g)(x) = f(x) - g(x)$$
$$= (3x + 4) - (2x - 5)$$
$$= x + 9$$

The domain of of $- g$ is $(-\infty, \infty)$.

$$(fg)(x) = f(x) \cdot g(x)$$
$$= (3x + 4)(2x - 5)$$
$$= 6x^2 - 7x - 20$$

The domain of fg is $(-\infty, \infty)$.

$$\left(\frac{f}{g}\right)(x) = \frac{f(x)}{g(x)}$$

$$= \frac{3x + 4}{2x - 5}$$

The domain is $\left(-\infty, \frac{5}{2}\right) \cup \left(\frac{5}{2}, \infty\right)$ since the denominator cannot be zero.

7. $f(x) = 2x^2 - 3x,\ g(x) = x^2 - x + 3$

$(f + g)(x) = f(x) + g(x)$
$$= (2x^2 - 3x) + (x^2 - x + 3)$$
$$= 3x^2 - 4x + 3$$

The domain of $f + g$ is $(-\infty, \infty)$.

$(f - g)(x) = f(x) - g(x)$
$$= (2x^2 - 3x) - (x^2 - x + 3)$$
$$= x^2 - 2x - 3$$

The domain of $f - g$ is $(-\infty, \infty)$.

$(fg)(x)$
$$= f(x) \cdot g(x)$$
$$= (2x^2 - 3x)(x^2 - x + 3)$$
$$= 2x^4 - 2x^3 + 6x^2 - 3x^3 + 3x^2 - 9x$$
$$= 2x^4 - 5x^3 + 9x^2 - 9x$$

The domain of fg is $(-\infty, \infty)$.

$$\left(\frac{f}{g}\right)(x) = \frac{f(x)}{g(x)}$$

$$= \frac{2x^2 - 3x}{x^2 - x + 3}$$

If $x^2 - x + 3 = 0$, then by the quadratic formula

$$x = \frac{1 \pm \sqrt{-11}}{2}.$$

The equation has no real solutions. There are no real numbers which make the denominator zero.
The domain of f/g is $(-\infty, \infty)$.

9. $f(x) = \sqrt{4x - 1},\ g(x) = \sqrt{x + 3}$

To find the domain of f we solve

$$4x - 1 \geq 0$$

$$x \geq \frac{1}{4}.$$

Domain of f is $[1/4, \infty)$.
To find the domain of g, we solve

$$x + 3 \geq 0$$

$$x \geq -3.$$

Domain of g is $[-3, \infty)$.
The intersection of the domain of f and the domain of g is

$$\left[\frac{1}{4}, \infty\right) \cap [-3, \infty) = \left[\frac{1}{4}, \infty\right)$$

$(f + g)(x) = f(x) + g(x)$
$$= \sqrt{4x - 1} + \sqrt{x + 3}$$

The domain of $(f + g)$ is $[1/4, \infty)$.

$(f - g)(x) = f(x) - g(x)$
$$= \sqrt{4x - 1} - \sqrt{x + 3}$$

The domain of $f - g$ is $[1/4, \infty)$.

$(fg)(x) = f(x) \cdot g(x)$
$$= \sqrt{4x - 1} \cdot \sqrt{x + 3}$$
$$= \sqrt{(4x - 1)(x + 3)}$$

The domain of fg is $[1/4, \infty)$.

$$\left(\frac{f}{g}\right)(x) = \frac{f(x)}{g(x)}$$

$$= \frac{\sqrt{4x - 1}}{\sqrt{x + 3}}$$

$$= \sqrt{\frac{4x - 1}{x + 3}}$$

The domain of f/g includes all real numbers in the intersection of the domains of f and g for which $g(x) \neq 0$. Since the intersection of the two domains is $[1/4, \infty)$ and the

only number for which g(x) = 0 is
-3, which is not in this interval,
the domain of f/g is also [1/4, ∞).

In Exercises 11–21, $f(x) = 5x^2 - 2x$ and
$g(x) = 6x + 4$.

11. $(f + g)(3) = f(3) + g(3)$
$$= 5(3)^2 - 2(3) + 6(3) + 4$$
$$= 45 - 6 + 18 + 4$$
$$= 61$$

13. $(fg)(4) = f(4) \cdot g(4)$
$$= [5(4)^2 - 2(4)] \cdot [6(4) + 4]$$
$$= 72(28)$$
$$= 2016$$

15. $\left(\dfrac{f}{g}\right)(-1) = \dfrac{f(-1)}{g(-1)}$
$$= \dfrac{5(-1)^2 - 2(-1)}{6(-1) + 4}$$
$$= \dfrac{5 + 2}{-2}$$
$$= -\dfrac{7}{2} \text{ or } -3.5$$

17. $(f - g)(m) = f(m) - g(m)$
$$= (5m^2 - 2m) - (6m + 4)$$
$$= 5m^2 - 2m - 6m - 4$$
$$= 5m^2 - 8m - 4$$

19. $(f \circ g)(2) = f[g(2)]$
$$= f(6 \cdot 2 + 4)$$
$$= f(16)$$
$$= 5 \cdot (16)^2 - 2(16)$$
$$= 1280 - 32 = 1248$$

21. $(g \circ f)(2) = g[f(2)]$
$$= g[5(2)^2 - 2(2)]$$
$$= g(16)$$
$$= 6(16) + 4 = 100$$

23. $f(1) + g(1) = 2 + 3 = 5$

25. $f(-2) \cdot g(4) = 0(2) = 0$

27. $(f \circ g)(2) = f[g(2)]$
$$= f(2) = 3$$

29. $(g \circ f)(-4) = g[f(-4)]$
$$= g(2) = 2$$

31. $(f \circ g)(2) = f[g(2)]$
$$= f(3) = 1$$

33. $(g \circ f)(3) = g[f(3)]$
$$= g(1) = 9$$

35. $(f \circ f)(4) = f[f(4)]$
$$= f(3) = 1$$

37. $(f \circ g)(1) = f[g(1)]$
$$= f(9)$$

However, f(9) cannot be determined
from the table given.

39. $f(x) = -6x + 9$, $g(x) = 5x + 7$
$(f \circ g)(x) = f[g(x)]$
$$= f(5x + 7)$$
$$= -6(5x + 7) + 9$$
$$= -30x - 42 + 9$$
$$= -30x - 33$$
$(g \circ f)(x) = g[f(x)]$
$$= g(-6x + 9)$$
$$= 5(-6x + 9) + 7$$
$$= -30x + 45 + 7$$
$$= -30x + 52$$

41. $f(x) = 4x^2 + 2x + 8$, $g(x) = x + 5$

$(f \circ g)(x)$

$= f[g(x)]$

$= f(x + 5)$

$= 4(x + 5)^2 + 2(x + 5) + 8$

$= 4(x^2 + 10x + 25) + 2x + 10 + 8$

$= 4x^2 + 40x + 100 + 2x + 18$

$= 4x^2 + 42x + 118$

$(g \circ f)(x)$

$= g[f(x)]$

$= g(4x^2 + 2x + 8)$

$= (4x^2 + 2x + 8) + 5$

$= 4x^2 + 2x + 13$

43. $f(x) = \dfrac{2}{x^4}$, $g(x) = 2 - x$

$(f \circ g)(x) = f[g(x)]$

$= f(2 - x)$

$= \dfrac{2}{(2 - x)^4}$

$(g \circ f)(x) = g[f(x)]$

$= g\left(\dfrac{2}{x^4}\right)$

$= 2 - \dfrac{2}{x^4}$

45. $f(x) = 9x^2 - 11x$, $g(x) = 2\sqrt{x + 2}$

$(f \circ g)(x)$

$= f[g(x)]$

$= f(2\sqrt{x + 2})$

$= 9(2\sqrt{x + 2})^2 - 11(2\sqrt{x + 2})$

$= 9[4(x + 2)] - 22\sqrt{x + 2}$

$= 36x + 72 - 22\sqrt{x + 2}$

$(g \circ f)(x)$

$= g[f(x)]$

$= g(9x^2 - 11x)$

$= 2\sqrt{(9x^2 - 11x) + 2}$

$= 2\sqrt{9x^2 - 11x + 2}$

47. $f(x) = 2x^3 - 1$, $g(x) = \sqrt[3]{\dfrac{x + 1}{2}}$

$(f \circ g)(x) = f[g(x)]$

$= f\left(\sqrt[3]{\dfrac{x + 1}{2}}\right)$

$= 2\left(\sqrt[3]{\dfrac{x + 1}{2}}\right)^3 - 1$

$= 2\left(\dfrac{x + 1}{2}\right) - 1$

$= (x + 1) - 1$

$= x$

$(g \circ f)(x) = g[f(x)]$

$= g(2x^3 - 1)$

$= \sqrt[3]{\dfrac{(2x^3 - 1) + 1}{2}}$

$= \sqrt[3]{\dfrac{2x^3}{2}}$

$= \sqrt[3]{x^3}$

$= x$

51. $(f \circ g)(x) = f[g(x)]$

When $g(x) > 0$, $f[g(x)] = 1$, from the graph of f.

When $g(x) = 0$, $f[g(x)] = 0$, from the graph of f.

When $g(x) < 0$, $f[g(x)] = -1$, from the graph of f.

Therefore, for x values where $g(x) > 0$, the graph of f ∘ g will be a portion of the horizontal line $y = 1$.

For x values where $g(x) = 0$, the graph of f ∘ g will be points on the x-axis.

For x values here $g(x) < 0$, the graph of f ∘ g will be a portion of the horizontal line $y = -1$.

See the answer graph in the back of the textbook.

53. $f(x) = 2x + 3$, $g(x) = \dfrac{x - 3}{2}$

$(f \circ g)(x) = f[g(x)]$

$\qquad = f\left(\dfrac{x - 3}{2}\right)$

$\qquad = 2\left(\dfrac{x - 3}{2}\right) + 3$

$\qquad = x - 3 + 3$

$\qquad = x$

$(g \circ f)(x) = g[f(x)]$

$\qquad = g(2x + 3)$

$\qquad = \dfrac{(2x + 3) - 3}{2}$

$\qquad = \dfrac{2x}{2}$

$\qquad = x$

55. $f(x) = \sqrt[3]{\dfrac{x + 1}{2}}$, $g(x) = 2x^3 - 1$

$(f \circ g)(x) = f[g(x)]$

$\qquad = f(2x^3 - 1)$

$\qquad = \sqrt[3]{\dfrac{(2x^3 - 1) + 1}{2}}$

$\qquad = \sqrt[3]{\dfrac{2x^3}{2}}$

$\qquad = \sqrt[3]{x^3}$

$\qquad = x$

$(g \circ f)(x) = g[f(x)]$

$\qquad = g\left(\sqrt[3]{\dfrac{x + 1}{2}}\right)$

$\qquad = 2\left(\sqrt[3]{\dfrac{x + 1}{2}}\right)^3 - 1$

$\qquad = 2\left(\dfrac{x + 1}{2}\right) - 1$

$\qquad = x + 1 - 1$

$\qquad = x$

In Exercises 57–61, we give only one of many possible ways.

57. $h(x) = (6x - 2)^2$

Let $f(x) = x^2$ and $g(x) = 6x - 2$.
Then $(f \circ g)(x) = f(6x - 2)$
$\qquad\qquad\qquad = (6x - 2)^2 = h(x)$.

59. $h(x) = \sqrt{x^2 - 1}$

Let $f(x) = \sqrt{x}$ and $g(x) = x^2 - 1$.
Then $(f \circ g)(x) = f(x^2 - 1)$
$\qquad\qquad\qquad = \sqrt{x^2 - 1} = h(x)$.

61. $h(x) = \sqrt{6x} + 12$

Let $f(x) = \sqrt{x} + 12$ and $g(x) = 6x$.
Then $(f \circ g)(x) = f(6x)$
$\qquad\qquad\qquad = \sqrt{6x} + 12 = h(x)$.

63. $Y_1 = 2x - 5$, $Y_2 = x^2$

(a) $X = 0$

$(Y_1 \circ Y_2)(0) = Y_1[Y_2(0)]$

$\qquad\qquad = Y_1(0)$

$\qquad\qquad = 2(0) - 5$

$\qquad\qquad = -5$

(b) $X = 1$

$(Y_1 \circ Y_2)(1) = Y_1[Y_2(1)]$

$\qquad\qquad = Y_1(1)$

$\qquad\qquad = 2(1) - 5$

$\qquad\qquad = -3$

(c) $X = 2$

$(Y_1 \circ Y_2)(2) = Y_1[Y_2(2)]$

$\qquad\qquad = Y_1(4)$

$\qquad\qquad = 2(4) - 5$

$\qquad\qquad = 3$

(d) X = 3

$$(Y_2 \circ Y_2)(3) = Y_1[Y_2(3)]$$
$$= Y_1(9)$$
$$= 2(9) - 5$$
$$= 13$$

65. f(x) = 12x, g(x) = 5280x

$$(f \circ g)(x) = f[g(x)]$$
$$= f(5280x)$$
$$= 12(5280x)$$
$$= 63,360x$$

The function f ∘ g computes the number of inches in x miles.

67. $A(x) = \frac{\sqrt{3}}{4}x^2$

(a) $A(2x) = \frac{\sqrt{3}}{4}(2x)^2$
$$= \frac{\sqrt{3}}{4}(4x^2)$$
$$= \sqrt{3}x^2$$

(b) $A(16) = A(2 \cdot 8)$
$$= \sqrt{3}(8)^2 \quad x = 8$$
$$= 64\sqrt{3} \text{ square units}$$

69. (a) r(t) = 4t
$$A(r) = \pi r^2$$
$$(A \circ r)(t) = A[r(t)]$$
$$= A(4t)$$
$$= \pi(4t)^2$$
$$= 16\pi t^2$$

(b) (A ∘ r)(t) defines the area of the leak in terms of the time t, in minutes.

(c) $A(3) = 16\pi(3)^2$
$$= 144\pi$$

The area is 144π sq ft.

71. Let x = the number of people less than 100 people that attend.

(a) x people fewer than 100 attend, so 100 - x people do attend.

$$N(x) = 100 - x$$

(b) The cost per person starts at $2 and increases by $.20 for each of the x people that do not attend. The total increase is $.20x, and the cost per person increases to $2 + $.20x.

$$G(x) = 2 + .2x$$

(c) $C(x) = N(x) \cdot G(x)$
$$= (100 - x)(2 + .2x)$$

(d) If 40 people attend,
$$x = 100 - 40 = 60.$$
$$C(60) = (100 - 60)[2 + .2(60)]$$
$$= (40)(14)$$
$$= 560$$

The total cost is $560.

Chapter 3   Review Exercises

1. $\{(-3, 6), (-1, 4), (8, 5)\}$

The domain is $\{-3, -1, 8\}$, the set of first elements.
The range is $\{6, 4, 5\}$, the set of second elements.

3. P(3, -1), Q(-4, 5)

$$d(P, Q) = \sqrt{(-4 - 3)^2 + [5 - (-1)]^2}$$
$$= \sqrt{(-7)^2 + 6^2}$$
$$= \sqrt{49 + 36}$$
$$= \sqrt{85}$$

$$\text{midpoint} = \left(\frac{3 + (-4)}{2}, \frac{-1 + 5}{2}\right)$$
$$= \left(-\frac{1}{2}, 2\right)$$

5. A(-6, 3), B(-6, 8)

$$d(A, B) = \sqrt{[-6 - (-6)]^2 + (8 - 3)^2}$$
$$= \sqrt{25} = 5$$

$$\text{midpoint} = \left(\frac{-6 + (-6)}{2}, \frac{3 + 8}{2}\right)$$
$$= \left(-6, \frac{11}{2}\right)$$

7. A(-1, 2), B(-10, 5), C(-4, k)

$$d(A, B) = \sqrt{[-1 - (-10)]^2 + (2 - 5)^2}$$
$$= \sqrt{90}$$

$$d(A, C) = \sqrt{[-4 - (-1)]^2 + (k - 2)^2}$$
$$= \sqrt{9 + (k - 2)^2}$$

$$d(B, C) = \sqrt{[-10 - (-4)]^2 + (5 - k)^2}$$
$$= \sqrt{36 + (k - 5)^2}$$

If segment AB is the hypotenuse,

$$(\sqrt{90})^2$$
$$= \left[\sqrt{9 + (k - 2)^2}\right]^2 + \left[\sqrt{36 + (k - 5)^2}\right]^2$$
$$90 = 9 + k^2 - 4k + 4 + 36 + k^2$$
$$- 10k + 25$$
$$0 = 2k^2 - 14k - 16$$
$$0 = k^2 - 7k - 8$$
$$0 = (k - 8)(k + 1)$$
$$k = 8 \quad \text{or} \quad k = -1.$$

If segment AC is the hypotenuse, the product of the slopes of lines AB and BC is -1 since the product of slopes of perpendicular lines is -1.

$$\left(\frac{5 - 2}{-10 + 1}\right) \cdot \left(\frac{k - 5}{-4 + 10}\right) = -1$$
$$\left(\frac{3}{-9}\right) \cdot \left(\frac{k - 5}{6}\right) = -1$$
$$\frac{k - 5}{-18} = -1$$
$$k - 5 = 18$$
$$k = 23$$

If segment BC is the hypotenuse, the product of the slopes of lines AB and AC is -1.

$$\left(\frac{3}{-9}\right) \cdot \left(\frac{k - 2}{-4 + 1}\right) = -1$$
$$\left(\frac{-1}{3}\right) \cdot \left(\frac{k - 2}{-3}\right) = -1$$
$$\frac{k - 2}{9} = -1$$
$$k - 2 = -9$$
$$k = -7$$

The possible values of k are -7, 23, 8, and -1.

9. Center (-2, 3), radius 15

$$(x - h)^2 + (y - k)^2 = r^2$$
$$[x - (-2)]^2 + (y - 3)^2 = 15^2$$
$$(x + 2)^2 + (y - 3)^2 = 225$$

11. Center (-8, -1), passing through (0, 16)

The radius is the distance from the center to any point on the circle. The distance between (-8, 1) and (0, 16) is

$$r = \sqrt{(-8 - 0)^2 + (1 - 16)^2}$$
$$= \sqrt{8^2 + 15^2}$$
$$= \sqrt{289} = 17.$$

The equation of the circle is

$$[x - (-8)]^2 + (y - 1)^2 = 17^2$$
$$(x + 8)^2 + (y - 1)^2 = 289.$$

13. $x^2 - 4x + y^2 + 6y + 12 = 0$

Complete the square on x and y to put the equation in center–radius form.

$$(x^2 - 4x + \quad) + (y^2 + 6y + \quad) = -12$$
$$(x^2 - 4x + 4) + (y^2 + 6y + 9) = -12 + 4 + 9$$
$$(x - 2)^2 + (y + 3)^2 = 1$$

The circle has center $(2, -3)$ and radius 1.

15. $2x^2 + 14x + 2y^2 + 6y + 2 = 0$
$$x^2 + 7x + y^2 + 3y + 1 = 0$$
$$(x^2 + 7x \quad) + (y^2 + 3y \quad) = -1$$
$$\left(x^2 + 7x + \frac{49}{4}\right) + \left(y^2 + 3y + \frac{9}{4}\right) = -1 + \frac{49}{4} + \frac{9}{4}$$
$$\left(x + \frac{7}{2}\right)^2 + \left(y + \frac{3}{2}\right)^2 = \frac{54}{4}$$

The circle has center $\left(-\frac{7}{2}, -\frac{3}{2}\right)$ and radius $\sqrt{\frac{54}{4}} = \frac{\sqrt{54}}{\sqrt{4}} = \frac{3\sqrt{6}}{2}.$

17. Find all possible values of x so that the distance between $(x, -9)$ and $(3, -5)$ is 6.

$$\sqrt{(3 - x)^2 + (-5 + 9)^2} = 6$$
$$\sqrt{9 - 6x + x^2 + 16} = 0$$
$$\sqrt{x^2 - 6x + 25} = 0$$
$$x^2 - 6x + 25 = 36$$
$$x^2 - 6x - 11 = 0$$

$$x = \frac{6 \pm \sqrt{36 - 4(1)(-11)}}{2}$$
$$= \frac{6 \pm \sqrt{36 + 44}}{2}$$
$$= \frac{6 \pm \sqrt{80}}{2}$$
$$= \frac{6 \pm 4\sqrt{5}}{2} = \frac{2(3 \pm 2\sqrt{5})}{2}$$
$$x = 3 + 2\sqrt{5} \quad \text{or} \quad x = 3 - 2\sqrt{5}$$

19. Find all points (x, y) with $x + y = 0$ so that (x, y) is 6 units from $(-2, 3)$.

$$6 = \sqrt{(x + 2)^2 + (y - 3)^2}$$
$$6 = \sqrt{(x + 2)^2 + (-x - 3)^2} \quad y = -x$$
$$36 = (x + 2)^2 + (-x - 3)^2$$
$$36 = x^2 + 4x + 4 + x^2 + 6x + 9$$
$$0 = 2x^2 + 10x - 23$$

$$x = \frac{-10 \pm \sqrt{100 - 4(2)(-23)}}{4}$$
$$= \frac{-10 \pm \sqrt{100 + 184}}{4}$$
$$= \frac{-10 \pm \sqrt{284}}{4}$$
$$= \frac{-10 \pm 2\sqrt{71}}{4}$$
$$x = \frac{-5 \pm \sqrt{71}}{2}$$

Since $x + y = 0$ or $y = -x$,

if $x = \dfrac{-5 + \sqrt{71}}{2}$, then $y = \dfrac{5 - \sqrt{71}}{2}$;

if $x = \dfrac{-5 - \sqrt{71}}{2}$, then $y = \dfrac{5 + \sqrt{71}}{2}$.

The points are

$$\left(\frac{-5 + \sqrt{71}}{2}, \frac{5 - \sqrt{71}}{2}\right),$$

and $\left(\dfrac{-5 - \sqrt{71}}{2}, \dfrac{5 + \sqrt{71}}{2}\right).$

21. This is not the graph of a function because a vertical line can intersect it in two points.
The domain of the relation is $(-\infty, \infty)$. The range is $[0, \infty)$.

23. This is the graph of a function. No vertical line will intersect the graph in more than one point. The domain of the function is $(-\infty, -2] \cup [2, \infty)$. The range is $[0, \infty)$.

25. This is the graph of a function. No vertical line will intersect the graph in more than one point. The domain of the function is $(-\infty, \infty)$. The range is $(-\infty, \infty)$.

27. The equation $x = \frac{1}{2}y^2$ does not define y as a function of x. For some values of x, there will be more than one value of y. For example, if $x = 8$,

$$8 = \frac{1}{2}y^2$$
$$y^2 = 16$$
$$y = \pm 4.$$

Therefore, the ordered pairs $(8, 4)$ and $(8, -4)$ would belong to the relation and the relation would not be a function.

29. The equation $y = \frac{-8}{x}$ defines y as a function of x because for every x in the domain, which is $(-\infty, 0) \cup (0, \infty)$, there will be exactly one value of y.

31. In the function $y = -4 + |x|$, we may use any real number for x. The domain is $(-\infty, \infty)$.

33. In the function $y = -\sqrt{\dfrac{5}{x^2 + 9}}$, we must have

$$\frac{5}{x^2 + 9} \geq 0.$$

However, this will be true for every real value of x. The domain is $(-\infty, \infty)$.

35. (a) As x is getting larger on the interval $[2, \infty)$, the value of y is increasing.

(b) As x is getting larger on the interval $(-\infty, -2]$, the value of y is decreasing.

37. $f(x) = 2x + 9$

$$\frac{f(x + h) - f(x)}{h}$$
$$= \frac{[2(x + h) - 9] - (2x + 9)}{h}$$
$$= \frac{2x + 2h - 9 - 2x - 9}{h}$$
$$= \frac{2h}{h}$$
$$= 2$$

39. (a) Because the graph contains vertical line segments, it is not the graph of a function.

(b) Oil prices were lowest in 1986 and highest in 1990.

(c) The lowest price was about $7, the highest price was about $27.50.

(d) The general trend of prices over the period shown on graph is down.

(e) The horizontal portions of the graph indicate a constant, stable price.

41. Find the slope of the line through $(8, 7)$ and $\left(\frac{1}{2}, -2\right)$.

$$m = \frac{y_2 - y_1}{x_2 - x_1}$$

$$= \frac{-2 - 7}{\frac{1}{2} - 8}$$

$$= \frac{-9}{-\frac{15}{2}}$$

$$= -9\left(-\frac{2}{15}\right)$$

$$= \frac{18}{15} = \frac{6}{5}$$

43. Find the slope of the line through $(5, 6)$ and $(5, -2)$.

$$m = \frac{y_2 - y_1}{x_2 - x_1}$$

$$= \frac{-2 - 6}{5 - 5} = \frac{-8}{0}$$

The slope is undefined.

45. Find the slope of the line

$$9x - 4y = 2.$$

Solve for y to put the equation in slope-intercept form.

$$-4y = -9x + 2$$

$$y = \frac{9}{4}x - \frac{1}{2}$$

$$m = \frac{9}{4}$$

(The slope can also be found by choosing two points on the line and using $m = \frac{y_2 - y_1}{x_2 - x_1}$.)

47. Find the slope of the line

$$x - 5y = 0.$$

Solve for y to put the equation in slope-intercept form.

$$-5y = x$$

$$y = \frac{1}{5}x$$

$$m = \frac{1}{5}$$

49. Two points on the graph are $(2, -4)$ and $(3, -7)$.

$$m = \frac{-7 - (-4)}{3 - 2}$$

$$= \frac{-3}{1} = -3$$

For Exercises 51–55, see the answer graphs in the back of the textbook.

51. $3x + 7y = 14$

$$7y = -3x + 14$$

$$y = -\frac{3}{7}x + 2$$

The graph is the line with slope $-3/7$ and y-intercept 2.

It may also be graphed using the x-intercept $14/3$ and y-intercept 2.

The domain and range are both $(-\infty, \infty)$.

53. $3y = x$

$y = \frac{1}{3}x$

The graph is the line with slope $1/3$ and y-intercept 0, which means that it passes through the origin. Use another point such as (3, 1) to complete the graph. The domain and range are both $(-\infty, \infty)$.

55. $x = -5$

The graph is the vertical line through $(-5, 0)$.
The domain is $\{-5\}$ and the range is $(-\infty, \infty)$.

57. Line through $(-2, 4)$ and $(1, 3)$

First find the slope.

$$m = \frac{3 - 4}{1 - (-2)} = -\frac{1}{3}$$

Now use the point-slope form with $(x_1, y_1) = (1, 3)$ and $m = -\frac{1}{3}$.

$$y - 3 = -\frac{1}{3}(x - 1)$$

$$3(y - 3) = -1(x - 1)$$

$$3y - 9 = -x + 1$$

$$x + 3y = 10 \quad \textit{Standard form}$$

59. x-intercept -3, y-intercept 5

Two points of the line are $(-3, 0)$ and $(0, 5)$.
First, find the slope.

$$m = \frac{5 - 0}{0 + 3} = \frac{5}{3}$$

The slope is $5/3$ and the y-intercept is 5. Write the equation in slope-intercept form.

$$y = \frac{5}{3}x + 5$$

Now rewrite the equation in standard form.

$$3y = 5x + 15$$

$$5x - 3y = -15$$

61. Line through (0, 5), perpendicular to $8x + 5y = 3$

First, find the slope of $8x + 5y = 3$.

$$8x + 5y = 3$$

$$5y = -8x + 3$$

$$y = -\frac{8}{5}x + \frac{3}{5}$$

Since the slope of $8x + 5y = 3$ is $-8/5$, the slope of a line perpendicular to it is $5/8$. Since $m = 5/8$ and $b = 5$, the equation is

$$y = \frac{5}{8}x + 5$$

$$8y = 5x + 40$$

$$-40 = -5x - 8y$$

or $5x - 8y = -40$. *Standard form*

63. Line through (3, -5), parallel to $y = 4$

This will be horizontal line through $(3, -5)$. Since y has the same value for all points on the line, $b = -5$. The equation is $y = -5$.

65.   Line through $(2, -4)$, $m = \dfrac{3}{4}$

First locate the point $(2, -4)$.
Since the slope is $3/4$, a change of
4 units horizontally (4 units to the
right) produces a change of 3 units
vertically (3 units up). This gives
a second point, $(6, -1)$, which can
be used to complete the graph.
See the answer graph in the back of
the textbook.

67.   Use the points $(.7, 1.4)$ and
$(5.3, 10.9)$.
The slope of the line is

$$m = \frac{10.9 - 1.4}{5.3 - .7}$$

$$= \frac{9.5}{4.6}$$

$$\approx 2.065.$$

Use the point-slope form with $m =$
2.065, $x_1 = .7$, and $y_1 = 1.4$.

$$y - y_1 = m(x - x_1)$$
$$y - 1.4 = 2.065(x - .7)$$
$$y - 1.4 = 2.065x - 1.4455$$
$$y = 2.065x - .0455$$
$$f(x) = 2.065x - .0455$$

Note: Using $m = 9.5/4.6$ and keeping
all digits in the calculator will
give a final answer of

$$f(x) = 2.065x - .0456.$$

This result will be used in Exercise
69.

69. $f(4.9) = 2.065(4.9) - .0456 \approx 10.1$

This represents approximately 10.1
million passengers, which agrees
favorably with the FAA prediction of
10.3 million.

For Exercises 71–79, see the answer
graphs in the back of the textbook.

71. $f(x) = -|x|$

The graph of $f(x) = -|x|$ is the re-
flection of the graph of $f(x) = |x|$
about the x-axis.

73. $f(x) = -|x| - 2$

Translate the graph in Exercise 71
down 2 units.

75. $f(x) = 2|x - 3| - 4$

Start with the graph of $f(x) = |x|$.
Stretch the graph vertically by a
factor of 2, translate it 3 units to
the right and translate it 4 units
down.

77. $f(x) = \left[\!\left[\frac{1}{2}x - 2\right]\!\right]$

For y to be 0, we need

$$0 \le \frac{1}{2}x - 2 < 1$$

$$2 \le \frac{1}{2}x < 3$$

$$4 \le x < 6.$$

Follow this pattern to graph the
step function.

79. $f(x) = \begin{cases} 3x + 1 & \text{if } x < 2 \\ -x + 4 & \text{if } x \geq 2 \end{cases}$

Graph the line $y = 3x + 1$ to the left of $x = 2$, and graph the line $y = -x + 4$ to the right of $x = 2$. The graph has an open circle at $(2, 7)$ and a closed circle at $(2, 2)$.

81. Since $x = 0$ corresponds to 1970, we may rewrite the table as follows.

x	Percent (as a decimal)
0	.05
5	.11
10	.17
15	.23
20	.23

Up to 1985, the percent changes by a constant amount each year. Using the points $(0, .05)$ and $(10, .17)$,

$$m = \frac{.17 - .05}{10 - 0}$$

$$= \frac{.12}{10}$$

$$= \frac{12}{1000}$$

$$m = \frac{3}{250}.$$

We see that $b = .05$ or $5/100$. Therefore, for this period,

$$y = \frac{3}{250}x + \frac{5}{100}.$$

Note that

$$f(15) = \frac{3}{250}(15) + \frac{5}{100} = .23.$$

From 1985 to 1990, the percent is constant, with equation $y = .23$.

Thus,

$$f(x) = \begin{cases} \frac{3}{250}x + \frac{5}{100} & \text{if } 0 \leq x \leq 15 \\ .23 & \text{if } 15 < x \leq 20. \end{cases}$$

83. $3y^2 - 5x^2 = 15$

Replace x with $-x$ to obtain

$$3y^2 - 5(-x)^2 = 15$$
$$3y^2 - 5x^2 = 15.$$

The result is the same as the original equation, so the graph is symmetric with respect to the y-axis.

Replace y with $-y$ to obtain

$$3(-y)^2 - 5x^2 = 15$$
$$3y^2 - 5x^2 = 15.$$

The result is the same as the original equation, so the graph is symmetric with respect to the x-axis.

Since the graph is symmetric with respect to the y-axis and x-axis, it must also be symmetric with respect to the origin.

85. $y^3 = x + 1$

Replace x with $-x$ to obtain

$$y^3 = -x + 1.$$

The result is not the same as the original equation, so the graph is not symmetric with respect to the y-axis.

Replace y with $-y$ to obtain

$$(-y)^3 = x + 1$$
$$-y^3 = x + 1$$
$$y^3 = -x - 1.$$

The result is not the same as the original equation, so the graph is not symmetric with respect to the x–axis.

Replace x with –x and y with –y to obtain

$$(-y)^3 = (-x) + 1$$
$$-y^3 = -x + 1$$
$$y^3 = x - 1.$$

The result is not the same as the original equation, so the graph is not symmetric with respect to the origin.

Therefore, the graph has none of the listed symmetries.

87. $|y| = -x$

Replace x with –x to obtain

$$|y| = -(-x)$$
$$|y| = x.$$

The result is not the same as the original equation, so the graph is not symmetric with respect to the y–axis.

Replace y with –y to obtain

$$|-y| = -x$$
$$|y| = -x.$$

The result is the same as the original equation, so the graph is symmetric with respect to the x–axis.

Replace x with –x and y with –y to obtain

$$|-y| = -(-x)$$
$$|y| = x.$$

The result is not the same as the original equation, so the graph is not symmetric with respect to the origin.

Therefore, the graph is symmetric with respect to the x–axis only.

89. $|x| = |y|$

Replace x with –x to obtain

$$|-x| = |y|$$
$$|x| = |y|.$$

The result is the same as the original equation, so the graph is symmetric with respect to the y–axis.

Replace y with –y to obtain

$$|x| = |-y|$$
$$|x| = |y|.$$

The result is the same as the original equation, so the graph is symmetric with respect to the x–axis.

Since the graph is symmetric with respect to the x–axis and with respect to the y–axis, it must also be symmetric with respect to the origin.

91.   To obtain the graph of h(x) = $|x| - 2$, translate the graph of f(x) = $|x|$ down 2 units.

93.   If the graph of f(x) = 3x – 4 is reflected about the x–axis, we obtain a graph whose equation is

$$y = -(3x - 4)$$
$$= -3x + 4.$$

95. If the graph of $f(x) = 3x - 4$ is reflected about the origin, every point (x, y) will be replaced by the point $(-x, -y)$. The equation for the graph will change from

$$y = 3x - 4$$

to

$$-y = 3(-x) - 4$$
$$-y = -3x - 4$$
$$y = 3x + 4.$$

For Exercises 97–105, $f(x) = 3x^2 - 4$ and $g(x) = x^2 - 3x - 4$.

97. $(f + g)(x)$
$$= f(x) + g(x)$$
$$= (3x^2 - 4) + (x^2 - 3x - 4)$$
$$= 4x^2 - 3x - 8$$

99. $(f - g)(4)$
$$= f(4) - g(4)$$
$$= (3 \cdot 4^2 - 4) - (4^2 - 3 \cdot 4 - 4)$$
$$= (48 - 4) - (16 - 12 - 4)$$
$$= 44 - 0$$
$$= 44$$

101. $(f + g)(2k)$
$$= f(2k) + g(2k)$$
$$= [3(2k)^2 - 4]$$
$$\quad + [(2k)^2 - 3(2k) - 4]$$
$$= (12k^2 - 4) + (4k^2 - 6k - 4)$$
$$= 16k^2 - 6k - 8$$

103. $\left(\dfrac{f}{g}\right)(x) = \dfrac{3x^2 - 4}{x^2 - 3x - 4}$

Since $x^2 - 3x - 4 = 0$ when $x = -1$ and division by 0 is undefined, $\left(\dfrac{f}{g}\right)(-1)$ is undefined.

105. $\left(\dfrac{f}{g}\right)(x) = \dfrac{3x^2 - 4}{x^2 - 3x - 4}$

$$= \dfrac{3x^2 - 4}{(x + 1)(x - 4)}$$

The expression is not undefined if $(x + 1)(x - 4) = 0$, that is, if $x = -1$ or $x = 4$.

Thus, the domain is the set of all real numbers except $x = -1$ and $x = 4$, or

$$(-\infty, -1) \cup (-1, 4) \cup (4, \infty).$$

For Exercises 107 and 109, $f(x) = \sqrt{x - 2}$ and $g(x) = x^2$.

107. $(f \circ g)(x) = f[g(x)]$
$$= f(x^2)$$
$$= \sqrt{x^2 - 2}$$

109. $(f \circ g)(-6) = f[g(-6)]$
$$= \sqrt{(-6)^2 - 2}$$
$$= \sqrt{34}$$

111. $(f \circ g)(2) = f[g(2)]$
$$= f(2)$$
$$= 1$$

113. $P(x) = 2x^2 + 1$

$f(a) = 3a + 2$

$(P \circ f)(a) = P[f(a)]$
$$= P(3a + 2)$$
$$= 2(3a + 2)^2 + 1$$
$$= 18a^2 + 24a + 9$$

115. Use the formula for the perimeter of a rectangle.

$$P = 2L + 2W$$
$$P(x) = 2(2x) + 2(x)$$
$$= 4x + 2x$$
$$P(x) = 6x$$

This is a linear function.

Chapter 3   Test

1. Since the line rises from left to right, the slope is positive.

2. $m = \dfrac{4 - 1}{3 - (-2)} = \dfrac{3}{5}$

3. We label the points A(-2, 1) and B(3, 4).

$$d(A, B) = \sqrt{[3 - (-2)]^2 + (4 - 1)^2}$$
$$= \sqrt{5^2 + 3^2} = \sqrt{34}$$

4. The midpoint has coordinates

$$M\left(\dfrac{-2 + 3}{2}, \dfrac{1 + 4}{2}\right) \text{ or } (1/2, 5/2).$$

5. Use the point-slope form with m = 3/5, $x_1 = -2$, and $y_1 = 1$.

$$y - 1 = \dfrac{3}{5}[x - (-2)]$$
$$y - 1 = \dfrac{3}{5}(x + 2)$$
$$5y - 5 = 3(x + 2)$$
$$5y - 5 = 3x + 6$$
$$-11 = 3x - 5y$$
$$3x - 5y = -11$$

6. Solve $3x - 5y = -11$ for y.

$$-5y = -3x - 11$$
$$y = \dfrac{3}{5}x + \dfrac{11}{5}$$

Therefore, the linear function is

$$f(x) = \dfrac{3}{5}x + \dfrac{11}{5}.$$

7. Point A has coordinates (5, -3).

(a) The equation of a vertical line through A is x = 5.

(b) The equation of a horizontal line through A is y = -3.

8. The slope of the graph of y = -3x + 2 is -3.

(a) A line parallel to the graph of y = -3x + 2 has a slope of -3.

$$y - 3 = -3(x - 2)$$
$$y - 3 = -3x + 6$$
$$y = -3x + 9$$

(b) A line perpendicular to the graph of y = -3x + 2 has a slope of 1/3.

$$y - 3 = \dfrac{1}{3}(x - 2)$$
$$y - 3 = \dfrac{1}{3}x - \dfrac{2}{3}$$
$$y = \dfrac{1}{3}x + \dfrac{7}{3}$$

9. Use the two points (0, 3) and (4, -13).

$$m = \dfrac{-13 - 3}{4 - 0} = \dfrac{-16}{4} = -4$$

From the point $(0, 3)$, we see that $b = 3$.
The equation that defines this function is

$$y = -4x + 3.$$

10. This is not the graph of a function because some vertical lines intersect it in more than one point.
The domain of the relation is $[0, 4]$.
The range is $[-4, 4]$.

11. This is the graph of a function because no vertical line intersects the graph in more than one point.
The domain of the function is $(-\infty, -1) \cup (-1, \infty)$.
The range is $(-\infty, 0) \cup (0, \infty)$.
As x is getting larger on the intervals $(-\infty, -1)$ and $(-1, \infty)$, the value of y is decreasing, so the function is decreasing on the intervals $(-\infty, -1)$ and $(-1, \infty)$. (The function is never increasing or constant.)

For Exercises 12–14, see the answer graphs in the back of the textbook.

12. To graph $y = |x - 2| - 1$, we translate the graph of $y = |x|$ 2 units to the right and 1 unit down.

13. $f(x) = [\![x + 1]\!]$

To get $y = 0$, we need
$$0 \le x + 1 < 1$$
$$-1 \le x < 0.$$

To get $y = 1$, we need
$$1 \le x + 1 < 2$$
$$0 \le x < 1.$$

Follow this pattern to graph the step function.

14. $f(x) = \begin{cases} 3 & \text{if } x < -2 \\ 2 - \frac{1}{2}x & \text{if } x \ge -2 \end{cases}$

For values of x with $x < -2$, we graph the horizontal line $y = 3$.
For values of x with $x \ge -2$, we graph the line with a slope of $-1/2$ and a y-intercept of 2. Two points on this line are $(-2, 3)$ and $(0, 2)$.

15. The graph of $y = -2\sqrt{x + 2} - 3$ can be obtained from the graph of $y = \sqrt{x}$ by translating the graph of $y = x$ 2 units to the left, stretching it by a factor of 2, reflecting it across the x-axis, and translating it 3 units down.

16. $3x^2 - y^2 = 3$

(a) Replace y with $-y$ to obtain
$$3x^2 - (-y)^2 = 3$$
$$3x^2 - y^2 = 3.$$

The result is the same as the original equation, so the graph is symmetric with respect to the x-axis.

(b) Replace x with $-x$ to obtain
$$3(-x)^2 - y^2 = 3$$
$$3x^2 - y^2 = 3.$$

The result is the same as the original equation, so the graph is symmetric with respect to the y-axis.

(c) Since the graph is symmetric with respect to the x-axis and with respect to the y-axis, it must also be symmetric with respect to the origin.

17. $f(x) = 2x^2 - 3x + 2$

$$f(-3) = 2(-3)^2 - 3(-3) + 2$$
$$= 2(9) + 9 + 2$$
$$= 29$$

18. $f(x) = 2x^2 - 3x + 2$

$\dfrac{f(x + h) - f(x)}{h}$

$$= \frac{[2(x+h)^2 - 3(x+h) + 2] - (2x^2 - 3x + 2)}{h}$$

$$= \frac{2(x^2 + 2xh + h^2) - 3x - 3h + 2 - 2x^2 + 3x - 2}{h}$$

$$= \frac{2x^2 + 4xh + 2h^2 - 3x - 3h + 2 - 2x^2 + 3x - 2}{h}$$

$$= \frac{4xh + 2h^2 - 3h}{h}$$

$$= \frac{h(4x + 2h - 3)}{h}$$

$$= 4x + 2h - 3$$

19. $f(x) = 2x^2 - 3x + 2$ and $g(x) = -2x + 1$

$(f \circ g)(x)$
$$= f[g(x)]$$
$$= f(-2x + 1)$$
$$= 2(-2x + 1)^2 - 3(-2x + 1) + 2$$
$$= 2(4x^2 - 4x + 1) + 6x - 3 + 2$$
$$= 8x^2 - 8x + 2 + 6x - 1$$
$$= 8x^2 - 2x + 1$$

20.   **(a)** If $x = 0$ represents 1982 and $x = 12$ represents 1994, then we have the two points (0, 1147) and (12, 4690).

$$m = \frac{4690 - 1147}{12 - 0}$$
$$= \frac{3543}{12}$$
$$= 295.25$$

Since the y-intercept is 1147,

$$y = 295.25x + 1147.$$

(b) For 1990, $x = 8$.
If $x = 8$,

$$y = 295.25(8) + 1147$$
$$y = 3509.$$

The predicted debt based on the model was $3509 billion. This is slightly higher than the actual debt.

CHAPTER 4 POLYNOMIAL AND RATIONAL FUNCTIONS

Section 4.1

1. $f(x) = (x + 3)^2 - 4$

   (a) Domain: $(-\infty, \infty)$

   Range: $[-4, \infty)$

   (b) Vertex: $(h, k) = (-3, -4)$

   (c) Axis: $x = -3$

   (d) To find the y-intercept, let $x = 0$.

   $$y = (0 + 3)^2 - 4$$
   $$y = 9 - 4$$
   $$y = 5$$

   y-intercept: 5

   (e) To find the x-intercepts, let $y = 0$.

   $$0 = (x + 3)^2 - 4$$
   $$(x + 3)^2 = 4$$
   $$x + 3 = \pm\sqrt{4} = \pm 2$$
   $$x = -3 \pm 2$$
   $$x = -5 \quad \text{or} \quad x = -1$$

   x-intercepts: -5 and -1

3. $f(x) = -2(x + 3)^2 + 2$

   (a) Domain: $(-\infty, \infty)$

   Range: $(-\infty, 2]$

   (b) Vertex: $(h, k) = (-3, 2)$

   (c) Axis: $x = -3$

   (d) To find the y-intercept, let $x = 0$.

   $$y = -2(0 + 3)^2 + 2$$
   $$y = -18 + 2$$
   $$y = -16$$

   y-intercept: -16

   (e) To find the x-intercepts, let $y = 0$.

   $$0 = -2(x + 3)^2 + 2$$
   $$(x + 3)^2 = 1$$
   $$x + 3 = \pm\sqrt{1} = \pm 1$$
   $$x = -3 \pm 1$$
   $$x = -4 \quad \text{or} \quad x = -2$$

   x-intercepts: -4 and -2

5. $f(x) = (x - 4)^2 - 3$

   Since $a > 0$, the parabola opens upward. The vertex is at $(4, -3)$. The correct graph, therefore, is **B**.

7. $f(x) = (x + 4)^2 - 3$

   Since $a > 0$, the parabola opens upward. The vertex is at $(-4, -3)$. The correct graph, therefore, is **D**.

9. For parts (a), (b), (c), and (d), see the answer graphs in the back of the textbook.

   (e) As the coefficient decreases in absolute value, the parabola becomes broader.

11. For parts (a), (b), (c), and (d), see the answer graphs in the back of the textbook.

    (e) Each of these graphs is a horizontal translation of the graph of $y = x^2$, so they differ only in the position of their vertices on the x-axis.

The graph of $y = (x - 2)^2$ is the same as that of $y = x^2$, but translated 2 units to the right. The graph of $y = (x + 1)^2$ is the same as that of $y = x^2$, but translated 1 unit to the left. The graph of $y = (x + 3)^2$ is the same as that of $y = x^2$, but translated 3 units to the left. The graph of $y = (x - 4)^2$ is the same as that of $y = x^2$, but translated 4 units to the right.

13. $y = a(x - h)^2 + k$

   (a) If $h < 0$ and $k < 0$, the vertex is in quadrant III.

   (b) If $h < 0$ and $k > 0$, the vertex is in quadrant II.

   (c) If $h > 0$ and $k < 0$, the vertex is in quadrant IV.

   (d) If $h > 0$ and $k > 0$, the vertex is in quadrant I.

For Exercises 15–25, see the answer graphs in the back of the textbook.

15. $f(x) = (x - 2)^2$

This equation is of the form $y = (x - h)^2$, with $h = 2$. The graph opens upward and has the same shape as that of $y = x^2$. It is a horizontal translation of the graph of $y = x^2$ 2 units to the right. The vertex is $(2, 0)$ and the axis is the vertical line $x = 2$. The domain and range can be seen on the graph. The

domain is $(-\infty, \infty)$. Since the smallest value of y is 0 and the graph opens upward, the range is $[0, \infty)$.

17. $f(x) = (x + 3)^2 - 4$

$y = [x - (-3)]^2 + (-4)$

This equation is of the form $y = (x - h)^2 + k$, with $h = -3$ and $k = -4$. The vertex is $(-3, -4)$. The graph opens upward and has the same shape as $y = x^2$. It is a translation of $y = x^2$ 3 units to the left and 4 units down. The axis is the vertical line $x = -3$. The domain is $(-\infty, \infty)$. Since the smallest value of y is -4 and the graph opens upward, the range is $[-4, \infty)$.

19. $f(x) = -2(x + 3)^2 + 2$

The vertex is $(-3, 2)$. The graph opens downward and is narrower than $y = x^2$. It is a translation of the graph of $y = -2x^2$ 3 units to the left and 2 units up. The axis is the vertical line $x = -3$. The domain is $(-\infty, \infty)$. Since the largest value of y is 2, the range is $(-\infty, 2]$.

21. $y = -\frac{1}{2}(x + 1)^2 - 3$

The vertex is $(-1, -3)$. The graph opens downward and is wider than $y = x^2$. It is a translation of the graph of $y = -\frac{1}{2}x^2$ 1 unit to the left and 3 units down.

The axis is the vertical line $x = -1$. The domain is $(-\infty, \infty)$. Since the largest value of y is -3, the range is $(-\infty, -3]$.

23. $f(x) = x^2 - 2x + 3$

$= (x^2 - 2x + 1) - 1 + 3$

$= (x - 1)^2 + 2$

The vertex is $(1, 2)$.
The graph opens upward and has the same shape as $y = x^2$. It is a translation of the graph of $y = x^2$ 1 unit to the right and 2 units up. The axis is the vertical line $x = 1$. The domain is $(-\infty, \infty)$. Since the smallest value of y is 2, the range is $[2, \infty)$.

25. $f(x) = 2x^2 - 4x + 5$

$= 2(x^2 - 2x \quad) + 5$

$= 2(x^2 - 2x + 1 - 1) + 5$

$= 2(x^2 - 2x + 1) - 2 + 5$

$= 2(x - 1)^2 + 3$

The vertex is $(1, 3)$. The graph opens upward and has the same shape as $y = 2x^2$. It is a translation of the graph of $y = 2x^2$ 1 unit to the right and 3 units up. The axis is the vertical line $x = 1$. The domain is $(-\infty, \infty)$. Since the smallest value of y is 3, the range is $[3, \infty)$.

27. The minimum value of $f(x)$ is $f(-3) = 3$.

29. There are no real solutions to the equation $f(x) = 1$ since the value of $f(x)$ is never less than 3.

31. Graph the functions $y = 3x^2 - 2$ and $y = 3(x^2 - 2)$ on the same screen. See the answer graph in the back of the textbook. We see that the graphs are two different parabolas. Since the graphs are not the same, the expressions $3x^2 - 2$ and $3(x^2 - 2)$ are not equivalent. This result can be confirmed algebraically:

$$3(x^2 - 2) = 3x^2 - 6 \neq 3x^2 - 2.$$

33. $a < 0$, $b^2 - 4ac = 0$

The correct choice is E. $a < 0$ indicates that the parabola opens downward, while $b^2 - 4ac = 0$ indicates that the graph has exactly one x-intercept.

35. $a < 0$, $b^2 - 4ac < 0$

The correct choice is D. $a < 0$ indicates that the parabola opens downward, while $b^2 - 4ac < 0$ indicates that the graph has no x-intercepts.

37. $a > 0$, $b^2 - 4ac > 0$

The correct choice is C. $a > 0$ indicates that the parabola opens upward, while $b^2 - 4ac > 0$ indicates that the graph has two x-intercepts.

39. Graph the function $f(x) = x^2 + 2x - 8$. From the vertex formula, the vertex is $(-1, -9)$ and the axis is $x = -1$. Use a table of values to find points on the graph. See the answer graph in the back of the textbook. From the graph, we see that the x-intercepts are -4 and 2.

40. The solution set of the inequality

$$x^2 + 2x - 8 < 0$$

consists of all x-values for which the graph of f lies *below* the x-axis. By examining the graph of $f(x) = x^2 + 2x - 8$ from Exercise 39, we see that the graph lies below the x-axis when x is between the -4 and 2. Thus, the solution set is the open interval $(-4, 2)$.

41. Graph $g(x) = -f(x) = -x^2 - 2x + 8$.

See the answer graph in the back of the textbook. The graph of g is obtained by reflecting the graph of f across the x-axis.

42. The solution set of the inequality

$$-x^2 - 2x - 8 > 0$$

consists of all x-values for which the graph of g lies *above* the x-axis. By examining the graph of $g(x) = -x^2 - 2x + 8$ from Exercise 41, we see that the graph lies above the x-axis when x is between -4 and 2. Thus, the solution set is the open interval $(-4, 2)$.

43. The two solution sets are the same, the open interval $(-4, 2)$.

45. The vertex of the parabola in the figure is $(1, 4)$ and the y-intercept is 2. The equation takes the form

$$f(x) = a(x - 1)^2 + 4.$$

When $x = 0$, $f(x) = 2$, so

$$2 = a(0 - 1)^2 + 4$$
$$2 = a + 4$$
$$a = -2.$$

The equation is

$$f(x) = -2(x - 1)^2 + 4.$$

This function may also be written as

$$f(x) = -2(x^2 - 2x + 1) + 4$$
$$= -2x^2 + 4x - 2 + 4$$
$$f(x) = -2x^2 + 4x + 2.$$

Graphing this function on a graphing calculator shows that the graph matches the equation.
See the answer graph in the back of the textbook.

47. In 1996, $x = 11$.

$$f(x) = 18.14x^2 + 234.03x + 3954$$
$$f(11) = 18.14(11)^2 + 234.03(11)$$
$$+ 3954$$
$$\approx 8723$$

The gross state product in 1996 would be approximately 8723 billion dollars.

49. $f(x) = -17x^2 + 44.6x + 2572$

The value of $f(x)$ will reach its peak when

$$x = -\frac{b}{2a}$$

$$= \frac{-44.6}{2(-17)} \approx 1.3$$

because this gives the x-value of the vertex of the parabola. Because $a = -17 < 0$, we know that the parabola opens downward, so the vertex will be the maximum point on the graph.

Therefore, the number of suicides reached its peak in 1986.

51. **(a)** Plot the 12 points $(2, 1563)$, $(3, 4647)$, ..., $(13, 361{,}509)$ on the same calculator screen. See the answer graph in the back of the textbook.

(b) One can see that the data are not linear but instead resemble the right half of a parabola opening upward. A quadratic function will model the data better than a linear function.

(c) Let the point $(2, 1563)$ be the vertex. Then,

$$f(x) = a(x - 2)^2 + 1563.$$

Next let the point $(13, 361{,}509)$ lie on the graph of the function. Solving for a results in

$$f(13) = a(13 - 2)^2 + 1563$$
$$= 361{,}509$$
$$a = \frac{359{,}946}{121} \approx 2974.76.$$

Thus, $f(x) = 2974.76(x - 2)^2 + 1563$. (Other choices will lead to other models.)

(d) Plotting the points together with $f(x)$, we see that there is a good fit. See the answer graph in the back of the textbook.

(e) $x = 19$ corresponds to the year 1999, and $x = 20$ corresponds to the year 2000.

$f(19) \approx 861{,}269$ and $f(20) \approx 965{,}385$.

In the year 2000, nearly 1 million people will have been diagnosed with AIDS since 1981.

(f) The number of new cases in the year 2000 will be approximately

$$f(20) - f(19) \approx 965{,}385 - 861{,}269$$
$$= 104{,}116.$$

53. $h(x) = -.5x^2 + 1.25x + 3$

(a) Find $h(x)$ when $x = 2$.

$$h(x) = -.5x^2 + 1.25x + 3$$
$$h(2) = -.5(2)^2 + 1.25(2) + 3$$
$$= 3.5.$$

When the distance from the base of the stump was 2 ft, the frog was 3.5 ft high.

(b) Find x when $h(x) = 3.25$.

$$3.25 = -.5x^2 + 1.25x + 3$$
$$0 = -.5x^2 + 1.25x - .25$$

Multiply by 100 to clear decimals.

$$0 = -50x^2 + 125x - 25$$

Divide by 25.

$$0 = -2x^2 + 5x - 1$$

Use the quadratic formula with $a = -2$, $b = 5$, and $c = -1$.

$$x = \frac{-5 \pm \sqrt{5^2 - 4(-2)(-1)}}{2(-2)}$$

$$= \frac{-5 \pm \sqrt{17}}{-4}$$

$$x = \frac{-5 + \sqrt{17}}{-4} \approx .21922$$

or $\quad x = \dfrac{-5 - \sqrt{17}}{4} \approx -2.2808$

The frog was 3.25 ft above the ground at approximately .2 sec (on the way up) and 2.3 sec (on the way down).

(c) Since the parabola opens downward, the vertex is the maximum point. Use the vertex formula to find the x-coordinate of the vertex of

$$h(x) = -.5x^2 + 1.25x + 3.$$

$$x = -\frac{b}{2a} = -\frac{1.25}{2(-.5)} = 1.25$$

The frog reached its highest point at 1.25 ft from the stump.

(d) The maximum height is the y-coordinate of the vertex.

$$h(x) = -.5x^2 + 1.25x + 3$$
$$y = h(1.25) = -.5(1.25)^2 + 1.25(1.25) + 3$$
$$= 3.78125$$

The maximum height reached by the frog was approximately 3.78 ft.

55. $A = x(320 - x)$

Find the vertex.

$$A = 320x - 2x^2$$
$$= -2x^2 + 320x$$
$$= -2(x^2 - 160x + 6400 - 6400)$$
$$= -2(x^2 - 160x + 6400) + 12,800$$
$$= -2(x - 80)^2 + 12,800$$

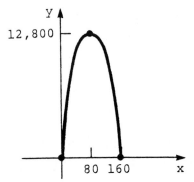

The vertex is (80, 12,800). The width, x, should be 80 ft in order to have maximum area of 12,800 sq ft. The length would be 320 - 2(80) = 160 ft.

57. $s(x) = -16x^2 + 80x + 100$
$$= -16(x^2 - 5x \quad) + 100$$
$$= -16(x^2 - 5x + 6.25) + 100$$
$$\quad + 100$$
$$= -16(x - 2.5)^2 + 200$$

The vertex of the parabola is (2.5, 200). Since $a = -16 < 0$, this is the maximum point.

(a) The ball will reach its maximum height after 2.5 sec.

(b) The maximum height is 200 ft.

59. $s(x) = -16x^2 + 200x + 50$

Complete the square to find the vertex.

$$s = -16(x^2 + 12.5x) + 50$$
$$= -16(x^2 + 12.5x + 39.0625)$$
$$\quad + 625 + 50$$
$$= -16(x - 6.25)^2 + 675$$

The vertex is $(6.25, 675)$. Since $a < 0$, this is the maximum point.

(a) The number of seconds to reach maximum height is 6.25 sec.
The maximum height is 675 ft.

(b) To find the time interval in which the rocket will be more than 300 ft above ground level, solve the inequality

$$-16x^2 + 200x + 50 > 300$$
$$-16x^2 + 200x - 250 > 0$$
$$-8x^2 + 100x - 125 > 0.$$

Solve the corresponding equation.

$$-8x^2 + 100x - 125 = 0$$

Use the quadratic formula with $a = -8$, $b = 100$, and $c = -125$.

$$x = \frac{-100 \pm \sqrt{100^2 - 4(-8)(-125)}}{2(-8)}$$
$$= \frac{-100 \pm \sqrt{6000}}{-16}$$
$$x = \frac{-100 + \sqrt{6000}}{-16} \approx 1.4$$
or $\quad x = \frac{-100 - \sqrt{6000}}{-16} \approx 11.1$

The values 1.4 and 11.1 divide the number line into three intervals: $(-\infty, 1.4)$, $(1.4, 11.1)$, and $(11.1, \infty)$.

Use a test point in each interval to determine where the expression $-8x^2 + 100x - 125$ is positive. We find that it is positive in the interval $(1.4, 1.11)$.
Therefore, the rocket will be more than 300 ft above the ground between 1.4 sec and 1.11 sec.
(The quadratic inequality could also be solved graphically by graphing $Y_1 = -8x^2 + 100x - 125$ on a graphing calculator and finding the interval where the graph lies above the x-axis.)

(c) To find the number of seconds it will take the toy rocket to hit the ground, let $s(x) = 0$ and solve for x.

$$0 = -16(x - 6.25)^2 + 675$$
$$-675 = -16(x - 6.25)^2$$
$$(x - 6.25)^2 = 42.1875$$
$$x = 6.25 \pm 6.5$$
$$x \approx 12.75$$

It will take approximately 12.75 sec. for the toy rocket to hit the ground.

61. $y = x^2 - 10x + c$

An x-intercept occurs where $y = 0$, or

$$0 = x^2 - 10x + c.$$

There will be exactly one x-intercept if this equation has exactly one solution, or if the discriminant is zero.

$$b^2 - 4ac = 0$$
$$(-10)^2 - 4(1)c = 0$$
$$100 = 4c$$
$$c = 25$$

63. $y = x^2 + bx + 9$

An x-intercept occurs where $y = 0$, or

$$0 = x^2 + bx + 9.$$

There will be exactly one x-intercept if this equation has exactly one solution, or if the discriminant is zero.

$$b^2 - 4ac = 0$$
$$b^2 - 4(1)(9) = 0$$
$$b^2 = 36$$
$$b = \pm 6$$

Thus, $x^2 + bx + 9$ has exactly one x-intercept when $b = 6$ or $b = -6$.

65. x-intercepts 1 and -2, and y-intercept 4

Every quadratic function may be written as

$$f(x) = ax^2 + bx + c.$$

Since the y-intercept is 4, if $x = 0$, $f(x) = 4$.

$$4 = a(0^2) + b(0) + c$$
$$c = 4$$

Since 1 and -2 are x-intercepts, $f(1) = 0$ and $f(-2) = 0$. First,

$$0 = a(1)^2 + b(1) + 4$$
$$0 = a + b + 4$$
$$-b - 4 = a. \qquad (1)$$

Then,

$$0 = a(-2)^2 + b(-2) + 4$$
$$0 = 4a - 2b + 4. \qquad (2)$$

Substitute for a in equation (2) and solve for b.

$$0 = 4(-b - 4) - 2b + 4$$
$$0 = -4b - 16 - 2b + 4$$
$$6b = -12$$
$$b = -2$$
$$a = -(-2) - 4 = -2$$

The required quadratic function is

$$f(x) = -2x^2 - 2x + 4.$$

67. $y = 3 + (x + 5)^2$

The graph of this equation is a parabola with vertex $(-5, 3)$ that opens upward. The smallest possible value of y is the value of y at the vertex, or $y = 3$.

(a) As a number decreases, its square root decreases. Since the smallest possible value of $3 + (x + 5)^2$ is 3, the smallest possible value of $\sqrt{3 + (x + 5)^2}$ is $\sqrt{3}$.

(b) As a number decreases, its reciprocal increases, since the smallest possible value of $3 + (x + 5)^2$ is 3, the largest possible value of $\dfrac{1}{3 + (x + 5)^2}$ is $\dfrac{1}{3}$.

Section 4.2

1. When the polynomial $f(x)$ is divided by $x - r$, the remainder is $f(r)$.

   The remainder theorem states if the polynomial $f(x)$ is divided by $x - k$, the remainder is $f(k)$. Therefore, this statement is true.

3. If $x^3 - 1$ is divided by $x + 1$, the remainder is 0.

   $$\frac{x^3 - 1}{x + 1}$$

   Use $x = -1$ since $x + 1 = x - (-1)$.

   $$\begin{array}{r|rrrr} -1 & 1 & 0 & 0 & -1 \\ & & -1 & 1 & -1 \\ \hline & 1 & -1 & 1 & -2 \end{array}$$

   The remainder is -2, not 0. Therefore, the statement is false.

5. $$\frac{x^3 + 4x^2 - 5x + 42}{x + 6}$$

   $$\begin{array}{r|rrrr} -6 & 1 & 4 & -5 & 42 \\ & & -6 & 12 & -42 \\ \hline & 1 & -2 & 7 & 0 \end{array}$$

   Thus,

   $$\frac{x^3 + 4x^2 - 5x + 42}{x + 6} = x^2 - 2x + 7.$$

7. $$\frac{4x^3 - 3x - 2}{x + 1}$$

   $$\begin{array}{r|rrrr} -1 & 4 & 0 & -3 & -2 \\ & & -4 & 4 & -1 \\ \hline & 4 & -4 & 1 & -3 \end{array}$$

   Thus,

   $$4x^3 - 3x - 2 = 4x^2 - 4x + 1 + \frac{-3}{x + 1}.$$

9. $$\frac{x^4 - 3x^3 - 4x^2 + 12x}{x - 3}$$

   $$\begin{array}{r|rrrrr} 3 & 1 & -3 & -4 & 12 & 0 \\ & & 3 & 0 & -12 & 0 \\ \hline & 1 & 0 & -4 & 0 & 0 \end{array}$$

   Thus,

   $$\frac{x^4 - 3x^3 - 4x^2 + 12x}{x - 3} = x^3 - 4x.$$

11. $$\frac{x^5 + 3x^4 + 2x^3 + 2x^2 + 3x + 1}{x + 2}$$

    $$\begin{array}{r|rrrrrr} -2 & 1 & 3 & 2 & 2 & 3 & 1 \\ & & -2 & -2 & 0 & -4 & 2 \\ \hline & 1 & 1 & 0 & 2 & -1 & 3 \end{array}$$

    Thus,

    $$\frac{x^5 + 3x^4 + 2x^3 + 2x^2 + 3x + 1}{x + 2}$$

    $$= x^4 + x^3 + 2x - 1 + \frac{3}{x + 2}.$$

13. $f(x) = 2x^3 + x^2 + x - 8;\ k = -1$

    Use synthetic division to write the polynomial in the form $f(x) = (x - k)q(x) + r$.

    $$\begin{array}{r|rrrr} -1 & 2 & 1 & 1 & -8 \\ & & -2 & 1 & -2 \\ \hline & 2 & -1 & 2 & -10 \end{array}$$

    $f(x) = (x + 1)(2x^2 - x + 2) - 10$

15. $f(x) = -x^3 + 2x^2 + 4;\ k = -2$

    $$\begin{array}{r|rrrr} -2 & -1 & 2 & 0 & 4 \\ & & 2 & -8 & 16 \\ \hline & -1 & 4 & -8 & 20 \end{array}$$

    $f(x) = (x + 2)(-x^2 + 4x - 8) + 20$

17. $f(x) = 4x^4 - 3x^3 - 20x^2 - x;\ k = 3$

$$
\begin{array}{r|rrrrr}
3 & 4 & -3 & -20 & -1 & 0 \\
  &   & 12 & 27 & 21 & 60 \\
\hline
  & 4 & 9 & 7 & 20 & 60
\end{array}
$$

$f(x)$
$\quad = (x - 3)(4x^3 + 9x^2 + 7x + 20) + 60$

19. $k = 3;\ f(x) = x^2 - 4x + 5$

$$
\begin{array}{r|rrr}
3 & 1 & -4 & 5 \\
  &   & 3 & -3 \\
\hline
  & 1 & -1 & 2
\end{array}
$$

$f(3) = 2$

21. $k = 2;\ f(x) = 2x^2 - 3x - 3$

$$
\begin{array}{r|rrr}
2 & 2 & -3 & -3 \\
  &   & 4 & 2 \\
\hline
  & 2 & 1 & -1
\end{array}
$$

$f(2) = -1$

23. $k = -1;\ f(x) = x^3 - 4x^2 + 2x + 1$

$$
\begin{array}{r|rrrr}
-1 & 1 & -4 & 2 & 1 \\
   &   & -1 & 5 & -7 \\
\hline
   & 1 & -5 & 7 & -6
\end{array}
$$

$f(-1) = -6$

25. $k = 3;\ f(x) = 2x^5 - 10x^3 - 19x^2 - 45$

$$
\begin{array}{r|rrrrrr}
3 & 2 & 0 & -10 & -19 & 0 & -45 \\
  &   & 6 & 18 & 24 & 15 & 45 \\
\hline
  & 2 & 6 & 8 & 5 & 15 & 0
\end{array}
$$

$f(3) = 0$

27. $k = -8;\ f(x) = x^6 + 7x^5 - 5x^4 + 22x^3$
$\qquad\qquad\qquad - 16x^2 + x + 19$

$$
\begin{array}{r|rrrrrrr}
-8 & 1 & 7 & -5 & 22 & -16 & 1 & 19 \\
   &   & -8 & 8 & -24 & 16 & 0 & -8 \\
\hline
   & 1 & -1 & 3 & -2 & 0 & 1 & 11
\end{array}
$$

$f(-8) = 11$

29. $k = 2 + i;\ f(x) = x^2 - 5x + 1$

$$
\begin{array}{r|rrr}
2 + i & 1 & -5 & 1 \\
      &   & 2 + i & -7 - i \\
\hline
      & 1 & -3 + i & -6 - i
\end{array}
$$

$f(2 + i) = -6 - i$

31. To determine if 3 is a zero of $f(x) = 2x^3 - 6x^2 - 9x + 4$, divide synthetically.

$$
\begin{array}{r|rrrr}
3 & 2 & -6 & -9 & 4 \\
  &   & 6 & 0 & -27 \\
\hline
  & 2 & 0 & -9 & -23
\end{array}
$$

No, 3 is not a zero of $f(x)$ because $f(3) = -23$.

33. To determine if -5 is a zero of $f(x) = x^3 + 7x^2 + 10x$, divide synthetically.

$$
\begin{array}{r|rrrr}
-5 & 1 & 7 & 10 & 0 \\
   &   & -5 & -10 & 0 \\
\hline
   & 1 & 2 & 0 & 0
\end{array}
$$

Yes, -5 is a zero of $f(x)$ because $f(-5) = 0$.

35. To determine if $2/5$ is a zero of $f(x) = 5x^4 + 2x^3 - x + 15$, divide synthetically.

$$
\begin{array}{r|rrrrr}
\frac{2}{5} & 5 & 2 & 0 & -1 & 15 \\
            &   & 2 & \frac{8}{5} & \frac{16}{25} & -\frac{18}{125} \\
\hline
            & 5 & 4 & \frac{8}{5} & -\frac{9}{25} & \frac{1857}{125}
\end{array}
$$

No, $2/5$ is not a zero of $f(x)$ because $f(2/5) = 1857/125$.

37. To determine if 2 − i is a zero of f(x) = x² + 3x + 4, divide synthetically.

$$2 - i \underline{\big|} \quad \begin{array}{ccc} 1 & 3 & 4 \\ & 2-i & 9-7i \\ \hline 1 & 5-i & 13-7i \end{array}$$

No, 2 − i is not a zero of f(x) because f(2 − i) = 13 − 7i.

39. 1 raised to *any* power is 1.

40. The result from Exercise 39 is 1. If we multiply 1 by a real number, the result is the real number. It is equal to the real number because 1 is the identity element for multiplication.

41. We can evaluate f(1) by adding the coefficients of f.

42. $f(x) = x^3 - 4x^2 + 9x - 6$

$$\begin{aligned} f(1) &= (1)^3 - 4(1)^2 + 9(1) - 6 \\ &= 1 - 4 + 9 - 6 \\ &= 0 \end{aligned}$$

The sum of the coefficients of f is

$$1 + (-4) + 9 + (-6) = 0.$$

The answers agree.

43. $f(x) = x^3 - 4x^2 + 9x - 6$

$$\begin{aligned} f(-x) &= (-x)^3 - 4(-x)^2 + 9(-x) - 6 \\ &= -x^3 - 4x^2 - 9x - 6 \end{aligned}$$

$$\begin{aligned} f(-1) &= (-1)^3 - 4(-1)^2 + 9(-1) - 6 \\ &= -1 - 4 - 9 - 6 \\ &= -20 \end{aligned}$$

44. $f(-x) = -x^3 - 4x^2 - 9x - 6$

The sum of the coefficients of f(−x) is

$$-1 + (-4) + (-9) + (-6) = -20.$$

f(−1) = −20

They are both −20.

Our conjecture is, to find f(−1), add the coefficients of f(−x).

Section 4.3

1. Given that x − 1 is a factor of f(x) = x⁶ − x⁴ + 2x² − 2, we are assured that f(1) = 0.

This statement is justified by the factor theorem; therefore, it it true.

3. For the function

$$f(x) = (x + 2)^4(x - 3),$$

2 is a zero of multiplicity 4.

To find the zero, set the factor equal to 0.

$$\begin{aligned} x + 2 &= 0 \\ x &= -2 \end{aligned}$$

2 is not a zero of the function; therefore, the statement is false. (It would be true to say that −2 is a zero of multiplicity 4.)

5. 4x² + 2x + 54; x − 4

Let f(x) = 4x² + 2x + 54. By the factor theorem, x − 4 will be a factor of f(x) only if f(4) = 0.

Use synthetic division and the re-mainder theorem.

$$\begin{array}{r|rrr} 4 & 4 & 2 & 54 \\ & & 16 & 72 \\ \hline & 4 & 18 & 126 \end{array}$$

Since the remainder is 126, $f(4) =$ 126, so $x - 4$ is not a factor of $f(x)$.

7. $x^3 + 2x^2 - 3$; $x - 1$

Let $f(x) = x^3 + 2x^2 - 1$.
By the factor theorem, $x - 1$ will be a factor of $f(x)$ only if $f(1) = 0$.

$$\begin{array}{r|rrrr} 1 & 1 & 2 & 0 & -3 \\ & & 1 & 3 & 3 \\ \hline & 1 & 3 & 3 & 0 \end{array}$$

Since $f(1) = 0$, $x - 1$ is a factor of $f(x)$.

9. $2x^4 + 5x^3 - 2x^2 + 5x + 6$; $x + 3$

Let $f(x) = 2x^4 + 5x^3 - 2x^2 + 5x + 6$.

$$\begin{array}{r|rrrrr} -3 & 2 & 5 & -2 & 5 & 6 \\ & & -6 & 3 & -3 & -6 \\ \hline & 2 & -1 & 1 & 2 & 0 \end{array}$$

Since $f(-3) = 0$, $x + 3$ is a factor of $f(x)$.

11. $f(x) = x^3 - x^2 - 4x - 6$; 3

Since 3 is a zero, first divide $f(x)$ by $x - 3$.

$$\begin{array}{r|rrrr} 3 & 1 & -1 & -4 & -6 \\ & & 3 & 6 & 6 \\ \hline & 1 & 2 & 2 & 0 \end{array}$$

This gives

$f(x) = (x - 3)(x^2 + 2x + 2)$.

Use the quadratic formula with $a = 1$, $b = 2$, and $c = 2$ to find the remaining two zeros.

$$\begin{aligned} x &= \frac{-2 \pm \sqrt{4 - 4(1)(2)}}{2(1)} \\ &= \frac{-2 \pm \sqrt{-4}}{2} = \frac{-2 \pm 2i}{2} \\ &= \frac{2(-1 \pm i)}{2} = -1 \pm i \end{aligned}$$

The remaining zeros are $-1 \pm i$.

13. $f(x) = 4x^3 + 6x^2 - 2x - 1$; $\frac{1}{2}$

Since $1/2$ is a zero, first divide $f(x)$ by $x - 1/2$.

$$\begin{array}{r|rrrr} \frac{1}{2} & 4 & 6 & -2 & -1 \\ & & 2 & 4 & 1 \\ \hline & 4 & 8 & 2 & 0 \end{array}$$

This gives

$$\begin{aligned} f(x) &= \left(x - \frac{1}{2}\right)(4x^2 + 8x + 2) \\ &= \left(x - \frac{1}{2}\right) \cdot 2(2x^2 + 4x + 1) \\ &= (2x - 1)(2x^2 + 4x + 1). \end{aligned}$$

Use the quadratic formula with $a = 2$, $b = 4$, and $c = 1$ to find the other two zeros.

$$\begin{aligned} x &= \frac{-4 \pm \sqrt{16 - 4(2)(1)}}{2(2)} \\ &= \frac{-4 \pm \sqrt{8}}{4} = \frac{-4 \pm 2\sqrt{2}}{4} \\ &= -1 \pm \frac{\sqrt{2}}{2}. \end{aligned}$$

(These zeros may also be written as $\frac{-2 \pm \sqrt{2}}{2}$).

15. $f(x) = x^4 + 5x^2 + 4$; $-i$

Since $-i$ is a zero, first divide $f(x)$ by $x + i$.

$$\begin{array}{r|rrrrr} -i & 1 & 0 & 5 & 0 & 4 \\ & & -i & -1 & -4i & -4 \\ \hline & 1 & -i & 4 & -4i & 0 \end{array}$$

By the conjugate zeros theorem, i is also a zero, so divide the quotient polynomial from the first synthetic division by i.

$$\begin{array}{r|rrrr} -i & 1 & -i & 4 & -4i \\ & & i & 0 & 4i \\ \hline & 1 & 0 & 4 & 0 \end{array}$$

The remaining zeros will be zeros of the new quotient polynomial, $x^2 + 4$. Find the remaining zeros by using the square root property.

$$x^2 + 4 = 0$$
$$x^2 = -4$$
$$x = \pm 2i.$$

The other zeros are i and $\pm 2i$.

17. Zeros of -3, 1, and 4; $f(2) = 30$

The polynomial function has the form

$$f(x) = a(x + 3)(x - 1)(x - 4)$$

for some real number a. To find a, use the fact that $f(2) = 30$.

$$f(2) = a(2 + 3)(2 - 1)(2 - 4) = 30$$
$$a(5)(1)(-2) = 30$$
$$-10a = 30$$
$$a = -3$$

Thus,

$$f(x) = -3(x + 3)(x - 1)(x - 4)$$
$$= -3(x^2 + 2x - 3)(x - 4)$$
$$= -3(x^3 - 2x^2 - 11x + 12)$$
$$= -3x^3 + 6x^2 + 33x - 36.$$

19. Zeros of -2, 1, and 0; $f(-1) = -1$

The polynomial function has the form

$$f(x) = a(x + 2)(x - 1)(x - 0)$$

for some real number a. To find a, use the fact that $f(-1) = -1$.

$$f(-1) = a(-1 + 2)(-1 - 1)(-1 - 0) = -1.$$
$$a(1)(-2)(-1) = -1$$
$$2a = -1$$
$$a = -\frac{1}{2}$$

Thus,

$$f(x) = -\frac{1}{2}(x + 2)(x - 1)x$$
$$= -\frac{1}{2}(x^2 + x - 2)x$$
$$= -\frac{1}{2}x^3 - \frac{1}{2}x^2 + x.$$

21. Zeros of 5, i, and $-i$; $f(2) = 5$

The polynomial function has the form

$$f(x) = a(x - 5)(x - i)(x + i)$$

for some real number a. To find a, use the fact that $f(2) = 5$.

$$f(2) = a(2 - 5)(2 - i)(2 + i) = 5$$
$$a(-3)(5) = 5$$
$$-15a = 5$$
$$a = -\frac{1}{3}$$

Thus,

$$f(x) = -\frac{1}{3}(x - 5)(x - i)(x + i)$$
$$= -\frac{1}{3}(x - 5)(x^2 + 1)$$
$$= -\frac{1}{3}(x^3 - 5x^2 + x - 5)$$
$$= -\frac{1}{3}x^3 + \frac{5}{3}x^2 - \frac{1}{3}x + \frac{5}{3}.$$

23. $f(x) = x^3 - 21x - 20$

Find the quotient

$$\frac{x^3 - 21x - 20}{x + 4}$$

$$\underline{-4} \begin{array}{cccc} 1 & 0 & -21 & -20 \\ & -4 & 16 & 20 \\ \hline 1 & -4 & -5 & 0 \end{array}$$

$g(x) = x^2 - 4x - 5$

24. See the answer graph in the back of the textbook.
The function g is a quadratic function. The x-intercepts of g are also x-intercepts of f.

25. $\dfrac{x^2 - 4x - 5}{x - 5}$

$$\underline{5} \begin{array}{ccc} 1 & -4 & -5 \\ & 5 & 5 \\ \hline 1 & 1 & 0 \end{array}$$

$h(x) = x + 1$

26. See the answer graph in the back of the textbook.
The function h is a linear function. The x-intercept of h is also an x-intercept of g.

27. $f(x) = 2x^3 - 3x^2 - 17x + 30; \ k = 2$

Since 2 is a zero of $f(x)$, $x - 2$ is a factor. Divide $f(x)$ by $x - 2$.

$$\underline{2} \begin{array}{cccc} 2 & -3 & -17 & 30 \\ & 4 & 2 & -30 \\ \hline 2 & 1 & -15 & 0 \end{array}$$

Thus,

$$f(x) = (x - 2)(2x^2 + x - 15)$$
$$= (x - 2)(2x - 5)(x + 3).$$

29. $f(x) = 6x^3 + 13x^2 - 14x + 3; \ k = -3$

Since -3 is a zero of $f(x)$, $x + 3$ is a factor. Divide $f(x)$ by $x + 3$.

$$\underline{-3} \begin{array}{cccc} 6 & 13 & -14 & 3 \\ & -18 & 15 & -3 \\ \hline 6 & -5 & 1 & 0 \end{array}$$

Thus,

$$f(x) = (x + 3)(6x^2 - 5x + 1)$$
$$= (x + 3)(3x - 1)(2x - 1).$$

31. $f(x) = 7x^3 + x$

To find the zeros, let $f(x) = 0$ and factor the binomial.
Set each factor equal to zero and solve for x.

$$7x^3 + x = 0$$
$$x(7x^2 + 1) = 0$$
$$x = 0$$

or

$$7x^2 + 1 = 0$$
$$7x^2 = -1$$
$$x^2 = -\frac{1}{7}$$
$$x = \pm\sqrt{-\frac{1}{7}} = \pm i\sqrt{\frac{1}{7}}$$
$$= \pm\frac{\sqrt{7}}{7}i$$

The zeros are 0 and $\pm\dfrac{\sqrt{7}}{7}i$.

33. $f(x) = 3(x - 2)(x + 3)(x^2 - 1)$

To find the zeros, let $f(x) = 0$.
Set each factor equal to zero and solve for x.

x - 2 = 0 or x + 3 = 0 or x² - 1 = 0

x = 2   x = -3   x² = 1

x = ±1

The zeros are 2, -3, 1, and -1.

35. $f(x) = (x^2 + x - 2)^5(x - 1 + \sqrt{3})^2$

To find the zeros, let $f(x) = 0$.
Set each factor equal to zero and
solve for x.

$$(x^2 + x - 2)^5 = 0$$

$$x^2 + x - 2 = 0$$

$$(x + 2)(x - 1) = 0$$

$$x + 2 = 0 \quad \text{or} \quad x - 1 = 0$$

x = -2, multiplicity 5;

x = 1, multiplicity 5

$$(x - 1 + \sqrt{3})^2 = 0$$

$$x - 1 + \sqrt{3} = 0$$

$$x = 1 - \sqrt{3},$$

multiplicity 2

The zeros are -2 (multiplicity 5),
1 (multiplicity 5) and $1 - \sqrt{3}$
(multiplicity 2).

In Exercises 37–45, find a polynomial of
lowest degree with real coefficients
having the given zeros. For each of these
exercises, other answers are possible.

37. 3 + i and 3 - i

$$f(x) = [x - (3 + i)][x - (3 - i)]$$

$$= (x - 3 - i)(x - 3 + i)$$

$$= [(x - 3) - i][(x - 3) + i]$$

$$= (x - 3)^2 - (i)^2$$

$$= x^2 - 6x + 9 - i^2$$

$$= x^2 - 6x + 9 + 1$$

$$= x^2 - 6x + 10$$

39. $1 + \sqrt{2}$, $1 - \sqrt{2}$, and 3

$$f(x) = [x - (1 + \sqrt{2})][x - (1 - \sqrt{2})]$$

$$\cdot (x - 3)$$

$$= (x - 1 - \sqrt{2})(x - 1 + \sqrt{2})$$

$$\cdot (x - 3)$$

$$= [(x - 1) - \sqrt{2}][(x - 1) + \sqrt{2}]$$

$$\cdot (x - 3)$$

$$= [(x - 1)^2 - (\sqrt{2})^2](x - 3)$$

Difference of two squares

$$= (x^2 - 2x + 1 - 2)(x - 3)$$

$$= (x^2 - 2x - 1)(x - 3)$$

$$= x^3 - 5x^2 + 5x + 3$$

41. -2 + i, -2 - i, 3, and -3

$$f(x) = [x - (-2 + i)][x - (-2 - i)]$$

$$\cdot (x - 3)(x + 3)$$

$$= [(x + 2) - i][(x + 2) + i]$$

$$\cdot (x - 3)(x + 3)$$

$$= [(x + 2)^2 - (i)^2](x - 3)(x + 3)$$

Difference of two squares

$$= (x^2 + 4x + 4 + 1)(x - 3)(x + 3)$$

$$= (x^2 + 4x + 5)(x^2 - 9)$$

$$= x^4 + 4x^3 - 4x^2 - 36x - 45$$

43. 2 and 3i

By the conjugate zeros theorem, -3i
must also be a zero.

$$f(x) = (x - 2)(x - 3i)[x - (-3i)]$$

$$= (x - 2)(x - 3i)(x + 3i)$$

$$= (x - 2)(x^2 + 9)$$

$$= x^3 - 2x^2 + 9x - 18$$

45. 1 + 2i, 2 (multiplicity 2)

By the conjugate zeros theorem,
1 - 2i must also be a zero.

$f(x) = (x - 2)^2[x - (1 + 2i)]$

$\cdot [x - (1 - 2i)]$

$= (x^2 - 4x + 4)[(x - 1) - 2i]$

$\cdot [(x - 1) + 2i]$

$= (x^2 - 4x + 4)[(x - 1)^2 - (2i)^2]$

$= (x^2 - 4x + 4)(x^2 - 2x + 1 + 4)$

$= (x^2 - 4x + 4)(x^2 - 2x + 5)$

$= x^4 - 6x^3 + 17x^2 - 28x + 20$

47. If -2 is a zero of multiplicity 2, then

$f(x) = x^4 + 2x^3 - 7x^2 - 20x - 12$

can be divided by $x + 2$, and the resulting quotient polynomial can be divided by $x + 2$ again. Each time the remainder should be 0. This is demonstrated in the following.

```
-2| 1    2   -7   -20   -12
        -2    0    14    12
   ─────────────────────────
     1    0   -7    -6     0
```

Divide the quotient polynomial by $x + 2$.

```
-2| 1    0   -7   -6
        -2    4    6
   ──────────────────
     1   -2   -3    0
```

Factor the quotient polynomial.

$x^2 - 2x - 3 = (x + 1)(x - 3)$

The remaining zeros are -1 and 3, and

$f(x) = (x + 2)^2(x + 1)(x - 3)$.

49. Prove that $\overline{c + d} = \overline{c} + \overline{d}$.
Let $c = a + bi$ and $d = m + ni$.
Work with the left-hand side.

$\overline{c + d} = \overline{(a + bi) + (m + ni)}$

$= \overline{(a + m) + (b + n)i}$

$= (a + m) - (b + n)i$

Work with right-hand side.

$\overline{c} + \overline{d} = \overline{a + bi} + \overline{m + ni}$

$= a - bi + m - ni$

$= (a + m) - (b + n)i$

Therefore,

$\overline{c + d} = \overline{c} + \overline{d}$.

51. Prove that $\overline{a} = a$ for any real number a.
Let $a = m + 0i$.
Work with left-hand side.

$\overline{a} = \overline{m + 0i}$

$= m - 0i$

$= m$

Work with right-hand side.

$a = m + 0i$

$= m$

Therefore, $\overline{a} = a$.

53. $f(x) = .86x^3 - 5.24x^2 + 3.55x + 7.84$

Enter this function into the calculator as

$Y_1 = .86 * x \wedge 3 - 5.24 * x \wedge 2$
$+ 3.55 * x + 7.84$.

solve $(Y_1, X, -5)$

$-.88$

solve $(Y_1, X, 1)$

2.12

solve $(Y_1, X, 6)$

4.86

The real zeros are $-.88$, 2.12 and 4.86.

55. $f(x) = 2.45x^4 - 3.22x^3 + .47x^2 - 6.54x + 3$

Enter this function into the calculator as

$Y_1 = 4 * x \wedge 4 + 8 * x \wedge 3 - 4 * x \wedge 2 + 4 * x + 1.$

solve $(Y_1, X, 1)$

.44

solve $(Y_1, X, 2)$

1.81

The real zeros are .44 and 1.81.

57. $f(x) = -\sqrt{7}x^3 + \sqrt{5}x + \sqrt{17}$

Enter this function into the calculator as

$Y_1 = -\sqrt{7} * x \wedge 3 + \sqrt{5} * x + \sqrt{17}.$

solve $(Y_1, X, 1)$

1.4

The only real zero is 1.40.

Section 4.4

1. $y = x^3 - 3x^2 - 6x + 8$

The range of an odd-degree polynomial is $(-\infty, \infty)$. The y-intercept of the graph is 8. The graph fitting these criteria is A.

3. Since graph C crosses the x-axis at one point, the graph has one real zero.

5. A polynomial of degree 3 can have at most 2 turning points. Graphs B and D have more than 2 turning points, so they cannot be graphs of cubic polynomial functions.

7. On the left side of the y-axis, graph A crosses the x-axis once, at -2. Therefore, graph A has one negative real zero.

For Exercises 9–17, see the answer graphs in the back of the textbook.

9. $f(x) = 2x^4$ is in the form $f(x) = ax^n$. $|a| = 2 > 1$, so the graph is narrower than $f(x) = x^4$.

11. $f(x) = -\frac{2}{3}x^5$ is in the form

$f(x) = ax^n$. $|a| = 2/3 < 1$, so the graph is broader than that of $f(x) = x^5$.

Since $a = -2/3$ is negative, the graph is the reflection of

$f(x) = \frac{2}{3}x^5$ about the x-axis.

13. $f(x) = \frac{1}{2}x^3 + 1$

The graph of $f(x) = \frac{1}{2}x^3 + 1$ looks like $y = x^3$ but is broader and is translated 1 unit up. The graph includes the points $(-2, -3)$,

$\left(-1, \frac{1}{2}\right)$, $(0, 1)$, $\left(1, \frac{3}{2}\right)$, and $(2, 5)$.

15. $f(x) = -(x + 1)^3$

The graph can be obtained by reflecting the graph of $f(x) = x^3$ about the x-axis and then translating it 1 unit to the left.

17. $f(x) = (x - 1)^4 + 2$

This graph has the same shape as $y = x^4$, but is translated 1 unit to the right and 2 units up.

19. (c) $f(x) = \dfrac{1}{x}$

does not define a polynomial function since polynomials never have variables in a denominator.

For Exercises 21–31, see the answer graphs in the back of the textbook.

21. $f(x) = 2x(x - 3)(x + 2)$

First set each of the three factors equal to 0 and solve the resulting equations to find the zeros of the function.

$2x = 0$ or $x - 3 = 0$ or $x + 2 = 0$
$x = 0$ or $x = 3$ or $x = -2$

The three zeros, 0, 3, and -2, divide the x-axis into four regions: $(-\infty, -2)$, $(-2, 0)$, $(0, 3)$ and $(3, \infty)$.

These intervals can be shown on a number line.

To find the sign of f(x) in each region, select a value of x in the region and determine by substitution whether the function values are positive or negative in that region.

Region	Test point	Value of f(x)	Sign of f(x)
$(-\infty, -2)$	-3	-36	Negative
$(-2, 0)$	-1	8	Positive
$(0, 3)$	1	-12	Negative
$(3, \infty)$	4	48	Positive

Sketch the graph.

23. $f(x) = x^2(x - 2)(x + 3)^2$

Set each factor equal to zero and solve the resulting equations to get the zeros 0, 2, and -3. The zeros divide the x-axis into four regions:

$(-\infty, -3)$, $(-3, 0)$, $(0, 2)$, and $(2, \infty)$.

Test a point in each region to find the sign of f(x) in that region.

Region	Test point	Value of f(x)	Sign of f(x)
$(-\infty, -3)$	-4	-96	Negative
$(-3, 0)$	-1	-12	Negative
$(0, 2)$	1	-16	Negative
$(2, \infty)$	3	324	Positive

Sketch the graph.

25. $f(x) = x^3 - x^2 - 2x$

$\qquad = x(x^2 - x - 2)$

$\qquad = x(x - 2)(x + 1)$

Set each factor equal to zero and solve the resulting equations to get the zeros 0, 2, and −1. The zeros divide the x−axis into four regions: $(-\infty, -1)$, $(-1, 0)$, $(0, 2)$, and $(2, \infty)$.

Test a point in each region to find the sign of $f(x)$ in that region.

Region	Test point	Value of $f(x)$	Sign of $f(x)$
$(-\infty, -1)$	−2	−8	Negative
$(-1, 0)$	$-\frac{1}{2}$	$\frac{5}{8}$	Positive
$(0, 2)$	1	−2	Negative
$(2, \infty)$	3	12	Positive

Sketch the graph.

27. $f(x) = (x + 2)(x - 1)(x + 1)$

Set each factor equal to zero and solve the resulting equations to get the zeros −2, 1, and −1. The zeros divide the x−axis into four regions: $(-\infty, -2)$, $(-2, -1)$, $(-1, 1)$ and $(1, \infty)$.

Test a point in each region to find the sign of $f(x)$ in that region.

Region	Test point	Value of $f(x)$	Sign of $f(x)$
$(-\infty, -2)$	−3	−8	Negative
$(-2, -1)$	$-\frac{3}{2}$	$\frac{5}{8}$	Positive
$(-1, 1)$	0	−2	Negative
$(1, \infty)$	2	12	Positive

Sketch the graph.

29. $f(x) = (3x - 1)(x + 2)^2$

Set each factor equal to zero and solve the resulting equations to get the zeros $\frac{1}{3}$ and −2. The zeros divide the x−axis into three regions:

$\qquad (-\infty, -2)$, $\left(-2, \frac{1}{3}\right)$, and $\left(\frac{1}{3}, \infty\right)$.

Test a point in each region to find the sign of $f(x)$ in that region.

Region	Test point	Value of $f(x)$	Sign of $f(x)$
$(-\infty, -2)$	−3	−10	Negative
$\left(-2, \frac{1}{3}\right)$	0	−4	Negative
$\left(\frac{1}{3}, \infty\right)$	1	18	Positive

Sketch the graph.

31. $f(x) = x^3 + 5x^2 - x - 5$

$\qquad = x^2(x + 5) - 1(x + 5)$

$\qquad = (x + 5)(x^2 - 1)$

$\qquad = (x + 5)(x + 1)(x - 1)$

Set each factor equal to zero, and solve the resulting equations to get the zeros −5, −1, and 1. The zeros divide the x−axis into four regions:

(−∞, −5), (−5, −1), (−1, 1), and (1, ∞).

Test a point in each region to find the sign of f(x) in that region.

Region	Test point	Value of f(x)	Sign of f(x)
(−∞, −5)	−6	−35	Negative
(−5, −1)	−2	9	Positive
(−1, 1)	0	−5	Negative
(1, ∞)	2	21	Positive

Sketch the graph.

33. $f(x) = 2x(x - 3)(x + 2)$

See the answer graph in the back of the textbook.

The graph from Exercise 21 is the same as this one.

35. $f(x) = (3x - 1)(x + 2)^2$

See the answer graph in the back of the textbook.

The graph from Exercise 29 is the same as this one.

37. $f(x) = 2x^2 - 7x + 4$; 2 and 3

Use synthetic division to find f(2) and f(3).

```
2| 2  -7   4
        4  -6
   ─────────────
   2  -3  -2
```

```
3| 2  -7   4
        6  -3
   ─────────────
   2  -1   1
```

Since $f(2) = -2$ is negative and $f(3) = 1$ is positive, there is a zero between 2 and 3.

39. $f(x) = 2x^3 - 5x^2 - 5x + 7$; 0 and 1

Use synthetic division to find f(0) and f(1).

```
0| 2  -5  -5   7
        0   0   0
   ──────────────────
   2  -5  -5   7
```

```
1| 2  -5  -5   7
        2  -3  -8
   ──────────────────
   2  -3  -8  -1
```

Since $f(0) = 7$ is positive and $f(1) = -1$ is negative, there is a zero between 0 and 1.

41. $f(x) = 2x^4 - 4x^2 + 4x - 8$; 1 and 2

Use synthetic division to find f(1) and f(2).

```
1| 2   0  -4   4  -8
        2   2  -2   2
   ──────────────────────
   2   2  -2   2  -6
```

```
2| 2   0  -4   4  -8
        4   8   8  24
   ──────────────────────
   2   4   4  12  16
```

Since $f(1) = -6$ is negative and $f(2) = 16$ is positive, there is a zero between 1 and 2.

43. $f(x) = 2x^2 - 7x + 4$

Enter f(x) as Y_1; then use the "solver" feature.

Solve $(Y_1, X, 2)$

                    2.7807764064

The real zero between 2 and 3 is 2.7807764064.

45. $f(x) = 2x^3 - 9x^2 + x + 20$

Enter $f(x)$ as Y_1; then use the "solver" feature.

solve $(Y_1, X, 2)$

                    2.19332495204

The real zero between 2 and 2.5 is 2.19332495204.

47. The graph shows that the zeros are -6, 2, and 5. The polynomial function has the form

$$f(x) = a(x + 6)(x - 2)(x - 5).$$

Since $(0, 30)$ is on the graph, $f(0) = 30$.

$$f(0) = a(0 + 6)(0 - 2)(0 - 5)$$
$$30 = 60a$$
$$\frac{1}{2} = a$$
$$a = .5$$

A cubic polynomial that has the graph shown is

$$f(x) = .5(x + 6)(x - 2)(x - 5).$$

49. $f(x) = 4x^3 - 3x^2 + 4x + 7$; no real zero greater than 1

Use the boundedness theorem. Divide synthetically by $x - 1$.

$$\begin{array}{r|rrrr}
1 & 4 & -3 & 4 & 7 \\
  &   & 4 & 1 & 5 \\
\hline
  & 4 & 1 & 5 & 12
\end{array}$$

Since $1 > 0$ and all numbers in the bottom row of the synthetic division are nonnegative, the boundedness theorem tells us that $f(x)$ has no real zero greater than 1.

51. $f(x) = x^4 + x^3 - x^2 + 3$; no real zero less than -2

Use the boundedness theorem. Divide synthetically by $x + 2$.

$$\begin{array}{r|rrrrr}
-2 & 1 & 1 & -1 & 0 & 3 \\
   &   & -2 & 2 & -2 & 4 \\
\hline
   & 1 & -1 & 1 & -2 & 7
\end{array}$$

Since $-2 < 0$ and the numbers in the bottom row of the synthetic division alternate in sign, the boundedness theorem tells us that $f(x)$ has no real zero less than -2.

53. $f(x) = x^3 + 3x^2 - 2x - 6$

The highest degree term is x^3, so the graph will have end behavior similar to the graph of $f(x) = x^3$, which is downward at the left and upward at the right. There is at least one real zero because the polynomial is of odd degree. There are at most three real zeros because the polynomial is third-degree. Using the fact that $f(0) = -6$, the end behavior, and the intermediate value theorem, we conclude that there must be either 0 or 2 negative zeros and 1 positive zero. We use synthetic division to search for these zeros, looking for places where $f(x) = 0$ or where the sign of $f(x)$ changes.

Find $f(1)$, $f(2)$, and so on by synthetic division.

Use the shortened form.

x				f(x)	
	1	3	-2	-6	
-3	1	0	-2	0	←Zero
-2	1	1	-4	2	
					←Zero
-1	1	2	-4	-2	
1	1	4	2	-4	
					←Zero
2	1	5	8	10	

f(x) has one zero between -1 and -2 and one zero between 1 and 2. To approximate these zeros to the nearest tenth, use synthetic division.

x				f(x)
	1	3	-2	-6
-1.5	1	1.5	-4.25	.375
-1.4	1	1.6	-4.24	-.064
1.4	1	4.4	4.16	-.176
1.5	1	4.5	4.75	1.125

Since f(-1.4) is nearer to zero than f(-1.5) is and f(1.4) is nearer to zero than f(1.5) is, the approximate zeros are -3.0, -1.4 and 1.4.

55. $f(x) = -2x^4 - x^2 + x + 5$

The highest degree term is $-x^4$, so the graph will have the same end behavior as the graph of $f(x) = -x^4$, which is downward at both the left and the right. Since f(0) = 5 > 0, the end behavior and the intermediate value theorem tell us that there must be at least one zero on each side of the y-axis, that is, at least one negative and one positive zero.

Search for approximate zeros using synthetic division.

x					f(x)	
	-2	0	-1	1	5	
-2	-2	4	-9	19	-33	
						←Zero
-1	-2	2	-3	4	1	
1	-2	-2	-3	-2	3	
						←Zero
2	-2	-4	-9	-17	-29	

f(x) has one zero between -1 and -2 and one zero between 1 and 2. To approximate these zeros to the nearest tenth, use synthetic division.

x					f(x)
	-2	0	1	1	5
-1.1	-2	2.2	-3.42	4.762	-.2382
-1	-2	2	-3	4	1
1.1	-2	-2.2	-3.42	-2.762	1.9618
1.2	-2	-2.4	-3.88	-3.656	.6128

Since f(-1.1) is nearer to zero than f(-1) is and f(1.2) is nearer to zero than f(1.2) is, the approximate zeros are -1.1 and 1.2.

For Exercises 59–63, see the answer graphs in the back of the textbook.

59. f(x) = x³ - 2x² - x + 1

Because the highest power term is x³, the left arrow points down and the right arrow points up.
Use synthetic division to locate the zeros.

x				f(x)	Ordered pair	
	1	-2	-1	1		
3	1	1	2	7	(3, 7)	←All positive
						←Zero
2	1	0	-1	-1	(2, -1)	
1	1	-1	-2	-1	(1, -1)	
						←Zero
0	1	-2	-1	1	(0, 1)	
						←Zero
-1	1	-3	2	-1	(-1, 1)	←Alternating signs

By the changes in sign of f(x), the polynomial has zeros between -1 and 0, 0 and 1, and 2 and 3. Each of these zeros is an x-intercept of the graph. Plot the points from the table and draw a continuous curve through them.

61. f(x) = -4x³ + 7x² - 2

Because the highest power term is -4x³, the left arrow points up and the right arrow points down.

Use synthetic division to locate the zeros.

x				f(x)	Ordered pair	
	-4	7	0	-2		
2	-4	-1	-2	-6	(2, -6)	
						←Zero
1	-4	3	3	1	(1, 1)	
						←Zero
0	-4	7	0	-2	(0, -2)	
						←Zero
-1	-4	11	-11	9	(-1, 9)	

By the changes in sign of f(x), the polynomial has zeros between 1 and 2, 0 and 1, and 0 and -1. Use the points from the table and draw a continuous curve through them.

63. f(x) = x⁴ - 5x² + 2

Because the highest power term is x⁴, both left and right arrows point up.
Use synthetic division to locate the zeros.

x					f(x)	Ordered pair	
	1	0	-5	0	2		
3	1	3	4	12	38	(3, 38)	
							←Zero
2	1	2	-1	-2	-2	(2, -2)	
1	1	1	-4	-4	-2	(1, -2)	
							←Zero
0	1	0	-5	0	2	(0, 2)	

By the changes in sign of f(x), the polynomial has zeros between 2 and 3, and 0 and 1. Because the graph is symmetric with respect to the y-axis, there are also zeros between −2 and −3, and 0 and −1. Use the points from the table to graph the function for x ≥ 0 and symmetry to graph the function for x < 0.

65.   **(a)** Using x = t, graph f(t) = $2.8 \times 10^{-4}t^3 - .011t^2 + .23t + .93$ and g(t) = 30 on the same calculator screen. See the answer graph in the back of the textbook.

(b) The graphs intersect at t ≈ 56.9. Since t = 0 corresponds to 1930, this would be during 1986.

(c) An increasing percentage of females have smoked during this time period. Smoking has been shown to increase the likelihood of lung cancer.

67.   g(x) = $-.006x^4 + .14x^3 - .05x^2 + .02x$

(a) See the answer graph in the back of the textbook.

(b) By using "maximum" in the CALC menu, we find that the greatest concentration is at 17.3 hours.

(c) Graph h(x) = 100 on the same screen as the graph of g(x). We want to determine the values of x for which g(x) > h(x).

Using "intersect" in the CALC menu, we find that the graphs intersect at x ≈ 11.4 and x ≈ 21.2. From the graph, we see that the graph of g(x) is above the graph of h(x) between the intersection points. Therefore, the river is polluted from 11.4 to 21.2 hours.

69.   **(a)** See the answer graph in the back of the textbook.

(b) From the graph, we see that all three functions approximate the data near 1986, but only the linear function

    (ii) f(x) = 1.088(x − 1986) + 8.6

approximates the data near 1994.

71.   f(x) = $.05\overline{3}x - .04\overline{3}$

(a) Let x = 10.

$$f(10) = .05\overline{3}(10) - .04\overline{3}$$
$$= .49$$

Assuming the function continues to model the situation, 49% of the patients will have AIDS after 10 years.

(b) Let f(x) = .5.

$$.5 = .05\overline{3}x - .043$$
$$x \approx 10.2$$

After approximately 10.2 years, half the patients will have AIDS.

73. **(a)** If the length of a pendulum increases, so does the period of oscillation T.

(b) There are a number of ways. One way is to realize that $k = L/T^n$ for some integer n. The ratio should be the constant k for each data point when the correct value of n is found.

(c) $k = \dfrac{L}{T^n}$ $\qquad k = \dfrac{L}{T^n}$

$k = \dfrac{1.0}{1.11^n}$ $\qquad k = \dfrac{2.0}{1.57^n}$

Try different values of n.
For n = 2,

$$k = \frac{1.0}{1.11^2} \approx .81$$

$$\text{and } k = \frac{2.0}{1.57^2} \approx .81.$$

When n = 2, k is constant.
$k \approx .81$

(d) $k \approx .81$, n = 2, L = 5

$$5 = .81T^2$$
$$6.1728 = T^2$$
$$T \approx 2.48$$

For a pendulum with length 5 ft, the value of T is 2.48 sec.

(e) If L = 2L, then

$$2L = kT^2$$
$$\sqrt{\frac{2L}{k}} = T.$$

T increases by a factor of $\sqrt{2} \approx 1.414$.

75. **(a)** Let x = the length;
20 − 2x = the width.

Both length and width must be positive, so

$$x > 0 \quad \text{and} \quad 20 - 2x > 0$$
$$-2x > -20$$
$$x < 10.$$

The restrictions on x are given by the inequality 0 < x < 10.

(b) $\qquad A(x) = x(20 - 2x)$
or $\quad A(x) = -2x^2 + 20x$

(c) The x-value of the vertex of the graph of this quadratic function is the maximum value of x.
Use the vertex formula with a = −2 and b = 2.

$$x = \frac{-b}{2a}$$
$$= \frac{-20}{2(-2)} = 5$$

The y-value of the vertex gives the maximum area.

$$A(x) = -2x^2 + 20x$$
$$A(5) = -2(5)^2 + 20(5)$$
$$= 50$$

The maximum cross-sectional area is 50 sq in.

(d) We must solve the quadratic inequality

$$-2x^2 + 20x < 40$$
$$\text{or } -2x^2 + 20x - 40 = 0.$$

Solve the corresponding quadratic equation

$$-2x^2 + 20x - 40 = 0$$
$$\text{or} \quad x^2 - 10x + 20 = 0.$$

Use the quadratic formula with $a = 1$, $b = -10$, and $c = 20$.

$$x = \frac{10 \pm \sqrt{10^2 - 4(20)}}{2}$$

$$\approx \frac{10 \pm 4.47}{2}$$

$$x \approx 2.76 \quad \text{or} \quad x \approx 7.24$$

The values 2.76 and 7.24 divide the number line into three intervals: $(-\infty, 2.76)$, $(2.76, 7.24)$, and $(7.24, \infty)$. However, in (a), we saw that in this problem x is restricted to $0 < x < 10$, that is, the open interval $(0, 10)$. Therefore, we need to consider the intervals $(0, 2.76)$, $(2.76, 7.24)$, and $(7.24, 10)$. Use a test point in each interval to determine where the expression $-2x^2 + 20x - 40$ is negative. We find that it is negative in the intervals $(0, 2.76)$ and $(7.24, 10)$. Therefore, the area of a cross section will be less than 40 square inches when x is between 0 and 2.76 or between 7.24 and 10.

77. Let x = the length of the hypotenuse.

(a) length of the leg = $x - 1$

(b) By the Pythagorean theorem,

$$a^2 + b^2 = c^2$$
$$a^2 + (x - 1)^2 = x^2$$
$$a^2 = x^2 - (x - 1)^2$$
$$a = \sqrt{x^2 - (x - 1)^2}$$

length of other leg = $\sqrt{x^2 - (x - 1)^2}$.

(c) $A = \frac{1}{2}bh$

$$84 = \frac{1}{2}(x - 1)(\sqrt{x^2 - (x - 1)^2})$$

Multiply by 2.

$$168 = (x - 1)(\sqrt{x^2 - (x - 1)^2})$$

Square both sides.

$$28,224 = (x - 1)^2[x^2 - (x - 1)^2]$$
$$28,224 = (x^2 - 2x + 1)$$
$$\qquad \cdot [x^2 - (x^2 - 2x + 1)]$$
$$28,224 = (x^2 - 2x + 1)(2x - 1)$$
$$28,224 = 2x^3 - 5x^2 + 4x - 1$$
$$2x^3 - 5x^2 + 4x - 28,225 = 0$$

(d) Solving this cubic equation graphically, we obtain x = 25. If x = 25,

$$x - 1 = 24,$$

and

$$\sqrt{x^2 - (x - 1)^2} = \sqrt{625 - 576}$$
$$= \sqrt{49} = 7.$$

The hypotenuse is 25 inches; the legs are 24 inches and 7 inches.

79. $y = 13,333.\overline{3}x^3 - 87,000x^2$
$\qquad + 122,666.\overline{6}x + 118,000$

See the answer graph in the back of the textbook.

(a) In 1991, x = 1.

$$y = 13,333.\overline{3}(1)^3 - 87,000(1)^2$$
$$\qquad + 122.666.\overline{6}(1) + 118,000$$
$$y = 167,000$$

In 1991, 167,000 students had bachelor's degrees.

(b) According to the bar graph, 122,000 students received bachelor's degrees in 1991. There is quite a large discrepancy.

Section 4.5

1. Graphs A, B, and C have a domain of $(-\infty, 3) \cup (3, \infty)$.

3. Graph A has a range of $(-\infty, 0) \cup (0, \infty)$.

5. Graph A has a single solution to the equation $f(x) = 3$.

7. Graphs A, C, and D have the x-axis as a horizontal asymptote.

For Exercises 9–13, see the answer graphs in the back of the textbook.

9. $f(x) = \dfrac{2}{x}$

Since $\dfrac{2}{x} = 2 \cdot \dfrac{1}{x}$, the graph of $f(x) = \dfrac{2}{x}$ will be similar to the graph of $f(x) = \dfrac{1}{x}$, except that each point will be twice as far from the x-axis. Just as with the graph of $f(x) = \dfrac{1}{x}$, $y = 0$ is the horizontal asymptote and $x = 0$ is the vertical asymptote. Sketch the graph.

11. $f(x) = \dfrac{1}{x + 2}$

Since $\dfrac{1}{x + 2} = \dfrac{1}{x - (-2)}$, the graph of $f(x) = \dfrac{1}{x + 2}$ will be similar to the graph of $f(x) = \dfrac{1}{x}$, except that each point will be translated 2 units to the left. Just as with $f(x) = \dfrac{1}{x}$, $y = 0$ is the horizontal asymptote, but this graph has $x = -2$ as its vertical asymptote. Sketch the graph.

13. $f(x) = \dfrac{1}{x} + 1$

The graph of this function will be similar to the graph of $f(x) = \dfrac{1}{x}$, except that each point will be translated 1 unit upward. Just as with $f(x) = \dfrac{1}{x}$, $x = 0$ is the vertical asymptote, but this graph has $y = 1$ as its horizontal asymptote. Sketch the graph.

15. $f(x) = \dfrac{3}{x - 5}$

To find the vertical asymptote, set the denominator equal to zero.

$$x - 5 = 0$$
$$x = 5$$

The equation of the vertical asymptote is

$$x = 5.$$

To find the horizontal asymptote, divide each term by the largest power of x in the expression.

$$f(x) = \frac{\dfrac{3}{x}}{\dfrac{x}{x} - \dfrac{5}{x}} = \frac{\dfrac{3}{x}}{1 - \dfrac{5}{x}}$$

As $|x| \to \infty$, $\frac{1}{x}$ approaches 0, f(x) approaches $\frac{0}{1-0} = 0$, so the equation of the horizontal asymptote is y = 0.

17. $f(x) = \dfrac{4 - 3x}{2x + 1}$

To find the vertical asymptote, set the denominator equal to zero.

$$2x + 1 = 0$$
$$x = -\frac{1}{2}$$

The equation of the vertical asymptote is

$$x = -\frac{1}{2}.$$

To find the horizontal asymptote, divide each term by the largest power of x in the expression.

$$f(x) = \frac{\dfrac{4}{x} - \dfrac{3x}{x}}{\dfrac{2x}{x} + \dfrac{1}{x}} = \frac{\dfrac{4}{x} - 3}{2 + \dfrac{1}{x}}$$

As $|x| \to \infty$, $\frac{1}{x}$ approaches 0, so f(x) approaches $\frac{0 - 3}{2 + 0} = -\frac{3}{2}$. The equation of the horizontal asymptote is

$$y = -\frac{3}{2}.$$

19. $f(x) = \dfrac{x^2 - 1}{x + 3}$

The vertical asymptote is x = -3, found by solving x + 3 = 0. Since the numerator is of degree exactly one more than the denominator, there is no horizontal asymptote, but there may be an oblique asymptote. To find it, divide the numerator by the denominator and disregard any remainder.

$$\begin{array}{r|rrr} -3 & 1 & 0 & -1 \\ & & -3 & 9 \\ \hline & 1 & -3 & 8 \end{array}$$

Thus,

$$f(x) = \frac{x^2 - 1}{x + 3} = x - 3 + \frac{8}{x + 3}.$$

The oblique asymptote is the line y = x - 3.

21. $f(x) = \dfrac{(x - 3)(x + 1)}{(x + 2)(2x - 5)}$

The vertical asymptotes are x = -2 and $x = \frac{5}{2}$, since these values make the denominator equal to 0. Multiply the factors in the numerator and denominator to get

$$f(x) = \frac{x^2 - 2x - 3}{2x^2 - x - 10}.$$

Thus, the horizontal asymptote, found by dividing the numerator and denominator by x^2, is $y = \frac{1}{2}$.

23. (a) $f(x) = \dfrac{1}{x^2 + 2}$

has a graph that does not have a vertical asymptote. This is because there is no real number that can make the denominator $x^2 + 2$ become 0.

25. (a) $f(x) = \dfrac{1}{(x - 2)^2}$

Notice that no matter what value x takes on $(x \neq 2)$, $(x - 2)^2$ will be greater than zero. That means $f(x) > 0$, which is graph C.

(b) $f(x) = \dfrac{1}{x - 2}$

Notice if $x > 2$, $f(x) > 0$. If $x < 2$, $x - 2 < 0$, and $f(x) < 0$. This is graph A.

(c) $f(x) = \dfrac{-1}{x - 2}$

Notice if $x > 2$, $f(x) < 0$. If $x < 2$, $f(x) > 0$. This is graph B.

(d) $f(x) = \dfrac{-1}{(x - 2)^2}$

Notice that no matter what value x takes on $(x \neq 2)$, $(x - 2)^2$ will be greater than zero. Thus $f(x) < 0$. This is graph D.

For Exercises 27–43, see the answer graphs in the textbook.

27. $f(x) = \dfrac{x + 1}{x - 4}$

The graph has a vertical asymptote where $x - 4 = 0$.

$$x - 4 = 0$$
$$x = 4$$

Since the degree of the numerator equals the degree of the denominator, the graph has a horizontal asymptote at $y = \dfrac{1}{1} = 1$.

The y-intercept is $f(0) = -1/4$. Any x-intercepts are found by solving $f(x) = 0$.

$$\dfrac{x + 1}{x - 4} = 0$$
$$x + 1 = 0$$
$$x = -1$$

The only x-intercept is -1. Find some additional points.

x	-3	-2	1	3	5	7
y	$\dfrac{2}{7}$	$\dfrac{1}{6}$	$-\dfrac{2}{3}$	-4	6	$\dfrac{8}{3}$

Use the asymptotes, intercepts, and these points to sketch the graph.

29. $f(x) = \dfrac{3x}{(x + 1)(x - 2)}$

The graph has vertical asymptotes when $(x + 1)(x - 2) = 0$, that is, when $x = -1$ and $x = 2$. Since the degree of the numerator is less than the degree of the denominator, the graph has a horizontal asymptote at $y = 0$ (the x-axis). The y-intercept is 0. The only x-intercept is also 0.

Find some additional points.

x	-3	-2	$-\dfrac{1}{2}$	$\dfrac{1}{2}$	1	3	4
y	$-\dfrac{9}{10}$	$-\dfrac{3}{2}$	$\dfrac{6}{5}$	$-\dfrac{2}{3}$	$-\dfrac{3}{2}$	$\dfrac{9}{4}$	$\dfrac{6}{5}$

Sketch the graph.

31. $f(x) = \dfrac{5x}{x^2 - 1}$

The graph has vertical asymptotes where $x^2 - 1 = 0$.

$$x^2 - 1 = 0$$
$$(x + 1)(x - 1) = 0$$
$$x + 1 = 0 \quad \text{or} \quad x - 1 = 0$$
$$x = -1 \quad \text{or} \quad x = 1$$

Since the degree of the numerator is less than the degree of the denominator, the graph has a horizontal asymptote at $y = 0$ (the x-axis). The y-intercept is 0. The only x-intercept is also 0.
Find some additional points.

y	-3	-2	$-\dfrac{1}{2}$	$\dfrac{1}{2}$	2	3
x	$-\dfrac{15}{8}$	$-\dfrac{10}{3}$	$\dfrac{10}{3}$	$-\dfrac{10}{3}$	$\dfrac{10}{3}$	$\dfrac{15}{8}$

Sketch the graph.

33. $f(x) = \dfrac{(x - 3)(x + 1)}{(x - 1)^2}$

The graph has a vertical asymptote where $(x - 1)^2 = 0$.

$$(x - 1)^2 = 0$$
$$x - 1 = 0$$
$$x = 1$$

Since the degree of the numerator is the same as the degree of the denominator $\left(f(x) = \dfrac{x^2 - 2x - 3}{x^2 - 2x + 1}\right)$, the graph has a horizontal asymptote at $y = \dfrac{1}{1} = 1$.

$$f(0) = \frac{(0 - 3)(0 + 1)}{(0 - 1)^2} = \frac{-3}{1} = -3,$$

so the y-intercept is -3.

The numerator $(x - 3)(x + 1) = 0$ when $x = 3$ or $x = -1$, so the x-intercepts are 3 and -1.
Using the asymptotes, the intercepts, and a few additional points, we sketch the graph.

35. $f(x) = \dfrac{x}{x^2 - 9}$

The graph has a vertical asymptote where $x^2 - 9 = 0$.

$$x^2 - 9 = 0$$
$$(x + 3)(x - 3) = 0$$
$$x + 3 = 0 \quad \text{or} \quad x - 3 = 0$$
$$x = -3 \quad \text{or} \quad x = 3$$

Since the degree of the numerator is less than the degree of the denominator, the graph has a horizontal asymptote at $y = 0$ (the x-axis). The y-intercept is 0 and the only x-intercept is also zero. Use the asymptotes, the intercepts, and some additional points to sketch the graph.

37. $f(x) = \dfrac{1}{x^2 + 1}$

Since $x^2 + 1$ has no real zeros, there are no vertical asymptotes. Since $f(x) = f(-x)$, the graph is symmetric with respect to the y-axis. Note that $f(x) > 0$ for all values of x, so the graph is entirely above the x-axis. Since the numerator has lower degree than the denominator, $y = 0$ is the horizontal asymptote.

$$f(0) = \frac{1}{0^2 + 1} = 1,$$

so the y-intercept is 1.
There are no x-intercepts.

x	-2	-1	0	1	2
y	$\frac{1}{5}$	$\frac{1}{2}$	1	$\frac{1}{2}$	$\frac{1}{5}$

Sketch the graph.

39. $f(x) = \dfrac{x^2 + 1}{x + 3}$

The vertical asymptote is x = -3.
Since the numerator is of degree
exactly one more than the denominator, there is no horizontal
asymptote, but there may be an
oblique asymptote.
Use synthetic division.

$$\begin{array}{r|rrr} -3 & 1 & 0 & 1 \\ & & -3 & 9 \\ \hline & 1 & -3 & 10 \end{array}$$

$$f(x) = \frac{x^2 + 1}{x + 3} = x - 3 + \frac{10}{x + 3}$$

The oblique asymptote is y = x - 3.

$$f(0) = \frac{0^2 + 1}{0 + 3} = \frac{1}{3},$$

so the y-intercept is $\frac{1}{3}$.

Note that f(x) can never be zero, so
the graph has no x-intercepts.

x	-4	-5	-2	-1	0	1	3
y	-17	-13	5	1	$\frac{1}{3}$	$\frac{1}{2}$	$\frac{5}{3}$

Sketch the graph.

41. $f(x) = \dfrac{x^2 + 2x}{2x - 1}$

f(x) has a vertical asymptote where
2x - 1 = 0.

$$2x - 1 = 0$$
$$2x = 1$$
$$x = \frac{1}{2}$$

Since the degree of the numerator
is one more than the degree of the
denominator, f(x) has an oblique
asymptote. Divide $x^2 + 2x$ by
2x - 1.

$$\begin{array}{r} \frac{1}{2}x + \frac{5}{4} \\ 2x - 1 \overline{\smash{)}x^2 + 2x } \\ x^2 - \frac{1}{2}x \\ \hline \frac{5}{2}x \\ \frac{5}{2}x - \frac{5}{4} \\ \hline \frac{5}{4} \end{array}$$

$$f(x) = \frac{x^2 + 2x}{2x - 1} = \frac{1}{2}x + \frac{5}{4} + \frac{\frac{5}{4}}{2x - 1}$$

The oblique asymptote is

$$y = \frac{1}{2}x + \frac{5}{4}.$$

If x = 0, f(x) = 0, so the
y-intercept is 0.
The numerator, $x^2 + 2x$, is equal to
0 when x = 0 or x = -2. Thus, the
graph has two x-intercepts, 0 and
-2.
Use asymptotes, the intercepts, and
a few additional points to sketch
the graph.

43. $f(x) = \dfrac{x^2 - 9}{x + 3}$

Since $x^2 - 9 = (x + 3)(x - 3)$,

$f(x) = \dfrac{(x + 3)(x - 3)}{x + 3}$

$\quad\;\; = x - 3 \quad (x \neq -3).$

The graph of this function will be the same as the graph of $y = x - 3$ (a straight line), with the exception of the point with x-value -3. A "hole" appears in the graph at $(-3, -6)$.

45. The graph has a vertical asymptote, $x = 2$, so $x - 2$ is in the denominator of the function. There is a "hole" in the graph at $x = -2$, so $x + 2$ is in the denominator and numerator also. The x-intercept is 3, so that when $f(x) = 0$, $x = 3$. This condition exists if $x - 3$ is a factor of the numerator.

Putting these conditions together, we have a possible function

$\qquad f(x) = \dfrac{(x - 3)(x + 2)}{(x - 2)(x + 2)}$

or $f(x) = \dfrac{x^2 - x - 6}{x^2 - 4}$.

47. The graph has vertical asymptotes at $x = 4$ and $x = 0$, so $x - 4$ and x are factors in the denominator of the function. The only x-intercept is 2, so that when $f(x) = 0$, $x = 2$. This condition exists if $x - 2$ is a factor of the numerator.

Putting these conditions together, we have a possible function

$f(x) = \dfrac{x - 2}{x(x - 4)}$ or $f(x) = \dfrac{x - 2}{x^2 - 4x}$.

49. $f(x) = \dfrac{x + 1}{x - 4}$

See the answer graph in the back of the textbook.
This graph resembles the hand-drawn graph for the same function in Exercise 27, but the calculator graph does not show the asymptotes.

51. $f(x) = \dfrac{(x - 3)(x + 1)}{(x - 1)^2}$

See the answer graph in the back of the textbook.
This graph resembles the hand-drawn graph for the same function in Exercise 33, but the calculator graph does not show the asymptotes.

55. **(a)** The vertical asymptote of $P(x) = \dfrac{x - 1}{x}$ is found by dividing each term by the largest power of x.

$\qquad P(x) = \dfrac{\dfrac{x}{x} - \dfrac{1}{x}}{\dfrac{x}{x}} = \dfrac{1 - \dfrac{1}{x}}{1}$

As $|x| \to \infty$, x approaches 0, so $P(x)$ approaches $\dfrac{1 - 0}{1} = 1$.

As x increases, the value of $P(x)$ approaches 1.

57. **(a)** Graph

$$y = d(x) = \frac{(8.71 \times 10^3)x^2 - (6.94 \times 10^4)x + (4.70 \times 10^5)}{(1.08)x^2 - (3.24 \times 10^2)x + (8.22 \times 10^4)}$$

and y = 300 on the same calculator screen.

See the answer graph in the back of the textbook.

The graphs intersect when x ≈ 52.1 miles per hour.

(b)

x	d(x)
20	34
25	56
30	85
35	121
40	164
45	215
50	273
55	340
60	415
65	499
70	591

(c) From the table, we can see that if the speed doubles, the stopping distance more than doubles. For example, d(35) ≈ 121 whereas d(70) ≈ 591. Generally, when the speed doubles, the stopping distance increases by a factor of 4.8–4.9 times.

(d) If the stopping distance doubled when the speed doubled, there would be a linear relationship between speed and distance. The graph would be linear and not curved like the graph of d.

59. The function and its rational approximation are graphed on the same calculator screen. From the graphs we can see that all of the rational approximations give excellent results on the interval [1, 10]. For (a)–(d), see the answer graphs in the back of the textbook.

(a) The rational approximation of

(a) $f_1(x) = \sqrt{x}$ is

(iii) $r_3(x) = \dfrac{10x^2 + 80x + 32}{x^2 + 40x + 80}$.

(b) The rational approximation of

(b) $f_2(x) = \sqrt{4x + 1}$ is

(ii) $r_2(x) = \dfrac{15x^2 + 75x + 33}{x^2 + 23x + 31}$.

(c) The rational approximation of

(c) $f_3(x) = \sqrt[3]{x}$ is

(iv) $r_4(x) = \dfrac{7x^3 + 42x^2 + 30x + 2}{2x^3 + 30x^2 + 42x + 7}$.

(d) The rational approximation of

(d) $f_4(x) = \dfrac{1 - \sqrt{x}}{1 + \sqrt{x}}$ is

(i) $r_1(x) = \dfrac{2 - 2x^2}{3x^2 + 10x + 3}$.

Chapter 4   Review Exercises

For Exercises 1 and 3, see the answer graphs in the back of the textbook.

1. $f(x) = 3(x + 4)^2 - 5$

The function has the form

$$f(x) = a(x - h)^2 + k$$

with $a = 3$, $h = -4$, and $k = -5$.
The graph is a parabola that opens upward.
Vertex: $(h, k) = (-4, -5)$
Axis: $x = -4$
Find the x-intercepts by letting $f(x) = 0$.

$$3(x + 4)^2 - 5 = 0$$
$$3(x + 4)^2 = 5$$
$$(x + 4)^2 = \frac{5}{3}$$
$$x + 4 = \pm\sqrt{\frac{5}{3}}$$
$$x = -4 \pm \sqrt{\frac{5}{3}} = -4 \pm \frac{\sqrt{15}}{3}$$
$$= \frac{-12 \pm \sqrt{15}}{3}$$

x-intercepts: $\dfrac{-12 \pm \sqrt{15}}{3}$

Find the y-intercept by letting $x = 0$.

$$f(0) = 3(0 + 4)^2 - 5$$
$$= 3(16) - 5$$
$$= 43$$

y-intercept: 43

3. $f(x) = -3x^2 - 12x - 1$

Complete the square to find the vertex.

$$f(x) = -3(x^2 + 4x \quad) - 1$$
$$= -3(x^2 + 4x + 4) + 12 - 1$$
$$= -3(x + 2)^2 + 11$$

The graph is a parabola that opens downward.
Vertex: $(h, k) = (-2, 11)$
Axis: $x = -2$
Find the x-intercepts by letting $f(x) = 0$.

$$-3(x + 2)^2 + 11 = 0$$
$$-3(x + 2)^2 = -11$$
$$(x + 2)^2 = \frac{11}{3}$$
$$x + 2 = \pm\sqrt{\frac{11}{3}}$$

$$x = -2 \pm \sqrt{\frac{11}{3}}$$

$$= -2 \pm \frac{\sqrt{33}}{3}$$

$$= -\frac{6 \pm \sqrt{33}}{3}$$

x–intercepts: $-\dfrac{6 \pm \sqrt{33}}{3}$

$f(0) = -3(0)^2 - 12(0) - 1 = -1$

Thus, the y–intercept is –1.

5. $f(x) = a(x - h)^2 + k;\ a > 0$

The graph is a parabola that opens upward. The y–coordinate of the lowest point of the graph is the y–coordinate of the vertex, k.

7. $f(x) = a(x - h)^2 + k,\ a > 0$

Find the y–intercept by letting $x = 0$.

$$f(0) = a(0 - h)^2 + k$$
$$= ah^2 + k$$

The y–intercept is $ah^2 + k$.

9. If a is positive, the graph of $y = ax^2 + bx + c$ is a parabola that opens upward, so the y–coordinate of the lowest point of the graph is the y–coordinate of the vertex, k, in the equation $y = a(x - h)^2 + k$. Complete the square to find the vertex.

$$y = ax^2 + bx + c$$
$$= a\left(x^2 + \frac{b}{a}x \quad \right) + c$$
$$= a\left(x^2 + \frac{b}{a}x + \frac{b^2}{4a^2}\right) + c - \frac{b^2}{4a}$$
$$y = a\left(x + \frac{b}{2a}\right)^2 + \left(c - \frac{b^2}{4a}\right)$$

The smallest value is $c - \dfrac{b^2}{4a}$.

11. **(a)** See the answer graph in the back of the textbook.

(b)

$$G(x) = 15 + 24x - 2x^2$$
$$= -2x^2 + 24x + 15$$
$$= -2(x^2 - 12x \quad) + 15$$
$$= -2(x^2 - 12x + 36) + 72 + 15$$
$$= -2(x - 6)^2 + 87$$

Vertex: $(h, k) = (6, 87)$

$x = 6$ represents June 6, so the maximum pollen measurement is 87 particles on June 6.

13. $f(x) = -2.64x^2 + 5.47x + 3.54$

The discriminant is $b^2 - 4ac$ in the standard quadratic equation $y = ax^2 + bx + c$.

$a = -2.64,\ b = 5.47,\ c = 3.54$

$$b^2 - 4ac = (5.47)^2 - 4(-2.64)(3.54)$$
$$= 67.3033$$

Because the discriminant is 67.3033, a positive number, there are two x–intercepts.

15. **(a)** $f(x) > 0$ on the open interval $(-.52, 2.59)$.

(b) $f(x) < 0$ on interval $(-\infty, -.52) \cup (2.59, \infty)$.

17. $f(x) = -2.64x^2 + 5.47x + 3.54$

Complete the square to find the vertex. (The vertex formula may also be used.)

$$f(x) = -2.64(x^2 - 2.07x\ \ \ \ ) + 3.54$$
$$= -2.64(x^2 - 2.07x + 1.07)$$
$$+ 2.83 + 3.54$$
$$= -2.64(x - 1.04)^2 + 6.37$$

The coordinates of the vertex are (1.04, 6.37).

19. $\dfrac{3x^3 + 8x^2 + 5x + 10}{x + 2}$

$$\begin{array}{r|rrrr} -2 & 3 & 8 & 5 & 10 \\ & & -6 & -4 & -2 \\ \hline & 3 & 2 & 1 & 8 \end{array}$$

The synthetic division shows that $q(x) = 3x^2 + 2x + 1$ and $r = 8$.

21. $f(x) = 2x^3 - 3x^2 + 7x - 12$; find $f(2)$.

$$\begin{array}{r|rrrr} 2 & 2 & -3 & 7 & -12 \\ & & 4 & 2 & 18 \\ \hline & 2 & 1 & 9 & 6 \end{array}$$

The synthetic division shows that $f(2) = 6$.

23. $f(x) = x^5 + 4x^2 - 2x - 4$; find $f(2)$.

$$\begin{array}{r|rrrrrr} 2 & 1 & 0 & 0 & 4 & -2 & -4 \\ & & 2 & 4 & 8 & 24 & 44 \\ \hline & 1 & 2 & 4 & 12 & 22 & 40 \end{array}$$

The synthetic division shows that $f(2) = 40$.

25. Zeros are -1, 4, and 7.

$$f(x) = (x + 1)(x - 4)(x - 7)$$
$$= (x + 1)(x^2 - 11x + 28)$$
$$= x^3 - 10x^2 + 17x + 28$$

27. Zeros are $\sqrt{3}$, $-\sqrt{3}$, 2, 3.

$$f(x) = (x - \sqrt{3})(x + \sqrt{3})(x - 2)(x - 3)$$
$$= (x^2 - 3)(x^2 - 5x + 6)$$
$$= x^4 - 5x^3 + 3x^2 + 15x - 18$$

29. Is -1 a zero of $f(x) = 2x^4 + x^3 - 4x^2 + 3x + 1$?

$$\begin{array}{r|rrrrr} -1 & 2 & 1 & -4 & 3 & 1 \\ & & -2 & 1 & 3 & -6 \\ \hline & 2 & -1 & -3 & 6 & -5 \end{array}$$

Since $f(-1) = -5 \neq 0$, -1 is not a zero of $f(x)$.

31. Find a polynomial function with real coefficients of degree 4 with 3, 1, and $-1 - 3i$ as zeros, and $f(2) = -36$.

Since $-1 - 3i$ is a zero, by the conjugate zeros theorem, $-1 + 3i$ is also a zero.

$$f(x) = a[x - (-1 - 3i)][x - (-1 + 3i)]$$
$$\cdot (x - 3)(x - 1)$$
$$= a[(x + 1) + 3i][(x + 1) - 3i]$$
$$\cdot (x - 3)(x - 1)$$
$$= a[(x + 1)^2 - (3i)^2](x^2 - 4x + 3)$$
$$= a(x^2 + 2x + 1 - 9i^2)$$
$$\cdot (x^2 - 4x + 3)$$
$$= a(x^2 + 2x + 10)(x^2 - 4x + 3)$$
$$f(x) = a(x^4 - 2x^3 + 5x^2 - 34x + 30)$$
$$f(2) = -36$$
$$= a(2^4 - 2 \cdot 2^3 + 5 \cdot 2^2 - 34 \cdot 2 + 30)$$
$$-36 = a(-18)$$
$$2 = a$$

The polynomial function is

$$f(x) = 2(x^4 - 2x^3 + 5x^2 - 34x + 30)$$
$$= 2x^4 - 4x^3 + 10x^2 - 68x + 60.$$

33. A fourth-degree polynomial function having exactly two distinct real zeros is any function such that the polynomial can be factored as

$$a(x - b)^2(x - c)^2,$$

where a, b, and c are real numbers. One example is

$$f(x) = 2(x - 1)^2(x - 3)^2.$$

See the answer graph in the back of the textbook.

35. $f(x) = x^4 - 3x^3 - 8x^2 + 22x - 24$; $1 - i$ is a zero.

Since $1 - i$ is a zero, $1 + i$ is also a zero.

$$[x - (1 - i)][x - (1 + i)]$$
$$= [(x - 1) + i][(x - 1) - i]$$
$$= (x - 1)^2 - i^2$$
$$= x^2 - 2x + 1 + 1$$
$$= x^2 - 2x + 2$$

Find $\dfrac{x^4 - 3x^3 - 8x^2 + 22x - 24}{x^2 - 2x + 2}$ by long division.
The quotient is $x^2 - x - 12 = (x - 4)(x + 3)$.
Since

$$f(x) = [x - (1 - i)][x - (1 + i)]$$
$$\cdot (x - 4)(x + 3),$$

all the zeros are $1 - i$, $1 + i$, 4, and -3.

37. $x - 4$ is a factor of $f(x) = x^3 - 2x^2 + sx + 4$ if 4 is a zero.

$$\begin{array}{r|rrrr} 4 & 1 & -2 & s & 4 \\ & & 4 & 8 & 4s + 32 \\ \hline & 1 & 2 & s + 8 & 4s + 36 \end{array}$$

$$4s + 36 = 0$$
$$4s = -36$$
$$s = -9$$

39. $\text{Volume}_{cube} = x^3$

The volume of the solid with the top sliced off is

$$V = x \cdot x \cdot (x - 2).$$

Since the volume is 32 cubic inches, we have the equation

$$32 = x^3 - 2x^2$$
$$0 = x^3 - 2x^2 - 32.$$

Graph $Y_1 = x^3 - 2x^2 - 32$ on a graphing calculator. The graph will show that 4 is the only real zero. Since the only real solution is 4, the dimensions of the original cube are 4 inches $\times$ 4 inches $\times$ 4 inches.

41. Show $f(x) = 3x^3 - 8x^2 + x + 2$ has zeros in $[-1, 0]$ and $[2, 3]$.

x				f(x)
	3	-8	1	2
-1	3	-11	12	-10
0	3	-8	1	2
2	3	-2	-3	-4
3	3	1	4	14

$f(-1) = -10$ and $f(0) = 2$. Since $f(-1) < 0$ and $f(0) > 0$, there is a zero between -1 and 0. $f(2) = -4 < 0$ and $f(3) = 14 > 0$, so there is a zero between 2 and 3.

45. $f(x) = x^4 - 4x^3 - 5x^2 + 14x - 15$

Graph this function in the window $[-10, 10]$ by $[-60, 60]$.
See the answer graph in the back of the textbook.
Using the calculator, we find that the real zeros are 4.58039972384 and -2.25883787095.

For Exercises 47–55, see the answer graphs in the back of the textbook.

47. $f(x) = x^4 - 3x^2 + 2$

$\quad\quad = (x^2 - 2)(x^2 - 1)$

$f(x) = (x + \sqrt{2})(x - \sqrt{2})(x + 1)(x - 1)$

Set each factor equal to zero and solve resulting equations to get the zeros $-\sqrt{2}$ (≈ -1.4), -1, 1, and $\sqrt{2}$ (≈ 1.4). The zeros divide the x-axis into five regions:

$(-\infty, -\sqrt{2})$, $(-\sqrt{2}, -1)$, $(-1, 1)$, $(1, \sqrt{2})$, and $(\sqrt{2}, \infty)$.

Test a point in each region to find the sign of $f(x)$ in that region.

Region	Test point	Value of $f(x)$	Sign of $f(x)$
$(-\infty, -\sqrt{2})$	-2	6	Positive
$(-\sqrt{2}, -1)$	-1.2	$-.25$	Negative
$(-1, 1)$	0	2	Positive
$(1, \sqrt{2})$	1.2	$-.25$	Negative
$(\sqrt{2}, \infty)$	2	6	Positive

Use the zeros and points from the table to sketch the graph. Note that the graph is symmetric with respect to the y-axis.

49. $f(x) = \dfrac{4}{x - 1}$

The vertical asymptote is $x = 1$.
The horizontal asymptote is $y = 0$.
The y-intercept is -4. There are no x-intercepts. Make a table of values.

x	-3	-1	0	2	3	5
y	-1	-2	-4	4	2	1

Use the asymptotes and these points to sketch the graph.

51. $f(x) = \dfrac{6x}{(x - 1)(x + 2)}$

The vertical asymptotes are $x = 1$ and $x = -2$. The horizontal asymptote is $y = 0$.
The y-intercept is 0. The only x-intercept is also 0. Use the asymptotes, the intercepts, and some additional points to sketch the graph.

53. $f(x) = \dfrac{x^2 + 4}{x + 2}$

The vertical asymptote is $x = -2$.
Since the degree of the numerator is exactly one more than the degree of the denominator, use synthetic division to find the oblique asymptote.

$$\begin{array}{r|rrr} -2 & 1 & 0 & 4 \\ & & -2 & 4 \\ \hline & 1 & -2 & 8 \end{array}$$

$$f(x) = \frac{x^2 + 4}{x + 2} = x - 2 + \frac{8}{x + 2}$$

The oblique asymptote is $y = x - 2$.
The y-intercept is 2. Since
$f(x) = 0$ has no real solutions, the
graph has no x-intercepts.
Use the asymptotes, the y-intercept,
and some additional points to sketch
the graph.

55. $f(x) = \frac{-2}{x^2 + 1}$

Since the denominator never becomes
0, there are no vertical asymptotes.
The horizontal asymptote is $y = 0$.
Since $f(x) = 0$ has no real solutions, the graph has no x-intercepts. When $x = 0$, $f(x) = -2$, so
the y-intercept is -2.
Notice that the value of y is always
negative and that the graph is symmetric with respect to the y-axis.

57. $C(x) = \frac{10x}{49(101 - x)}$

(a) Graph $C(x)$ in the window
[0, 101] by [0, 10]. See the answer
graph in the back of the textbook.

(b) To find the cost of removing 95%
of the pollutant, find $C(95)$.

$$C(95) = \frac{10(95)}{49(101 - 95)}$$
$$\approx 3.23$$

To earn 95 points, an owner would
expect to pay approximately 3.23
thousand dollars.

59. **(a)** See the answer graph in the back
of the textbook.

(b) The graph has a vertical asymptote $x = 3$, so $x - 3$ is in the
denominator of the function. The
x-intercepts are 2 and 4, so that
when $f(x) = 0$, $x = 2$ or $x = 4$.
This would exist if $x - 2$ and $x - 4$
were factors of the numerator.
The horizontal asymptote is $y = 1$,
so the numerator and denominator
have the same degree. Since the
numerator will have degree 2, we
must make the denominator also have
degree 2.
Putting these conditions together,
we have a possible function

$$f(x) = \frac{(x - 2)(x - 4)}{(x - 3)^2}.$$

Chapter 4   Test

1. $f(x) = -2x^2 + 6x - 3$
$$= -2(x^2 - 3x \quad) - 3$$
$$= -2\left(x^2 - 3x + \frac{9}{4}\right) + \frac{9}{2} - 3$$
$$= -2\left(x - \frac{3}{2}\right)^2 + \frac{3}{2}$$

Vertex: $(h, k) = \left(\frac{3}{2}, \frac{3}{2}\right)$

Axis: $x = \frac{3}{2}$

To find the x-intercepts, let
$f(x) = 0$.

$$0 = -2\left(x - \frac{3}{2}\right)^2 + \frac{3}{2}$$

$$-\frac{3}{2} = -2\left(x - \frac{3}{2}\right)^2$$

$$\frac{3}{4} = \left(x - \frac{3}{2}\right)^2$$

$$\pm\sqrt{\frac{3}{4}} = x - \frac{3}{2}$$

$$x = \frac{3}{2} \pm \sqrt{\frac{3}{4}} = \frac{3}{2} \pm \frac{\sqrt{3}}{2}$$

x-intercepts: $\dfrac{3 \pm \sqrt{3}}{2}$

To find the y-intercept, let x = 0.

$$f(0) = -2(0)^2 + 6(0) - 3$$

$$= -3$$

y-intercept: −3

Domain: $(-\infty, \infty)$

Range: $\left(-\infty, \frac{3}{2}\right]$

See the answer graph in the back of the textbook.

2. $f(x) = 5334.6x^2 - 23.040x + 617,519.1$

(a) In 1986, x = 1.

$$f(1) = 5334.6(1)^2 - 23,040(1)$$
$$+ 617,519.1$$
$$= 599,814$$

(b) Complete the square to find the vertex, which is the maximum point.

$$f(x) = 5334.6x^2 - 23.040x$$
$$+ 617,519.1$$
$$= 5334.6(x^2 - 4.31897x)$$
$$+ 617,519.1$$
$$= 5334.6(x^2 - 4.31897x + 4.66338)$$
$$- 24877.3 + 617,519.1$$
$$= 5334.6(x - 2)^2 + 592,642$$

The minimum number of degrees occurred when x = 2 in 1987. The minimum value is 592,642.

3. $\dfrac{3x^3 + 4x^2 - 9x + 6}{x + 2}$

$$
\begin{array}{r|rrrr}
-2 & 3 & 4 & -9 & 6 \\
   &   & -6 & 4 & 10 \\
\hline
   & 3 & -2 & -5 & 16
\end{array}
$$

$q(x) = 3x^2 - 2x - 5;\ r = 16$

4. $\dfrac{2x^3 - 11x^2 + 28}{x - 5}$

$$
\begin{array}{r|rrrr}
5 & 2 & -11 & 0 & 28 \\
  &   & 10 & -5 & -25 \\
\hline
  & 2 & -1 & -5 & 3
\end{array}
$$

$q(x) = 2x^2 - x - 5;\ r = 3$

5. $f(x) = 2x^3 - 9x^2 + 4x + 8;\ k = 5$

$$
\begin{array}{r|rrrr}
5 & 2 & -9 & 4 & 8 \\
  &   & 10 & 5 & 45 \\
\hline
  & 2 & 1 & 9 & 53
\end{array}
$$

$f(5) = 53$

6. $6x^4 - 11x^3 - 35x^2 + 34x + 24;\ x - 3$

Let

$f(x) = 6x^4 - 11x^3 - 35x^2 + 34x + 24.$

By the factor theorem, x − 3 will be a factor of f(x) only if f(3) = 0.

$$
\begin{array}{r|rrrrr}
3 & 6 & -11 & -35 & 34 & 24 \\
  &   & 18 & 21 & -42 & -24 \\
\hline
  & 6 & 7 & -14 & -8 & 0
\end{array}
$$

Since f(3) = 0, x − 3 is a factor of f(x). The other factor is

$6x^3 + 7x^2 - 14x - 8.$

7. Find all zeros of f(x), given that

   f(x) = 2x³ − x² − 13x − 6, and −2 is

   a zero.

   Since −2 is a zero, first divide

   f(x) by x + 2.

   ```
   -2| 2  -1  -13  -6
      |    -4   10   6
      ─────────────────
        2  -5   -3   0
   ```

   This gives

   2x² − 5x − 3 = (2x + 1)(x − 3).

   The zeros of f(x) are −2, −1/2, and

   3.

8. Zeros of −1, 2 and i; f(3) = 80

   By the conjugate zeros theorem, −1

   is also a zero.

   The polynomial has the form

   f(x) = a(x + 1)(x − 2)(x − i)(x + i).

   Use the condition f(3) = 80 to find

   a.

   80 = a(3 + 1)(3 − 2)(3 − i)(3 + i)

   80 = a(4)(1)(10)

   80 = 40a

   a = 2

   Thus,

   f(x) = 2(x + 1)(x − 2)(x − i)(x + i)

   = 2(x² − x − 2)(x² + 1)

   = 2(x⁴ − x³ − x² − x − 2)

   = 2x⁴ − 2x³ − 2x² − 2x − 4.

10. **(a)** f(x) = x³ − 5x² + 2x + 7; 1 and 2

    ```
    1| 1  -5   2   7
     |     1  -4  -2
     ─────────────────
       1  -4  -2   5
    ```

    ```
    2| 1  -5   2   7
     |     2  -6  -8
     ─────────────────
       1  -3  -4  -1
    ```

By the intermediate value theorem,

since f(1) = 5 > 0 and f(2) =

−1 < 0, there must be at least one

real zero between 1 and 2.

(b) The real zeros of

   f(x) = x³ − 5x² + 2x + 7

are 4.09376345695, 1.83703814322,

and −.930801600173.

11. See the answer graph in the back of

    the textbook.

    To obtain the graph of f₂, shift the

    graph of f₁ 5 units to the left,

    stretch by a factor of 2, reflect

    across the x−axis, and shift 3 units

    up.

12. f(x) = −x⁷ + x − 4

    Since f(x) is of odd degree and the

    sign of aₙ is negative, the left

    arrow points up and the right arrow

    points down.

    The correct graph is C.

For Exercises 13 and 14, see the answer
graphs in the back of the textbook.

13. f(x) = (3 − x)(x + 2)(x + 5)

    To find the zeros, set each factor

    equal to 0 and solve the resulting

    equations. We obtain the zeros −5,

    −2, and 3. The zeros divide the

    x−axis into four regions: (−∞, −5),

    (−5, −2), (−2, 3), and (3, 0). Test

    a point in each region to find the

    sign of f(x) in that region.

Region	Test point	Value of $f(x)$	Sign of $f(x)$
$(-\infty, -5)$	-6	36	Positive
$(-5, -2)$	-3	-12	Negative
$(-2, 3)$	0	30	Positive
$(3, 0)$	4	-54	Negative

Use the zeros and the points from the table to sketch the graph.

14. $f(x) = 2x^4 - 8x^3 + 8x^2$
$\qquad\quad = 2x^2(x^2 - 4x + 4)$
$f(x) = 2x^2(x - 2)^2$

By setting each factor equal to zero and solving the resulting equations, we see that the zeros are 0 and 2 (each of multiplicity two). The zeros divide the x-axis into three regions: $(-\infty, 0)$, $(0, 2)$, and $(2, \infty)$. Test a point in each region.

Region	Test point	Value of $f(x)$	Sign of $f(x)$
$(-\infty, 0)$	-1	18	Positive
$(0, 2)$	1	2	Positive
$(2, \infty)$	3	18	Positive

(We can also see from the equation that $f(x)$ can never be negative.) Use the zeros and the points from the table to sketch the graph.

15. The zeros are -3 and 2; $f(0) = 24$

Since the left arrow points up and the right arrow points down, the polynomial is of odd degree.

$f(x) = a(x - 2)^2(x + 3)$
$24 = a(0 - 2)^2(0 + 3)$
$24 = 12a$
$a = 2$

The polynomial function is

$f(x) = 2(x - 2)^2(x + 3).$

16. **(a)** $f(t) = 1.06t^3 - 24.6t^2 + 180t$

Let $t = 2$.

$f(2) = 1.06(2)^3 - 24.6(2)^2 + 180(2)$
$\qquad = 270.08$

(b) Graph the function

$f(x) = 1.06x^3 - 24.6x^2 + 180x$

in the window $[0, 15]$ by $[0, 600]$.

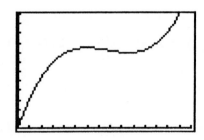

From the graph we see that the amount of change is increasing from $t = 0$ to $t = 5.9$ and from $t = 9.5$ to $t = 15$ and decreasing from $t = 5.9$ to $t = 9.5$.

For Exercises 17–19, see the answer graph in the back of the textbook.

17. $f(x) = \dfrac{3x - 1}{x - 2}$

To find any vertical asymptotes, solve the equation $x - 2 = 0$.

There is one vertical asymptote, x = 2.

$$f(x) = \frac{3x - 1}{x - 2} = \frac{3 - \frac{1}{x}}{1 - \frac{5}{x}},$$

so, as $|x| \to \infty$, $f(x) \to 3$.

The horizontal asymptote is y = 3.

$$f(0) = \frac{3 \cdot 0 - 1}{0 - 2} = \frac{1}{2},$$

so the y-intercept is $\frac{1}{2}$.

The only zero of the numerator is $\frac{1}{3}$, so $\frac{1}{3}$ is the only x-intercept.

Sketch the graph.

18. $f(x) = \frac{x^2 - 1}{x^2 - 9}$

$$= \frac{(x + 1)(x - 1)}{(x + 3)(x - 3)}$$

There are vertical asymptotes at x = -3 and x = 3.

Since the numerator and denominator have the same degree, divide by x^2 to find the horizontal asymptote.

$$f(x) = \frac{1 - \frac{1}{x^2}}{1 - \frac{9}{x^2}} \to \frac{1}{1} \text{ as } |x| \to \infty,$$

so y = 1 is a horizontal asymptote.

The y-intercept is $\frac{1}{9}$.

The x-intercepts are -1 and 1.

Sketch the graph.

19. $f(x) = \frac{x^2 - 16}{x + 4}$

$f(x) = \frac{x^2 - 16}{x + 4}$ can be rewritten as

$$f(x) = \frac{(x + 4)(x - 4)}{x + 4},$$

which equals x - 4 when x ≠ -4. The graph of f(x) is the line y = x - 4 with a "hole" at (-4, -8) since f(x) is undefined at x = -4.

20. $f(x) = \frac{2x^2 + x - 6}{x - 1}$

(a) To find the oblique asymptote, divide:

$$\frac{2x^2 + x - 6}{x - 1} = 2x + 3 - \frac{3}{x - 1}.$$

For very large values of $|x|$, $\frac{3}{x - 1}$ is close to 0, so the line y = 2x + 3 is the oblique asymptote.

(b) To find the x-intercepts, let f(x) = 0.

$$0 = \frac{2x^2 + x - 6}{x - 1}$$

$$2x^2 + x - 6 = 0$$

$$(2x - 3)(x + 2) = 0$$

$$x = \frac{3}{2} \quad \text{or} \quad x = -2$$

x-intercepts: -2, 3/2

(c) To find the y-intercept, let x = 0.

$$f(x) = \frac{2(0)^2 + 0 - 6}{0 - 1}$$

$$= 6$$

y-intercept: 6

(d) To find the vertical asymptote, set the denominator equal to zero and solve for x.

$$x - 1 = 0$$
$$x = 1$$

The equation of the vertical asymptote is x = 1.

(e) Use the information from (a)-(d) and a few additional points to graph the function. See the answer graph in the back of the textbook.

CHAPTER 5 EXPONENTIAL AND LOGARITHMIC FUNCTIONS

Section 5.1

1. In order for a function to have an inverse, it must be *one-to-one*.

3. If f and g are inverses, then $(f \circ g)(x) = x$, and $(g \circ f)(x) = x$.

5. If the point (a, b) lies on the graph of f, and f has an inverse, then the point (*b*, *a*) lies on the graph of f^{-1}.

7. If the function f has an inverse, then the graph of f^{-1} may be obtained by reflecting the graph of f across the line with equation $y = x$.

9. This is a one-to-one function since every horizontal line intersects the graph in no more than one point.

11. This is a one-to-one function since every horizontal line intersects the graph in no more than one point.

13. This is not a one-to-one function since there is a horizontal line that intersects the graph in more than one point. (Here it intersects the curve at an infinite number of points.)

15. $y = (x - 2)^2$

If $x = 0$, $y = 4$.

If $x = 4$, $y = 4$.

Thus, there exist two distinct x-values that lead to the same y-value. Thus, the function is not one-to-one.

17. $y = \sqrt{36 - x^2}$

Both $x = 6$ and $x = -6$ lead to the same y-value, 0. Thus, the function is not one-to-one.

19. $y = 2x^3 + 1$

If $x_1 \neq x_2$, then

$$x_1^3 \neq x_2^3$$
$$2x_1^3 \neq 2x_2^3$$
$$2x_1^3 + 1 \neq 2x_2^3 + 1$$
$$f(x_1) \neq f(x_2),$$

so the function is one-to-one.

21. $y = \dfrac{1}{x + 2}$

If $x_1 \neq x_2$, then

$$x_1 + 2 \neq x_2 + 2,$$
$$\frac{1}{x_1 + 2} \neq \frac{1}{x_2 + 2}$$
$$f(x_1) \neq f(x_2),$$

so the function is one-to-one.

23. Yes, they are inverses, because, for the ordered pairs shown, the values of x and y in the second function are the reverse of the values of x and y in the first function.

24. If $f(g(x)) = x$ and $g(f(x)) = x$, then f and g are *inverse* functions.

25. (a) $6 + (x - 6) = x$, so addition and *subtraction* are inverse functions.

    (b) $4 \cdot (x \div 4) = x$, so multiplication and *division* are inverse operations.

26. When the appropriate operation is applied to an inverse, the result is the *identity* for that operation.

27. The inverse operation of tying your shoelaces would be untying your shoelaces, since untying "undoes" tying.

28. The inverse operation of starting a car would be stopping a car, since stopping "undoes" starting.

29. The inverse operation of entering a room would be leaving a room, since leaving "undoes" entering.

30. The inverse operation of climbing the stairs would be descending the stairs, since descending "undoes" climbing.

31. The inverse operation of taking off in an airplane would be landing in an airplane, since landing "undoes" taking off.

32. The inverse operation of filling a cup would be emptying a cup, since emptying "undoes" filling.

33. These functions are inverses since their graphs are symmetric with respect to the line $y = x$.

35. These functions are not inverses since their graphs are not symmetric with respect to the line $y = x$.

37. $f(x) = 2x + 4$, $g(x) = \frac{1}{2}x - 2$

$$(f \circ g)(x) = f[g(x)]$$
$$= 2\left(\frac{1}{2}x - 2\right) + 4$$
$$= x - 4 + 4$$
$$= x$$

$$(g \circ f)(x) = g[f(x)]$$
$$= \frac{1}{2}(2x + 4) - 2$$
$$= x + 2 - 2$$
$$= x$$

Since $(f \circ g)(x) = x$ and $(g \circ f)(x) = x$, these functions are inverses.

39. $f(x) = \dfrac{2}{x + 6}$, $g(x) = \dfrac{6x + 2}{x}$

$$(f \circ g)(x) = f[g(x)]$$
$$= f\left(\frac{6x + 2}{x}\right)$$
$$= \frac{2}{\dfrac{6x + 2}{x} + 6}$$
$$= \frac{2}{\dfrac{6x + 2 + 6x}{x}}$$
$$= \frac{2}{1} \cdot \frac{x}{12x + 2}$$

$$= \frac{2x}{12x + 2}$$

$$= \frac{x}{6x + 1} \neq x$$

Since $(f \circ g)(x) \neq x$, the functions are not inverses. It is not necessary to check $(g \circ f)(x)$.

41. $f(x) = x^2 + 3$, domain $[0, \infty)$,
    $g(x) = \sqrt{x - 3}$, domain $[3, \infty)$

$$(f \circ g)(x) = f[g(x)]$$
$$= f(\sqrt{x - 3})$$
$$= (\sqrt{x - 3})^2 + 3$$
$$= x$$

$$(g \circ f)(x) = g[f(x)]$$
$$= g(x^2 + 3)$$
$$= \sqrt{x^2 + 3 - 3}$$
$$= \sqrt{x^2}$$
$$= |x|$$
$$= x, \text{ since } x \geq 0$$

Thus, these functions are inverses.

43. $f(x) = -|x + 5|$, domain $[-5, \infty)$,
    $g(x) = |x - 5|$, domain $[5, \infty)$

For the interval $[-5, \infty)$,

$$|x + 5| = x + 5.$$

For the interval $[5, \infty)$,

$$|x - 5| = x - 5.$$

Thus, we have

$$(f \circ g)(x) = f[g(x)]$$
$$= f(x - 5)$$
$$= -(x - 5 + 5)$$
$$= -x$$
$$\neq x.$$

Thus, these functions are not inverses. Another way to see this is to note that the range of $f(x)$ is $(-\infty, 0]$, which is not the same as the domain of $g(x)$, which is $[5, \infty)$.

For Exercises 45–49, see the answer graphs in the textbook.

45. Draw the mirror image of the original graph across the line $y = x$.

47. Carefully draw the mirror image of the original graph across the line $y = x$.

49. To graph the inverse, first draw the line $y = x$ and then draw the mirror image of the graph of the original functions across $y = x$. The graph of the inverse will be another line that also passes through $(0, 0)$.

51. To find $f^{-1}(4)$, find the point with y-coordinate equal to 4. That point is $(4, 4)$. The graph of f^{-1} contains $(4, 4)$. Hence $f^{-1}(4) = 4$.

53. To find $f^{-1}(0)$, find the point with y-coordinate equal to 0. That point is $(2, 0)$. The graph of f^{-1} contains $(0, 2)$. Hence $f^{-1}(0) = 2$.

55. $f^{-1}(-3) = -2$, since the point on the graph of f that has y-coordinate equal to -3 is -2.

For Exercises 57–65, see the answer graphs in the textbook.

57. $y = 3x - 4$

Solve for x.

$y + 4 = 3x$

$\dfrac{y + 4}{3} = x = f^{-1}(y)$

Exchange x and y.

$\dfrac{x + 4}{3} = y$

$f^{-1}(x) = \dfrac{x + 4}{3}$

The graph of the original function is a line with slope 3 and y-intercept −4. Since $f^{-1}(x) = \dfrac{x + 4}{3} = \dfrac{1}{3}x + \dfrac{4}{3}$, the graph of the inverse function is a line with slope $\dfrac{1}{3}$ and y-intercept $\dfrac{4}{3}$.

59. $y = x^3 + 1$

$y - 1 = x^3$

$\sqrt[3]{y - 1} = x = f^{-1}(y)$

Exchange x and y.

$\sqrt[3]{x - 1} = y$

$f^{-1}(x) = \sqrt[3]{x - 1}$

Plot points to graph these functions.

x	−1	0	1
f(x)	0	1	2

x	0	1	2
$f^{-1}(x)$	−1	0	1

61. $y = x^2$

This is not a one-to-one function since $(2)^2 = 4$ and $(-2)^2 = 4$. Thus, the function has no inverse function.

63. $y = \dfrac{1}{x}$

$xy = 1$

$x = \dfrac{1}{y}$

Exchange x and y.

$y = \dfrac{1}{x}$

$f^{-1}(x) = \dfrac{1}{x}$

Observe that this function is its own inverse. Plot points to draw the graph.

x	−2	−1	$-\dfrac{1}{2}$	$\dfrac{1}{2}$	1	2
y	$-\dfrac{1}{2}$	−1	−2	2	1	$\dfrac{1}{2}$

65. $f(x) = \sqrt{6 + x}$

$y = f(x)$ is one-to-one, so it has an inverse function.

Solve for x.

$$y = \sqrt{6 + x}$$
$$y^2 = 6 + x, \ y \geq 0$$
$$y^2 - 6 = x, \ y \geq 0$$

Exchange x and y.

$$y = x^2 - 6, \ x \geq 0$$
$$f^{-1}(x) = x^2 - 6, \ x \geq 0$$

The domain of f is [−6, ∞) and the range is [0, ∞).

The domain of f⁻¹ is [0, ∞) and the range is [−6, ∞).

Note that the domain of f⁻¹ is the range of f and the range of f⁻¹ is the domain of f.

x	−6	−5	−2	3
f(x)	0	1	2	3

x	0	1	2	3
f⁻¹(x)	−6	−5	−2	3

67. Find $f^{-1}(7)$ if $f(x) = x^2 + 5x$ for $x \geq -5/2$.

Find the values of x that correspond to y = 7.

$$7 = x^2 + 5x$$

$$0 = x^2 + 5x - 7$$

$$x = \frac{-5 \pm \sqrt{25 - 4(1)(-7)}}{2}$$

$$x = \frac{-5 \pm \sqrt{53}}{2}$$

$$x = \frac{-5 + \sqrt{53}}{2} \approx 1.14$$

$$\text{or} \quad x = \frac{-5 - \sqrt{53}}{2} \approx -6.14$$

Since $x \geq -5/2$, we must use $\frac{-5 + \sqrt{53}}{2}$.

Therefore, $f\left(\frac{-5 + \sqrt{53}}{2}\right) = 7$, so

$$f^{-1}(7) = \frac{-5 + \sqrt{53}}{2} = 1.14.$$

69. $f^{-1}(1000)$ represents the number of dollars required to build 1000 cars.

71. $$B = \frac{k}{d}$$

$$Bd = k$$

$$d = \frac{k}{B}$$

73. If a line has slope a, the slope of its reflection in the line y = x will be the reciprocal of a, which is 1/a.

75. Use a graphing calculator to graph $f(x) = 6x^3 + 11x^2 - x - 6$ using the window [−3, 2] by [−10, 10]. The horizontal line test will show that this function is not one-to-one.

77. Use a graphing calculator to graph $f(x) = \frac{x - 5}{x + 3}$ using the window [−8, 8] by [−6, 8]. The horizontal line test will show that this function is one-to-one. Find the equation of f⁻¹.

$$y = \frac{x - 5}{x + 3}$$

$$y(x + 3) = x - 5$$

$$yx + 3y = x - 5$$

$$yx - x = -5 - 3y$$

$$x(y - 1) = -5 - 3y$$

$$x = \frac{-5 - 3y}{y - 1} = f^{-1}(y)$$

$$f^{-1}(x) = \frac{-5 - 3x}{x - 1}$$

Graph $f^{-1}(x)$ on the same screen where you have graphed $f(x)$. Note that the two graphs are symmetric with respect to the line $y = x$.

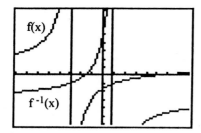

Section 5.2

1. $f(x) = a^x$ is a one-to-one function. Therefore, an inverse function exists for f.

2. Since $f(x) = a^x$ has an inverse, the graph of $f^{-1}(x)$ will be the reflection of f across the line $y = x$. See the answer graph in the back of the textbook.

3. Since $f(x) = a^x$ has an inverse, we find it as follows.

$$y = a^x$$
$$x = a^y$$

4. If $a = 10$, the equation for $f^{-1}(x)$ will be given by

$$x = 10^y.$$

5. If $a = e$, the equation for $f^{-1}(x)$ will be given by

$$x = e^y.$$

For Exercises 7–13, see the answer graphs in the back of the textbook.

7. **(a)** $f(x) = 3^{-x} - 2$

This graph is obtained by translating the graph of $f(x) = 3^{-x}$ down 2 units.

x	−2	−1	0	1	2
y	7	1	−1	$-\dfrac{5}{3}$	$-\dfrac{17}{9}$

(b) $f(x) = 3^{-x} + 4$

This graph is obtained by translating the graph of $f(x) = 3^{-x}$ up 4 units.

x	−2	−1	0	1	2	3
y	13	7	5	$4\dfrac{1}{3}$	$4\dfrac{1}{9}$	$4\dfrac{1}{27}$

(c) $f(x) = 3^{-x-2}$

This equation may also be written as $f(x) = 3^{-(x+2)}$. The graph is obtained by translating the graph of $f(x) = 3^{-x}$ 2 units to the left.

x	−5	−3	−2	−1	0	1	2
y	27	3	1	$\dfrac{1}{3}$	$\dfrac{1}{9}$	$\dfrac{1}{27}$	$\dfrac{1}{81}$

(d) $f(x) = 3^{-x+4}$

This equation may also be written as $f(x) = 3^{-(x-4)}$. This graph is obtained by translating the graph of $f(x) = 3^{-x}$ 4 units to the right.

9. $f(x) = 3^x$

Make a table of values.

x	-2	-1	0	1	2
y	$\frac{1}{9}$	$\frac{1}{3}$	1	3	9

Plot these points and draw a smooth curve through them. This is an increasing function. The domain is $(-\infty, \infty)$ and the range is $(0, \infty)$. The x-axis is a horizontal asymptote.

11. $f(x) = \left(\frac{3}{2}\right)^x$

The domain is $(-\infty, \infty)$ and the range is $(0, \infty)$. Make a table of values.

x	-2	-1	0	1	2
y	$\frac{4}{9}$	$\frac{2}{3}$	1	$\frac{3}{2}$	$\frac{9}{4}$

13. $f(x) = 2^{|x|}$

Make a table of values.

x	-3	-2	-1	0	1	2	3
y	8	4	2	1	2	4	8

Notice that for $x < 0$, $|x| = -x$, so the graph is the same as that of $f(x) = 2^{-x}$. For $x \geq 0$, $|x| = x$, so the graph is the same as that of $f(x) = 2^x$. Since $|-x| = |x|$, the graph is symmetric with respect to the y-axis.

15. $a = 2.3$

Since graph A is increasing, $a > 1$. Since graph A is the middle of the three increasing graphs, the value of a must be the middle of the three values of a greater than 1.

17. $a = .75$

Since graph C is decreasing, $0 < a < 1$. Since graph C decreases at the slowest rate of the three decreasing graphs, the value of a must be the closest to 1 of the three values of a less than 1.

19. $a = .31$

Since graph E is decreasing, $0 < a < 1$. Since graph E decreases at the fastest rate of the three decreasing graphs, the value of a must be the closest to 0 of the three values of a less than 1.

21. For $a > 1$, the value of $f(x) = a^x$ increases as x increases. For $0 < a < 1$, the value of $f(x) = a^x$ decreases as x decreases.

25. If the graph of the exponential function $f(x) = a^x$ contains the point $(3, 8)$, we have

$$a^3 = 8$$
$$(a^3)^{1/3} = 8^{1/3}$$
$$a = 2.$$

Thus, the equation which satisfies the given condition is

$$f(x) = 2^x.$$

27. $f(t) = 3^{2t+3}$

$\qquad = 3^{2t}3^3$

$\qquad = 27 \cdot 3^{2t}$

$\qquad = 27 \cdot (3^2)^t$

$\qquad = 27 \cdot 9^t$

31. $\quad 125^r = 5$

Write both sides as powers of 5.

$(5^3)^r = 5^1$

$\quad 5^{3r} = 5^1$

$\qquad 3r = 1 \quad$ *Property* (*b*)

$\qquad r = \dfrac{1}{3}$

Solution set: $\quad \left\{\dfrac{1}{3}\right\}$

33. $\quad \left(\dfrac{2}{3}\right)^x = \dfrac{9}{4}$

$\quad \left(\dfrac{2}{3}\right)^x = \left(\dfrac{3}{2}\right)^2$

$\quad \left(\dfrac{2}{3}\right)^x = \left(\dfrac{2}{3}\right)^{-2}$

$\qquad x = -2 \quad$ *Property* (*b*)

Solution set: $\quad \{-2\}$

35. $\quad 5^{2p+1} = 25$

$\quad 5^{2p+1} = 5^2$

$\quad 2p + 1 = 2 \quad$ *Property* (*b*)

$\qquad 2p = 1$

$\qquad p = \dfrac{1}{2}$

Solution set: $\quad \left\{\dfrac{1}{2}\right\}$

37. $\qquad \dfrac{1}{81} = k^{-4}$

$\qquad\quad 81^{-1} = k^{-4}$

$\quad [(\pm 3)^4]^{-1} = k^{-4}$

$\qquad (\pm 3)^{-4} = k^{-4}$

$\quad [(\pm 3)^{-4}]^{-1/4} = (k^{-4})^{-1/4}$

$\qquad\qquad \pm 3 = k$

Solution set: $\{-3, 3\}$

39. $\qquad z^{5/2} = 32$

Raise both sides to the 2/5 power.

$(z^{5/2})^{2/5} = (32)^{2/5}$

$\qquad z = (32)^{2/5}$

$\qquad\quad = (32^{1/5})^2$

$\qquad\quad = 2^2$

$\qquad\quad = 4$

Solution set: $\{4\}$

41. $\quad 32^t = 16^{1-t}$

Write both sides as powers of 2.

$(2^5)^t = (2^4)^{1-t}$

$\quad 2^{5t} = 2^{4-4t}$

$\quad 5t = 4 - 4t \quad$ *Property* (*b*)

$\quad 9t = 4$

$\qquad t = \dfrac{4}{9}$

Solution set: $\quad \left\{\dfrac{4}{9}\right\}$

43. $\quad \left(\dfrac{2}{3}\right)^{k-1} = \left(\dfrac{81}{16}\right)^{k+1}$

$\quad \left(\dfrac{2}{3}\right)^{k-1} = \left[\left(\dfrac{3}{2}\right)^4\right]^{(k+1)}$

$\quad \left(\dfrac{2}{3}\right)^{k-1} = \left[\left(\dfrac{2}{3}\right)^{-4}\right]^{(k+1)}$

$\quad k - 1 = -4k - 4$

$\qquad 5k = -3$

$\qquad k = -\dfrac{3}{5}$

Solution set: $\quad \left\{-\dfrac{3}{5}\right\}$

For Exercises 45 and 47, see the answer graphs in the textbook.

45. $f(x) = \dfrac{e^x + e^{-x}}{2}$

Graph this function in the standard viewing window.

47. $f(x) = x^2 \cdot 2^{-x}$

This function may be graphed in the standard viewing window, but a better picture of the graph is obtained by using the window $[-2, 8]$ by $[-2, 5]$.

49. $P = 56{,}780$, $t = \dfrac{23}{4}$, $m = 4$, $r = .053$

$$A = P\left(1 + \frac{r}{m}\right)^{tm}$$

$$= 56{,}780\left(1 + \frac{.053}{4}\right)^{(23/4)(4)}$$

$$= 56{,}780(1.01325)^{23}$$

$$= 76{,}855.95$$

The future value is $76,855.95.

51. Use the compound interest formula to find the present value P if the future value $A = 45{,}678.93$, $r = .096$, $m = 12$, and $t = \dfrac{11}{12}$.

$$A = P\left(1 + \frac{r}{m}\right)^{tm}$$

$$45{,}678.93 = P\left(1 + \frac{.096}{12}\right)^{(11/12)(12)}$$

$$45{,}678.93 = P(1.008)^{11}$$

$$P = \frac{45{,}678.93}{(1.008)^{11}}$$

$$P = 41{,}845.63$$

The present value is $41,845.63.

53. Use the compound interest formula to find r if $P = \$1200$, $A = \$1780$, $m = 4$, and $t = 5$.

$$A = P\left(1 + \frac{r}{m}\right)^{tm}$$

$$1780 = 1200\left(1 + \frac{r}{4}\right)^{5(4)}$$

$$1780 = 1200\left(1 + \frac{r}{4}\right)^{20}$$

$$1.4833 = \left(1 + \frac{r}{4}\right)^{20}$$

Now, take the 20th root of both sides. This can be done on a calculator by using an exponential key to find

$$(1.4833)^{1/20} = (1.4833)^{.05}$$

$$\approx 1.01991.$$

$$1.01991 = 1 + \frac{r}{4}$$

$$.01991 = \frac{r}{4}$$

$$.07964 = r$$

The interest rate, to the nearest tenth, is 8.0%.

55. (a) $T = 50{,}000(1 + .06)^n$

After 4 years, $n = 4$.

$$T = 50{,}000(1 + .06)^4$$

$$\approx 63{,}000$$

The total population after 4 years is about 63,000.

(b) $T = 30{,}000(1 + .12)^n$

After 3 years, $n = 3$.

$$T = 30{,}000(1 + .12)^3$$

$$\approx 42{,}000$$

There would be about 42,000 deer after 3 years.

(c) $T = 45,000(1 + .08)^n$

After 5 years, n = 5.

$$T = 45,000(1 + .08)^5$$
$$\approx 66,000$$

We can expect about 66,000 − 45,000 = 21,000 additional deer after 5 years.

57. $A(t) = 500e^{-.032t}$

(a) $A(4) = 500e^{-.032(4)}$
$$= 500e^{-.128}$$
$$\approx 500(.8799)$$
$$\approx 440$$

After 4 yr, about 440 g will remain.

(b) $A(8) = 500e^{-.032(8)}$
$$= 500e^{-.256}$$
$$\approx 500(.7741)$$
$$\approx 387$$

After 8 yr, about 387 g will remain.

(c) $A(20) = 500e^{-.032(20)}$
$$= 500e^{-.64}$$
$$\approx 500(.5273)$$
$$\approx 264$$

After 20 yr, about 264 g will remain.

(d) The domain is [0, ∞) and the graph passes through (4, 440), (8, 387), and (20, 264).
See the answer graph in the textbook.

59. **(a)** T is a linear function, not an exponential function.

(b) The graph of T(R) passes through the points (0, 0) and (20, 20.6). The slope between these points is

$$m = \frac{20.6 - 0}{20 - 0} = 1.03.$$

Since the slope is 1.03 and the y-intercept is 0, the equation is

$$T(R) = 1.03R.$$

(c) $T(5) = 1.03(5) = 5.15$

When R = 5 w/m², the global temperature increase is 5.15°F.

61. $x = 2^x$

Use a graphing calculator to graph the line f(x) = x and the exponential function g(x) = 2ˣ on the same screen. These two graphs do not intersect, so the given equation has no solution.
Solution set: ∅

63. $6^{-x} = 1 - x$

Use a graphing calculator to graph the exponential function f(x) = 6⁻ˣ and the line g(x) = 1 − x on the same screen. These two graphs intersect in two points whose coordinates may be found by using the "intersect" option in the CALC menu. The x-coordinates of these intersection points are the solutions of the given equation: x = 0 and x ≈ .73.
Solution set: {0, .73}

65. Graph $f(x) = (1 + 1/x)^x$ and $y = 2.71828$ on the same screen using the window $[1, 25]$ by $[0, 3]$. As x gets large, $f(x)$ approaches the line $y = 2.71828$.

Since 2.71828 is a decimal approximation for e, this graph illustrates that as x increases, the value of $(1 + 1/x)^x$ approaches e.

Section 5.3

1. $y = \log_a x$ if and only if $x = a^y$.

3. The statement $\log_5 125 = 3$ tells us that 3 is the power of 5 that equals 125.

5. $3^4 = 81$ is equivalent to
$$\log_3 81 = 4.$$

7. $\left(\frac{2}{3}\right)^{-3} = \frac{27}{8}$ is equivalent to
$$\log_{2/3} \left(\frac{27}{8}\right) = -3.$$

9. $\log_6 36 = 2$ is equivalent to
$$6^2 = 36.$$

11. $\log_{\sqrt{3}} 81 = 8$ is equivalent to
$$(\sqrt{3})^8 = 81.$$

15. Let $y = \log_5 25$.
Write the equation in exponential form.
$$5^y = 25$$
$$5^y = 5^2$$
$$y = 2$$

Thus, $\log_5 25 = 2$.

17. Let $y = \log_{10} .001$.
$$10^y = .001$$
$$10^y = (10)^{-3}$$
$$y = -3$$

Thus, $\log_{10} .001 = -3$.

19. Let $y = \log_4 \left(\frac{\sqrt[3]{4}}{2}\right)$.
$$4^y = \frac{\sqrt[3]{4}}{2}$$
$$(2^2)^y = \frac{\sqrt[3]{2^2}}{2}$$
$$2^{2y} = \frac{2^{2/3}}{2}$$
$$2^{2y} = 2^{(2/3) - 1}$$
$$2^{2y} = 2^{-1/3}$$
$$2y = -\frac{1}{3}$$
$$y = -\frac{1}{6}$$

Thus, $\log_4 \frac{\sqrt[3]{4}}{2} = -\frac{1}{6}$.

21. $2^{\log_2 9} = 9$

This is true by the theorem on inverses.

23. $x = \log_2 32$

Write the equation in exponential form.

$$2^x = 32$$
$$2^x = 2^5$$
$$x = 5$$

Solution set: $\{5\}$

25. $\log_x 25 = 2$
$$x^{-2} = 25$$
$$(x^{-2})^{-1/2} = (25)^{-1/2}$$
$$x = \frac{1}{25^{1/2}}$$
$$x = \frac{1}{5}$$

Solution set: $\left\{\frac{1}{5}\right\}$

For Exercises 29–35, see the answer graphs in the textbook.

29. $f(x) = \log_{1/2} x$

(a) $f(x) = (\log_{1/2} x) - 2$

This is the graph of $f(x) = \log_{1/2} x$ translated down 2 units.

(b) $f(x) = \log_{1/2} (x - 2)$

This is the graph of $f(x) = \log_{1/2} x$ translated to the right 2 units.
The graph has a vertical asymptote at $x = 2$.

(c) $f(x) = |\log_{1/2} (x - 2)|$

This is the same graph as (b) with the part below the x-axis reflected about the x-axis.

31. $f(x) = \log_3 x$

Write $y = \log_3 x$ in exponential form as $x = 3^y$ to find ordered pairs that satisfy the equation. It is easier to choose values for y and find the corresponding values of x.
Make a table of values.

x	$\frac{1}{9}$	$\frac{1}{3}$	1	3	9
y	-2	-1	0	1	2

The graph can also be found by reflecting the graph of $f(x) = 3^x$ about the line $y = x$. The graph has the y-axis as a vertical asymptote. Sketch the graph.

33. $f(x) = \log_{1/2} (1 - x)$
Make a table of values.

x	-7	-3	-1	0	$\frac{1}{2}$	$\frac{3}{4}$	$\frac{7}{8}$
1 - x	8	4	2	1	$\frac{1}{2}$	$\frac{1}{4}$	$\frac{1}{8}$
y	-3	-2	-1	0	1	2	3

The graph has a vertical asymptote at $x = 1$. Sketch the graph.

35. $f(x) = \log_3 (x - 1)$

To graph the function, translate the graph of $f(x) = \log_3 x$ (from Exercise 31) 1 unit to the right. The vertical asymptote will be $x = 1$.

37. Graph $y = \log_{10} x^2$ and $y = 2 \log_{10} x$ on separate viewing screens, using the window $[-5, 5]$ by $[-3, 3]$ for each. See the answer graphs in the back of the textbook. The graphs of $y = \log_{10} x^2$ and $y = 2 \log_{10} x$ are not the same because the domains are not the same. The domain of $y = \log_{10} x^2$ is $(-\infty, 0) \cup (0, \infty)$, while the domain of $y = 2 \log_{10} x$ is $(0, \infty)$.

39. $f(x) = \log_2 (2x)$

The graph will be similar to that of $f(x) = \log_2 x$ (graph E) but will increase more rapidly. Note that while the x-intercept for $f(x) = \log_2 x$ is 1, the x-intercept for $f(x) = \log_2 2x$ will be $1/2$, since $\log_2 2(1/2) = \log_2 1 = 0$. The correct graph is D.

41. $f(x) = \log_2 \left(\frac{x}{2}\right)$

$\quad\quad = \log_2 \frac{1}{2}x$

The graph will be similar to that of $f(x) = \log_2 x$ but will increase less rapidly. The x-intercept is 2 since $\log_2 (2/2) = \log_2 1 = 0$. The correct graph is C.

43. $f(x) = \log_2 (-x)$

The graph of this function is the reflection of the graph of $f(x) = \log_2 x$ about the y-axis. Note that the x-intercept is -1 since $\log_2 [-(-1)] = \log_2 1 = 0$. The correct graph is A.

45. $f(x) = x^2 \log_{10} x$

Graph $f(x) = x^2 \log x$ in the window $[-1, 5]$ by $[-5, 2]$. See the answer graph in the back of the textbook.

47. $2^{-x} = \log_{10} x$

Graph $y_1 = 2^{-x}$ and $y_2 = \log_{10} x$ on a graphing calculator. The x-coordinate of the intersection point will be the solution of the original equation. Using the "intersect" option in the CALC menu, we find that solution is $x \approx 1.87$. Solution set: $\{1.87\}$

48. If x and y are positive numbers,

$\log_a \frac{x}{y} = \log_a x - \log_a y.$

49. Since $\log_2 \left(\frac{x}{4}\right) = \log_2 x - \log_2 4$ by the quotient rule, the graph of $y = \log_2 \left(\frac{x}{4}\right)$ can be obtained by shifting the graph of $y = \log_2 x$ down $\log_2 4 = 2$ units.

50. Graph $f(x) = \log_2 \left(\frac{x}{4}\right)$ and $g(x) = \log_2 x$ on the same axes. See the answer graph in the textbook. The graph of f is 2 units below the graph of g. This supports the answer in Exercise 49.

51. If $x = 4$, $\log_2 \frac{x}{4} = 0$; since $\log_2 x = 2$

and $\log_2 4 = 2$, $\log_2 x - \log_2 4 = 0$.

By the quotient rule,

$$\log_2 \left(\frac{x}{4}\right) = \log_2 x - \log_2 4.$$

Both sides should equal 0. Since $2 - 2 = 0$, they do.

53. $\log_3 \left(\frac{4p}{q}\right)$

$= \log_3 4p - \log_3 q$
    Logarithm of a quotient

$= \log_3 4 + \log_3 p - \log_3 q$
    Logarithm of a product

55. $\log_2 \left(\frac{2\sqrt{3}}{5}\right)$

$= \log_2 2\sqrt{3} - \log_2 5$
    Logarithm of a quotient

$= \log_2 2 + \log_2 \sqrt{3} - \log_2 5$
    Logarithm of a product

$= 1 + \frac{1}{2} \log_2 3 - \log_2 5$
    Logarithm of a power

57. $\log_6 (7m + 3q)$

Since this is a sum, none of the logarithm properties apply, so this expression cannot be simplified.

59. $\log_p \sqrt[3]{\dfrac{m^5 n^4}{t^2}}$

$= \frac{1}{3} \log_p \left(\frac{m^5 n^4}{t^2}\right)$

$= \frac{1}{3}(\log_p m^5 + \log_p n^4 - \log_p t^2)$

$= \frac{1}{3}(5 \log_p m + 4 \log_p n - 2 \log_p t)$

$= \frac{5}{3} \log_p m + \frac{4}{3} \log_p n - \frac{2}{3} \log_p t$

$= \frac{1}{3}(5 \log_p m + 4 \log_p n - 2 \log_p t)$

61. $(\log_b k - \log_b a) - \log_b a$

$= \log_b \frac{k}{m} - \log_b a$
    Logarithm of a quotient

$= \log_b \left(\frac{k}{ma}\right)$
    Logarithm of a quotient

63. $\frac{1}{2} \log_y p^3 q^4 - \frac{2}{3} \log_y p^4 q^3$

$= \log_y (p^3 q^4)^{1/2} - \log_y (p^3 q^4)^{2/3}$

$= \log_y \frac{(p^3 q^4)^{1/2}}{(p^4 q^3)^{2/3}}$

$= \log_y \frac{p^{3/2} \cdot q^2}{p^{8/3} \cdot q^2}$

$= \log_y p^{-7/6}$

65. $\log_b (2y + 5) - \frac{1}{2} \log_b (y + 3)$

$= \log_b \frac{2y + 5}{(y + 3)^{1/2}}$

$= \log_b \left(\frac{2y + 5}{\sqrt{y + 3}}\right)$

67. $\log_{10} 12 = \log_{10} (3 \cdot 2^2)$

$= \log_{10} 3 + 2 \cdot \log_{10} 2$

$= .4771 + 2(.3010)$

$= 1.0791$

69. $\log_{10} \left(\frac{20}{27}\right)$

$= \log_{10} 20 - \log_{10} 27$

$= \log_{10} 2 \cdot 10 - \log_{10} 3^3$

$= \log_{10} 2 + \log_{10} 10 - 3 \log_{10} 3$

$\approx .3010 + 1 - 3(.4771)$

$= -.1303$

71. Since f is a logarithmic function,

$$f(27) = f(3^3)$$
$$= 3f(3)$$
$$= 3(2)$$
$$= 6.$$

73. Prove: $\log_a \frac{x}{y} = \log_a x - \log_a y$.

Let $m = \log_a x$ and $n = \log_a y$.

Changing to exponential form,

$$a^m = x \text{ and } a^n = y.$$

Then,

$$\frac{x}{y} = \frac{a^m}{a^n}$$

$$\frac{x}{y} = a^{m-n}.$$

Changing to logarithmic form,

$$\log_a \frac{x}{y} = m - n$$

$$\log_a \frac{x}{y} = \log_a x - \log_a y.$$

Section 5.4

1. $\log 43 = 1.6335$

To find this value, enter 43 and press the log key.

3. $\log .014 = -1.8539$

5. $\ln 580 = 6.3630$

To find this value, enter 580 and press the ln key.

7. $\ln .7 = -.3567$

9. The graph of $y = \log x$ has coordinates $x = 8$, $y = .90308999$. Thus,

$$\log 8 \approx .90308999.$$

11. $\log_3 4$ is the logarithm to the base 3 of 4.

($\log_4 3$ would be the logarithm to the base 4 of 3.)

12. The exact value of $\log_3 9$ is 2 since $3^2 = 9$.

13. The exact value of $\log_3 27$ is 3 since $3^3 = 27$.

14. $\log_3 16$ must lie between 2 and 3. Because the function defined by $y = \log_3 x$ is increasing and $9 < 16 < 27$, we have

$$\log_3 9 < \log_3 16 < \log_3 27.$$

15. By the change-of-base theorem,

$$\log_3 16 = \frac{\log 16}{\log 3} = \frac{\ln 16}{\ln 3}$$
$$\approx 2.523719014.$$

This value is between 2 and 3, as predicted in Exercise 14.

16. The exact value of $\log_5 (1/5)$ is -1 since $5^{-1} = \frac{1}{5}$.

The exact value of $\log_5 1$ is 0 since $5^0 = 1$.

17. $\log_5 .68$ must lie between -1 and 0. Since the function defined by $y = \log_5 x$ is increasing and

$$\frac{1}{5} = .2 < .68 < 1,$$

we must have

$$\log_5 .2 < \log_5 .68 < \log_5 1.$$

By the change-of-base theorem,

$$\log_5 .68 = \frac{\log .68}{\log 5} = \frac{\ln .68}{\ln 5}$$

$$\approx -.239625573.$$

This value is between -1 and 0, as predicted above.

19. To find $\log_9 12$, use the change-of-base theorem with $a = 9$, $b = e$, and $x = 12$.

$$\log_a x = \frac{\log_b x}{\log_b a}$$

$$\log_9 12 = \frac{\ln 12}{\ln 9}$$

$$\approx \frac{2.4849}{2.19722}$$

$$\approx 1.13$$

21. $\log_{1/2} 3 = \dfrac{\ln 3}{\ln .5}$

$$\approx \frac{1.0986}{-.6931}$$

$$\approx -1.58$$

23. $\log_{200} 175 = \dfrac{\ln 175}{\ln 200}$

$$\approx \frac{5.1648}{5.2983}$$

$$\approx .97$$

25. $\log_{5.8} 12.7 = \dfrac{\ln 12.7}{\ln 5.8}$

$$\approx \frac{2.5416}{1.7579}$$

$$\approx 1.45$$

27. The table for $Y_1 = \log_3 (4 - x)$ shows "ERROR" for $X \geq 4$. This is because the function is undefined for $X \geq 4$. The domain of $y = \log_a X$ is $X > 0$, which means that for Y_1, the domain is $4 - X > 0$, or $X < 4$.

29. To graph $f(x) = \log_x 5$, use the change-of-base theorem; enter $y_1 = (\log 5)/(\log x)$. Graph this function in the window $[-1, 5]$ by $[-3, 3]$. See the answer graph in the back of the textbook.
The vertical line simulates an asymptote at $x = 1$. The base must be greater than 0 and not equal to 1.

31. Grapefruit, 6.3×10^{-4}

$$pH = -\log [H_3O^+]$$
$$= -\log (6.3 \times 10^{-4})$$
$$= -(\log 6.3 + \log 10^{-4})$$
$$= -(.7793 - 4)$$
$$= -.7993 + 4$$
$$pH = 3.2$$

The answer is rounded to the nearest tenth because it is customary to round pH values to the nearest tenth.
The pH of grapefruit is 3.2.

33. Limes, 1.6×10^{-2}

$$pH = -\log [H_3O^+]$$
$$= -\log (1.6 \times 10^{-2})$$
$$= -(\log 1.6 + \log 10^{-2})$$
$$= -(.2041 - 2)$$
$$= -(-1.7959)$$

pH = 1.8

The pH of limes is 1.8.

35. Soda pop, 2.7

$$\text{pH} = -\log\,[H_3O^+]$$
$$2.7 = -\log\,[H_3O^+]$$
$$-2.7 = \log\,[H_3O^+]$$
$$[H_3O^+] = 2.0 \times 10^{-3}$$

37. Beer, 4.8

$$\text{pH} = -\log\,[H_3O^+]$$
$$4.8 = -\log\,[H_3O^+]$$
$$-4.8 = \log\,[H_3O^+]$$
$$[H_3O^+] = 10^{-4.8}$$
$$[H_3O^+] = 1.6 \times 10^{-5}$$

39. Let r = the decibel rating of a sound.

$$r = 10 \cdot \log_{10} \frac{I}{I_0}$$

(a) $r = 10 \cdot \log_{10} \dfrac{100 \cdot I_0}{I_0}$

$$= 10 \cdot \log_{10} 100$$
$$= 10 \cdot 2$$
$$= 20$$

(b) $r = 10 \cdot \log_{10} \dfrac{1000 \cdot I_0}{I_0}$

$$= 10 \cdot \log_{10} 1000$$
$$= 10 \cdot 3$$
$$= 30$$

(c) $r = 10 \cdot \log_{10} \dfrac{100,000 \cdot I_0}{I_0}$

$$= 10 \cdot 5$$
$$= 50$$

(d) $r = 10 \cdot \log_{10} \dfrac{1,000,000 \cdot I_0}{I_0}$

$$= 10 \cdot 6$$
$$= 60$$

41. Let r = the Richter scale rating of an earthquake.

$$r = \log_{10}\left(\frac{I}{I_0}\right)$$

(a) $r = \log_{10} \dfrac{1000 \cdot I}{I_0}$

$$= \log_{10} 1000$$
$$= 3$$

(b) $r = \log_{10} \dfrac{1,000,000 \cdot I_0}{I_0}$

$$= \log_{10} 1,000,000$$
$$= 6$$

(c) $r = \log_{10} \dfrac{100,000,000 \cdot I_0}{I_0}$

$$= \log_{10} 100,000,000$$
$$= 8$$

43. **(a)** $8.3 = \log_{10} \dfrac{I}{I_0}$

$$\frac{I}{I_0} = 10^{8.3}$$
$$I = 10^{8.3} \cdot I_0$$
$$I = 200,000,000 I_0$$

The magnitude was about $200,000,000 I_0$.

(b) $7.1 = \log_{10} \dfrac{I}{I_0}$

$$\frac{I}{I_0} = 10^{7.1}$$
$$I = 10^{7.1} \cdot I_0$$
$$I = 13,000,000 I_0$$

The magnitude was about $13,000,000 I_0$.

(c) $\dfrac{200,000,000 \cdot I_0}{13,000,000 \cdot I_0} \approx 15.38$

The 1906 earthquake had a magnitude more than 15 times greater than the 1989 earthquake.

45. $S_n = a \ln \left(1 + \dfrac{n}{a}\right)$

$= .36 \ln \left(1 + \dfrac{n}{.36}\right)$

(a) $S(100) = .36 \ln \left(1 + \dfrac{100}{.36}\right)$

$= .36 \ln 278.77$

$= (.36)(5.6304)$

≈ 2

(b) $S(200) = .36 \ln \left(1 + \dfrac{200}{.36}\right)$

$= .36 \ln (556.555)$

$= .36(6.322)$

≈ 2

(c) $S(150) = .36 \ln \left(1 + \dfrac{150}{.36}\right)$

$= .36 \ln 417.666$

$= .36(6.0347)$

≈ 2

(d) $S(10) = .36 \ln \left(1 + \dfrac{10}{.36}\right)$

$= .36 \ln 28.777$

$= .36(3.3596)$

≈ 1

47. The index of diversity M for 2 species is given by

$H = -[P_1 \log_2 P_1 + P_2 \log_2 P_2]$.

$P_1 = \dfrac{50}{100}$

$= .5$

$P_2 = \dfrac{50}{100}$

$= .5$

Substituting into the formula gives

$H = -[.5 \log_2 .5 + .5 \log_2 .5]$.

Since $\log_2 .5 = \log_2 \dfrac{1}{2} = -1$, we have

$H = -[.5(-1) + .5(-1)]$

$= -(-1)$

$= 1$

The index of diversity is 1.

49. $g(x) = e^x$

(a) By the theorem on inverses,

$g(\ln 3) = e^{\ln 3}$

$= 3.$

(b) By the theorem on inverses,

$g[\ln (5^2)] = e^{\ln 5^2}$

$= 5^2 = 25$

(c) By the theorem on inverses,

$g[\ln \left(\dfrac{1}{e}\right)] = e^{\ln(1/e)}$

$= \dfrac{1}{e}.$

51. $f(x) = \ln x$

(a) By the theorem on inverses,

$f(e^5) = \ln e^5$

$= 5.$

(b) By the theorem on inverses,

$f(e^{\ln 3}) = \ln e^{\ln 3}$

$= \ln 3.$

(c) By the theorem on inverses,

$f(e^{2 \ln 3}) = \ln e^{2 \ln 3}$

$= 2 \ln 3$

or $\ln 9.$

53. $f(x) = -266 + 72 \ln x$

In the year 2000, $x = 100$.

$$f(100) = -266 + 72 \ln 100$$
$$\approx 66$$

Thus, the number of visitors in the year 2000 will be about 66 million. Beyond 1993, we must assume that the rate of increase continues to be logarithmic.

55. From Example 6,

$$T(R) = 1.03R$$

and

$$R = k \ln\left(\frac{C}{C_0}\right).$$

By substitution, we have

$$T(k) = 1.03k \ln\left(\frac{C}{C_0}\right).$$

Since $10 \le k \le 16$ and $\frac{C}{C_0} = 2$, the range for

$$T = 1.03k \ln\left(\frac{C}{C_0}\right)$$

will be between

$$T = 1.03(10) \ln 2 \approx 7.1$$

and

$$T = 1.03(16) \ln 2 \approx 11.4.$$

The predicted increased global temperature due to the greenhouse effect from a doubling of the carbon dioxide in the atmosphere is between 7°F and 11°F.

57. **(a)** The table of natural logarithms takes the following form. Let $x = \ln D$ and $y = \ln P$ for each planet.

Planet	ln D	ln P
Mercury	$-.94$	-1.43
Venus	$-.33$	$-.48$
Earth	0	0
Mars	.42	.64
Jupiter	1.65	2.48
Saturn	2.26	3.38
Uranus	2.95	4.43
Neptune	3.40	5.10

Plot this data in the window $[-2, 4]$ by $[-2, 6]$. See the answer graph in the back of the textbook.
From the plot, the data appear to be linear.

(b) Choose two points from the table showing $\ln D$ and $\ln P$ and find the equation through them. If we use $(0, 0)$, representing Earth and $(3.40, 5.10)$ representing Neptune, we obtain

$$m = \frac{5.10 - 0}{3.40 - 0} = 1.5.$$

Since the y-intercept is 0, the equation is

$$y = 1.5x$$

or $\ln P = 1.5 \ln D$.

Since the points lie approximately but not exactly on a line, a slightly different equation will be found if a different pair of points is used.

(c) For Pluto, D = 39.5, so

$$\ln P = 1.5 \ln D$$
$$= 1.5 \ln 39.5$$

Then

$$P = e^{1.5 \ln 39.5}$$
$$= e^{\ln 39.5^{1.5}}$$
$$= (39.5)^{1.5}$$
$$\approx 248.3.$$

The linear equation predicts that the period of the planet Pluto is 248.3 years, which is very close to the true value of 248.5 years.

Section 5.5

5. $3^x = 6$

Take base e (natural) logarithms of both sides.

$$\ln 3^x = \ln 6$$
$$x \ln 3 = \ln 6 \quad \textit{Logarithm of a power}$$
$$x = \frac{\ln 6}{\ln 3} \quad \textit{Divide by ln 3}$$
$$\approx \frac{1.7918}{1.0986}$$
$$\approx 1.631$$

Solution set: $\{1.631\}$

7. $6^{1-2k} = 8$

Take base 10 (common) logarithms of both sides. (This exercise can also be done using natural logarithms.)

$$\log 6^{1-2k} = \log 8$$
$$(1 - 2k) \log 6 = \log 8$$
$$1 - 2k = \frac{\log 8}{\log 6}$$
$$2k = 1 - \frac{\log 8}{\log 6}$$
$$k = \frac{1}{2}\left(1 - \frac{\log 8}{\log 6}\right)$$
$$\approx \frac{1}{2}\left(1 - \frac{.9031}{.7782}\right)$$
$$\approx -.080$$

Solution set: $\{-.080\}$

9.
$$e^{k-1} = 4$$
$$\ln e^{k-1} = \ln 4$$
$$k - 1 = \ln 4 \quad \textit{Theorem on inverses}$$
$$k = \ln 4 + 1$$
$$= 1.3863 + 1$$
$$= 2.386$$

Solution set: $\{2.386\}$

11.
$$2e^{5a+2} = 8$$
$$e^{5a+2} = 4$$
$$\ln e^{5a+2} = \ln 4$$
$$5a + 2 = \ln 4 \quad \textit{Theorem on inverses}$$
$$5a = \ln 4 - 2$$
$$a = \frac{1}{5}(\ln 4 - 2)$$
$$\approx \frac{1}{5}(1.3863 - 2)$$
$$\approx -.123$$

Solution set: $\{-.123\}$

13. $2^x = -3$ has no solution since 2 raised to any power is positive.

Solution set: $\emptyset$

15. $e^{2x} \cdot e^{5x} = e^{14}$

$\quad\quad e^{7x} = e^{14}$ *Product rule for exponents*

$\quad\quad 7x = 14$ *Property 1*

$\quad\quad x = 2$

Solution set: $\{2\}$

17. $100(1 + .02)^{3+n} = 150$

$\quad\quad (1.02)^{3+n} = 1.5$

$(3 + n) \log 1.02 = \log 1.5$
$\quad\quad\quad$ *Logarithm of a power*

$\quad\quad 3 + n = \dfrac{\log 1.5}{\log 1.02}$

$\quad\quad 3 + n \approx \dfrac{.1761}{.0086}$

$\quad\quad 3 + n \approx 20.475$

$\quad\quad n \approx 17.475$

Solution set: $\{17.475\}$

19. $\log (t - 1) = 1$

$\quad\quad t - 1 = 10^1$
$\quad\quad\quad$ *Exponential form*

$\quad\quad t - 1 = 10$

$\quad\quad t = 11$

Solution set: $\{11\}$

21. $\ln (y + 2) = \ln (y - 7) + \ln 4$

$\ln (y + 2) = \ln [(y - 7) \cdot 4]$
$\quad\quad\quad$ *Logarithm of a product*

$\quad\quad y + 2 = 4(y - 7)$ *Property 2*

$\quad\quad y + 2 = 4y - 28$

$\quad\quad -3y = -30$

$\quad\quad y = 10$

Solution set: $\{10\}$

23. $\ln (5 + 4y) - \ln (3 + y) = \ln 3$

$\quad\quad \ln \dfrac{5 + 4y}{3 + y} = \ln 3$
$\quad\quad\quad$ *Logarithm of a quotient*

$\quad\quad \dfrac{5 + 4y}{3 + y} = 3$
$\quad\quad$ *Property 2*

$\quad\quad 5 + 4y = 3(3 + y)$

$\quad\quad 5 + 4y = 9 + 3y$

$\quad\quad y = 4$

Solution set: $\{4\}$

25. $2 \ln (x - 3) = \ln (x + 5) + \ln 4$

$\ln (x - 3)^2 = \ln [4(x + 5)]$
$\quad\quad\quad$ *Logarithm of a power; logarithm of a product*

$\quad\quad (x - 3)^2 = 4(x + 5)$ *Property 2*

$\quad\quad x^2 - 6x + 9 = 4x + 20$

$\quad\quad x^2 - 10x - 11 = 0$

$(x - 11)(x + 1) = 0$

$\quad x - 11 = 0$ or $x + 1 = 0$

$\quad\quad x = 11$ or $\quad\quad x = -1$

If $x = -1$, $x - 3 = -4$, so -1 is not in the domain of $\ln (x - 3)$ and cannot be used.

Solution set: $\{11\}$

27. $\log_3 (a - 3) = 1 + \log_3 (a + 1)$

$\log_3 (a - 3) = \log_3 3 + \log_3 (a + 1)$
$\quad\quad\quad$ $\log_3 3 = 1$

$\log_3 (a - 3) = \log_3 [3(a + 1)]$
$\quad\quad\quad$ *Logarithm of a product*

$\quad\quad a - 3 = 3(a + 1)$
$\quad\quad\quad$ *Property 2*

$\quad\quad a - 3 = 3a + 3$

$\quad\quad -2a = 6$

$\quad\quad a = -3$

−3 is not in the domain of $\log_3 (a − 3)$ or $\log_3 (a + 1)$ and therefore cannot be used.

Solution set: ∅

29. $\ln e^x − \ln e^3 = \ln e^5$

   $x − 3 = 5$ *Theorem on inverses*

   $x = 8$

Solution set: $\{8\}$

31. $\log_2 \sqrt{2y^2} − 1 = \dfrac{1}{2}$

   $\log_2 \sqrt{2y^2} = \dfrac{3}{2}$ *Add 1*

   $2^{3/2} = \sqrt{2y^2}$ *Change to exponential form*

   $(2^{3/2})^2 = (\sqrt{2y^2})^2$ *Square both sides*

   $2^3 = 2y^2$

   $8 = 2y^2$

   $4 = y^2$

   $\pm 2 = y$

Since the solution involves squaring both sides, both proposed solutions must be checked in the original equation. Both answers check.

Solution set: $\{-2, 2\}$

33. $\log z = \sqrt{\log z}$

   $(\log z)^2 = (\sqrt{\log z})^2$ *Square both sides*

   $(\log z)^2 = \log z$

   $(\log z)^2 − \log z = 0$

   $\log z(\log z − 1) = 0$ *Factor out log z*

$\log z = 0$ or $\log z − 1 = 0$

                         $\log z = 1$

$10^0 = z$ $10^1 = z$

$1 = z$ or $10 = z$

Since the work involves squaring both sides, both proposed solutions must be checked in the original equation. Both answers check.

Solution set: $\{1, 10\}$

37. $I = \dfrac{E}{R}(1 − e^{-Rt/2})$ for t

   $I = \dfrac{E}{R} − \dfrac{E}{R}e^{-Rt/2}$

   $I − \dfrac{E}{R} = -\dfrac{E}{R}e^{-Rt/2}$

   $\dfrac{RI − E}{R}\left(-\dfrac{R}{E}\right) = e^{-Rt/2}$

   $\dfrac{E − RI}{E} = e^{-Rt/2}$

   $\ln\left(\dfrac{E − RI}{E}\right) = \ln e^{-Rt/2}$

   $\ln\left(\dfrac{E − RI}{E}\right) = -\dfrac{Rt}{2}$

   $-\dfrac{2}{R}\ln\left(1 − \dfrac{RI}{E}\right) = t$

39. $p = a + \dfrac{k}{\ln x}$ for x

   $p − a = \dfrac{k}{\ln x}$

   $(p − a)\ln x = k$

   $\ln x = k/(p − a)$

To solve for x, change this equation from logarithmic to exponential form.

   $x = e^{k/(p-a)}$

42. $(e^x)^2 - 4e^x + 3 = 0$

$(e^x - 1)(e^x - 3) = 0$

43. $(e^x - 1)(e^x - 3) = 0$

Set each factor to 0 and solve.

$e^x - 1 = 0$ or $e^x - 3 = 0$

$e^x = 1$ $\qquad$ $e^x = 3$

$\ln e^x = \ln 1$ $\qquad$ $\ln e^x = \ln 3$

$x = 0$ or $\qquad$ $x = \ln 3$

Solution set: $\{0, \ln 3\}$

44. Graph $y = e^{2x} - 4e^x + 3$ on a graph-ing calculator, using the window $[-5, 5]$ by $[-5, 10]$. See the answer graph in the back of the textbook. The graph intersects the x-axis at 0 and $1.099 \approx \ln 3$.

45. From the graph, we see that the intervals where $y > 0$ are $(-\infty, 0)$ and $(\ln 3, \infty)$, so the solution set of the inequality

$$e^{2x} - 4e^x + 3 < 0$$

is $(-\infty, 0) \cup (\ln 3, \infty)$.

46. From the graph we see that $y < 0$ on the interval $(0, \ln 3)$, so the solu-tion set of the inequality

$$e^{2x} - 4e^x + 3 < 0$$

is $(0, \ln 3)$.

47. $e^x + \ln x = 5$

Graph $y_1 = e^x + \ln x$

and $\quad y_2 = 5$

on the same screen.

Using the "intersect" option in the CALC menu, we find that the two graphs intersect at approximately $(1.52, 5)$. The x-coordinate of this point is the solution of the equa-tion.

Solution set: $\{1.52\}$

49. $2e^x + 1 = 3e^{-x}$

Graph $y_1 = 2e^x + 1$

and $\quad y_2 = 3e^{-x}$

on the same screen. The two curves intersect at the point $(0, 3)$. The x-coordinate of this point is the solution of the equation.

Solution set: $\{0\}$

51. $\log x = x^2 - 8x + 14$

Graph $y_1 = \log x$ and

$\qquad y_2 = x^2 - 8x + 14$.

The intersection points are at $x = 2.45$ and $x = 5.66$.

Solution set: $\{2.45, 5.66\}$

53. $f(x) = e^{3x+1}$

$\quad y = e^{3x+1}$

To solve for x, take natural loga-rithms on both sides.

$$\ln y = \ln e^{3x+1}$$
$$\ln y = 3x + 1$$
$$3x = \ln y - 1$$
$$x = \frac{1}{3}(\ln y - 1) = f^{-1}(y)$$

Exchange x and y.

$$f^{-1}(x) = \frac{1}{3}(\ln x - 1)$$

The domain of f^{-1} is $(0, \infty)$ and the range is $(-\infty, \infty)$.

55. $\log_3 x > 3$

$\log_3 x - 3 > 0$

Graph $Y_1 = \log_3 x - 3$.

The graph is positive in the interval $(27, \infty)$, so the solution to the original inequality is $(27, \infty)$.

57. $d = 10 \log \dfrac{I}{I_0}$

For 89 decibels, we have

$$89 = 10 \log \frac{I}{I_0}$$

$$8.9 = \log \frac{I}{I_0}.$$

Change this equation to exponential form.

$$\frac{I}{I_0} = 10^{8.9}$$

$$I = 10^{8.9} I_0$$

For 86 decibels, we have

$$86 = 10 \log \frac{I}{I_0}$$

$$8.6 = \log \frac{I}{I_0}$$

$$\frac{I}{I_0} = 10^{8.6}$$

$$I = 10^{8.6} I_0.$$

To compare these intensities, find their ratio.

$$\frac{10^{8.9} I_0}{10^{8.6} I_0} = \frac{10^{8.9}}{10^{8.6}} \approx 2$$

From this calculation, we see that 89 decibels is about twice as loud as 86 decibels, for a 100% increase.

59. $A = P\left(1 + \dfrac{r}{m}\right)^{tm}$

To solve for t, substitute $A = 2063.40$, $P = 1786$, $r = .116$, and $m = 12$.

$$2063.40 = 1786\left(1 + \frac{.116}{12}\right)^{(t)(12)}$$

$$\frac{2063.40}{1786} = \left(1 + \frac{.116}{12}\right)^{12t}$$

$$\ln \frac{2063.40}{1786} = 12t \ln \left(1 + \frac{.116}{12}\right)$$

$$\frac{\ln \dfrac{2063.40}{1786}}{12 \ln \left(1 + \dfrac{.116}{23}\right)} = t$$

$$1.25 \approx t$$

To the nearest hundredth, t = 1.25 yr.

61. $A = P\left(1 + \dfrac{r}{m}\right)^{tm}$

$$20,000 = 16,000\left(1 + \frac{r}{4}\right)^{(5.25)(4)}$$

$$20,000 = 16,000\left(1 + \frac{r}{4}\right)^{21}$$

$$1.25 = \left(1 + \frac{r}{4}\right)^{21}$$

$$\sqrt[21]{1.25} = 1 + \frac{r}{4}$$

$$4\left(\sqrt[21]{1.25} - 1\right) = r$$

$$.0427 \approx r$$

The interest rate is about 4.27%.

63. $f(x) = 6.2(10)^{-12}(1.4)^x$

Find x when $f(x) = 2000$.

$$2000 = 6.2(10)^{-12}(1.4)^x$$

$$322.6 \times 10^{12} = (1.4)^x$$

$$\log (322.6 \times 10^{12}) = x \log 1.4$$

$$99 \approx x$$

Software exports will double their 1997 value in 1999.

65. **(a)** $\ln(1 - P) = -.0034 - .0053T$

Change this equation to exponential form.

$$1 - P = e^{-.0034 - .0053T}$$

$$P(T) = 1 - e^{-.0034 - .0034T}$$

(b) See the answer graph in the textbook.

From the graph one can see that inititally there is a rapid reduction of carbon dioxide emissions. However, after a while there is little benefit in raising taxes further.

(c) $P(T) = 1 - e^{-.0034 - .0053T}$

$$P(60) = 1 - e^{-.0034 - .0053(60)}$$

$$\approx .275 \text{ or } 27.5\%$$

The reduction in carbon emissions from a tax of $60 per ton of carbon is 27.5%.

(d) We must determine T when P = .05.

$$P(T) = 1 - e^{-.0034 - .0053T} = .5$$

$$.5 = 1 - e^{-.0034 - .0053T}$$

$$.5 = e^{-.0034 - .0053T}$$

$$\ln .5 = -.0034 - .0053T$$

$$T = \frac{\ln .5 + .0034}{-.0053} \approx 130.14$$

The value T = $130.14 will give a 50% reduction in carbon emissions.

Section 5.6

1. $A(t) = 500e^{-.032t}$

(a) $t = 4$

$$A(4) = 500e^{-.032(4)}$$

$$\approx 440$$

After 4 years, about 440 g remain.

(b) $t = 8$

$$A(8) = 500e^{-.032(8)}$$

$$\approx 387$$

After 8 years, about 387 g remain.

(c) $t = 20$

$$A(20) = 500e^{-.032(20)}$$

$$\approx 264$$

After 20 years, about 264 g remain.

(d) Find t when $A(t) = 250$.

$$250 = 500e^{-.032t}$$

$$.5 = e^{-.032t}$$

$$\ln .5 = \ln e^{-.032t}$$

$$-.6931 = -.032t$$

$$21.66 \approx t$$

The half-life is about 21.66 yr.

3. $A(t) = A_0\, e^{-.00043t}$

Find t when $A(t) = .5\, A_0$.

$$.5\, A_0 = A_0\, e^{-.00043t}$$

$$.5 = e^{-.00043t}$$

$$-.6931 = -.00043t$$

$$1611.97 \approx t$$

The half-life is about 1611.97 yr.

7. $p(h) = 86.3 \ln h - 680$

   (a) $h = 3000$

   $P(3000) = 86.3 \ln 3000 - 680$
   ≈ 11

   At 3000 ft, about 11% of moisture falls as snow.

   (b) $h = 4000$

   $P(4000) = 86.3 \ln 4000 - 680$
   ≈ 36

   At 4000 ft, about 36% of moisture falls as snow.

   (c) $h = 7000$

   $P(7000) = 86.3 \ln 7000 - 680$
   ≈ 84

   At 7000 ft, about 84% of moisture falls as snow.

9. $A(t) = A_0 e^{kt}$

   $k \approx -(\ln 2)(1/5700)$

   Solve $A(t) = \frac{1}{3}A_0$ for t.

   $$\frac{1}{3}A_0 = A_0 e^{-(\ln 2)(1/5700)t}$$

   $$\frac{1}{3} = e^{-(\ln 2)(1/5700)t}$$

   $$.3333 \approx e^{-(\ln 2)(1/5700)t}$$

   $$\ln .3333 \approx -(\ln 2)(1/5700)t$$

   $$-1.0986 \approx -.0001t$$

   $$9000 \approx t$$

   The Egyptian died about 9000 yr ago.

11. $y = y_0 e^{-(\ln 2)(1/5700)t}$

    or

    $y \approx y_0 e^{-.0001216t}$

Find t when $y = .15y_0$.

$$.15y_0 = y_0 e^{-.0001216t}$$

$$.15 = e^{-.0001216t}$$

$$\ln .15 = -.0001216t$$

$$\frac{\ln .15}{-.0001216} = t$$

$$16,000 \approx t$$

The paintings are about 16,000 yr old.

13. $A(t) = T_0 + Ce^{-kt}$

    Substitute $T_0 = 20$, $C = 100$, and $k = .1$ into the formula, and solve $A(t) = 25$ for t.

    $$25 = 20 + 100e^{-.1t}$$

    $$5 = 100e^{-.1t}$$

    $$.05 = e^{-.1t}$$

    $$\ln .05 = -.1t$$

    $$-2.9957 \approx -.1t$$

    $$30 \approx t$$

    It will take about 30 min.

15. **(a)** $A = P\left(1 + \frac{r}{m}\right)^{tm}$

    $$5000 = 1000\left(1 + \frac{.035}{4}\right)^{4t}$$

    $$5 = (1 + .00875)^{4t}$$

    $$\ln 5 = 4t \ln 1.00875$$

    $$46.2 \approx t$$

    It will take about 46.2 yr if interest is compounded quarterly.

    (b) $A = Pe^{rt}$

    $$5000 = 1000e^{.035t}$$

    $$5 = e^{.035t}$$

    $$\ln 5 = .035t$$

    $$46.0 \approx t$$

It will take about 46.0 yr if interest is compounded continuously.

17. **(a)**
$$A = P\left(1 + \frac{r}{m}\right)^{tm}$$
$$2P = P\left(1 + \frac{.025}{4}\right)^{4t}$$
$$2 = (1.00625)^{4t}$$
$$\ln 2 = 4t \ln 1.00625$$
$$27.81 \approx t$$

The doubling time is about 27.81 yr if interest is compounded quarterly.

(b)
$$A = Pe^{rt}$$
$$2P = Pe^{.025t}$$
$$2 = e^{.025t}$$
$$\ln 2 = .025t$$
$$27.73 \approx t$$

The doubling time is about 27.73 yr if interest is compounded continuously.

19. $A = Pe^{rt}$

$A = 80,000, \quad P = 60,000, \quad r = .0675$

$$80,000 = 60,000e^{.0675t}$$
$$1.3333 \approx e^{.0675t}$$

Take natural logarithms of both sides.

$$\ln 1.3333 \approx .0675t$$
$$4.3 \approx t$$

With the continuous compounding plan, it will take about 4.3 yr for Ms. Youngman's $60,000 to grow to $80,000.

21. Graph $Y_1 = 5(1.04)^{2x}$ and
$$Y_2 = 5e^{.08x}.$$

When $x = 25$, $Y_1 \approx 35.533417$
and $\qquad Y_2 \approx 36.94528.$

Graph $Y_1 = 5\left(1 + \frac{.08}{12}\right)^{12x}$.

When $x = 25$, $Y_1 \approx 36.70088$.
(The value of Y_2 is unchanged).

23. $f(t) = 625e^{.0516t}$

In 1996, $t = 17$.

$$f(16) = 625e^{.0516(17)}$$
$$\approx 1503$$

Based on this model, in 1996, there will be about 1503 thousand, or 1.503 million cesarean section deliveries.

25.

$$G(t) = \frac{MG_0}{G_0 + (M - G_0)e^{-kMt}}$$

$$G(t) = \frac{(2500)(100)}{100 + (2500 - 100)e^{-.0004(2500)t}}$$

$$G(t) = \frac{250,000}{100 + 2400e^{-t}}$$

Graph this function. See the answer graph in the back of the textbook.

27. The graph of $G(t)$ intersects the line $Y = 1000$ at $t \approx 2.8$. Thus, it takes about 2.8 decades for the population to reach 1000.
Now solve $G(t) = 1000$ analytically.

$$\frac{250,000}{100 + 2400e^{-t}} = 1000$$

$$250,000 = 1000(100 + 2400e^{-t})$$

$$250 = 100 + 2400e^{-t}$$

$$.0625 = e^{-t}$$

$$\ln .0625 = \ln e^{-t}$$

$$-t = \ln .0625$$

$$t = -\ln .0625 \text{ (exact)}$$

$$t \approx 2.7726$$

29. $A(t) = 1757e^{.0264t}$

In 1996, $t = 11$.

$$A(11) = 1757e^{.0264(11)}$$

$$\approx 2349$$

About 2349 million, or 2.349 billion books will be sold in 1996.

31. $A(t) = 34e^{.04t}$

We want to find the year in which the CPI will be 150.

$$150 = 34e^{.04t}$$

$$\frac{150}{34} = e^{.04t}$$

$$\ln \left(\frac{150}{34}\right) = \ln e^{.04t}$$

$$\ln \left(\frac{150}{34}\right) = .04t$$

$$t = \frac{\ln (150/34)}{.04}$$

$$t \approx 37$$

Since $t = 37$ corresponds to the year $1960 + 37 = 1997$, costs will be 50% higher in 1997 than in 1987.

33. $S(t) = 50,000e^{-.1t}$

Find t when $S(t) = 25,000$.

$$25,000 = 50,000e^{-.1t}$$

$$.5 = e^{-.1t}$$

$$\ln .5 = -.1t$$

$$6.9 \approx t$$

It will take about 6.9 yr for sales to fall to half the initial sales.

35.
$$A = Pe^{rt}$$

$$2P = Pe^{.06t}$$

$$2 = e^{.06t}$$

$$\ln 2 = \ln e^{.06t}$$

$$\ln 2 = .06t \ln e$$

$$\ln 2 = .06t$$

$$\frac{\ln 2}{.06} = t$$

$$11.6 \approx t$$

It will take about 11.6 yr before twice as much electricity is needed.

37. $L = 9 + 2e^{.15t}$

(a) In 1982, $t = 0$.

$$L = 9 + 2e^{.15(0)} = 11$$

(b) In 1986, $t = 4$.

$$L = 9 + 2e^{.15(4)} \approx 12.6$$

(c) In 1992, $t = 10$.

$$L = 9 + 2e^{.15(10)} \approx 18.0$$

(d) Graph $y_1 = 9 + 2e^{.15x}$ in the window $[0, 10]$ by $[0, 30]$. See the answer graph in the back of the textbook.

(e) According to this equation, living standards are increasing, but at a slow rate.

39.
$$n = -7600 \log r$$

$$-\frac{n}{7600} = \log r$$

$$10^{-n/7600} = r$$

(a) $n = 1000$

$$r = 10^{-1000/7600} \approx .74$$

(b) $n = 2500$

$$r = 10^{-2500/7600} \approx .47$$

41. Let A denote the maximum amount of Social Security tax.

$$A = P(1 + r)^t$$

$$4681.80 = 2791.80(1 + r)^{10}$$

$$(1 + r)^{10} = \frac{4681.80}{2791.80}$$

$$1 + r = \left(\frac{4681.80}{2791.80}\right)^{1/10}$$

$$r = \left(\frac{4681.80}{2791.80}\right)^{1/10} - 1$$

$$\approx .0531 \text{ or } 5.31\%.$$

Social Security taxes rose at a faster rate than consumer prices.

43. Let K be the total number of executives, managers, and administrators in retail trade. The number of Hispanic executives, managers, and administrators is .048K. If this number increases at a rate of 3%, we must determine when it will reach .09K. That is, we must determine when

$$.048K(1.03)^t = .09K.$$

Solve this equation.

$$(1.03)^t = \frac{.09}{.048}$$

$$t \ln 1.03 = \left(\frac{.09}{.048}\right)$$

$$t = \frac{\ln (.09/.048)}{\ln 1.03} \approx 21.3$$

Since 1995 + 21 = 2016, this will occur in the year 2016.

Chapter 5 Review Exercises

1. $f(x) = x^3 - 3$

$f(x)$ is one-to-one, so it has an inverse function.
Let $y = f(x)$ and solve for x.

$$y + 3 = x^3$$

$$\sqrt[3]{y + 3} = x = f^{-1}(y)$$

Exchange x and y.

$$f^{-1}(x) = \sqrt[3]{x + 3}$$

3. $f^{-1}(\$50,000)$ represents the number of years after 1992 required for the investment to reach $50,000.

5. The two graphs are reflections of each other across the line $y = x$; thus, they are inverses of each other.

7. $f(x) = \log_{2/3} x$ defines a decreasing function since the base, 2/3, is between 0 and 1.

9. $y = e^x$

The point $(0, 1)$ is on the graph since $e^0 = 1$, so the correct choice must be either A or D.
Since the base is e and $e > 1$, $y = e^x$ is an increasing function, and so the correct choice must be A.

11. $y = (.3)^x$

The point $(0, 1)$ is on the graph since $(.3)^0 = 1$, so the correct choice must be either A or D.
Since the base is $.3$ and $0 < .3 < 1$, $y = (.3)^x$ is a decreasing function, and so the correct choice must be D.

13. $100^{1/2} = 10$ is written in logarithmic form as $\log_{100} 10 = \frac{1}{2}$.

15. $\left(\frac{3}{4}\right)^{-1} = \frac{4}{3}$ is written in logarithmic form as $\log_{3/4} \left(\frac{4}{3}\right) = -1$.

17. $e^{2.4849} = 12$ is written in logarithmic form as $\log_e 12 = 2.4849$, or $\ln 12 = 2.4849$.

19. Since the graph is decreasing, $a < 1$.

21. The range of f is $(0, \infty)$.

23. To graph $y = f^{-1}(x)$, reflect $y = f(x)$ across the line $y = x$. See the answer graph in the back of the textbook.

25. $\log_{10} .001 = -3$ is written in exponential form as
$$10^{-3} = .001.$$

27. $\log 3.45 = .537819$ is written in exponential form as
$$10^{.537819} = 3.45.$$

29. $\log_9 27 = \frac{3}{2}$ is written in exponential form as
$$9^{3/2} = 27.$$

31. Let $f(x) = \log_a x$ be the required function. Then
$$f(81) = 4$$
$$\log_a 81 = 4$$
$$a^4 = 81$$
$$a^4 = 3^4$$
$$a = 3.$$

The base is 3.

33. $\log_3 \left(\frac{mn}{5r}\right)$

$= \log_3 mn - \log_3 5r$
    Logarithm of a quotient

$= \log_3 m + \log_3 n - \log_3 5 - \log_3 r$
    Logarithm of a product

35. $\log_5 (x^2 y^4 \sqrt[5]{m^3 p})$

$= \log_5 x^2 y^4 (m^3 p)^{1/5}$

$= \log_5 x^2 + \log_5 y^4 + \log_5 (m^3 p)^{1/5}$
    Logarithm of a product

$= 2 \log_5 x + 4 \log_5 y$

$\quad + \frac{1}{5}(\log_5 m^3 p)$
    Logarithm of a power

$$= 2 \log_5 x + 4 \log_5 y$$
$$+ \frac{1}{5}(\log_5 m^3 + \log_5 p)$$
Logarithm of a product
$$= 2 \log_5 x + 4 \log_5 y$$
$$+ \frac{1}{5}(3 \log_5 m + \log_5 p)$$
Logarithm of a power

37. The correct statement is

$$\log_5 125 - \log_5 25 = \log_5 \left(\frac{125}{25}\right)$$
$$= \log_5 5$$
$$= 1.$$

39. $\log .0411 = -1.386$

41. $\ln 144,000 = 11.878$

43. $\log_{2/3} \left(\frac{5}{8}\right)$

$$= \frac{\log \frac{5}{8}}{\log \frac{2}{3}}$$
$$\approx \frac{-.20412}{-.17609}$$
$$\approx 1.159$$

45. $y = y_0 e^{kt}$

When $t = 44$,

$$y = 2y_0.$$
$$2y_0 = y_0 e^{44k}$$
$$2 = e^{44k}$$

When $t = 22$,

$$y = y_0 e^{22k}$$
$$= y_0 (e^{44k})^{1/2}$$
$$= y_0 (2)^{1/2}$$
$$= 2^{1/2} y_0.$$

In 22 yr, the population will increase by a factor of $2^{1/2} \approx 1.4$.

47.
$$8^k = 32$$
$$(2^3)^k = 2^5$$
$$2^{3k} = 2^5$$
$$3k = 5$$
$$k = \frac{5}{3}$$

Solution set: $\left\{\frac{5}{3}\right\}$

49. $10^{2r-3} = 17$

Take common logarithms of both sides.

$$\log 10^{2r-3} = \log 17$$
$$(2r - 3) \log 10 = \log 17$$
$$2r - 3 = \log 17$$
$$2r = \log 17 + 3$$
$$r = \frac{\log 17 + 3}{2}$$
$$r \approx 2.115$$

Solution set: $\{2.115\}$

51. $\log_{64} y = \frac{1}{3}$

$$64^{1/3} = y \quad \text{Change to exponential form}$$
$$4 = y$$

Solution set: $\{4\}$

53. $\log_{16} \sqrt{x + 1} = \frac{1}{4}$

$$16^{1/4} = \sqrt{x + 1}$$
Change to exponential form
$$(16^{1/4})^2 = (\sqrt{x + 1})^2$$
Square both sides
$$16^{1/2} = x + 1$$
$$4 = x + 1$$
$$3 = x$$

The solution must be checked in the original equation since both sides were squared.

Solution set: $\{3\}$

55. $\ln [\ln (e^{-x})] = \ln 3$

$\ln (-x) = \ln 3$ *ln 3^{-x} = -x by theorem on inverses*

$-x = 3$

$x = 3$

Solution set: $\{-3\}$

57. Substitute P = 48,000, A = 58,344, r = .05, and m = 2 into the formula.

$$A = P\left(1 + \frac{r}{m}\right)^{tm}$$

$$58,344 = 48,000\left(1 + \frac{.05}{2}\right)^{(t)(2)}$$

$$58,344 = 48,000(1.025)^{2t}$$

$$1.2155 = (1.025)^{2t}$$

$$\ln 1.2155 = \ln (1.025)^{2t}$$

$$\ln 1.2155 = 2t \ln 1.025$$

$$\frac{\ln 1.2155}{2 \ln 1.025} = t$$

$$4.0 \approx t$$

$48,000 will increase to $58,344 in about 4.0 yr.

59. $A = P\left(1 + \frac{r}{m}\right)^{tm}$

First, substitute P = 12,000, r = .05, t = 8, and m = 1 into the formula.

$$A = 12,000\left(1 + \frac{.05}{1}\right)^{8(1)}$$

$$= 12,000(1.05)^8$$

$$\approx 12,000(1.4775)$$

$$\approx 17,729.47$$

After the first 8 yr, there would be $17,729.47 in the account.

To finish off the 14-year period, substitute P = 17,729.47, r = .06, t = 6, and m = 1 into the original formula.

$$A = 17,729.47\left(1 + \frac{.06}{1}\right)^{6(1)}$$

$$= 17,729.47(1.06)^6$$

$$\approx 17,729.47(1.4185)$$

$$\approx 25,149.59$$

At the end of the 14-year period, $25,149.59 would be in the account.

61. $A(t) = (5 \times 10^{12})e^{-.04t}$

(a) Use t = 1990 − 1970 = 20.

$$A(20) = (5 \times 10^{12})e^{-.04(20)}$$

$$= (5 \times 10^{12})e^{-.8}$$

$$\approx (5 \times 10^{12})(.4493)$$

$$= (5 \times .4493) \times 10^{12}$$

$$\approx 2.2 \times 10^{12}$$

In 1990, about 2.2×10^{12} tons of coal were available, according to this formula.

(b) Solve $A(t) = \frac{1}{2}A(0)$ for t.

Note that $\frac{1}{2}A(0) = \frac{1}{2}(5 \times 10^{12})$.

$$\frac{1}{2}(5 \times 10^{12}) = (5 \times 10^{12})e^{-.04t}$$

$$.5 = e^{-.04t}$$

$$\ln .5 = -.04t$$

$$-.6931 \approx -.04t$$

$$17 \approx t$$

According to this formula, coal reserves were half of their 1970 levels 17 yr after 1970, in 1987.

63. Graph $Y_1 = x^2$ and $Y_2 = 2^x$ on the same screen. Since the range of both functions is $(0, \infty)$, there is no need to include Quadrants III and IV in the viewing window. A good choice for the window is $[-5, 5]$ by $[0, 20]$.

The graph shows that the two curves intersect in three points. Two of these points are $(2, 4)$ and $(4, 16)$, as given in the exercise. To find the coordinates of the third point, use the "intersect" option in the CALC menu. The calculator displays the coordinates

$$(-.7666647, .58777476).$$

65. Graph $Y_3 = 3^{x+4} - 27^{x+1}$ in the window $[0, 3]$ by $[-20, 60]$. See the answer graph in the back of the textbook.

The x-intercept is .5, which agrees with the x-value found in Exercise 64.

67. To graph

$$Y_1 = \log_2 x + \log_2 (x + 2) - 3,$$

use the change-of-base theorem with base 10 to write the function as

$$Y_1 = \frac{\log x}{\log 2} + \frac{\log (x + 2)}{\log 2} - 3.$$

69. **(a)** Compute the natural logarithm of P for each value of P given in the table in the textbook.

x	ln P
0	6.921
1000	6.801
2000	6.678
3000	6.553
4000	6.425
5000	6.293
6000	6.157
7000	6.019
8000	5.878
9000	5.730
10,000	5.580

Plot this data on a graphing calculator using the window $[-1000, 11,000]$ by $[5, 7]$ with Xscl = 1000 and Yscl = 1.

See the answer graph in the textbook.

There appears to be a linear relationship.

(b) $P = Ce^{kx}$

Take natural logarithms on both sides.

$$\ln P = \ln Ce^{kx}$$
$$\ln P = \ln C + \ln e^{kx}$$
$$\ln P = kx + \ln C$$

Thus,

$$y = \ln P = ax + b$$

where $a = k$ and $b = \ln C$, which is a linear function.

71. **(a)** Plot the year on the x-axis and the number of processors on the y-axis. Let x = 0 correspond to the year 1971. See the answer graph in the textbook.

(b) The data are clearly not linear and do not level off like a logarithmic function. The data are increasing at a faster rate as x increases. Of the three choices, the exponential function (iii) will describe this data best.

(c) Let x = 0 correspond to the year 1971. We can require that the exponential function pass through the points

(0, 2300) and (24, 5,500,000).

We want to find a function of the form

$$f(x) = ae^{bx}.$$

First, find the value of a.

$$f(0) = ae^0 = 2300$$
$$a = 2300$$

Now find the value of b.

$$f(24) = 2300e^{24b} = 5,500,000$$
$$e^{24b} = \frac{5,500,000}{2300}$$
$$e^{24b} \approx 2391$$
$$24b \approx \ln 2391$$
$$b \approx \frac{\ln 2391}{24}$$
$$b \approx .3241$$

Thus, we obtain the exponential function

$$f(x) = 2300e^{.3241x}.$$

(If a different pair of points from the table is used, a slightly different function will be found.) See the answer graph in the textbook.

(d) We can predict the number of transistors on a chip in the year 2000 by evaluating

$$f(29) = 2300e^{.3241(29)} \approx 28,000,000.$$

Chapter 5   Test

1. $25^{2x-1} = 125^{x+1}$
$$(5^2)^{2x-1} = (5^3)^{x+1}$$
$$5^{4x-2} = 5^{3x+3}$$
$$4x - 2 = 3x + 3$$
$$x = 5$$

Solution set: $\{5\}$

2. $a^2 = b$ is written in logarithmic form as

$$\log_a b = 2.$$

3. $e^c = 4.82$

is written in logarithmic form as

$$\ln 4.82 = c.$$

4. $\log_3 \sqrt{27} = \frac{3}{2}$ is written in exponential form as

$$3^{3/2} = \sqrt{27}.$$

5. $\ln 5 = a$ is written in exponential form as

$$e^a = 5.$$

6. $y = \log_a x$

$\log_a a = 1$, so $(a, 1)$ is a point on the graph.

$\log_a 1 = 0$, so $(1, 0)$ is a point on the graph.

For Exercises 7 and 8, see the answer graphs in the textbook.

7. $y = (1.5)^{x+2}$

x	-4	-3	-2	-1	0	1	2
x + 2	-2	-1	0	1	2	3	4
y	.444	.667	1	1.5	2.25	3.375	5.0625

8. $y = \log_{1/2} x$

x	$\frac{1}{16}$	$\frac{1}{8}$	$\frac{1}{4}$	$\frac{1}{2}$	1	2	4	8	16
y	4	3	2	1	0	-1	-2	-3	-4

9. $\log_7 \left(\dfrac{x^2 \sqrt[4]{y}}{z^3} \right)$

$= \log_7 x^2 + \log_7 \sqrt[4]{y} - \log_7 z^3$

$= 2 \log_7 x + \dfrac{1}{4} \log_7 y - 3 \log_7 z$

10. $\ln 2300 = 7.741$

11. $\log_{2.7} 94.6 = \dfrac{\ln 94.6}{\ln 2.7}$

$\approx \dfrac{4.5497}{.99325}$

≈ 4.581

12. $\log_a (2x - 3) = -1$

$\log_a t$ is defined when t is in the interval $(0, \infty)$, but it is undefined when t is in the interval $(-\infty, 0]$.

$\log_a (2x - 3)$ will be undefined when $2x - 3$ is in the interval $(-\infty, 0]$, which corresponds to

$2x - 3 \leq 0$

$2x \leq 3$

$x \leq \dfrac{3}{2}.$

The numbers in the interval $\left(-\infty, \dfrac{3}{2}\right]$ cannot be solutions of the given equation.

14. $8^{2w-4} = 100$

Take natural logarithms of both sides.

$(2w - 4) \ln 8 = \ln 100$

$2w - 4 = \dfrac{\ln 100}{\ln 8}$

$\approx \dfrac{4.6052}{2.0794}$

$2w - 4 \approx 2.2147$

$2w \approx 6.2147$

$w \approx 3.107$

Solution set: $\{3.107\}$

15. $\log_3 (m + 2) = 2$

$3^2 = m + 2$

$9 = m + 2$

$7 = m$

Solution set: $\{7\}$

16. $\ln x - 4 \ln 3 = \ln \left(\dfrac{5}{x}\right)$

$\ln \left(\dfrac{x}{3^4}\right) = \ln \left(\dfrac{5}{x}\right)$

$\dfrac{x}{81} = \dfrac{5}{x}$

$x^2 = 405$

$x = 20.125$

(Since the domain of ln x is $(0, \infty)$, only the positive square root of 405 can be a solution.)

Solution set: $\{20.125\}$

17. $A(t) = 600e^{-.05t}$

(a) $A(12) = 600e^{-.05(12)}$

$= 600e^{-.6}$

$\approx 600(.5488)$

≈ 329.3

The amount of radioactive material present after 12 days is about 329.3 g.

(b) Solve $A(t) = \frac{1}{2} A_0$ for t.

Note that $\frac{1}{2}A_0 = \frac{1}{2}A(0) = \frac{1}{2}(600) = 300$.

$300 = 600e^{-.05t}$

$.5 = e^{-.05t}$

$\ln 5 = -.05t$

$-.691 \approx -.05t$

$13.9 \approx t$

The half-life of the material is about 13.9 days.

18. $v(t) = 176(1 - e^{-.18t})$

Find the time t at which $v(t) = 147$.

$147 = 176(1 - e^{-.18t})$

$\frac{147}{176} = 1 - e^{-.18t}$

$e^{-.18t} = 1 - \frac{147}{176}$

$e^{-.18t} = \frac{29}{176}$

$\ln e^{-.18t} = \ln \frac{29}{176}$

$-.18t \approx \ln .16477$

$t \approx \frac{\ln .16477}{-.18}$

$\approx \frac{-1.8}{-.18} = 10$

It will take the skydiver about 10 sec to attain the speed of 147 ft per sec (100 mph).

19. $A = P\left(1 + \frac{r}{m}\right)^{tm}$

$18,000 = 5000\left(1 + \frac{.068}{12}\right)^{12t}$

$3.6 = \left(1 + \frac{.068}{12}\right)^{12t}$

$\log 3.6 = 12t \log \left(1 + \frac{.068}{12}\right)$

$t = \frac{\log 3.6}{12 \log (1 + .068/12)}$

$t \approx 18.9$

It will take about 18.9 yr.

20. (a) The population of New York (in millions) can be approximated by

$y = 18.2(1.001)^x$,

while the population of Florida can be approximated by

$y = 14.0(1.017)^x$,

where $x = 0$ corresponds to July, 1994.

(b) Graph the two functions from (a) on the same screen. A good choice for the viewing window is [0, 25] by [13, 22]. Use the "intersect" option in the CALC menu to find the coordinates of the intersection point.

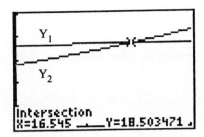

The graphs intersect at the point
(16.5, 18.5).

Since x = 0 corresponds to July,
1994, x = 16.5 corresponds to 16.5
years after July, 1994. Since 16
years after July, 1994 is July,
2010, we see that 16.5 years after
July, 1994 is January, 2011. We
estimate that Florida's population
will exceed New York's in January,
2011.

Cumulative Review Exercises (Chapter 1–5)

1. Use the distributive property of real numbers to rewrite the sum $6x - 3x^2$ as a product.

2. Evaluate $\dfrac{|-x^2|}{-x^2}$ if x is a nonzero real number.

3. Multiply the two polynomials $x^3 - 3x^2 + 1$ and $3x^2 + x - 4$.

4. Write out the binomial expansion of $(3m - 2n)^4$.

5. Find the product $\dfrac{x^2 + 7x + 10}{x^2 + x - 2} \cdot \dfrac{x^2 + 2x - 15}{x^2 + 5x - 24}$.

6. Simplify $\dfrac{4^{1/2} \cdot 4^{-2}}{4^3}$.

7. Write $\sqrt{4x^2 + 12xy + 9y^2}$ without a radical.

8. Find i^{-5}.

9. Solve the equation $\dfrac{4}{5 - x} - \dfrac{5}{2 + x} = 1$.

10. A rectangle is twice as long as it is wide. Find the dimensions of the rectangle if the perimeter is 24 inches.

11. Solve the equation $5 + \dfrac{2}{x} - \dfrac{3}{x^2} = 0$.

12. In 1991, Rick Mears won the (500-mi) Indianapolis 500 race. His speed (rate) was 100 mph faster than that of the 1911 winner, Ray Harroun. Mears completed the race in 3.74 hr less than Harroun. Find Mears's rate to the nearest whole number of miles per hour.

13. Solve $\sqrt{5x + 1} + \sqrt{7 - x} = 6$.

14. When the brakes of a car are applied, the speed that the car was traveling is proportional to the square root of the distance that the car travels before coming to a stop. Under certain conditions, a car moving at 60 mph will travel 18 m after the brakes are applied. Find a formula giving the stopping distance in terms of the speed.

15. Solve the inequality $5 \le \dfrac{1}{x - 2}$.

16. Solve the inequality $|3x - 5| > 2$.

17. Find the midpoint of the line segment with endpoints $(5, -3)$ and $(-4, 7)$.

18. Find the center and radius of the circle $x^2 - 6x + y^2 + 4y = 3$.

Exercises 19 through 21 refer to the function $f(x)$ whose graph is shown below.

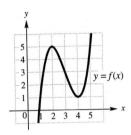

19. Find $f(2)$.

20. Solve $f(x) = 1$ for x.

21. Determine the interval(s) of the domain of $f(x)$ for which the function is decreasing.

22. Graph the equation $-x + 2y = 2$.

23. Write the equation in standard form of the line through $(-2, 5)$ and $(3, 7)$.

24. Find the equation in standard form of the line that passes through the point $(6, 4)$ and is perpendicular to the line $4x + 3y = 7$.

For Exercises 25 and 26, let $f(x) = 3\sqrt{x} - 1$; $g(x) = \begin{cases} 2x + 14 & \text{if } x \leq 2 \\ x^2 & \text{if } x > 2 \end{cases}$.

25. Find $(f + g)(1)$.

26. Find $f(g(3))$.

27. Without graphing, determine whether the graph of the equation $y = 2x^3 - 3x$ is symmetric with respect to the x-axis, the y-axis, the origin, or none of these.

28. The graph below is the result of applying two translations to the graph of $y = |x|$. Describe the translations and then give the equation of the graph.

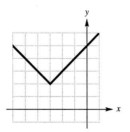

29. Find $f \circ g$ and $g \circ f$ where $f(x) = \sqrt{x - 1}$ and $g(x) = x^2 + 1$.

30. The convention for men's hat sizes varies from country to country. For instance, a hat of size 7 in the United States has size 6 7/8 in England and size 56 in France. The function

$$f(x) = \frac{1}{8}x$$

converts from French sizes to U.S. sizes, and the function

$$g(x) = 8x + 1$$

converts from English sizes to French sizes. Determine the function $f(g(x))$ and give its interpretation.

31. The graph of the parabola below has an equation of the form $y = a(x - h)^2 + k$, where a is 1 or −1. Determine a, h, and k.

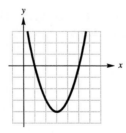

32. Graph the parabola $y = -\frac{1}{2}x^2 - 2x + 1$.

33. Use synthetic division to divide $2x^3 - 5x^2 + 4$ by $x - 3$.

34. Find the remainder when the polynomial $x^{98} + 2x^{49} - x^3$ is divided by $x - i$.

35. Find a polynomial $f(x)$ of degree 3 with zeros of 3, −2, and 5 for which $f(1) = 8$.

36. What is the least possible degree of the polynomial with real coefficients having 2, $3 + 4i$, and $-2 + 3i$ as zeros?

37. Graph the polynomial function $f(x) = .2x(x - 3)^2(x + 2)$.

38. Show that the polynomial $f(x) = x^4 + x^3 - 4x^2 - x + 3$ has no real zero greater than 2.

39. Sketch the graph of the rational function $f(x) = \dfrac{3(x^2 + x - 2)}{x^2}$.

40. Find an equation for the rational function graph below,

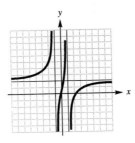

41. Find the inverse of the function $f(x) = \dfrac{5 - x}{4x}$.

42. Let $f(x) = \dfrac{x + 1}{x - 1}$. Show that $f(f(x)) = x$. What can we conclude about $f^{-1}(x)$?

43. Solve $81^x = 3^{2-x}$.

44. Find the present value of \$20,000 if interest is 4% compounded quarterly for two years.

45. Find $5^{\log_5 12}$.

46. Write $\log_2 \dfrac{3\sqrt{2}}{4}$ as a sum, difference, or product of logarithms. Simplify the result if possible.

47. Use natural logarithms to find $\log_3 8$.

48. Solve $\log (2 + 3x) - \log (4 + x) = \log 2$.

49. Solve $5e^{2x-3} = 10$.

50. A round table hanging in Winchester Castle (in England) was alleged to belong to King Arthur, who lived in the 5th century. A chemical analysis recently showed that the table had 91% of the amount of radiocarbon present in living wood. How old is the table? (Note: The decay constant of radiocarbon is $\approx .00012$.)

51. A parcel of land purchased for \$10,000 in 1990 was valued at \$14,000 in 1997. At what rate, compounded continuously, did the land appreciate? If the land continues to appreciate at the same rate, in what year will it be worth \$20,000?

Solutions to Cumulative Review Exercises (Chapters 1–5)

1. Use the distributive property to factor out the common factor of 3x.

$$6x - 3x^2 = 3x \cdot 2 - 3x \cdot x$$
$$= 3x(2 - x)$$

2. If x is a nonzero real number,

$$\frac{|-x^2|}{-x^2} = \frac{x^2}{-x^2} = -1.$$

3. Multiply using the distributive property.

$$(x^3 - 3x^2 + 1)(3x^2 + x - 4)$$
$$= 3x^5 + x^4 - 4x^3 - 9x^4 - 3x^3$$
$$+ 12x^2 + 3x^2 + x - 4$$
$$= 3x^5 - 8x^4 - 7x^3 + 15x^2 + x - 4$$

4. To expand $(3m - 2n)^4$, use the fourth row of Pascal's tringle.

$(3m - 2n)^4$
$= (3m)^4 + 4(3m)^3(-2n)^1 + 6(3m)^2(-2n)^2$
$\quad + 4(3m)^1(-2n)^3 + (-2n)^4$
$= 81m^4 + 4(27m^3)(-2n) + 6(9m^2)(4n^2)$
$\quad + 4(3m)(-8n^3) + 16m^4$
$= 81m^4 - 216m^3n + 216m^2n^2 - 96mn^3 + 16m^4$

5. $\dfrac{x^2 + 7x + 10}{x^2 + x - 2} \cdot \dfrac{x^2 + 2x - 15}{x^2 + 5x - 24}$

$$= \frac{(x + 2)(x + 5)}{(x + 2)(x - 1)} \cdot \frac{(x + 5)(x - 3)}{(x + 8)(x - 3)}$$

$$= \frac{(x + 5)(x + 5)}{(x - 1)(x + 8)}$$

$$= \frac{(x + 5)^2}{(x - 1)(x + 8)}$$

6. $\dfrac{4^{1/2} \cdot 4^{-2}}{4^3} = \dfrac{4^{1/2-2}}{4^3} = \dfrac{4^{-3/2}}{4^3} = 4^{-3/2-3}$

$$= 4^{-9/2} = (2^2)^{-9/2} = 2^{-9}$$

$$= \frac{1}{2^9} = \frac{1}{512}$$

7. $\sqrt{4x^2 + 12xy + 9y^2} = \sqrt{(2x + 3y)^2}$
$$= |2x + 3y|$$

8. $i^{-5} = \dfrac{1}{i^5} = \dfrac{1}{i} \cdot \dfrac{i}{i} = \dfrac{i}{-1} = -i$

9. $\dfrac{4}{5 - x} - \dfrac{5}{2 + x} = 1$

Multiply by the LCD, $(5 - x)(2 + x)$.

$$(5 - x)(2 + x) \cdot \frac{4}{5 - x} - (5 - x)(2 + x)$$

$$\cdot \frac{5}{2 + x} = (5 - x)(2 + x) \cdot 1$$

$$4(2 + x) - 5(5 - x) = (5 - x)(2 + x)$$
$$8 + 4x - 25 + 5x = 10 + 3x - x^2$$
$$x^2 + 6x - 27 = 0$$
$$(x + 9)(x - 3) = 0$$
$$x + 9 = 0 \quad \text{or} \quad x - 3 = 0$$
$$x = -9 \text{ or} \qquad x = 3$$

Solution set: $\{-9, 3\}$

10. Let x = width of the rectangle. Then 2x = length of the rectangle.

Use the formula for the perimeter of a rectangle.

$$P = 2L + 2W$$
$$24 = 2(2x) + 2(x)$$
$$24 = 4x + 2x$$
$$24 = 6x$$
$$4 = x$$

If x = 4, then 2x = 8.

The dimensions of the rectangle are 8 in. by 4 in.

11. $5 + \dfrac{2}{x} - \dfrac{3}{x^2} = 0$

Multiply by the LCD, x^2.

$$x^2 \cdot 5 + x^2 \cdot \dfrac{2}{x} - x^2 \cdot \dfrac{3}{x^2} = x^2 \cdot 0$$

$$5x^2 + 2x - 3 = 0$$

$$(5x - 3)(x + 1) = 0$$

$$5x - 3 = 0 \quad \text{or} \quad x + 1 = 0$$

$$5x = 3$$

$$x = \dfrac{3}{5} \quad \text{or} \quad\quad x = -1$$

Solution set: $\left\{-1,\ \dfrac{3}{5}\right\}$

12.   Let x = Mears's rate.

Then $x - 100$ = Harroun's rate.

Summarize the given table.

	d	r	t
Mears	500	x	$\dfrac{500}{x}$
Harroun	500	x − 100	$\dfrac{500}{x - 100}$

Harroun's time − Mears's time = 3.74

$$\dfrac{500}{x - 100} - \dfrac{500}{x} = 3.74$$

$$x(x - 100) \cdot \dfrac{500}{x - 100} - x(x - 100) \cdot \dfrac{500}{x}$$
$$= x(x - 100) \cdot 3.74$$

$$500(x) - 500(x - 100) = 3.74x(x - 100)$$

$$500x - 500x + 50{,}000 = 3.74x^2 - 374x$$

$$0 = 3.74x^2 - 374x$$
$$- 50{,}000$$

Use the quadratic formula, with $a = 3.74$, $b = -374$, $c = -50{,}000$.

$$x = \dfrac{374 \pm \sqrt{(374)^2 - 4(3.74)(-50{,}000)}}{2(3.74)}$$

Using a calculator,

$$x \approx 176 \quad \text{or} \quad x \approx -76.$$

Discard −76 since speed cannot be negative. Mears's speed is about 176 mph.

13. $\sqrt{5x + 1} + \sqrt{7 - x} = 6$

$$\sqrt{5x + 1} = 6 - \sqrt{7 - x}$$

Square both sides.

$$(\sqrt{5x + 1})^2 = (6 - \sqrt{7 - x})^2$$

$$5x + 1 = 36 - 12\sqrt{7 - x}$$
$$+ 7 - x$$

$$6x - 42 = -12\sqrt{7 - x}$$

Divide by 6.

$$x - 7 = -2\sqrt{7 - x}$$

Square both sides again.

$$(x - 7)^2 = (-2\sqrt{7 - x})^2$$

$$x^2 - 14x + 49 = 28 - 4x$$

$$x^2 - 10x + 21 = 0$$

$$(x - 3)(x - 7) = 0$$

$$x - 3 = 0 \quad \text{or} \quad x - 7 = 0$$

$$x = 3 \quad \text{or} \quad\quad c = 7$$

Check $x = 3$.

$$\sqrt{5(3) + 1} + \sqrt{7 - 3} = 6 \quad ?$$

$$\sqrt{16} + \sqrt{4} = 6 \quad ?$$

$$4 + 2 = 6 \quad\quad \textit{True}$$

Check x = 7.

$$\sqrt{5(7) + 1} + \sqrt{7 - 7} = 6 \quad ?$$
$$\sqrt{36} + \sqrt{0} = 6 \quad ?$$
$$6 + 0 = 6 \qquad \textit{True}$$

Both of the proposed solutions check.

Solution set: $\{3, 7\}$

14. Let s = the speed of the car,
    d = the distanced traveled.

We have

$$s = k\sqrt{d}$$
$$\text{or} \quad ks^2 = d$$

for some constant k.

Since d = 18 when s = 60,

$$k(60)^2 = 18$$
$$k = \frac{18}{60^2}$$
$$k = .005.$$

The formula is $d = .005s^2$.

15. $5 \leq \dfrac{1}{x - 2}$

We cannot solve this inequality by multiplying by the LCD, x - 2, because the sign of x - 2 depends on the value of x. If x - 2 were negative, we would have to reverse the inequality sign.

Solve the inequality using the sign graph method.

$$5 \leq \frac{1}{x - 2}$$
$$5 - \frac{1}{x - 2} \leq 0$$
$$\frac{5(x - 2) - 1}{x - 2} \leq 0$$
$$\frac{5x - 11}{x - 2} \leq 0$$

To draw a sign graph, first solve the equations

$$5x - 11 = 0 \quad \text{and} \quad x - 2 = 0,$$

getting the solutions

$$x = \frac{11}{5} \quad \text{and} \qquad x = 2.$$

Use the values 2 and 11/5 to divide the number line into three intervals: $(-\infty, 2)$, $(2, 11/5)$, and $(11/5, \infty)$.

Because x - 2 is negative in the interval $(-\infty, 2)$ and positive in the interval $(2, \infty)$, while 5x - 11 is negative in the interval $(-\infty, 11/5)$ and positive in the interval $(11/5, \infty)$, a sign graph will show that $\dfrac{5x - 11}{x - 2}$ is negative only in the interval $(2, 11/5)$. Because the inequality sign is "$\leq$", the endpoint 11/5 is included in the solution. However, since if x = 2, the denominator x - 2 would be 0, the endpoint 2 cannot be included. Therefore, the solution set is $(2, 11/5]$.

16. $|3x - 5| > 2$

$3x - 5 < -2$ or $3x - 5 > 2$

$3x < 3$ $\qquad\qquad$ $3x > 7$

$x < 1$ or $\qquad$ $x > \dfrac{7}{3}$

Solution set: $(-\infty, 1) \cup \left(\dfrac{7}{3}, \infty\right)$

17. Using the midpoint formula, the coordinates of M are

$\left(\dfrac{5 + (-4)}{2}, \dfrac{-3 + 7}{2}\right) = \left(\dfrac{1}{2}, 2\right).$

18. $x^2 - 6x + y^2 + 4y = 3$

Complete the square on x and on y.

$x^2 - 6x + 9 + y^2 + 4y + 4 = 3 + 9 + 4$

$(x - 3)^2 + (y + 2)^2 = 16$

The center is at $(3, -2)$ and the radius is 4.

19. From the graph, $f(2) = 5$.

20. From the graph, $f(x) = 1$ when

$x = 1$ or $x = 4$

Solution set: $\{1, 4\}$

21. The function is decreasing on the interval $[2, 4]$.

22. $-x + 2y = 2$

$2y = x + 2$

$y = \dfrac{1}{2}x + 1$

The slope is 1/2 and the y-intercept is 1. The line may be graphed by using the slope and y-intercept or by using the two intercepts: x-intercept = -2, y-intercept = 1.

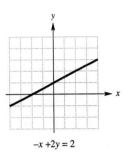

$-x + 2y = 2$

23. $(-2, 5)$ and $(3, 7)$

Find the slope first,

$m = \dfrac{7 - 5}{3 - (-2)} = \dfrac{2}{5}$

Use the slope and the point $(3, 7)$ in the point–slope form to find the equation of the line.

$y - 7 = \dfrac{2}{5}(x - 3)$

$5(y - 7) = 2(x - 3)$

$5y - 35 = 2x - 6$

$-29 = 2x - 5y$

$2x - 5y = -29$

24. Find the slope of the given line.

$4x + 3y = 7$

$3y = -4x + 7$

$y = -\dfrac{4}{3}x + \dfrac{7}{3}$

The slope is $-4/3$; the slope of the desired perpendicular line is $3/4$. Use the slope and the point $(6, 4)$ to find the equation.

$y - 4 = \dfrac{3}{4}(x - 6)$

$4(y - 4) = 3(x - 6)$

$4y - 16 = 3x - 18$

$2 = 3x - 4y$

$3x - 4y = 2$

25. First, find $f(1)$ and $g(1)$.

$$f(1) = 3\sqrt{1} - 1 = 2$$
$$g(1) = 2(1) + 14 = 16$$

$$(f + g)(1) = f(1) + g(1)$$
$$= 2 + 16$$
$$= 18$$

26. $g(3) = 3^2 = 9$

$$f(g(3)) = f(9) = 3\sqrt{9} - 1$$
$$= 9 - 1$$
$$= 8$$

27. $y = 2x^3 - 3x$

Replace x with $-x$.

$$y = 2(-x)^3 - 3(-x) = -2x^3 + 3x$$

This is not equivalent to the original equation, so the graph is not symmetric with respect to the x-axis.

Replace y with $-y$.

$$-y = 2x^3 - 3x$$
$$y = -2x^3 + 3x$$

This is not equivalent to the original equation, so the graph is not symmetric with respect to the y-axis.

Replace x with $-x$ and y with $-y$.

$$-y = 2(-x)^3 - 3(-x)$$
$$-y = -2x^3 + 3x$$
$$y = 2x^3 - 3x$$

This is equivalent to the original equation, so the graph is symmetric with respect to the origin.

28. The graph of $y = |x|$ is translated 3 units to the left and 2 units up. The equation is

$$y = |x + 3| + 2.$$

29. $f(x) = \sqrt{x - 1}$, $g(x) = x^2 + 1$

$$(f \circ g)(x) = f(g(x)) = f(x^2 + 1)$$
$$= \sqrt{x^2 + 1 - 1}$$
$$= \sqrt{x^2}$$
$$= |x|$$

$$(g \circ f)(x) = g(f(x)) = g(\sqrt{x - 1})$$
$$= (\sqrt{x - 1})^2 + 1$$
$$= x - 1 + 1$$
$$= x$$

30. $f(x) = \frac{1}{8}x$, $g(x) = 8x + 1$

$$h(x) = g(g(x)) = f(8x + 1)$$
$$= \frac{1}{8}(8x + 1)$$
$$= x + \frac{1}{8}$$

$h(x)$ converts from British sizes to U.S. sizes.

31. The vertex is $(3, -4)$ so $h = 3$, $k = -4$, giving the equation

$$y = a(x - 3)^2 - 4.$$

Using the point $(1, 0)$, substitute to find a.

$$0 = a(1 - 3)^2 - 4$$
$$0 = 4a - 4$$
$$1 = a$$

Therefore, $a = 1$, $h = 3$, and $k = -4$.

32. $y = -\frac{1}{2}x^2 - 2x + 1$

Complete the square to find the vertex.

$$y = -\frac{1}{2}(x^2 + 4x + 4) + 2 + 1$$

$$y = -\frac{1}{2}(x + 2)^2 + 3$$

The parabola opens down, and has a vertex of $(-2, 3)$.

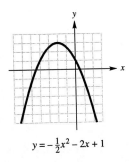

$y = -\frac{1}{2}x^2 - 2x + 1$

33. Use synthetic division to divide $2x^3 - 5x^2 + 4$ by $x - 3$.

```
3 |  2   -5    0    4
   |       6    3    9
   ‾‾‾‾‾‾‾‾‾‾‾‾‾‾‾‾‾‾
      2    1    3   13
```

The result is $2x^2 + x + 3 + \dfrac{13}{x - 3}$.

34. When $f(x) = x^{98} + 2x^{49} - x^3$ is divided by $x - i$, the remainder is $f(i)$.

$$f(i) = i^{98} + 2i^{49} - i^3$$
$$= i^2 + 2i - i^3$$
$$= -1 + 2i + i$$
$$= -1 + 3i$$

The remainder is $-1 + 3i$.

35. The polynomial has the form

$$f(x) = a(x - 3)(x + 2)(x - 5).$$

Since $f(1) = 8$,

$$8 = a(1 - 3)(1 + 2)(1 - 5)$$
$$8 = a(-2)(3)(-4)$$
$$8 = 24a$$
$$\frac{1}{3} = a.$$

Thus,

$$f(x) = \frac{1}{3}(x - 3)(x + 2)(x - 5)$$
$$= \frac{1}{3}(x^2 - x - 6)(x - 5)$$
$$= \frac{1}{3}(x^3 - 6x^2 - x + 30)$$
$$f(x) = \frac{1}{3}x^3 - 2x^2 - \frac{1}{3}x + 10$$

36. The complex numbers $3 - 4i$ and $-2 - 3i$ must also be zeros, because of the conjugate zeros theorem. Therefore, the polynomial has at least 5 zeros, and must be at least of degree 5.

37. $f(x) = .2x(x - 3)^2(x + 2)$

The zeros are 0, 3, and -2. These three zeros divide the x-axis into four regions.

Region	Test point	Value of $f(x)$	Sign of $f(x)$
$(-\infty, -2)$	-3	21.6	Positive
$(-2, 0)$	-1	-3.2	Negative
$(0, 3)$	1	2.4	Positive
$(3, \infty)$	4	4.8	Positive

Plot the three zeros and the test points and connect them with a smooth curve.

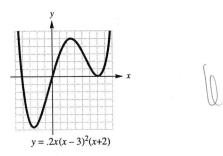

$y = .2x(x-3)^2(x+2)$

38. $f(x) = x^4 + x^3 - 4x^2 - x + 3$

Use synthetic division to divide $P(x)$ by $x - 2$.

```
2⌋  1   1   -4   -1    3
         2    6    4    6
   ─────────────────────────
    1   3    2    3    9
```

Since all the numbers in the bottom row are nonnegative, by the boundedness theorem, $f(x)$ has no zeros greater than 2.

39. $f(x) = \dfrac{3(x^2 + x - 2)}{x^2}$

$\qquad = \dfrac{3(x + 2)(x - 2)}{x^2}$

The vertical asymptote is $x = 0$. Rewrite the function as

$$f(x) = \frac{3x^2 + 3x - 6}{x^2}.$$

The horizontal asymptote is 3, since the numerator and denominator have the same degree and $\dfrac{a_n}{b_n} = \dfrac{3}{1} = 3$.

Since $f(0)$ is undefined, there is no y-intercept.

Solve $f(x) = 0$.

$$\frac{3(x + 2)(x - 1)}{x^2} = 0$$

$$3(x + 2)(x - 1) = 0$$

$$x = -2 \quad \text{or} \quad x = 1$$

The x-intercepts are −2 and 1. To determine whether the graph intersects its horizontal asymptote, which is $y = 3$, solve the equation $f(x) = 3$.

$$f(x) = \frac{3x^2 + 3x - 6}{x^2} = 3$$

$$3x^2 + 3x - 6 = 3x^2$$

$$3x - 6 = 0$$

$$x = 2$$

The graph intersects its horizontal asymptote at (2, 3).

The vertical asymptote and x-intercepts divide the x-axis into 4 regions.

Region	Test point	Value of $f(x)$	Sign of $f(x)$
$(-\infty, -2)$	−3	1.33	Positive
$(-2, 0)$	−1	−6	Negative
$(0, 1)$	.5	−15	Negative
$(1, \infty)$	2	3	Positive

Use this information to plot the graph.

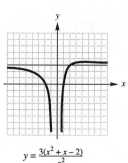

$y = \dfrac{3(x^2 + x - 2)}{x^2}$

40. The vertical asymptotes are $x = -1$ and $x = 1$, so the denominator contains the factors $(x + 1)$ and $(x - 1)$. The x-intercepts are 0 and 2, so the numerator contains the factors x and $(x - 2)$. The function is

$$f(x) = \frac{ax(x - 2)}{(x + 1)(x - 1)}.$$

Since the horizontal asymptote is $y = 2$, the value of a is 2, giving

$$f(x) = \frac{2x(x - 2)}{(x + 1)(x - 1)}.$$

41. $f(x) = \dfrac{5 - x}{4x}$

Let $y = f(x)$ and solve for x.

$$y = \frac{5 - x}{4x}$$

$$4xy = 5 - x$$

$$4xy + x = 5$$

$$x(4y + 1) = 5$$

$$x = \frac{5}{4y + 1} = f^{-1}(y)$$

Exchange x and y.

$$f^{-1}(x) = \frac{5}{4x + 1}$$

42. $f(x) = \dfrac{x + 1}{x - 1}$

$$f(f(x)) = f\!\left(\frac{x + 1}{x - 1}\right)$$

$$= \frac{\dfrac{x + 1}{x - 1} + 1}{\dfrac{x + 1}{x - 1} - 1}$$

$$= \frac{x + 1 + x - 1}{x - 1} \div \frac{x + 1 - (x - 1)}{x - 1}$$

$$= \frac{2x}{x - 1} \cdot \frac{x - 1}{2}$$

$$= x$$

Since $f(f(x)) = x$, $f(x)$ is its own inverse, that is, $f^{-1}(x) = f(x)$.

43.
$$81^x = 3^{2-x}$$
$$(3^4)^x = 3^{2-x}$$
$$3^{4x} = 3^{2-x}$$
$$4x = 2 - x$$
$$5x = 2$$
$$x = \frac{2}{5}$$

Solution set: $\left\{\dfrac{2}{5}\right\}$

44.
$$A = P\!\left(1 + \frac{r}{m}\right)^{tm}$$
$$20{,}000 = P\!\left(1 + \frac{.04}{4}\right)^{2(4)}$$
$$20{,}000 = P(1.01)^8$$
$$P = \frac{20{,}000}{(1.01)^8} = 18{,}469.66.$$

The present value is \$18,469.66.

45. $5^{\log_5 12} = 12$, by the theorem on inverses.

46.
$$\log_2 \frac{3\sqrt{2}}{4} = \log_2 3\sqrt{2} - \log_2 4$$
$$= \log_2 3 + \log_2 \sqrt{2} - 2$$
$$= \log_2 3 = \log_2 2^{1/2} - 2$$
$$= \log_2 3 + \frac{1}{2}\log_2 2 - 2$$
$$= \log_2 3 + \frac{1}{2} - 2$$
$$= \log_2 3 - \frac{3}{2}$$

47. By the change-of-base theorem,

$$\log_3 8 = \frac{\ln 8}{\ln 3} \approx 1.8927893.$$

48. $\log(2 + 3x) - \log(4 + x) = \log 2$

$$\log\left(\frac{2 + 3x}{4 + x}\right) = \log 2$$

$$\frac{2 + 3x}{4 + x} = 2$$

$$2 + 3x = 2(4 + x)$$

$$2 + 3x = 8 + 2x$$

$$x = 6$$

Solution set: $\{6\}$

Since $t = 0$ corresponds to the year 1990, $t = 14$ corresponds to the year 2004.

The land will be worth \$20,000 in the year 2004.

49. $5e^{2x-3} = 10$

$e^{2x-3} = 2$

$2x - 3 = \ln 2$

$2x = \ln 2 + 3$

$x = .5(\ln 2 + 3)$

Solution set: $\{.5(\ln 2 + 3)\}$

50. $y = y_0\, e^{-.00012t}$

$y = .91y_0$, therefore,

$.91y_0 = y_0\, e^{-.00012t}$

$.91 = e^{-.00012t}$

$\ln .91 = -.00012t$

$790 \approx t.$

The table is about 790 years old.

51. Let $t = 0$ represent the year 1990.

$$A = Pe^{rt}$$

$$14{,}000 = 10{,}000\, e^{r(7)}$$

$$1.4 = e^{7r}$$

$$\ln 1.4 = 7r$$

$$.048 \approx r$$

The interest rate is about 4.8%.
Find t when $A = 20{,}000$.

$$20{,}000 = 10{,}000e^{.048t}$$

$$2 = e^{.048t}$$

$$\ln 2 = .048t$$

$$14 \approx t$$

CHAPTER 6 SYSTEMS OF EQUATIONS AND INEQUALITIES

Section 6.1

1. From the graph, we see that two-piece aluminum can production equaled three-piece steel can production in approximately 1976.

3. The solution of the system containing the graphs of the two-piece steel production and the three-piece steel production is (1979, 10 billion).

5. If an equation were determined that modeled the production of total beverage cans produced, then x would represent the *year* and y would represent the *number of cans produced* (*in billions*).

7. $x - 5y = 8$ (*1*)

   $x = 6y$ (*2*)

   Substitute 6y for x in equation (1).

   $$6y - 5y = 8$$
   $$y = 8$$

   To find x, substitute 8 for y in equation (2).

   $$x = 6(8) = 48$$

   Solution set: $\{(48, 8)\}$

9. $6x - y = 5$ (*1*)

   $y = 11x$ (*2*)

   Substitute 11x for y in equation (1).

   $$6x - 11x = 5$$
   $$-5x = 5$$
   $$x = -1$$

   To find x, substitute -1 for x in equation (2).

   $$y = 11(-1) = -11$$

   Solution set: $\{(-1, -11)\}$

11. $7x - y = -10$ (*1*)

    $3y - x = 10$ (*2*)

    Solve equation (1) for y.

    $$y = 7x + 10 \quad (3)$$

    Substitute 7x + 10 for y in equation (2) and solve for x.

    $$3(7x + 10) - x = 10$$
    $$21x + 30 - x = 10$$
    $$20x = -20$$
    $$x = -1$$

    To find y, substitute -1 for x in equation (3).

    $$y = 7(-1) + 10 = 3$$

    Solution set: $\{(-1, 3)\}$

13. $-2x = 6y + 18$ (*1*)

    $-29 = 5y - 3x$ (*2*)

    Solve equation (1) for x by dividing both sides by -2.

    $$x = -3y - 9$$

Substitute this expression for x in equation (2).

$$-29 = 5y - 3(-3y - 9)$$
$$-29 = 5y + 9y + 27$$
$$-56 = 14y$$
$$y = -4$$

If $y = -4$, $x = -3(-4) - 9 = 3$.
Solution set: $\{(3, -4)\}$

15. $3y = 5x + 6$ (1)
    $x + y = 2$ (2)

Solve equation (2) for y by subtracting x from both sides.

$$y = 2 - x$$

Substitute this expression for y in equation (1).

$$3(2 - x) = 5x + 6$$
$$6 - 3x = 5x + 6$$
$$-3x = 5x$$
$$0 = 8x$$
$$0 = x$$

If $x = 0$, $y = 2 - 0 = 2$.
Solution set: $\{(0, 2)\}$

17. $4x + 2y = 6$ (1)
    $5x - 2y = 12$ (2)

Add equations (1) and (2) and solve for x.

$$
\begin{array}{r}
4x + 2y = 6 \\
\underline{5x - 2y = 12} \\
9x = 18 \\
x = 2
\end{array}
$$

Substitute $x = 2$ in equation (1) and solve for y.

$$4(2) + 2y = 6$$
$$8 + 2y = 6$$
$$2y = -2$$
$$y = -1$$

Solution set: $\{(2, -1)\}$

19. $3x - y = -4$ (1)
    $x + 3y = 12$ (2)

Multiply equation (2) by -3 and add the result to equation (1).

$$
\begin{array}{r}
3x - y = -4 \\
\underline{-3x - 9y = -36} \\
-10y = -40 \\
y = 4
\end{array}
$$

$x + 3(4) = 12$ Let $y = 4$ in (2)
$x = 0$

Solution set: $\{(0, 4)\}$

21. $4x + 3y = -1$ (1)
    $2x + 5y = 3$ (2)

Multiply equation (2) by -2 and add to equation (1).

$$
\begin{array}{r}
4x + 3y = -1 \\
\underline{-4x - 10y = -6} \\
-7y = -7 \\
y = 1
\end{array}
$$

$2x + 5(1) = 3$ Let $y = 1$ in (2)
$x = -1$

Solution set: $\{(-1, 1)\}$

23. $12x - 5y = 9$ *(1)*

$3x - 8y = -18$ *(2)*

$12x - 5y = 9$

$\underline{-12x + 32y = 72}$

$27y = 81$ *(3)*   *Add (1) to -4 times (2)*

$y = 3$ *Solve (3) for y*

$3x - 8(3) = -18$ *Let y = 3 in (2)*

$3x = 6$

$x = 2$

Solution set: $\{(2, 3)\}$

25. $\dfrac{x}{2} + \dfrac{y}{3} = 4$ *(1)*

$\dfrac{3x}{2} + \dfrac{3y}{2} = 15$ *(2)*

To clear denominators, multiply equation (1) by -6 and equation (2) by 2. Add the resulting equations.

$-3x - 2y = -24$

$\underline{3x + 3y = 30}$

$y = 6$

Substitute 6 for y in equation (1).

$\dfrac{x}{2} + \dfrac{6}{3} = 4$

$\dfrac{x}{2} + 2 = 4$

$\dfrac{x}{2} = 2$

$x = 4$

Solution set: $\{(4, 6)\}$

27. $9x - 2y = 30$ *(1)*

$\dfrac{5x}{2} + \dfrac{2y}{3} = 12$ *(2)*

Multiply equation (2) by 6 to clear denominators.

$9x - 2y = 30$ *(1)*

$15x + 4y = 72$ *(3)*

Multiply equation (1) by 2 and then add the result to equation (3).

$18x - 4y = 60$

$\underline{15x + 4y = 72}$

$33x = 132$

$x = 4$

Substitute this into equation (1).

$9(4) - 2y = 30$

$36 - 2y = 30$

$-2y = -6$

$y = 3$

Solution set: $\{(4, 3)\}$

29. $\dfrac{2x - 1}{3} + \dfrac{y + 2}{4} = 4$ *(1)*

$\dfrac{x + 3}{2} - \dfrac{x - y}{3} = 3$ *(2)*

Multiply equation (1) by 12 and equation (2) by 6 to clear denominators.

$4(2x - 1) + 3(y + 2) = 48$

$3(x + 3) - 2(x - y) = 18$

Remove parentheses and combine like terms.

$8x + 3y = 46$ *(3)*

$x + 2y = 9$ *(4)*

Multiply equation (4) by -8 and then add the result to equation (3).

$8x + 3y = 46$

$\underline{-8x - 16y = -72}$

$-13y = -26$

$y = 2$

Substitute this value into equation (4).

$$x + 2(2) = 9$$
$$x = 5$$

Solution set: $\{(5, 2)\}$

31. $\sqrt{3}x - y = 5$ (1)
$100x - y = 9$ (2)

We solve each equation for y and use

$$y_1 = \sqrt{3}x - 5$$
$$y_2 = -100x + 9.$$

Graph these two functions on the same screen. One suitable choice for the viewing window is [-1, 3] by [-10, 5]. Using the "intersect" option in the CALC menu, we find that the coordinates of the intersection point are approximately (.138, -4.762).

Solution set: $\{(.138, -4.762)\}$

33. $.2x + \sqrt{2}y = 1$ (1)
$\sqrt{5}x + .7y = 1$ (2)

We solve each equation for y and use

$$y_1 = \frac{1 - .2x}{\sqrt{2}}$$
$$y_2 = \frac{1 - \sqrt{5}x}{.7}.$$

Graph these two functions on the same screen. One suitable choice for the viewing window is [-2, 2] by [-2, 2]. Using the "intersect" option in the CALC menu, we find that the coordinates of intersection point are approximately (.236, .674).

Solution set: $\{(.236, .674)\}$

35. $9x - 5y = 1$ (1)
$-18x + 10y = 1$ (2)

Multiply equation (1) by 2 and add the result to equation (2).

$$18x - 10y = 2$$
$$\underline{-18x + 10y = 1}$$
$$0 = 3$$

This is a false statement. The solution set is $\emptyset$, and the system is inconsistent.

37. $4x - y = 9$ (1)
$-8x + 2y = -18$ (2)

Multiply equation (1) by 2 and add the result to equation (2).

$$8x - 2y = 18$$
$$\underline{-8x + 2y = -18}$$
$$0 = 0$$

This is a true statement. The equations are dependent. We will express the solution set with y as the arbitrary variable. Solve equation (1) for x.

$$4x - y = 9$$
$$x = \frac{y + 9}{4}$$

Solution set: $\left\{\left(\frac{y + 9}{4}, y\right)\right\}$

39. $3x - 4y = 1$ (1)
$2x + 3y = 12$ (2)

To eliminate y, multiply both sides of equation (1) by 3 and both sides of equation (2) by 4 to get equations (3) and (4).

$$9x - 12y = 3 \quad (3)$$
$$\underline{8x + 12y = 48} \quad (4)$$
$$17x \quad = 51$$

Thus, y has been eliminated.

41.
$$x + y + z = 2 \quad (1)$$
$$2x + y - z = 5 \quad (2)$$
$$x - y + z = -2 \quad (3)$$

Eliminate z first.
Add equations (1) and (2) to get

$$3x + 2y = 7. \quad (4)$$

Add equations (2) and (3) to get

$$3x = 3$$
$$x = 1.$$

$$3(1) + 2y = 7 \quad Let\ x = 1\ in\ (4)$$
$$2y = 4$$
$$y = 2$$

$$1 + 2 + z = 2 \quad Let\ x = 1,\ y = 2$$
$$in\ (1)$$
$$z = -1$$

Check by substituting x = 1, y = 2, and z = -1 in the three original equations.

Solution set: $\{(1, 2, -1)\}$

43.
$$x + 3y + 4z = 14 \quad (1)$$
$$2x - 3y + 2z = 10 \quad (2)$$
$$3x - y + z = 9 \quad (3)$$

Eliminate y first. Add equations (1) and (2) to get

$$3x + 6z = 24. \quad (4)$$

Multiply equation (3) by 3 and add the result to equation (1).

$$x + 3y + 4z = 14$$
$$\underline{9x - 3y + 3z = 27}$$
$$10x + \quad 7z = 41$$

This gives the new system

$$3x + 6z = 24 \quad (4)$$
$$10x + 7z = 41. \quad (5)$$

Multiply equation (4) by 10 and equation (5) by -3 and add.

$$30x + 60z = 240$$
$$\underline{-30x - 21z = -123}$$
$$39z = 117$$
$$z = 3$$

$$3x + 6(3) = 24 \quad Let\ z = 3\ in\ (4)$$
$$3x = 6$$
$$x = 2$$

$$2 + 3y + 4(3) = 14 \quad Let\ x = 2,$$
$$z = 3\ in\ (1)$$
$$3y = 0$$
$$y = 0$$

Solution set: $\{(2, 0, 3)\}$

45.
$$x + 4y - z = 6 \quad (1)$$
$$2x - y + z = 3 \quad (2)$$
$$3x + 2y + 3z = 16 \quad (3)$$

Eliminate z first. Add equations (1) and (2) to get

$$3x + 3y = 9 \quad or$$
$$x + y = 3. \quad (4)$$

Multiply equation (1) by 3 and add the result to equation (3).

$$3x + 12y - 3z = 18$$
$$\underline{3x + 2y + 3z = 16}$$
$$6x + 14y = 34 \quad \text{or}$$
$$3x + 7y = 17 \quad (5)$$

The new system is

$$x + y = 3 \quad (4)$$
$$3x + 7y = 17. \quad (5)$$

Multiply equation (3) by -3 and add the result to equation (7).

$$-3x - 3y = -9$$
$$\underline{3x + 7y = 17}$$
$$4y = 8$$
$$y = 2$$

$$x + 2 = 3 \quad \textit{Let } y = 2 \textit{ in (4)}$$
$$x = 1$$

$$1 + 4(2) - z = 6 \quad \textit{Let } x = 1,$$
$$ y = 2 \textit{ in (1)}$$
$$9 - z = 6$$
$$z = 3$$

Solution set: $\{(1, 2, 3)\}$

47. $5x + y - 3z = -6 \quad (1)$
$2x + 3y + z = 5 \quad (2)$
$-3x - 2y + 4z = 3 \quad (3)$

Eliminate y first.
Multiply equation (1) by 2 and add to equation (3).

$$10x + 2y - 6z = -12$$
$$\underline{-3x - 2y + 4z = 3}$$
$$7x - 2z = -9 \quad (4)$$

Multiply equation (1) by -3 and add to equation (2).

$$-15x - 3y + 9z = 18$$
$$\underline{2x + 3y + z = 5}$$
$$-13x + 10z = 23 \quad (5)$$

This gives the new system

$$7x - 2z = -9 \quad (4)$$
$$-13x + 10z = 23. \quad (5)$$

Multiply equation (4) by 5 and add to equation (5).

$$35x - 10z = -45$$
$$\underline{-13x + 10z = 23}$$
$$22x = -22$$
$$x = -1$$

$$7(-1) - 2z = -9 \quad \textit{Let } x = -1 \textit{ in (4)}$$
$$-2z = -2$$
$$z = 1$$

$$5(-1) + y - 3(1) = -6 \quad \textit{Let } x = -1,$$
$$ z = 1 \textit{ in (1)}$$
$$y - 8 = -6$$
$$y = 2$$

Solution set: $\{(-1, 2, 1)\}$

49. $x - 3y - 2z = -3 \quad (1)$
$3x + 2y - z = 12 \quad (2)$
$-x - y + 4z = 3 \quad (3)$

Eliminate x first. Add equations (1) and (3).

$$-4 + 2z = 0 \quad (4)$$

Multiply equation (3) by 3 and add to equation (2).

$$-3x - 3y + 12z = 9$$
$$\underline{3x + 2y - z = 12}$$
$$-2y + 11z = 21 \quad (5)$$

This gives the new system

$$-4y + 2z = 0 \quad (4)$$
$$-y + 11z = 21. \quad (5)$$

Multiply equation (5) by -4 and add to equation (4).

$$\begin{array}{r} -4y + 2z = 0 \\ \underline{4y - 44z = -84} \\ -42z = -84 \\ z = 2 \end{array}$$

$$-4y + 2(2) = 0 \quad \textit{Let } z = 2 \textit{ in } (4)$$
$$-4y = -4$$
$$y = 1$$

$$x - 3(1) - 2(2) = -3 \quad \textit{Let } y = 1,$$
$$\textit{z = 2 in } (1)$$
$$x - 7 = -3$$
$$x = 4$$

Solution set: $\{(4, 1, 2)\}$

51.
$$\begin{array}{llr} 2x + 6y - z = 6 & (1) \\ 4x - 3y + 5z = -5 & (2) \\ 6x + 9y - 2z = 11 & (3) \end{array}$$

Eliminate y first. Add equation (1) to 2 times equation (2).

$$\begin{array}{r} 2x + 6y - z = 6 \\ \underline{8x - 6y + 10z = -10} \\ 10x \quad\quad + 9z = -4 \quad (4) \end{array}$$

Add equation (3) to 3 times equation (2).

$$\begin{array}{r} 6x + 9y - 2z = 11 \\ \underline{12x - 9y + 15z = -15} \\ 18x \quad\quad + 13z = -4 \quad (5) \end{array}$$

This gives the new system

$$\begin{array}{lr} 10x + 9z = -4 & (4) \\ 18x + 13z = -4. & (5) \end{array}$$

Multiply equation (4) by 9 and equation (5) by -5 and add the results.

$$\begin{array}{r} 90x + 81z = -36 \\ \underline{-90x - 65z = 20} \\ 16z = -16 \\ z = -1 \end{array}$$

$$10x + 9(-1) = -4 \quad \textit{Let } z = -1 \textit{ in } (4)$$
$$10x = 5$$
$$x = \frac{1}{2}$$

$$2\left(\frac{1}{2}\right) + 6y - (-1) = 6 \quad \textit{Let } x = 1/2,$$
$$\textit{z = -1 in } (1)$$
$$6y + 2 = 6$$
$$6y = 4$$
$$y = \frac{2}{3}$$

Solution set: $\left\{\left(\frac{1}{2}, \frac{2}{3}, -1\right)\right\}$

53.
$$x + y + z = 4 \quad\quad (1)$$

(a)
$$\begin{array}{lr} x + 2y + z = 5 & (2) \\ 2x - y + 3z = 4 & (3) \end{array}$$

Equations (1), (2), and (3) form a system having exactly one solution, namely $(4, 1, -1)$.
(There are other equations that would do the same.)

(b)
$$\begin{array}{lr} x + y + z = 5 & (4) \\ 2x - y + 3z = 4 & (5) \end{array}$$

Equations (1), (4), and (5) form a system having no solution, since no ordered triple can satisfy equations (1) and (4) simultaneously.
(There are other equations that would do the same.)

(c) $2x + 2y + 2z = 8$ *(6)*

$\ \ 2x - \ y + 3z = 4$ *(7)*

Equations (1), (6), and (7) form a system having infinitely many solutions, since all the ordered triples that satisfy equation (1) will also satisfy equation (6).

(There are other equations that would do the same.)

55. $3x + \ 5y - \ z = -2$ *(1)*

$\ \ \ 4x - \ \ y + 2z = \ \ 1$ *(2)*

$\ -6x - 10y + 2z = \ \ 0$ *(3)*

We first eliminate z. Multiply equation (1) by 2 and add the result to equation (2).

$6x + 10y - 2z = -4$ *(1)*

$\underline{4x - \ \ y + 2z = \ \ 1}$ *(2)*

$10x + \ 9y = -3$ *(4)*

Multiply equation (2) by -1 and add the result to equation (3).

$-4x + \ \ y - 2z = -1$ *(2)*

$\underline{-6x - 10y + 2z = \ \ 0}$ *(3)*

$-10x - \ 9y = -1$ *(5)*

We now have the system

$10x + 9y = -3$ *(4)*

$-10x - 9y = -1.$ *(5)*

Adding these equations, we obtain

$$0 = -4,$$

which is a false statement.
The solution set is Ø, and the system is inconsistent.

57. $5x - 3y + z = 1$ *(1)*

$\ \ 2x + \ y - z = 4$ *(2)*

Geometrically, the solution is the intersection of two nonparallel planes, which is a line. There will be infinitely many ordered triples in the solution set. To describe these ordered triples, we proceed as follows. We will express the solution set with z as the arbitrary variable.

We first eliminate y. Multiply equation (2) by 3 and add the result to equation (1).

$5x - 3y + \ z = \ \ 1$

$\underline{6x + 3y - 3z = 12}$

$11x - 2z = 13$

Solve this equation for x in terms of z.

$$11x = 2z + 13$$

$$x = \frac{2}{11}z + \frac{13}{11}$$

Substitute this expression into equation (2); then solve for y in terms of z.

$$2\left(\frac{2}{11}z + \frac{13}{11}\right) + y - z = 4$$

$$\frac{4}{11}z + \frac{26}{11} + y - z = 4$$

$$y = 4 - \frac{4}{11}z - \frac{26}{11} + z$$

$$y = \frac{44}{11} - \frac{26}{11} - \frac{4}{11}z + \frac{11}{11}z$$

$$y = \frac{18}{11} + \frac{7}{11}z$$

The solution set is

$$\left\{\left(\frac{2}{11}z + \frac{13}{11}, \frac{7}{11}z + \frac{18}{11}, z\right)\right\}.$$

59.

$$x + y + z = 6 \quad (1)$$
$$2x - y - z = 3 \quad (2)$$

The two equations represent two non-parallel planes. The system will have infinitely many solutions, which we will express with z as the arbitrary variable.

We add the equations to obtain

$$3x = 9$$
$$x = 3.$$

Substitute this value into equation (1); then solve for y in terms of x.

$$3 + y + z = 6$$
$$y + z = 3$$
$$y = 3 - z$$

The solution set is $\{(3, 3 - z, z)\}.$

61. Since $y = ax + b$ and the line passes through $(-2, 1)$ and $(-1, -2)$, we have the following equations.

$$1 = a(-2) + b$$
$$-2 = a(-1) + b$$

This becomes the following system:

$$-2a + b = 1 \quad (1)$$
$$-a + b = -2. \quad (2)$$

Multiply equation (1) by -1 and add the result to equation (2).

$$2a - b = -1$$
$$\underline{-a + b = -2}$$
$$a = -3$$

Substitute this value into equation (1).

$$-2(-3) + b = 1$$
$$6 + b = 1$$
$$b = -5$$

The equation is $y = -3x - 5.$

63. Since $y = ax^2 + b + c$ and the parabola passes through the points $(2, 3)$, $(-1, 0)$, and $(-2, 2)$, we have the equations

$$3 = a(2)^2 + b(2) + c$$
$$0 = a(-1)^2 + b(-1) + c$$
$$2 = a(-2)^2 + b(-2) + c.$$

This becomes the following system.

$$4a + 2b + c = 3 \quad (1)$$
$$a - b + c = 0 \quad (2)$$
$$4a - 2b + c = 2 \quad (3)$$

First, we will eliminate c. Multiply equation (2) by -1 and add the result to equation (1).

$$4a + 2b + c = 3$$
$$\underline{-a + b - c = 0}$$
$$3a + 3b = 3 \quad (4)$$

Multiply equation (2) by -1 and add the result to equation (3).

$$-a + b - c = 0$$
$$\underline{4a - 2b + c = 2}$$
$$3a - b = 2 \quad (5)$$

We solve the system

$$3a + 3b = 3 \quad (4)$$
$$3a - b = 2 \quad (5)$$

by multiplying equation (4) by −1 and then adding the result to equation (5).

$$-3a - 3b = -3$$
$$\underline{3a - b = 2}$$
$$-4b = -1$$
$$b = \frac{1}{4}$$

Substitute this value into equation (5).

$$3a - \left(\frac{1}{4}\right) = 2$$
$$3a = \frac{9}{4}$$
$$a = \frac{3}{4}$$

Substitute this value into equation (1).

$$4\left(\frac{3}{4}\right) + 2\left(\frac{1}{4}\right) + c = 3$$
$$3 + \frac{1}{2} + c = 3$$
$$c = -\frac{1}{2}$$

The equation of the parabola is

$$y = \frac{3}{4}x^2 + \frac{1}{4}x - \frac{1}{2}.$$

65. Since $y = ax^2 + bx + c$ and the parabola passes through the points $(-2, -3.75)$, $(4, -3.75)$, and $(-1, -1.25)$, we have the equations

$$-3.75 = a(-2)^2 + b(-2) + c$$
$$-3.75 = a(4)^2 + b(4) + c$$
$$-1.25 = a(-1)^2 + b(-1) + c.$$

This becomes the following system.

$$4a - 2b + c = -3.75 \quad (1)$$
$$16a + 4b + c = -3.75 \quad (2)$$
$$a - b + c = -1.25 \quad (3)$$

First, we will eliminate c. Multiply equation (2) by −1 and add the result to equation (1).

$$4a - 2b + c = -3.75$$
$$\underline{-16a - 4b - c = 3.75}$$
$$-12a - 6b = 0 \quad (4)$$

Multiply equation (2) by −1 and add the result to equation (3).

$$-16a - 4b - c = 3.75$$
$$\underline{a - b + c = -1.25}$$
$$-15a - 5b = 2.50 \quad (5)$$

We now solve the system

$$-12a - 6b = 0 \quad (4)$$
$$-15a - 5b = 2.50. \quad (5)$$

Solve equation (4) for b.

$$-6b = 12a$$
$$b = -2a$$

Substitute this expression into equation (5).

$$-15a - 5(-2a) = 2.50$$
$$-5a = 2.50$$
$$a = -.5$$

Since $b = -2a$, $b = -2(-.5) = 1$. Substitute into equation (1).

$$4(-.5) - 2(1) + c = -3.75$$
$$-4 + c = -3.75$$
$$c = .25$$

The equation of the parabola is

$$y = -.5x^2 + x + .25$$

$$\text{or} \quad y = -\frac{1}{2}x^2 + x + \frac{1}{4}.$$

67. Since

$$x^2 + y^2 + ax + by + c = 0$$

and the circle passes through the points $(2, 1)$, $(-1, 0)$, and $(3, 3)$, we have the equations

$$(2)^2 + (1)^2 + a(2) + b(1) + c = 0$$
$$(-1)^2 + (0)^2 + a(-1) + b(0) + c = 0$$
$$(3)^2 + (3)^2 + a(3) + b(3) + c = 0.$$

This becomes the following system.

$$2a + b + c = -5 \quad (1)$$
$$-a + c = -1 \quad (2)$$
$$3a + 3b + c = -18 \quad (3)$$

First, we eliminate b by multiplying equation (1) by -3 and adding the result to equation (3).

$$-6a - 3b - 3c = 15$$
$$\underline{3a + 3b + c = -18}$$
$$-3a - 2c = -3 \quad (4)$$

We use this equation with equation (2) to form the system

$$-a + c = -1 \quad (2)$$
$$-3a - 2c = -3. \quad (4)$$

We eliminate c by multiplying equation (2) by 2 and adding the result to equation (4).

$$-2a + 2c = -2$$
$$\underline{-3a - 2c = -3}$$
$$-5a = -5$$
$$a = 1$$

We substitute this value into equation (2).

$$-(1) + c = -1$$
$$c = 0$$

Substitute into equation (1).

$$2(1) + b + (0) = -5$$
$$b = -7$$

The equation of the circle is

$$x^2 + y^2 + x - 7y = 0.$$

69. $p = 16 - \frac{5}{4}q$

(a) $p = 16 - \frac{5}{4} \cdot 0$

$$= 16$$

The price is $16.

(b) $p = 16 - \frac{5}{4} \cdot 4$

$$= 16 - 5$$
$$= 11$$

The price is $11.

(c) $p = 16 - \frac{5}{4} \cdot 8$

$$= 16 - 10$$
$$= 6$$

The price is $6.

(d) $\quad 6 = 16 - \frac{5}{4}q$

$$-10 = -\frac{5}{4}q$$
$$-40 = -5q$$
$$8 = q$$

The demand is 8 units.

(e) $11 = 16 - \frac{5}{4}q$

$-5 = -\frac{5}{4}q$

$-20 = -5q$

$4 = q$

The demand is 4 units.

(f) $16 = 16 - \frac{5}{4}q$

$0 = -\frac{5}{4}q$

$0 = q$

The demand is 0 units.

(g) See the graph of $p = 16 - \frac{5}{4}q$ in the textbook.

(h) $p = \frac{3}{4}q$

$0 = \frac{3}{4}q$

$0 = q$

(i) $10 = \frac{3}{4}q$

$\frac{4}{3}(10) = q$

$\frac{40}{3} = q$

(j) $20 = \frac{3}{4}q$

$\frac{4}{3}(20) = q$

$\frac{80}{3} = q$

(k) See the graph of $p = \frac{3}{4}q$ in the textbook.

(1) To find the equilibrium supply, solve the system

$p = 16 - \frac{5}{4}q$ (*1*)

$p = \frac{3}{4}q.$ (*2*)

The value of q will give the equilibrium supply.

$\frac{3}{4}q = 16 - \frac{5}{4}q$

$4\left(\frac{3}{4}q\right) = 4(16) - 4\left(\frac{5}{4}q\right)$

$3x = 64 - 5q$

$8q = 64$

$q = 8$

The equilibrium supply is 8.

(m) To find p, substitute q = 8 into equation (2).

$p = \frac{3}{4}(8)$

$= 6$

The equilibrium price is $6.

71. supply: $p = \frac{2}{5}q$

demand: $p = 100 - \frac{2}{5}q$

(a) See the graph in the textbook.

(b) The equilibrium demand may be found from reading the graph. The lines intersect at (125, 50), so the equilibrium demand is the x-coordinate of this point, 125. To solve algebraically, solve the system of the two given equations by the substitution method.

$$p = \frac{2}{5}q \qquad (1)$$

$$p = 100 - \frac{2}{q} \qquad (2)$$

$$\frac{2}{5}q = 100 - \frac{2}{q}$$

$$5\left(\frac{2}{5}q\right) = 5(100) - 5\left(\frac{2}{5}q\right)$$

$$2q = 500 - 2q$$

$$4q = 500$$

$$q = 125$$

The equilibrium demand is 125.

(c) The equilibrium price may be found by reading the graph. This price is the y-coordinate of the intersection point of the two lines, 50.
To find p algebraically, substitute q = 125 into equation (1).

$$p = \frac{2}{5}(125)$$

$$= 50$$

The equilibrium price is $50.

73. Let x = the number of $3.00 gallons;
y = the number of $4.50 gallons;
z = the number of $9.00 gallons.

One equation is y = 2x, or
2x − y = 0.

$$x + y + z = 300 \qquad (1)$$
$$2x - y = 0 \qquad (2)$$
$$3.00x + 4.50y + 9.00z = 6.00(300) \qquad (3)$$

Eliminate z.

$$-9x - 9y - 9z = -2700 \quad \textit{Multiply} \\ \textit{(1) by -9}$$
$$\underline{3x + 4.50y + 9z = 1800 \quad (3)}$$
$$-6x - 4.50y = -900 \quad (4)$$

Use equations (2) and (4) to solve for y .

$$6x - 3y = 0 \quad \textit{Multiply (2) by 3}$$
$$\underline{-6x - 4.50y = -900 \qquad (4)}$$
$$-7.50y = -900$$
$$y = 120$$

Now solve for x and z.

$$2x - 120 = 0 \quad \textit{Let y = 120 in (2)}$$
$$2x = 120$$
$$x = 60$$

$$60 + 120 + z = 300 \quad \textit{Let x = 60 and}$$
$$z = 120 \qquad \textit{y = 120 in (1)}$$

She should use 60 gal of the $3.00 water, 120 gal of the $4.50 water, and 120 gal of the $9.00 water.

75. Let x = the length of the shortest side;

y = the length of the medium side;

z = the length of the longest side.

$$z = y + 11$$
$$y = x + 3$$
$$x + y + z = 59$$

Rewrite these equations.

$$-y + z = 11 \quad (1)$$
$$-x + y = 3 \quad (2)$$
$$x + y + z = 59 \quad (3)$$

First eliminate x.

$$-x + y \qquad = 3 \quad (2)$$
$$\underline{x + y + z = 59 \quad (3)}$$
$$2y + z = 62 \quad (4)$$

Use equations (1) and (4) to solve for z.

$$-2y + 2z = 22 \qquad \textit{Multiply (1) by 2}$$
$$\underline{2y + z = 62 \qquad (4)}$$
$$3z = 84$$
$$z = 28$$

Now solve for x and y.

$$-y + 28 = 11 \quad \textit{Let z = 28 in (1)}$$
$$-y = -17$$
$$y = 17$$

$$-x + 17 = 3 \quad \textit{Let y = 17 in (2)}$$
$$-x = -14$$
$$x = 14$$

The lengths of the sides of the triangle are 14 inches, 17 inches, and 28 inches.

77. Let x = the amount invested at 5%;

　　y = the amount invested at 4.5%;

　　z = the amount invested at 3.75%.

$z = x + y - 20,000$ may be rewritten as $x + y - z = 20,000$.

$$x + y - z = 20,000 \quad (1)$$
$$x + y + z = 100,000 \quad (2)$$
$$.05x + .045y + .0375z = 4450 \quad (3)$$

First eliminate z. Add equations (1) and (2).

$$x + y - z = 20,000 \qquad (1)$$
$$\underline{x + y + z = 100,000 \qquad (2)}$$
$$2x + 2y = 120,000$$
$$\text{or} \quad x + y = 60,000 \qquad (4)$$

Multiply equation (1) by .0375 and add the result to equation (3).

$$.0375x + .0375y - .0375z = 750$$
$$\underline{.05x + .045y + .0375z = 4450}$$
$$.0875x + .0825y = 5200 \qquad (5)$$

Use equations (4) and (5) to solve for x.

Multiply equation (4) by $-.0825$ and add the result to equation (5).

$$-.0825x - .0825y = -4950$$
$$\underline{.0875x + .0825y = 5200}$$
$$.005x = 250$$
$$x = 50,000$$

Now solve for y and z.

$$50,000 + y = 60,000 \quad \textit{Let x = 50,000}$$
$$\textit{in (4)}$$
$$y = 10,000$$

$$50,000 + 10,000 + z = 100,000$$
$$\textit{Let x = 50,000}$$
$$\textit{and y = 10,000}$$
$$\textit{in (2)}$$
$$z = 40,000$$

The amounts invested were $50,000 at 5%, $10,000 at 4.5%, and $40,000 at 3.75%.

79. Since

$$C = at^2 + bt + c$$

and we have the ordered pairs (0, 315), (15, 325), and (30, 352), we have the equations

$$315 = a(0)^2 + b(0) + c$$
$$325 = a(15)^2 + b(15) + c$$
$$352 = a(30)^2 + b(30) + c.$$

This becomes the following system:

$$c = 315 \quad (1)$$
$$225a + 15b + c = 325 \quad (2)$$
$$900a + 30b + c = 352. \quad (3)$$

Since c = 315, we substitute this value into equations (2) and (3) to obtain

$$225a + 15b + 315 = 325$$
$$900a + 30b + 315 = 352.$$

This leads to the following system:

$$225a + 15b = 10 \quad (4)$$
$$900a + 30b = 37. \quad (5)$$

We eliminate b by multiplying equation (4) by -2 and adding the result to equation (5).

$$-450a - 30b = -20$$
$$\underline{900a + 30b = 37}$$
$$450a = 17$$
$$a = \frac{17}{450}$$

Substitute this value into equation (5).

$$900\left(\frac{17}{450}\right) + 30b = 37$$
$$34 + 30b = 37$$
$$30b = 3$$
$$b = \frac{1}{10}$$

The constants are a = 17/450, b = 1/10, and c = 315.
The relationship is

$$C = \frac{17}{450}t^2 + \frac{1}{10}t + 315.$$

(b) Since t = 0 corresponds to 1958, the amount of carbon dioxcide will be double its 1958 level when

$$\frac{17}{450}t^2 + \frac{1}{10}t + 315 = 2(315).$$

Solve this equation.

$$\frac{17}{450}t^2 + \frac{1}{10}t - 315 = 0$$
$$17t^2 + 45t - 141{,}750 = 0$$
$$(17t + 1575)(t - 90) = 0$$
$$t = -\frac{1575}{17} \quad \text{or} \quad t = 90$$

We reject the first proposed solution because time cannot be negative.
If t = 90, the year is 1958 + 90 or 2048.

81.
$$\frac{5}{x} + \frac{15}{y} = 16$$
$$\frac{5}{x} + \frac{4}{y} = 5$$

Let t = 1/x and u = 1/y.
The system becomes

$$5t + 15u = 16$$
$$5t + 4u = 5.$$

82. 5t + 15u = 16 \quad (1)
5t + 4u = 5 \quad (2)

We eliminate t by multiplying equation (2) by -1 and adding the result to equation (1).

$$5t + 15u = 16$$
$$\underline{-5t - 4u = -5}$$
$$11u = 11$$
$$u = 1$$

Substitute this value into equation (1).

$$5t + 15(1) = 16$$
$$5t = 1$$
$$t = \frac{1}{5}$$

Thus, $t = 1/5$ and $u = 1$.

83. Since $t = \frac{1}{5}$,

$$\frac{1}{x} = \frac{1}{5} \text{ and } x = 5.$$

Since $u = 1$,

$$\frac{1}{y} = 1 \text{ and } y = 1.$$

The solution for the given system is $x = 5$, $y = 1$.

84. $\frac{5}{x} + \frac{15}{y} = 16$

Multiply both sides by the LCD, xy.

$$xy\left(\frac{5}{x}\right) + xy\left(\frac{15}{y}\right) = xy(16)$$
$$5y + 15x = 16xy$$
$$5y - 16xy = -15x$$

Factor out x on the left.

$$y(5 - 16x) = -15x$$
$$y = \frac{-15x}{5 - 16x}$$

85. $\frac{5}{x} + \frac{4}{y} = 5$

Multiply both sides by xy.

$$xy\left(\frac{5}{x}\right) + xy\left(\frac{4}{y}\right) = xy(5)$$
$$5y + 4x = 5xy$$
$$5y - 5xy = -4x$$

Factor out y on the left.

$$y(5 - 5x) = -4x$$
$$y = \frac{-4x}{5 - 5x}$$

86. Use

$$Y_1 = \frac{-15x}{5 - 16x}$$

and

$$Y_2 = \frac{-4x}{5 - 5x}$$

in the viewing window $[0, 10]$ by $[0, 2]$. The point of intersection is $(5, 1)$.

87. $\frac{2}{x} + \frac{1}{y} = \frac{3}{2}$

$$\frac{3}{x} - \frac{1}{y} = 1$$

Let $t = \frac{1}{x}$ and $u = \frac{1}{y}$. The system becomes

$$2t + u = \frac{3}{2} \quad (1)$$
$$3t - u = 1. \quad (2)$$

Add the equations to eliminate u.

$$5t = \frac{5}{2}$$
$$t = \frac{1}{2}$$

Substitute this value into equation (1).

$$2\left(\frac{1}{2}\right) + u = \frac{3}{2}$$
$$1 + u = \frac{3}{2}$$
$$u = \frac{1}{2}$$

Since $t = \frac{1}{2}$, $\frac{1}{x} = \frac{1}{2}$ and $x = 2$.

Since $u = \frac{1}{2}$, $\frac{1}{y} = \frac{1}{2}$ and $y = 2$.

Solution set: $\{(2, 2)\}$

89.
$$\frac{1}{x} + \frac{1}{y} - \frac{1}{z} = \frac{1}{4}$$

$$\frac{2}{x} - \frac{1}{y} + \frac{3}{z} = \frac{9}{4}$$

$$-\frac{1}{x} - \frac{2}{y} + \frac{4}{z} = 1$$

Let $r = 1/x$, $s = 1/y$, and $t = 1/z$.
The system becomes

$$r + s - t = \frac{1}{4} \quad (1)$$

$$2r - s + 3t = \frac{9}{4} \quad (2)$$

$$-r - 2s + 4t = 1. \quad (3)$$

Eliminate s by adding equations (1) and (2).

$$r + s - t = \frac{1}{4}$$

$$\underline{2r - s + 3t = \frac{9}{4}}$$

$$3r \quad + 2t = \frac{5}{2} \quad (4)$$

Eliminate s again by multiplying equation (2) by -2 and adding the result to equation (3).

$$-4r + 2s - 6t = -\frac{9}{2}$$

$$\underline{-r - 2s + 4t = 1}$$

$$-5r \quad - 2t = -\frac{7}{2} \quad (5)$$

We now solve the system

$$3r + 2t = \frac{10}{4} \quad (4)$$

$$-5r - 2t = -\frac{7}{2}. \quad (5)$$

Adding equations (4) and (5), we obtain

$$-2r = -1$$

$$r = \frac{1}{2}.$$

Substitute this value into equation (4).

$$3\left(\frac{1}{2}\right) + 2t = \frac{5}{2}$$

$$2t = 1$$

$$t = \frac{1}{2}$$

Substitute these values into equation (1).

$$\left(\frac{1}{2}\right) + s - \left(\frac{1}{2}\right) = \frac{1}{4}$$

$$s = \frac{1}{4}.$$

Since $r = \frac{1}{2}$, $\frac{1}{x} = \frac{1}{2}$ and $x = 2$.

Since $s = \frac{1}{4}$, $\frac{1}{y} = \frac{1}{4}$ and $y = 4$.

Since $t = \frac{1}{2}$, $\frac{1}{z} = \frac{1}{2}$ and $z = 2$.

Solution set: $\{(2, 4, 2)\}$

Section 6.2

1. $x^2 = y - 1 \quad (1)$
    $y = 3x + 5 \quad (2)$

The graph shows that the points of intersection are $(-1, 2)$ and $(4, 17)$. First consider $(-1, 2)$. We substitute into equation (1).

$$(-1)^2 = 2 - 1 \quad ?$$

$$1 = 1 \qquad \qquad \textit{True}$$

Then we substitute into equation (2).

$$2 = 3(-1) + 5 \quad ?$$
$$2 = 2 \qquad\qquad \textit{True}$$

Next consider (4, 17). We substitute into equation (2).

$$(4)^2 = 17 - 1 \quad ?$$
$$16 = 16 \qquad\qquad \textit{True}$$

Then we substitute into equation (2).

$$17 = 3(4) + 5 \quad ?$$
$$17 = 17 \qquad\qquad \textit{True}$$

Thus, (-1, 2) and (4, 17) are solutions of the system.

3. $x^2 + y^2 = 5 \quad (1)$
$-3x + 4y = 2 \quad (2)$

The graph shows that the points of intersection are (-2, -1) and (38/25, 41/25).
First consider (-2, -1). We substitute into equation (1).

$$(-2)^2 + (-1)^2 = 5 \quad ?$$
$$5 = 5 \qquad \textit{True}$$

Then we substitute into equation (2).

$$-3(-2) + 4(-1) = 2 \quad ?$$
$$2 = 2 \qquad \textit{True}$$

Next consider (38/25, 41/25). We substitute into equation (1).

$$\left(\frac{38}{25}\right)^2 + \left(\frac{41}{25}\right)^2 = 5 \quad ?$$
$$\frac{1444}{625} + \frac{1681}{625} = 5 \quad ?$$
$$\frac{3125}{625} = 5 \quad ?$$
$$5 = 5 \qquad \textit{True}$$

Then we substitute into equation (2).

$$-3\left(\frac{38}{25}\right) + 4\left(\frac{41}{25}\right) = 2 \quad ?$$
$$-\frac{114}{25} + \frac{164}{25} = 2 \quad ?$$
$$\frac{50}{25} = 2 \quad ?$$
$$2 = 2 \qquad \textit{True}$$

Thus, (-2, -1) and (38/25, 41/25) are solutions of the system.

5. $y = x^2 \qquad (1)$
$x + y = 2 \quad (2)$

The screens show that the points of intersection are (-2, 4) and (1, 1).
First consider (-2, 4). We substitute into equation (1).

$$4 = (-2)^2 \quad ?$$
$$4 = 4 \qquad\qquad \textit{True}$$

Then we substitute into equation (2).

$$(-2) + 4 = 2 \quad ?$$
$$2 = 2 \qquad\qquad \textit{True}$$

Next consider (1, 1). We substitute into equation (1).

$$1 = (1)^2 \quad ?$$
$$1 = 1 \qquad\qquad \textit{True}$$

Then we substitute into equation (2).

$$(1) + (1) = 2 \quad ?$$
$$2 = 2 \qquad \textit{True}$$

Thus, $(-2, 4)$ and $(1, 1)$ are solutions of the system.

7. The system

$$x^2 - y = 4$$
$$x + y = -2$$

cannot have more than two solutions because a parabola and a line cannot intersect in more than two points.

9. $y = x^2 \qquad (1)$
$x + y = 2 \qquad (2)$

Use the substitution method.
Solve equation (2) for y.

$$y = 2 - x$$

Substitute this result into equation (1).

$$2 - x = x^2$$
$$x^2 + x - 2 = 0$$
$$(x + 2)(x - 1) = 0$$
$$x = -2 \quad \text{or} \quad x = 1$$

If $x = -2$, then $y = (-2)^2 = 4$.
If $x = 1$, then $y = 1^2 = 1$.

Solution set: $\{(-2, 4), (1, 1)\}$

11. $y = (x - 1)^2 \qquad (1)$
$x - 3y = -1 \qquad (2)$

Substitute $(x - 1)^2$ for y in equation (2).

$$x - 3(x - 1)^2 = -1$$
$$x - 3(x^2 - 2x + 1) = -1$$
$$x - 3x^2 + 6x - 3 = -1$$
$$-3x^2 + 7x - 2 = 0$$
$$3x^2 - 7x + 2 = 0$$
$$(3x - 1)(x - 2) = 0$$
$$x = \frac{1}{3} \quad \text{or} \quad x = 2$$

If $x = \frac{1}{3}$, then

$$y = \left(\frac{1}{3} - 1\right)^2$$
$$= \left(-\frac{2}{3}\right)^2 = \frac{4}{9}.$$

If $x = 2$, then

$$y = (2 - 1)^2$$
$$= 1^2 = 1.$$

Solution set: $\left\{\left(\frac{1}{3}, \frac{4}{9}\right), (2, 1)\right\}$

13. $y = x^2 + 4x \qquad (1)$
$2x - y = -8 \qquad (2)$

Substitute $x^2 + 4x$ for y in equation (2).

$$2x - (x^2 + 4x) = -8$$
$$2x - x^2 - 4x = -8$$
$$x^2 + 2x - 8 = 0$$
$$(x - 2)(x + 4) = 0$$
$$x = 2 \quad \text{or} \quad x = -4$$

If $x = 2$, then

$$y = 2^2 + 4(2)$$
$$= 4 + 8 = 12.$$

If x = -4, then

$$y = (-4)^2 + 4(-4)$$
$$= 0.$$

Solution set: $\{(2, 12), (-4, 0)\}$

15. $3x^2 + 2y^2 = 5$ (1)
 $x - y = -2$ (2)

Solve equation (2) for x.

$$x = y - 2$$

Substitute y - 2 for x in equation (1).

$$3(y - 2)^2 + 2y^2 = 5$$
$$3(y^2 - 4y + 4) + 2y^2 = 5$$
$$2y^2 - 12y + 12 + 2y^2 = 5$$
$$5y^2 - 12y + 7 = 0$$
$$(5y - 7)(y - 1) = 0$$
$$y = \frac{7}{5} \quad \text{or} \quad y = 1$$

If $y = \frac{7}{5}$, then,

$$x = y - 2$$
$$= \frac{7}{5} - 2$$
$$= -\frac{3}{5}.$$

If y = 1, then

$$x = 1 - 2 = -1.$$

Solution set: $\left\{\left(-\frac{3}{5}, \frac{7}{5}\right), (-1, 1)\right\}$

17. $x^2 + y^2 = 8$ (1)
 $x^2 - y^2 = 0$ (2)

Use the addition method.
Add equations (1) and (2).

$$x^2 + y^2 = 8$$
$$\underline{x^2 - y^2 = 0}$$
$$2x^2 \qquad = 8$$
$$x^2 = 4$$
$$x = \pm 2$$

If x = 2, then

$$2^2 - y^2 = 0$$
$$4 - y^2 = 0$$
$$y^2 = 4$$
$$y = \pm 2.$$

If x = -2, then

$$(-2)^2 - y^2 = 0$$
$$4 - y^2 = 0$$
$$y^2 = 4$$
$$y = \pm 2.$$

Solution set:
$\{(2, 2), (2, -2), (-2, 2), (-2, -2)\}$

19. $5x^2 - y^2 = 0$ (1)
 $3x^2 + 4y^2 = 0$ (2)

Multiply equation (1) by 4 and add to equation (2).

$$20x^2 - 4y^2 = 0$$
$$\underline{3x^2 + 4y^2 = 0}$$
$$23x^2 \qquad = 0$$
$$x = 0$$

If x = 0,

$$5(0)^2 - y^2 = 0,$$
$$y = 0.$$

Solution set: $\{(0, 0)\}$

21. $3x^2 + y^2 = 3$ (1)

$4x^2 + 5y^2 = 26$ (2)

Multiply equation (1) by −5 and add to equation (2).

$$-15x^2 - 5y^2 = -15$$
$$\underline{4x^2 + 5y^2 = 26}$$
$$-11x^2 = 11$$
$$x^2 = -1$$
$$x = \pm i$$

If $x = \pm i$,

$$3(\pm i)^2 + y^2 = 3$$
$$-3 + y^2 = 3$$
$$y^2 = 6$$
$$y = \pm\sqrt{6}.$$

Solution set:

$$\{(i, \sqrt{6}), \ (-i, \sqrt{6}), \ (i, -\sqrt{6}), \ (-i, -\sqrt{6})\}$$

23. $2x^2 + 3y^2 = 5$ (1)

$3x^2 - 4y^2 = -1$ (2)

Multiply equation (1) by 4 and equation (2) by 3 and add.

$$8x^2 + 12y^2 = 20$$
$$\underline{9x^2 - 12y^2 = -3}$$
$$17x^2 = 17$$
$$x^2 = 1$$
$$x = \pm 1$$

If $x = \pm 1$,

$$2(\pm 1)^2 + 3y^2 = 5$$
$$3y^2 = 3$$
$$y^2 = 1$$
$$y = \pm 1.$$

Solution set:

$$\{(1, 1), \ (1, -1), \ (-1, 1), \ (-1, -1)\}$$

25. $2x^2 + 2y^2 = 20$ (1)

$4x^2 + 4y^2 = 30$ (2)

We eliminate x^2 by multiplying equation (1) by −2 and adding the result to equation (2).

$$-4x^2 - 4y^2 = -40$$
$$\underline{4x^2 + 4y^2 = 30}$$
$$0 = -10$$

This is a false statement.
The solution set is ∅.

27. $2x^2 - 3y^2 = 8$ (1)

$6x^2 + 5y^2 = 24$ (2)

$-6x^2 + 9y^2 = -24$ *Multiply (1) by −3*

$\underline{6x^2 + 9y^2 = 24}$

$$18y^2 = 0$$
$$y^2 = 0$$
$$y = 0$$

If $y = 0$, then

$$2x^2 - 3 \cdot 0 = 8$$
$$2x^2 = 8$$
$$x^2 = 4$$
$$x = \pm 2.$$

Solution set: $\{(2, 0), \ (-2, 0)\}$

29. $xy = 8$ (1)

$3x + 2y = -16$ (2)

Use the substitution method.
Solve equation (1) for y.

$$y = \frac{8}{x} \qquad (3)$$

Substitute $\frac{8}{x}$ for y in equation (2) and solve for x.

$$3x + 2 \cdot \frac{8}{x} = -16$$

$$3x^2 + 16 = -16x \quad \textit{Multiply by x}$$

$$3x^2 + 16x + 16 = 0$$

$$(x + 4)(3x + 4) = 0$$

$$x = -4 \quad \text{or} \quad x = -\frac{4}{3}$$

If $x = -4$, then

$$-4 \cdot y = 8$$

$$y = -2.$$

If $x = -\frac{4}{3}$, then

$$-\frac{4}{3} \cdot y = 8$$

$$y = -6.$$

Solution set: $\left\{(-4, -2), \left(-\frac{4}{3}, -6\right)\right\}$

31.
$$-5xy + 2 = 0 \qquad (1)$$
$$x - 15y = 5 \qquad (2)$$

Solve equation (1) for y.

$$-5xy + 2 = 0$$

$$-5xy = -2$$

$$y = \frac{2}{5x} \quad (3)$$

Substitute $\frac{2}{5x}$ for y in equation (2) and solve for x.

$$x - 15\left(\frac{2}{5x}\right) = 5$$

$$x - \frac{6}{x} = 5$$

$$x^2 - 6 = 5x \quad \textit{Multiply by x}$$

$$x^2 - 5x - 6 = 0$$

$$(x - 6)(x + 1) = 0$$

$$x = 6 \quad \text{or} \quad x = -1$$

Substitute $x = 6$ and $x = -1$ into equation (3) to find the corresponding value of y.

If $x = 6$, then

$$y = \frac{2}{5 \cdot 6}$$

$$= \frac{1}{15}.$$

If $x = -1$, then

$$y = \frac{2}{5(-1)}$$

$$= -\frac{2}{5}.$$

Solution set: $\left\{\left(6, \frac{1}{15}\right), \left(-1, -\frac{2}{5}\right)\right\}$

33.
$$5x^2 - 2y^2 = 6 \qquad (1)$$
$$xy = 2 \qquad (2)$$

Solve equation (2) for y.

$$y = \frac{2}{x} \quad (3)$$

Substitute $\frac{2}{x}$ for y in equation (1) and solve for x.

$$5x^2 - 2\left(\frac{2}{x}\right)^2 = 6$$

$$5x^2 - \frac{8}{x^2} = 6$$

$$5x^4 - 8 = 6x^2 \quad \textit{Multiply by } x^2$$

$$5x^4 - 6x^2 - 8 = 0$$

$$(5x^2 + 4)(x^2 - 2) = 0$$

$$5x^2 + 4 = 0$$

$$x^2 = -\frac{4}{5}$$

$$x = \pm\frac{2}{\sqrt{5}}i = \pm\frac{2\sqrt{5}}{5}i$$

or

$$x^2 - 2 = 0$$
$$x^2 = 2$$
$$x = \pm\sqrt{2}.$$

Substitute each value of x into equation (3) to find the corresponding value of y.

If $x = \dfrac{2}{\sqrt{5}}i$,

$$y = \frac{2\sqrt{5}}{2i}$$
$$= \frac{\sqrt{5}}{i} \cdot \frac{i}{i}$$
$$= -i\sqrt{5}.$$

If $x = -\dfrac{2}{\sqrt{5}}i$,

$$y = -\frac{2\sqrt{5}}{2i}$$
$$= -\frac{\sqrt{5}}{i} \cdot \frac{i}{i}$$
$$= i\sqrt{5}.$$

If $x = \sqrt{2}$,

$$y = \frac{2}{\sqrt{2}}$$
$$= \sqrt{2}.$$

If $x = -\sqrt{2}$,

$$y = -\sqrt{2}.$$

Solution set:

$$\left\{ \left(\frac{2\sqrt{5}}{5}i, -i\sqrt{5}\right), \left(-\frac{2\sqrt{5}}{5}i, i\sqrt{5}\right), \left(\sqrt{2}, \sqrt{2}\right), \left(-\sqrt{2}, -\sqrt{2}\right) \right\}$$

35.

$$3x^2 + xy + 3y^2 = 7 \qquad (1)$$
$$x^2 + y^2 = 2 \qquad (2)$$

$$\begin{array}{rl} 3x^2 + xy + 3y^2 = & 7 \\ -3x^2 \qquad\quad - 3y^2 = & -6 \quad \text{\textit{Multiply (2)}} \\ \hline xy \qquad\quad = & 1 \quad \text{\textit{by -3}} \end{array}$$

$$y = \frac{1}{x} \qquad (3)$$

Substitute $\dfrac{1}{x}$ for y in equation (2) and solve for x.

$$x^2 + \frac{1}{x^2} = 2$$
$$x^4 + 1 = 2x^2 \quad \text{\textit{Multiply by } } x^2$$
$$x^4 - 2x^2 + 1 = 0$$
$$(x^2 - 1)(x^2 - 1) = 0$$
$$x^2 - 1 = 0$$
$$x^2 = 1$$
$$x = \pm 1$$

Substitute each value of x into equation (3) to find the corresponding value of y.

If $x = 1$,

$$y = \frac{1}{1}$$
$$y = 1.$$

If $x = -1$,

$$y = \frac{1}{-1}$$
$$y = -1.$$

Solution set: $\{(1, 1), (-1, -1)\}$

37.

$$3x^2 + 2xy - y^2 = 9 \qquad (1)$$
$$x^2 - xy + y^2 = 9 \qquad (2)$$

This system can be solved using a combination of the addition and substitution methods.

First add equations (1) and (2).

$$3x^2 + 2xy - y^2 = 9 \qquad (1)$$
$$\underline{x^2 - xy + y^2 = 9} \qquad (2)$$
$$4x^2 + xy = 18 \qquad (3)$$

Solve equation (3) for y.

$$xy = 18 - 4x^2$$
$$y = \frac{18 - 4x^2}{x} \qquad (4)$$

Substitute $\dfrac{18 - 4x^2}{x}$ for y in

equation (2) and solve for x.

$$x^2 - x\left(\frac{18 - 4x^2}{x}\right) + \left(\frac{18 - 4x^2}{x}\right)^2 = 9$$

$$x^2 - 18 + 4x^2 + \frac{324 - 144x^2 + 16x^4}{x^2} = 9$$

$$x^4 - 18x^2 + 4x^4 + 324 - 144x^2 + 16x^4 = 9x^2$$

Multiply by x^2

$$21x^4 - 171x^2 + 324 = 0$$

Combine terms

$$7x^4 - 57x^2 + 108 = 0$$

Divide by 3

$$(7x^2 - 36)(x^2 - 3) = 0$$

$$7x^2 - 36 = 0$$

$$7x^2 = 36$$

$$x^2 = \frac{36}{7}$$

$$x = \pm\frac{6}{\sqrt{7}} \quad \text{or} \quad x^2 - 3 = 0$$

$$= \pm\frac{6\sqrt{7}}{7} \qquad x^2 = 3$$

$$x = \pm\sqrt{3}$$

Substitute each value of x into equation (4) to find the corresponding value of y.

If $x = \dfrac{6\sqrt{7}}{7}$,

$$y = \frac{18 - 4\left(\frac{36}{7}\right)}{\frac{6\sqrt{7}}{7}}$$

$$= -\frac{3\sqrt{7}}{7}.$$

If $x = -\dfrac{6\sqrt{7}}{7}$, $y = \dfrac{3\sqrt{7}}{7}$.

If $x = \sqrt{3}$,

$$y = \frac{18 - 4 \cdot 3}{\sqrt{3}}$$

$$= 2\sqrt{3}.$$

If $x = -\sqrt{3}$, $y = -2\sqrt{3}$.

Solution set:

$$\left\{\left(\frac{6\sqrt{7}}{7}, -\frac{3\sqrt{7}}{7}\right), \left(-\frac{6\sqrt{7}}{7}, \frac{3\sqrt{7}}{7}\right), \right.$$
$$\left. (\sqrt{3}, 2\sqrt{3}), (-\sqrt{3}, -2\sqrt{3})\right\}$$

39.
$$2x + |y| = 4 \qquad (1)$$
$$x^2 + y^2 = 5 \qquad (2)$$

Solve equation (1) for y.

$$|y| = -2x + 4$$
$$y = -2x + 4 \quad \text{or} \quad (3)$$
$$y = 2x - 4 \qquad (4)$$

Substitute $-2x + 4$ for y in equation (2) and solve for x.

$$x^2 + (-2x + 4)^2 = 5$$
$$x^2 + 4x^2 - 16x + 16 = 5$$
$$5x^2 - 16x + 11 = 0$$
$$(5x - 11)(x - 1) = 0$$
$$x = \frac{11}{5} \quad \text{or} \quad x = 1$$

Substituting $2x - 4$ for y in equation (2) leads to the same solutions.

Solve equation (2) for y.

$$x^2 + y^2 = 5$$
$$y = \pm\sqrt{5 - x^2} \quad (5)$$

Substitute each value of x into equation (5) to find the corresponding values of y.

If $x = 1$,

$$y = \pm\sqrt{5 - 1}$$
$$= \pm 2.$$

If $x = \frac{11}{5}$,

$$y = \pm\sqrt{5 - \left(\frac{11}{5}\right)^2}$$
$$= \sqrt{\frac{125 - 121}{25}}$$
$$= \pm\frac{2}{5}.$$

Check $\left(\frac{11}{5}, \pm\frac{2}{5}\right)$ in equation (2).

$$2\left(\frac{11}{5}\right) + \left|\pm\frac{2}{5}\right| = \frac{22}{5} + \frac{2}{5}$$
$$= \frac{24}{5} \neq 4$$

$\left(\frac{11}{5}, \pm\frac{2}{5}\right)$ does not check, but

$(1, \pm 2)$ does check because

$$2(1) + |\pm 2| = 2 + 2 = 4.$$

Solution set: $\{(1, 2), (1, -2)\}$

41. Shift the graph of $y = |x|$ one unit to the right to obtain the graph of $y = |x - 1|$.

42. Shift the graph of $y = x^2$ four units down to obtain the graph of $y = x^2 - 4$.

43. If $x - 1 \geq 0$,

$$|x - 1| = x - 1.$$

Thus, if $x \geq 1$,

$$|x - 1| = x - 1.$$

If $x - 1 < 0$,

$$|x - 1| = -(x - 1).$$

Thus, if $x < 1$,

$$|x - 1| = 1 - x.$$

Therefore,

$$y = \begin{cases} x - 1 & \text{if } x \geq 1 \\ 1 - x & \text{if } x < 1. \end{cases}$$

44. $x^2 - 4 = x - 1$ if $x \geq 1$.
$x^2 - 4 = 1 - x$ if $x < 1$.

45. $x^2 - 4 = x - 1$ if $x \geq 1$
$x^2 - x - 3 = 0$

$$x = \frac{-(-1) \pm \sqrt{(-1)^2 - 4(1)(-3)}}{2(1)}$$

$$x = \frac{1 \pm \sqrt{13}}{2}$$

$$\frac{1 + \sqrt{13}}{2} \approx 2.3 \geq 1$$

but $\dfrac{1 - \sqrt{13}}{2} \approx -1.3 \not\geq 1.$

Therefore,

$$x = \frac{1 + \sqrt{13}}{2}.$$

46. If $y = |x - 1|$ and $x = \dfrac{1 + \sqrt{13}}{2}$,

$$y = \left| \dfrac{1 + \sqrt{13}}{2} - 1 \right|$$

$$y = \left| \dfrac{1 + \sqrt{13}}{2} - \dfrac{2}{2} \right|$$

$$y = \left| \dfrac{-1 + \sqrt{13}}{2} \right|$$

$$y = \dfrac{-1 + \sqrt{13}}{2},$$

since $\dfrac{-1 + \sqrt{13}}{2} \geq 0$.

One solution of the system is

$$\left(\dfrac{1 + \sqrt{13}}{2}, \dfrac{-1 + \sqrt{13}}{2} \right).$$

47. $x^2 - 4 = 1 - x$ if $x < 1$

$x^2 + x - 5 = 0$

$$x = \dfrac{-(1) \pm \sqrt{(1)^2 - 4(1)(-5)}}{2(1)}$$

$$x = \dfrac{-1 \pm \sqrt{21}}{2}$$

$$\dfrac{-1 + \sqrt{21}}{2} \approx 1.79 \nless 1$$

but $\dfrac{-1 - \sqrt{21}}{2} \approx -2.79 < 1.$

Therefore,

$$x = \dfrac{-1 - \sqrt{21}}{2}.$$

48. If $y = |x - 1|$ and $x = \dfrac{-1 - \sqrt{21}}{2}$,

$$y = \left| \dfrac{-1 - \sqrt{21}}{2} - 1 \right|$$

$$y = \left| \dfrac{-1 - \sqrt{21}}{2} - \dfrac{2}{2} \right|$$

$$y = \left| \dfrac{-3 - \sqrt{21}}{2} \right|$$

$$y = \dfrac{3 + \sqrt{21}}{2},$$

since $\dfrac{-3 - \sqrt{21}}{2} < 0.$

Another solution of the system is

$$\left(\dfrac{-1 - \sqrt{21}}{2}, \dfrac{3 + \sqrt{21}}{2} \right).$$

49. Using the results of Exercises 46 and 48, the solution set of the original system is

$$\left\{ \left(\dfrac{1 + \sqrt{13}}{2}, \dfrac{-1 + \sqrt{13}}{2} \right), \left(\dfrac{-1 - \sqrt{21}}{2}, \dfrac{3 + \sqrt{21}}{2} \right) \right\}.$$

50. The displays at the bottom of the screens correspond to approximations of the x- and y-values in the ordered pair solutions.

51. $y = \log (x + 5)$

$y = x^2$

Use $y_1 = \log (x + 5)$ and $y_2 = x^2$. Using a graphing calculator, we find that the two curves intersect in two points whose coordinates are approximately $(-.79, .62)$ and $(.88, .77)$. Solution set:

$$\{(-.79, .62), (.88, .77)\}$$

53. $y = e^{x+1}$

$2x + y = 3$

Use $y_1 = e^{x+1}$ and $y_2 = 3 - 2x$. Using a graphing calculator, find that the curve and the line intersect in one point whose coordinates are approximately $(.06, 2.88)$. Solution set: $\{(.06, 2.88)\}$

55. $y = \sqrt[3]{x - 4}$

$x^2 + y^2 = 6$

Since $x^2 + y^2 = 6$, $y = \pm\sqrt{6 - x^2}$.
Use $y_1 = \sqrt{6 - x^2}$, $y_2 = -\sqrt{6 - x^2}$, and
$y_3 = (x - 4) \wedge (1/3)$.

Using a graphing calculator, we find
that the graph of $y = \sqrt[3]{x - 4}$ intersects the circle (the graph of
$x^2 + y^2 = 6$) in two points. The
approximate coordinates of these
points are $(-1.68, -1.78)$ and
$(2.12, -1.24)$.

Solution set:

$\{(-1.68, -1.78), (2.12, -1.24)\}$

57. Let x = one number;

y = the other number.

$x + y = 17$ (1)

$xy = 42$ (2)

Solve equation (1) for y.

$y = 17 - x$

Substitute this into equation (2).

$x(17 - x) = 42$

$17x - x^2 = 42$

$0 = x^2 - 17x + 42$

$0 = (x - 3)(x - 14)$

$x = 3$ or $x = 14$

Using equation (1), if $x = 3$

$3 + y = 17$

$y = 14.$

If $x = 14$,

$14 + y = 17$

$y = 3.$

The two numbers are 3 and 14.

59. Let x = one number;

y = the other number.

$x^2 + y^2 = 100$ (1)

$x^2 - y^2 = 28$ (2)

We add the equations to eliminate
y^2.

$2x^2 = 128$

$x^2 = 64$

$x = \pm 8$

Substitute -8 for x in equation (1).

$(-8)^2 + y^2 = 100$

$64 + y^2 = 100$

$y^2 = 36$

$y = \pm 6$

Similarly, if we substitute 8 for x
in equation (1), we obtain $y = \pm 6$.
The two numbers are -8 and 6, -8 and
-6, 8 and 6, or 8 and -6.

61. Let x and y represent the numbers.

$\dfrac{x}{y} = \dfrac{9}{2}$ (1)

$xy = 162$ (2)

Rewrite (1) as $x = \dfrac{9}{2}y$, and substi-

tute $\dfrac{9}{2}y$ for x in (2).

$\left(\dfrac{9}{2}y\right)y = 162$

$\dfrac{9}{2}y^2 = 162$

$y^2 = 36$

$y = \pm 6$

If $y = 6$, $x = \dfrac{9}{2}(6) = 27.$

If $y = -6$, $x = \dfrac{9}{2}(-6) = -27.$

The two numbers are either 6 and 27, or −6 and −27.

63. If the system

$$3x - 2y = 9 \quad (1)$$
$$x^2 + y^2 = 25 \quad (2)$$

has a solution, the line and the circle intersect.
Solve equation (1) for x.

$$3x = 9 + 2y$$
$$x = 3 + \frac{2}{3}y \quad (3)$$

Substitute into equation (2).

$$\left(3 + \frac{2}{3}y\right)^2 + y^2 = 25$$

$$9 + 4y + \frac{4}{9}y^2 + y^2 = 25$$

$$\frac{13}{9}y^2 + 4y - 16 = 0$$

$$13y^2 + 36y - 144 = 0$$

Use the quadratic formula.

$$y = \frac{-36 \pm \sqrt{(36)^2 - 4(13)(144)}}{2(13)}$$

$$= \frac{-36 \pm \sqrt{8784}}{26}$$

$$y \approx 2.2 \quad \text{or} \quad y \approx -4.989$$

Substitute into equation (3) to find x.

If y = 2.22,

$$x = 3 + \frac{2}{3}(2.22)$$

$$= 4.48.$$

If y = −4.989,

$$x = 3 + \frac{2}{3}(-4.989)$$

$$= -.326.$$

Thus, the circle and the line do intersect, in fact twice, at (4.48, 2.22) and at (−3.26, −4.99).

65.

$$x + 2y = b \quad (1)$$
$$x^2 + y^2 = 9 \quad (2)$$

Solve equation (1) for x.

$$x = b - 2y$$

Substitute b − 2y for x in equation (2).

$$(b - 2y)^2 + y^2 = 9$$
$$b^2 - 4by + 4y^2 + y^2 = 9$$
$$b^2 - 4by + 5y^2 - 9 = 0$$
$$5y^2 - 4b \cdot y + (b^2 - 9) = 0$$

This equation will have a unique solution when the discriminant is 0.

$$(-4b)^2 - 4(5)(b^2 - 9) = 0$$
$$16b^2 - 20b^2 + 180 = 0$$
$$-4b^2 = -180$$
$$b^2 = 45$$

$$b = \pm\sqrt{45} \quad \text{or} \quad \pm3\sqrt{5}$$

The line x + 2y = b will touch the circle $x^2 + y^2 = 9$ in only one point if $b = \pm3\sqrt{5}$.

67. **(a)** The emission of carbon is increasing with time. The carbon emissions from the former USSR and Eastern Europe have surpassed the emissions of Western Europe.

(b) They were equal in 1962 or 1963 when the levels were approximately 400 million metric tons.

(c) W = E

$375(1.008)^{(t-1950)}$

$\qquad = 260(1.038)^{(t-1950)}$

$\log \left[375(1.008)^{(t-1950)} \right]$

$\qquad = \log \left[260(1.038)^{(t-1950)} \right]$

$\log 375 + (t - 1950) \log 1.008$

$\qquad = \log 260 + (t - 1950) \log 1.038$

$\log 375 - \log 260$

$\qquad = (t - 1950) \log 1.038$

$\qquad\qquad - (t - 1950) \log 1.008$

$\log 375 - \log 260$

$\qquad = (\log 1.038 - \log 1.008)$

$\qquad\qquad \cdot (t - 1950)$

$\dfrac{\log 375 - \log 260}{\log 1.038 - \log 1.008}$

$\qquad = t - 1950$

$t = 1950 + \dfrac{\log 375 - \log 260}{\log 1.038 - \log 1.008}$

$t \approx 1962.49$

If t = 1962.49,

$W = 375(1.008)^{(1962.49-1950)} \approx 14.24.$

In 1962, the emission levels were equal and were approximately 414 million metric tons.

Section 6.3

1. $\begin{bmatrix} 2 & 4 \\ 4 & 7 \end{bmatrix}$; -2 times row 1 added to row 2

Using the third row transformation, the matrix is changed to

$\begin{bmatrix} 2 & 4 \\ 4 + (-2)(2) & 7 + (-2)(4) \end{bmatrix}$

$= \begin{bmatrix} 2 & 4 \\ 0 & -1 \end{bmatrix}.$

3. $\begin{bmatrix} 1 & 5 & 6 \\ -2 & 3 & -1 \\ 4 & 7 & 0 \end{bmatrix}$; 2 times row 1 added to row 2

Using the third row transformation, the matrix is changed to

$\begin{bmatrix} 1 & 5 & 6 \\ -2 + 2(1) & 3 + 2(5) & -1 + 2(6) \\ 4 & 7 & 0 \end{bmatrix}$

$= \begin{bmatrix} 1 & 5 & 6 \\ 0 & 13 & 11 \\ 4 & 7 & 0 \end{bmatrix}.$

5. $2x + 3y = 11$

$\quad x + 2y = 8$

The augmented matrix is

$\begin{bmatrix} 2 & 3 & | & 11 \\ 1 & 2 & | & 8 \end{bmatrix}.$

7. $2x + y + z = 3$

$3x - 4y + 2z = -7$

$\quad x - y + z = 2$

has the augmented matrix

$\begin{bmatrix} 2 & 1 & 1 & | & 3 \\ 3 & -4 & 2 & | & -7 \\ 1 & 1 & 1 & | & 2 \end{bmatrix}.$

9. $\begin{bmatrix} 3 & 2 & 1 & | & 1 \\ 0 & 2 & 4 & | & 22 \\ -1 & -2 & 3 & | & 15 \end{bmatrix}$

is associated with the system

$3x + 2y + z = 1$

$\qquad 2y + 4z = 22$

$-x - 2y + 3z = 15.$

11. $\begin{bmatrix} 1 & 0 & 0 & | & 2 \\ 0 & 1 & 0 & | & 3 \\ 0 & 0 & 1 & | & -2 \end{bmatrix}$

is associated with the system

$$x = 2$$
$$y = 3$$
$$z = -2.$$

13. The augmented matrix $\begin{bmatrix} 1 & 1 & 0 & | & 3 \\ 0 & 2 & 1 & | & -4 \\ 1 & 0 & -1 & | & 5 \end{bmatrix}$

is associated with the following system of equations:

$$x + y = 3$$
$$2y + z = -4$$
$$x - z = 5.$$

15. $x + y = 5$
    $x - y = -1$

has the augmented matrix

$\begin{bmatrix} 1 & 1 & | & 5 \\ 1 & -1 & | & -1 \end{bmatrix}.$

$\begin{bmatrix} 1 & 1 & | & 5 \\ 0 & -2 & | & -6 \end{bmatrix}$ $-1R1 + R2$

$\begin{bmatrix} 1 & 1 & | & 5 \\ 0 & 1 & | & 3 \end{bmatrix}$ $-\frac{1}{2}R2$

$\begin{bmatrix} 1 & 0 & | & 2 \\ 0 & 1 & | & 3 \end{bmatrix}$ $-1R2 + R1$

Solution set: $\{(2, 3)\}$

17. $x + y = -3$
    $2x - 5y = -6$

$\begin{bmatrix} 1 & 1 & | & -3 \\ 2 & -5 & | & -6 \end{bmatrix}$

$\begin{bmatrix} 1 & 1 & | & -3 \\ 0 & -7 & | & 0 \end{bmatrix}$ $-2R1 + R2$

$\begin{bmatrix} 1 & 1 & | & -3 \\ 0 & 1 & | & 0 \end{bmatrix}$ $-\frac{1}{7}R2$

$\begin{bmatrix} 1 & 0 & | & -3 \\ 0 & 1 & | & 0 \end{bmatrix}$ $-1R2 + R1$

Solution set: $\{(-3, 0)\}$

19. $6x + y - 5 = 0$
    $5x + y - 3 = 0$

Rewrite the system as

$$6x + y = 5$$
$$5x + y = 3.$$

$\begin{bmatrix} 6 & 1 & | & 5 \\ 5 & 1 & | & 3 \end{bmatrix}$

$\begin{bmatrix} 1 & \frac{1}{6} & | & \frac{5}{6} \\ 5 & 1 & | & 3 \end{bmatrix}$ $\frac{1}{6}R1$

$\begin{bmatrix} 1 & \frac{1}{6} & | & \frac{5}{6} \\ 0 & \frac{1}{6} & | & -\frac{7}{6} \end{bmatrix}$ $-5R1 + R2$

$\begin{bmatrix} 1 & \frac{1}{6} & | & \frac{5}{6} \\ 0 & 1 & | & -7 \end{bmatrix}$ $6R2$

$\begin{bmatrix} 1 & 0 & | & 2 \\ 0 & 1 & | & -7 \end{bmatrix}$ $-\frac{1}{6}R2 + R1$

Solution set: $\{(2, -7)\}$

21. $4x - y - 3 = 0$
    $-2x + 3y - 1 = 0$

Rewrite the system as

$$4x - y = 3$$
$$-2x + 3y = 1.$$

$\begin{bmatrix} 4 & -1 & | & 3 \\ -2 & 3 & | & 1 \end{bmatrix}$

$\begin{bmatrix} 1 & -\frac{1}{4} & | & \frac{3}{4} \\ -2 & 3 & | & 1 \end{bmatrix}$ $\frac{1}{4}R1$

$\begin{bmatrix} 1 & -\frac{1}{4} & | & \frac{3}{4} \\ 0 & \frac{5}{2} & | & \frac{5}{2} \end{bmatrix}$ $2R1 + R2$

$$\begin{bmatrix} 1 & -\frac{1}{4} & \Big| & \frac{3}{4} \\ 0 & 1 & \Big| & 1 \end{bmatrix} \quad \frac{2}{5}R2$$

$$\begin{bmatrix} 1 & 0 & \Big| & 1 \\ 0 & 1 & \Big| & 1 \end{bmatrix} \quad \frac{1}{4}R2 + R1$$

Solution set: $\{(1, 1)\}$

23. $x + y - z = 6$

$2x - y + z = -9$

$x - 2y + 3z = 1$

$$\begin{bmatrix} 1 & 1 & -1 & \Big| & 6 \\ 2 & -1 & 1 & \Big| & -9 \\ 1 & -2 & 3 & \Big| & 1 \end{bmatrix}$$

$$\begin{bmatrix} 1 & 1 & -1 & \Big| & 6 \\ 0 & -3 & 3 & \Big| & -21 \\ 0 & -3 & 4 & \Big| & -5 \end{bmatrix} \quad \begin{matrix} -2R1 + R2 \\ -1R1 + R3 \end{matrix}$$

$$\begin{bmatrix} 1 & 1 & -1 & \Big| & 6 \\ 0 & -3 & 3 & \Big| & -21 \\ 0 & 0 & 1 & \Big| & 16 \end{bmatrix} \quad -1R2 + R3$$

$$\begin{bmatrix} 1 & 1 & -1 & \Big| & 6 \\ 0 & 1 & -1 & \Big| & 7 \\ 0 & 0 & 1 & \Big| & 16 \end{bmatrix} \quad -\frac{1}{3}R2$$

$$\begin{bmatrix} 1 & 1 & 0 & \Big| & 22 \\ 0 & 1 & 0 & \Big| & 23 \\ 0 & 0 & 1 & \Big| & 16 \end{bmatrix} \quad \begin{matrix} R3 + R1 \\ R3 + R2 \end{matrix}$$

$$\begin{bmatrix} 1 & 0 & 0 & \Big| & -1 \\ 0 & 1 & 0 & \Big| & 23 \\ 0 & 0 & 1 & \Big| & 16 \end{bmatrix} \quad -1R2 + R1$$

Solution set: $\{(-1, 23, 16)\}$

25. $x - z = -3$

$y + z = 9$

$x + z = 7$

$$\begin{bmatrix} 1 & 0 & -1 & \Big| & -3 \\ 0 & 1 & 1 & \Big| & 9 \\ 1 & 0 & 1 & \Big| & 7 \end{bmatrix}$$

$$\begin{bmatrix} 1 & 0 & -1 & \Big| & -3 \\ 0 & 1 & 1 & \Big| & 9 \\ 0 & 0 & 2 & \Big| & 10 \end{bmatrix} \quad -1R1 + R3$$

$$\begin{bmatrix} 1 & 0 & -1 & \Big| & -3 \\ 0 & 1 & 1 & \Big| & 9 \\ 0 & 0 & 1 & \Big| & 5 \end{bmatrix} \quad \frac{1}{2}R3$$

$$\begin{bmatrix} 1 & 0 & -1 & \Big| & -3 \\ 0 & 1 & 0 & \Big| & 4 \\ 0 & 0 & 1 & \Big| & 5 \end{bmatrix} \quad -1R3 + R2$$

$$\begin{bmatrix} 1 & 0 & 0 & \Big| & 2 \\ 0 & 1 & 0 & \Big| & 4 \\ 0 & 0 & 1 & \Big| & 5 \end{bmatrix} \quad R3 + R1$$

Solution set: $\{(2, 4, 5)\}$

27. $y = -2x - 2z + 1$

$x = -2y - z + 2$

$z = x - y$

Rewrite the system as

$2x + y + 2z = 1$

$x + 2y + z = 2$

$x - y - z = 0.$

$$\begin{bmatrix} 2 & 1 & 2 & \Big| & 1 \\ 1 & 2 & 1 & \Big| & 2 \\ 1 & -1 & -1 & \Big| & 0 \end{bmatrix}$$

$$\begin{bmatrix} 1 & 2 & 1 & \Big| & 2 \\ 2 & 1 & 2 & \Big| & 1 \\ 1 & -1 & -1 & \Big| & 0 \end{bmatrix} \quad R1 \leftrightarrow R2$$

$$\begin{bmatrix} 1 & 2 & 1 & \Big| & 2 \\ 0 & -3 & 0 & \Big| & -3 \\ 0 & -3 & -2 & \Big| & -2 \end{bmatrix} \quad \begin{matrix} -2R1 + R2 \\ -1R1 + R3 \end{matrix}$$

$$\begin{bmatrix} 1 & 2 & 1 & \Big| & 2 \\ 0 & 1 & 0 & \Big| & 1 \\ 0 & -3 & -2 & \Big| & -2 \end{bmatrix} \quad -\frac{1}{3}R2$$

$$\begin{bmatrix} 1 & 0 & 1 & \Big| & 0 \\ 0 & 1 & 0 & \Big| & 1 \\ 0 & 0 & -2 & \Big| & 1 \end{bmatrix} \quad \begin{matrix} -2R2 + R1 \\ 3R2 + R3 \end{matrix}$$

$$\begin{bmatrix} 1 & 0 & 1 & \Big| & 0 \\ 0 & 1 & 0 & \Big| & 1 \\ 0 & 0 & 1 & \Big| & -\frac{1}{2} \end{bmatrix} \quad -\frac{1}{2}R3$$

$$\begin{bmatrix} 1 & 0 & 0 & \bigm| & \frac{1}{2} \\ 0 & 1 & 0 & \bigm| & 1 \\ 0 & 0 & 1 & \bigm| & -\frac{1}{2} \end{bmatrix} \qquad -1R3 + R1$$

Solution set: $\left\{\left(\frac{1}{2},\ 1,\ -\frac{1}{2}\right)\right\}$

29. $2x - y + 3z = 0$
$\quad\ \ x + 2y - z = 5$
$\quad\qquad\quad 2y + z = 1$

$$\begin{bmatrix} 2 & -1 & 3 & \bigm| & 0 \\ 1 & 2 & -1 & \bigm| & 5 \\ 0 & 2 & 1 & \bigm| & 1 \end{bmatrix}$$

$$\begin{bmatrix} 1 & -\frac{1}{2} & \frac{3}{2} & \bigm| & 0 \\ 1 & 2 & -1 & \bigm| & 5 \\ 0 & 2 & 1 & \bigm| & 1 \end{bmatrix} \quad \frac{1}{2}R1$$

$$\begin{bmatrix} 1 & -\frac{1}{2} & \frac{3}{2} & \bigm| & 0 \\ 0 & \frac{5}{2} & -\frac{5}{2} & \bigm| & 5 \\ 0 & 2 & 1 & \bigm| & 1 \end{bmatrix} \quad -1R1 + R2$$

$$\begin{bmatrix} 1 & -\frac{1}{2} & \frac{3}{2} & \bigm| & 0 \\ 1 & 1 & -1 & \bigm| & 2 \\ 0 & 2 & 1 & \bigm| & 1 \end{bmatrix} \quad \frac{2}{5}R2$$

$$\begin{bmatrix} 1 & -\frac{1}{2} & \frac{3}{2} & \bigm| & 0 \\ 0 & 1 & -1 & \bigm| & 2 \\ 0 & 0 & 3 & \bigm| & -3 \end{bmatrix} \quad -2R2 + R3$$

$$\begin{bmatrix} 1 & -\frac{1}{2} & \frac{3}{2} & \bigm| & 0 \\ 0 & 1 & -1 & \bigm| & 2 \\ 0 & 0 & 1 & \bigm| & -1 \end{bmatrix} \quad \frac{1}{3}R3$$

$$\begin{bmatrix} 1 & 0 & 1 & \bigm| & 1 \\ 0 & 1 & -1 & \bigm| & 2 \\ 0 & 0 & 1 & \bigm| & -1 \end{bmatrix} \quad \frac{1}{2}R2 + R1$$

$$\begin{bmatrix} 1 & 0 & 0 & \bigm| & 2 \\ 0 & 1 & 0 & \bigm| & 1 \\ 0 & 0 & 1 & \bigm| & -1 \end{bmatrix} \quad \begin{matrix} -1R3 + R1 \\ R3 + R2 \end{matrix}$$

Solution set: $\{(2,\ 1,\ -1)\}$

31. $3x + 5y - z + 2 = 0$
$\quad\ \ 4x - y + 2z - 1 = 0$
$\quad -6x - 10y + 2z = 0$

Rewrite the system as
$\quad 3x + 5y - z = -2$
$\quad 4x - y + 2z = 1$
$\quad -6x - 10y + 2z = 0.$

$$\begin{bmatrix} 3 & 5 & -1 & \bigm| & -2 \\ 4 & -1 & 2 & \bigm| & 1 \\ -6 & -10 & 2 & \bigm| & 0 \end{bmatrix}$$

$$\begin{bmatrix} 1 & \frac{5}{3} & -\frac{1}{3} & \bigm| & -\frac{2}{3} \\ 4 & -1 & 2 & \bigm| & 1 \\ -6 & -10 & 2 & \bigm| & 0 \end{bmatrix} \quad \frac{1}{3}R1$$

$$\begin{bmatrix} 1 & \frac{5}{3} & -\frac{1}{3} & \bigm| & -\frac{2}{3} \\ 0 & -\frac{23}{3} & \frac{10}{3} & \bigm| & \frac{11}{3} \\ 0 & 0 & 0 & \bigm| & -4 \end{bmatrix} \quad \begin{matrix} -4R1 + R2 \\ 6R1 + R3 \end{matrix}$$

The last row indicates that there is
no solution.
The solution set is Ø.

33. $x - 8y + z = 4$
$\quad 3x - y + 2z = -1$

$$\begin{bmatrix} 1 & -8 & 1 & \bigm| & 4 \\ 3 & -1 & 2 & \bigm| & -1 \end{bmatrix}$$

$$\begin{bmatrix} 1 & -8 & 1 & \bigm| & 4 \\ 0 & 23 & -1 & \bigm| & -13 \end{bmatrix} \quad -3R1 + R2$$

$$\begin{bmatrix} 1 & -8 & 1 & \bigm| & 4 \\ 0 & 1 & -\frac{1}{23} & \bigm| & -\frac{13}{23} \end{bmatrix} \quad \frac{1}{23}R2$$

$$\begin{bmatrix} 1 & 0 & \frac{15}{23} & \bigm| & -\frac{12}{23} \\ 0 & 1 & -\frac{1}{23} & \bigm| & -\frac{13}{23} \end{bmatrix} \quad 8R2 + R1$$

This matrix is equivalent to the following system:

$$x + \frac{15}{23}z = -\frac{12}{23}$$

$$y - \frac{1}{23}z = -\frac{13}{23}.$$

This system has infinitely many solutions. We will express the solution set with z as the arbitrary variable.
Therefore,

$$x = -\frac{15}{23}z - \frac{12}{23}$$

$$\text{and} \quad y = \frac{1}{23}z - \frac{13}{23}.$$

Solution set:

$$\left\{ \left(-\frac{15}{23}z - \frac{12}{23}, \frac{1}{23}z - \frac{13}{23}, z \right) \right\}$$

35.
$$x - y + 2z + w = 4$$
$$y + z = 3$$
$$z - w = 2$$

$$\begin{bmatrix} 1 & -1 & 2 & 1 & | & 4 \\ 0 & 1 & 1 & 0 & | & 3 \\ 0 & 0 & 1 & -1 & | & 2 \end{bmatrix}$$

$$\begin{bmatrix} 1 & 0 & 3 & 1 & | & 7 \\ 0 & 1 & 1 & 0 & | & 3 \\ 0 & 0 & 1 & -1 & | & 2 \end{bmatrix} \quad 1R2 + R1$$

$$\begin{bmatrix} 1 & 0 & 0 & 4 & | & 1 \\ 0 & 1 & 0 & 1 & | & 1 \\ 0 & 0 & 1 & -1 & | & 2 \end{bmatrix} \quad \begin{matrix} -3R3 + R1 \\ -1R3 + R2 \end{matrix}$$

This gives the following system:

$$x + 4w = 1$$
$$y + w = 1$$
$$z - w = 2.$$

This system has an infinite number of solutions. We will write the solution set with w as the arbitrary variable.

$$x = 1 - 4w$$
$$y = 1 - w$$
$$z = 2 + w$$

Solution set:
$$\{(1 - 4w, 1 - w, 2 + w, w)\}$$

39.
$$.268x + y = 9.814$$
$$x - .329y = 1.414$$

$$\begin{bmatrix} .268 & 1 & | & 9.814 \\ 1 & -.329 & | & 1.414 \end{bmatrix}$$

Using a graphing calculator capable of performing row operations, we obtain the following solution set:
$$\{(4.267, 8.671)\}.$$

41.
$$.3x + 2.7y - \sqrt{2}z = 3$$
$$\sqrt{7}x - 20y + 12z = -2$$
$$4x + \sqrt{3}y - 1.2z = \frac{3}{4}$$

$$\begin{bmatrix} .3 & 2.7 & -\sqrt{2} & | & 3 \\ \sqrt{7} & -20 & 12 & | & -2 \\ 4 & \sqrt{3} & -1.2 & | & \frac{3}{4} \end{bmatrix}$$

Using a graphing calculator capable of performing row operations, we obtain the following solution set:
$$\{(.571, 7.041, 11.442)\}.$$

43.
$$\frac{1}{(x - 1)(x + 1)} = \frac{A}{x - 1} + \frac{B}{x + 1}$$

Add the rational expression on the right.

$$\frac{1}{(x - 1)(x + 1)} = \frac{A(x + 1) + B(x - 1)}{(x - 1)(x + 1)}$$

Since the denominators are equal, the numerators must be equal.

$$1 = A(x + 1) + B(x - 1)$$

$$1 = Ax + A + Bx - B$$

Collect like terms.

$$1 = (A + B)x + (A - B)$$

Equating the coefficients of like powers of x gives the following system of equations.

$$A + B = 0 \quad (1)$$

$$A - B = 1 \quad (2)$$

This system can be solved by any of the methods studied so far. We use the elimination method.

$$
\begin{array}{r}
A + B = 0 \\
\underline{A - B = 1} \\
2A \qquad = 1 \\
A = \dfrac{1}{2} \\
B = -\dfrac{1}{2}
\end{array}
$$

45.

$$\frac{x}{(x - a)(x + a)} = \frac{A}{x - a} + \frac{B}{x + a}$$

$$\frac{x}{(x - a)(x + a)} = \frac{A(x + a) + B(x - a)}{(x - a)(x + a)}$$

$$x = A(x + a) + B(x - a)$$

$$x = Ax + Aa + Bx - Ba$$

$$x = (A + B)x + (A - B)a$$

$$
\begin{array}{r}
A + B = 1 \\
\underline{A - B = 0} \\
2A \qquad = 1 \\
A = \dfrac{1}{2} \\
B = \dfrac{1}{2}
\end{array}
$$

47. Let x = the number of days worked by the husband;

  y = the number of days worked by the wife.

$$x + y = 72$$

$$56x + 64y = 4352$$

$$\begin{bmatrix} 1 & 1 & | & 72 \\ 56 & 64 & | & 4352 \end{bmatrix}$$

$$\begin{bmatrix} 1 & 1 & | & 72 \\ 0 & 8 & | & 320 \end{bmatrix} \quad -56R1 + R2$$

$$\begin{bmatrix} 1 & 1 & | & 72 \\ 0 & 1 & | & 40 \end{bmatrix} \quad \frac{1}{8}R2$$

$$\begin{bmatrix} 1 & 0 & | & 32 \\ 0 & 1 & | & 40 \end{bmatrix} \quad -1R2 + R1$$

From the final matrix, we have x = 32 and y = 40. The husband worked 32 days, and the wife worked 40 days.

49. Let x = number of cubic centimeters of the 2% solution;

  y = number of cubic centimeters of the 7% solution.

$$x + y = 40$$

$$.02x + .07y = .032(40)$$

$$\begin{bmatrix} 1 & 1 & | & 40 \\ .02 & .07 & | & 1.28 \end{bmatrix}$$

$$\begin{bmatrix} 1 & 1 & | & 40 \\ 0 & .05 & | & .48 \end{bmatrix} \quad -.02R1 + R2$$

$$\begin{bmatrix} 1 & 1 & | & 40 \\ 0 & 1 & | & 9.6 \end{bmatrix} \quad \frac{1}{.05}R2$$

$$\begin{bmatrix} 1 & 0 & | & 30.4 \\ 0 & 1 & | & 9.6 \end{bmatrix} \quad -1R2 + R1$$

Solution set: $\{(30.4, 9.6)\}$

The chemist should mix 30.4 cm³ of the 2% solution with 9.6 cm³ of the 7% solution.

53. Let x = number of grams of food A;

y = number of grams of food B;

z = number of grams of food C.

From the given information, we have

$$x + y + z = 400$$

$$x = \frac{1}{3}y$$

$$x + z = 2y.$$

The system becomes

$$x + y + z = 400$$

$$3x - y = 0$$

$$x - 2y + z = 0.$$

$$\begin{bmatrix} 1 & 1 & 1 & | & 400 \\ 3 & -1 & 0 & | & 0 \\ 1 & -2 & 1 & | & 0 \end{bmatrix}$$

$$\begin{bmatrix} 1 & 1 & 1 & | & 400 \\ 0 & -4 & -3 & | & -1200 \\ 0 & -3 & 0 & | & -400 \end{bmatrix} \begin{matrix} \\ -3R1 + R2 \\ -1R1 + R3 \end{matrix}$$

$$\begin{bmatrix} 1 & 1 & 1 & | & 400 \\ 0 & 1 & \frac{3}{4} & | & 300 \\ 0 & -3 & 0 & | & -400 \end{bmatrix} \begin{matrix} \\ -\frac{1}{4}R2 \\ \\ \end{matrix}$$

$$\begin{bmatrix} 1 & 0 & \frac{1}{4} & | & 100 \\ 0 & 1 & \frac{3}{4} & | & 300 \\ 0 & 0 & \frac{9}{4} & | & 500 \end{bmatrix} \begin{matrix} -1R2 + R1 \\ \\ 3R2 + R3 \end{matrix}$$

$$\begin{bmatrix} 1 & 0 & \frac{1}{4} & | & 100 \\ 0 & 1 & \frac{3}{4} & | & 300 \\ 0 & 0 & 1 & | & \frac{2000}{9} \end{bmatrix} \begin{matrix} \\ \\ \frac{4}{9}R3 \end{matrix}$$

$$\begin{bmatrix} 1 & 0 & 0 & | & \frac{400}{9} \\ 0 & 1 & 0 & | & \frac{400}{3} \\ 0 & 0 & 1 & | & \frac{2000}{9} \end{bmatrix} \begin{matrix} -\frac{1}{4}R3 + R1 \\ -\frac{3}{4}R3 + R2 \\ \\ \end{matrix}$$

From the last matrix, we have x =
400/9 ≈ 44.4, y = 400/3 ≈ 133.3, and
z = 2000/9 ≈ 222.2.

The diet should include 44.4 g of
food A, 133.3 g of food B, and
222.2 g of food C.

55. **(a)** A height of 6'11" is 83".

If W = 7.46H − 374,

$$W = 7.46(83) - 374$$

$$W = 245.18.$$

Using this equation, the predicted
weight is approximately 245 pounds.
If W = 7.93H − 405,

$$W = 7.93(83) - 405$$

$$W = 253.19.$$

Using this equation, the predicted
weight is approximately 253 pounds.

(b) For the model W = 7.46H − 374, a
1-inch increase in height results in
a 7.46-pound increase in weight.
For the model W = 7.93H − 405, a 1-
inch increase in height results in a
7.93-pound increase in weight.
In each case, the change is given by
the slope of the line that is the
graph of the given equation.

(c) W − 7.46H = −374

W − 7.93H = −405

Solve this system by the Gauss-
Jordan method.

$$\begin{bmatrix} 1 & -7.46 & | & -374 \\ 1 & -7.93 & | & -405 \end{bmatrix}$$

$$\begin{bmatrix} 1 & -7.46 & | & -374 \\ 0 & -.47 & | & -31 \end{bmatrix} \quad -1R1 + R2$$

$$\begin{bmatrix} 1 & -7.46 & | & -374 \\ 0 & 1 & | & 65.957 \end{bmatrix} \quad -\frac{1}{.47}R2$$

$$\begin{bmatrix} 1 & 0 & | & 118.043 \\ 0 & 1 & | & 65.957 \end{bmatrix} \quad 7.46R2 + R1$$

From the last matrix, we have $W \approx$ 118 and $H \approx 66$.

The two models agree at a height of 66 inches and a weight of 118 pounds.

57. $F = a + bA + cP + dW$

Substituting the values, we have the following system of equations:

$$a + 871b + 11.5c + 3d = 239$$
$$a + 847b + 12.2c + 2d = 234$$
$$a + 685b + 10.6c + 5d = 192$$
$$a + 969b + 14.2c + 1d = 343.$$

58.   The augmented matrix is

$$\begin{bmatrix} 1 & 871 & 11.5 & 3 & | & 239 \\ 1 & 847 & 12.2 & 2 & | & 234 \\ 1 & 685 & 10.6 & 5 & | & 192 \\ 1 & 969 & 14.2 & 1 & | & 343 \end{bmatrix}.$$

Using a graphing calculator capable of performing row operations, the solution we obtain is

$$a \approx -715.457, \ b \approx .34756,$$
$$c \approx 48.6585, \text{ and } d \approx 30.71951.$$

59.   Using these values,

$$F = -715.457 + .34756A + 48.6585P$$
$$+ 30.71951W.$$

60.   Using $A = 960$, $P = 12.6$, and $W = 3$ we obtain

$$F = -715.457 + .34756(960) + 48.6585(12.6)$$
$$+ 30.71951(3)$$
$$F = 323.45623 \approx 323$$

Therefore, the predicted fawn count is approximately 323.

Section 6.4

1.   If

$$\begin{bmatrix} w & x \\ y & z \end{bmatrix} = \begin{bmatrix} 3 & 2 \\ -1 & 4 \end{bmatrix},$$

then $w = 3$, $x = 2$, $y = -1$, and $z = 4$.

3.   If

$$\begin{bmatrix} 2 & 5 & 6 \\ 1 & m & n \end{bmatrix} = \begin{bmatrix} z & y & w \\ 1 & 8 & -2 \end{bmatrix},$$

then $2 = z$, $5 = y$, $6 = w$, $m = 8$, and $n = -2$.   Therefore, $m = 8$, $n = -2$, $w = 6$, $y = 5$, and $z = 2$.

5. $\begin{bmatrix} a+2 & 3z+1 & 5m \\ 8k & 0 & 3 \end{bmatrix} + \begin{bmatrix} 3a & 2z & 5m \\ 2k & 5 & 6 \end{bmatrix}$

$$= \begin{bmatrix} 10 & -14 & 80 \\ 10 & 5 & 9 \end{bmatrix}$$

Add the matrices on the left.

$$\begin{bmatrix} 4a+2 & 5z+1 & 10m \\ 10k & 5 & 9 \end{bmatrix} = \begin{bmatrix} 10 & -14 & 80 \\ 10 & 5 & 9 \end{bmatrix}$$

For these two matrices to be equal, corresponding elements must be equal, so we have

$4a + 2 = 10$	$5z + 1 = -14$	$10m = 80$
$4a = 8$	$5z = -15$	$m = 8$
$a = 2$	$z = -3$	

$$10k = 10$$
$$k = 1.$$

Thus, $a = 2$, $z = -3$, $m = 8$, and $k = 1$.

7. $\begin{bmatrix} -4 & 8 \\ 2 & 3 \end{bmatrix}$

This matrix has 2 rows and 2 columns, so it is a 2 × 2 square matrix.

9. $\begin{bmatrix} -6 & 8 & 0 & 0 \\ 4 & 1 & 9 & 2 \\ 3 & -5 & 7 & 1 \end{bmatrix}$

This matrix has 3 rows and 4 columns, so it is a 3 × 4 matrix.

11. $\begin{bmatrix} 2 \\ 4 \end{bmatrix}$

This matrix has 2 rows and 1 column, so it is a 2 × 1 column matrix.

15. $\begin{bmatrix} 6 & -9 & 2 \\ 4 & 1 & 3 \end{bmatrix} + \begin{bmatrix} -8 & 2 & 5 \\ 6 & -3 & 4 \end{bmatrix}$

$= \begin{bmatrix} 6+(-8) & -9+2 & 2+5 \\ 4+6 & 1+(-3) & 3+4 \end{bmatrix}$

$= \begin{bmatrix} -2 & -7 & 7 \\ 10 & -2 & 7 \end{bmatrix}$

17. $\begin{bmatrix} -6 & 8 \\ 0 & 0 \end{bmatrix} - \begin{bmatrix} 0 & 0 \\ -4 & -2 \end{bmatrix}$

$= \begin{bmatrix} -6 & 8 \\ 0 & 0 \end{bmatrix} + \begin{bmatrix} 0 & 0 \\ 4 & 2 \end{bmatrix}$

$= \begin{bmatrix} -6 & 8 \\ 4 & 2 \end{bmatrix}$

19. $\begin{bmatrix} 3x+y & x-2y & 2x \\ 5x & 3y & x+y \end{bmatrix}$

$+ \begin{bmatrix} 2x & 3y & 5x+y \\ 3x+2y & x & 2x \end{bmatrix}$

$= \begin{bmatrix} 5x+y & x+y & 7x+y \\ 8x+2y & x+3y & 3x+y \end{bmatrix}$

21. $\begin{bmatrix} 3 \\ 2 \end{bmatrix} + \begin{bmatrix} 2 & 3 \end{bmatrix}$

These two matrices are not the same size, so they cannot be added.

In Exercises 23–27,

$A = \begin{bmatrix} -2 & 4 \\ 0 & 3 \end{bmatrix}$ and $B = \begin{bmatrix} -6 & 2 \\ 4 & 0 \end{bmatrix}$.

23. $2A = 2\begin{bmatrix} -2 & 4 \\ 0 & 3 \end{bmatrix} = \begin{bmatrix} -4 & 8 \\ 0 & 6 \end{bmatrix}$

25. $2A - B = 2\begin{bmatrix} -2 & 4 \\ 0 & 3 \end{bmatrix} - \begin{bmatrix} -6 & 2 \\ 4 & 0 \end{bmatrix}$

$= \begin{bmatrix} -4 & 8 \\ 0 & 6 \end{bmatrix} + \begin{bmatrix} 6 & -2 \\ -4 & 0 \end{bmatrix}$

$= \begin{bmatrix} 2 & 6 \\ -4 & 6 \end{bmatrix}$

27. $-A + \frac{1}{2}B = -\begin{bmatrix} -2 & 4 \\ 0 & 3 \end{bmatrix} + \frac{1}{2}\begin{bmatrix} -6 & 2 \\ 4 & 0 \end{bmatrix}$

$= \begin{bmatrix} 2 & -4 \\ 0 & -3 \end{bmatrix} + \begin{bmatrix} -3 & 1 \\ 2 & 0 \end{bmatrix}$

$= \begin{bmatrix} -1 & -3 \\ 2 & -3 \end{bmatrix}$

29. $\begin{bmatrix} 1 & 2 \\ 3 & 4 \end{bmatrix}\begin{bmatrix} -1 \\ 7 \end{bmatrix}$

$= \begin{bmatrix} 1(-1)+2(7) \\ 3(-1)+4(7) \end{bmatrix}$

$= \begin{bmatrix} 13 \\ 25 \end{bmatrix}$

31. $\begin{bmatrix} 3 & -4 & 1 \\ 5 & 0 & 2 \end{bmatrix}\begin{bmatrix} -1 \\ 4 \\ 2 \end{bmatrix}$

$= \begin{bmatrix} 3(-1)+(-4)(4)+1(2) \\ 5(-1)+0(4)+2(2) \end{bmatrix}$

$= \begin{bmatrix} -17 \\ -1 \end{bmatrix}$

33. $\begin{bmatrix} 5 & 2 \\ -1 & 4 \end{bmatrix} \begin{bmatrix} 3 & -2 \\ 1 & 0 \end{bmatrix} = \begin{bmatrix} 5(3) + 2(1) & 5(-2) + 2(0) \\ -1(3) + 4(-1) & -1(-2) + 4(0) \end{bmatrix}$

$= \begin{bmatrix} 17 & -10 \\ 1 & 2 \end{bmatrix}$

35. $\begin{bmatrix} 2 & 2 & -1 \\ 3 & 0 & 1 \end{bmatrix} \begin{bmatrix} 0 & 2 \\ -1 & 4 \\ 0 & 2 \end{bmatrix}$

$= \begin{bmatrix} 2(0) + 2(-1) + (-1)(0) & 2(2) + 2(4) + (-1)(2) \\ 3(0) + 0(-1) + 1(0) & 3(2) + 0(4) + 1(2) \end{bmatrix} = \begin{bmatrix} -2 & 10 \\ 0 & 8 \end{bmatrix}$

37. $\begin{bmatrix} -1 & 2 & 0 \\ 0 & 3 & 2 \\ 0 & 1 & 4 \end{bmatrix} \begin{bmatrix} 2 & -1 & 2 \\ 0 & 2 & 1 \\ 3 & 0 & -1 \end{bmatrix}$

$\begin{bmatrix} (-1)(2) + 2(0) + 0(3) & (-1)(-1) + 2(2) + 0(0) & (-1)(2) + 2(1) + 0(-1) \\ 0(2) + 3(0) + 2(3) & 0(-1) + 3(2) + 2(0) & 0(2) + 3(1) + 2(-1) \\ 0(2) + 1(0) + 4(3) & 0(-1) + 1(2) + 4(0) & 0(2) + 1(1) + 4(-1) \end{bmatrix}$

$= \begin{bmatrix} -2 & 5 & 0 \\ 6 & 6 & 1 \\ 12 & 2 & -3 \end{bmatrix}$

39. $\begin{bmatrix} -2 & 4 & 1 \end{bmatrix} \begin{bmatrix} 3 & -2 & 4 \\ 2 & 1 & 0 \\ 0 & -1 & 4 \end{bmatrix}$

$= \begin{bmatrix} -2(3) + 4(2) + 1(0) & (-2)(-2) + 4(1) + 1(-1) & (-2)(4) + 4(0) + 1(4) \end{bmatrix}$

$= \begin{bmatrix} 2 & 7 & -4 \end{bmatrix}$

41. $\begin{bmatrix} -3 & 0 & 2 & 1 \\ 4 & 0 & 2 & 6 \end{bmatrix} \begin{bmatrix} -4 & 2 \\ 0 & 1 \end{bmatrix}$

It is not possible to find this product because the number of columns of the first matrix (four) is not equal to the number of rows of the second matrix (two).

For Exercises 43–49,

$$A = \begin{bmatrix} 4 & -2 \\ 3 & 1 \end{bmatrix}, \; B = \begin{bmatrix} 5 & 1 \\ 0 & -2 \\ 3 & 7 \end{bmatrix}, \text{ and } C = \begin{bmatrix} -5 & 4 & 1 \\ 0 & 3 & 6 \end{bmatrix}.$$

43. Using the matrix capabilities of a graphing calculator,

$$[B][A] = \begin{bmatrix} 23 & -9 \\ -6 & -2 \\ 33 & 1 \end{bmatrix}.$$

45. Using the matrix capabilities of a graphing calculator,

$$[B][C] = \begin{bmatrix} -25 & 23 & 11 \\ 0 & -6 & -12 \\ -15 & 33 & 45 \end{bmatrix}.$$

47. It is not possible to find the product [A][B] because the size of A is 2 × 2 and the size of B is 3 × 2.

49. Using the matrix capabilities of a graphing calculator,

$$[A]^2 = \begin{bmatrix} 10 & -10 \\ 15 & -5 \end{bmatrix}.$$

51. Since

$$A + B = \begin{bmatrix} 6 & 12 & 0 \\ -10 & -4 & 11 \end{bmatrix},$$

$$A = \begin{bmatrix} 6 & 12 & 0 \\ -10 & -4 & 11 \end{bmatrix} - B$$

$$A = \begin{bmatrix} 6 & 12 & 0 \\ -10 & -4 & 11 \end{bmatrix} - \begin{bmatrix} 4 & 6 & -5 \\ -6 & 3 & 2 \end{bmatrix} = \begin{bmatrix} 2 & 6 & 5 \\ -4 & -7 & 9 \end{bmatrix}.$$

53. $AB = \begin{bmatrix} -2 & 4 \\ 1 & 3 \end{bmatrix}\begin{bmatrix} -2 & 1 \\ 3 & 6 \end{bmatrix} = \begin{bmatrix} -2(-2)+4(3) & -2(1)+4(6) \\ 1(-2)+3(3) & 1(1)+3(6) \end{bmatrix} = \begin{bmatrix} 16 & 22 \\ 7 & 19 \end{bmatrix}$

55. $AC = \begin{bmatrix} -2 & 4 \\ 1 & 3 \end{bmatrix}\begin{bmatrix} 5 & -2 & 1 \\ 0 & 3 & 7 \end{bmatrix} = \begin{bmatrix} -2(5)+4(0) & -2(-2)+4(3) & -2(1)+4(7) \\ 1(5)+3(0) & 1(-2)+3(3) & 1(1)+3(7) \end{bmatrix}$

$$= \begin{bmatrix} -10 & 16 & 26 \\ 5 & 7 & 22 \end{bmatrix}$$

57. From Exercise 53, $AB = \begin{bmatrix} 16 & 22 \\ 7 & 19 \end{bmatrix}$.

From Exercise 54, $BA = \begin{bmatrix} 5 & -5 \\ 0 & 30 \end{bmatrix}$.

For the given matrices A and B, $AB \neq BA$.

From Exercise 55,

$$AC = \begin{bmatrix} -10 & 16 & 26 \\ 5 & 7 & 22 \end{bmatrix}.$$

From Exercise 56, CA is not possible.

These two examples illustrate that matrix multiplication is not commutative.

59. The given information may be written as the 3×2 matrix

$$\begin{bmatrix} 100 & 150 \\ 125 & 50 \\ 175 & 200 \end{bmatrix},$$

or as the 2×3 matrix

$$\begin{bmatrix} 100 & 125 & 175 \\ 150 & 50 & 200 \end{bmatrix}.$$

61. **(a)** The sales figure information may be written as the 3×3 matrix

$$\begin{bmatrix} 50 & 100 & 30 \\ 10 & 90 & 50 \\ 60 & 120 & 40 \end{bmatrix}.$$

(b) The income per gallon information may be written as the 3×1 matrix

$$\begin{bmatrix} 12 \\ 10 \\ 15 \end{bmatrix}.$$

(If the matrix in part (a) had been written with its rows and columns interchanged, then this income per gallon information would be written instead as a 1×3 matrix.)

(c) $\begin{bmatrix} 50 & 100 & 30 \\ 10 & 90 & 50 \\ 60 & 120 & 40 \end{bmatrix} \begin{bmatrix} 12 \\ 10 \\ 15 \end{bmatrix} = \begin{bmatrix} 2050 \\ 1770 \\ 2520 \end{bmatrix}$

(This result may be written as a 1×3 matrix instead.)

(d) $2050 + 1770 + 2520 = 6340$

The total daily income from the three locations is $6340.

63. **(a)** The word HELP would be written as H = 8, E = 5, L = 12, P = 16 or 8 5 12 16 in numbers.

Then, $B = \begin{bmatrix} 8 & 12 \\ 5 & 16 \end{bmatrix}$. The word would be coded by taking the product AB.

$AB = \begin{bmatrix} 2 & 1 \\ -5 & -2 \end{bmatrix} \begin{bmatrix} 8 & 12 \\ 5 & 16 \end{bmatrix} = \begin{bmatrix} 21 & 40 \\ -50 & -92 \end{bmatrix}$

Scaling the numbers results in

$$21 - 0(26) = 21$$
$$40 - 1(26) = 14$$
$$-50 + 2(26) = 2$$
$$-92 + 4(26) = 12.$$

The new coded matrix is then written as $C = \begin{bmatrix} 21 & 14 \\ 2 & 12 \end{bmatrix}$. The entries of C are written as 21 2 14 12, which become UBNL. Thus, HELP is coded as the word UBNL.

(b) The word LETTER would be written as 12 5 20 20 5 18 in numbers.

$$B = \begin{bmatrix} 12 & 20 & 5 \\ 5 & 20 & 18 \end{bmatrix}$$

Then, the word would be coded by taking the product AB.

$$AB = \begin{bmatrix} 2 & 1 \\ -5 & -2 \end{bmatrix} \begin{bmatrix} 12 & 20 & 5 \\ 5 & 20 & 18 \end{bmatrix}$$

$$= \begin{bmatrix} 29 & 60 & 28 \\ -70 & -140 & -61 \end{bmatrix}$$

$$29 - 1(26) = 3$$
$$60 - 2(26) = 8$$
$$28 - 1(26) = 2$$
$$-70 + 3(26) = 8$$
$$-140 + 6(26) = 16$$
$$-61 + 2(26) = 17$$

The new coded matrix is then written as $C = \begin{bmatrix} 3 & 8 & 2 \\ 8 & 16 & 17 \end{bmatrix}$. The entries of C are written as 3 8 8 16 2 17, which becomes CHHPBQ.

65. $A^2 = AA$

$$= \begin{bmatrix} .8 & .2 \\ .4 & .6 \end{bmatrix} \begin{bmatrix} .8 & .2 \\ .4 & .6 \end{bmatrix}$$

$$= \begin{bmatrix} .72 & .28 \\ .56 & .44 \end{bmatrix}$$

$$.72 + .28 = 1$$
$$.56 + .44 = 1$$

Since the sum of the entries in each row is 1, A^2 is also a stochastic matrix.

67. $A = \begin{bmatrix} .1 & .6 & .3 \\ .3 & .4 & .3 \\ .7 & .2 & .1 \end{bmatrix}$

Use the matrix capabilities of a graphing calculator to find successive powers of A.

$$A^2 = AA = \begin{bmatrix} .40 & .36 & .24 \\ .36 & .40 & .24 \\ .20 & .52 & .28 \end{bmatrix}$$

$$.40 + .36 + .24 = 1$$
$$.36 + .40 + .24 = 1$$
$$.20 + .52 + .28 = 1$$

Therefore, A^2 is also a stochastic matrix.

$$A^3 = AA^2 = \begin{bmatrix} .316 & .432 & .252 \\ .324 & .424 & .252 \\ .372 & .384 & .244 \end{bmatrix}$$

$$A^4 = AA^3 = \begin{bmatrix} .3376 & .4128 & .2496 \\ .336 & .4144 & .2496 \\ .3232 & .4256 & .2512 \end{bmatrix}$$

$$A^5 = AA^4 = \begin{bmatrix} .33232 & .4176 & .25008 \\ .33264 & .41728 & .25008 \\ .33584 & .4144 & .24976 \end{bmatrix}$$

$$A^6 = AA^5 = \begin{bmatrix} .333568 & .416448 & .249984 \\ .333504 & .416512 & .249984 \\ .332736 & .417216 & .250048 \end{bmatrix}$$

The rows appear to be getting closer and closer to $[.\overline{3} \quad .41\overline{6} \quad .25]$ or $\begin{bmatrix} \frac{1}{3} & \frac{5}{12} & \frac{1}{4} \end{bmatrix}$.

69. $(A + B)^T = \left(\begin{bmatrix} a & b \\ c & d \end{bmatrix} + \begin{bmatrix} m & n \\ p & q \end{bmatrix} \right)^T$

$$= \begin{bmatrix} a+m & b+n \\ c+p & d+q \end{bmatrix}^T$$

$$= \begin{bmatrix} a+m & c+p \\ b+n & d+q \end{bmatrix}$$

$$A^T + B^T = \begin{bmatrix} a & c \\ b & d \end{bmatrix} + \begin{bmatrix} m & p \\ n & q \end{bmatrix}$$

$$= \begin{bmatrix} a+m & c+p \\ b+n & d+q \end{bmatrix}$$

Therefore,

$$(A + B)^T = A^T + B^T.$$

71. $A + B = \begin{bmatrix} a_{11} & a_{12} \\ a_{21} & a_{22} \end{bmatrix} + \begin{bmatrix} b_{11} & b_{12} \\ b_{21} & b_{22} \end{bmatrix} = \begin{bmatrix} a_{11} + b_{11} & a_{12} + b_{12} \\ a_{21} + b_{21} & a_{22} + b_{22} \end{bmatrix}$

Since addition of real numbers is commutative, we can rewrite the last matrix.

$A + B = \begin{bmatrix} b_{11} + a_{11} & b_{12} + a_{12} \\ b_{21} + a_{21} & b_{22} + a_{22} \end{bmatrix} = \begin{bmatrix} b_{11} & b_{12} \\ b_{21} & b_{22} \end{bmatrix} + \begin{bmatrix} a_{11} & a_{12} \\ a_{21} & a_{22} \end{bmatrix}$

Therefore,

$$A + B = B + A.$$

The statement is true.

73. $A + B = \begin{bmatrix} a_{11} & a_{12} \\ a_{21} & a_{22} \end{bmatrix} + \begin{bmatrix} b_{11} & b_{12} \\ b_{21} & b_{22} \end{bmatrix} = \begin{bmatrix} a_{11} + b_{11} & a_{12} + b_{12} \\ a_{21} + b_{21} & a_{22} + b_{22} \end{bmatrix}$

This is a 2×2 matrix, so the statement is true.

75. Let $-A = \begin{bmatrix} -a_{11} & -a_{12} \\ -a_{21} & -a_{22} \end{bmatrix}$.

Then,
$A + (-A) = \begin{bmatrix} a_{11} & a_{12} \\ a_{21} & a_{22} \end{bmatrix} + \begin{bmatrix} -a_{11} & -a_{12} \\ -a_{21} & -a_{22} \end{bmatrix} = \begin{bmatrix} a_{11} + (-a_{11}) & a_{12} + (-a_{12}) \\ a_{21} + (-a_{21}) & a_{22} + (-a_{22}) \end{bmatrix} = \begin{bmatrix} 0 & 0 \\ 0 & 0 \end{bmatrix} = O$

Similarly, $-A + A = O$.

Therefore, the statement is true.

77. $A(B + C) = \begin{bmatrix} a_{11} & a_{12} \\ a_{21} & a_{22} \end{bmatrix} \left(\begin{bmatrix} b_{11} & b_{12} \\ b_{21} & b_{22} \end{bmatrix} + \begin{bmatrix} c_{11} & c_{12} \\ c_{21} & c_{22} \end{bmatrix} \right)$

$= \begin{bmatrix} a_{11} & a_{12} \\ a_{21} & a_{22} \end{bmatrix} \begin{bmatrix} b_{11} + c_{11} & b_{12} + c_{12} \\ b_{21} + c_{21} & b_{22} + c_{22} \end{bmatrix}$

$= \begin{bmatrix} a_{11}(b_{11} + c_{11}) + a_{12}(b_{21} + c_{21}) & a_{11}(b_{12} + c_{12}) + a_{12}(b_{22} + c_{22}) \\ a_{21}(b_{11} + c_{11}) + a_{22}(b_{21} + c_{21}) & a_{21}(b_{12} + c_{12}) + a_{22}(b_{22} + c_{22}) \end{bmatrix}$

$= \begin{bmatrix} (a_{11}b_{11} + a_{12}b_{21}) + (a_{11}c_{11} + a_{11}c_{21}) & (a_{11}b_{12} + a_{12}b_{22}) + (a_{11}c_{12} + a_{12}c_{22}) \\ (a_{21}b_{11} + a_{22}b_{21}) + (a_{21}c_{11} + a_{22}c_{21}) & (a_{21}b_{12} + a_{22}b_{22}) + (a_{21}c_{12} + a_{22}c_{22}) \end{bmatrix}$

$= \begin{bmatrix} a_{11}b_{11} + a_{12}b_{21} & a_{11}b_{12} + a_{12}b_{22} \\ a_{21}b_{11} + a_{22}b_{21} & a_{21}b_{12} + a_{22}b_{22} \end{bmatrix} + \begin{bmatrix} a_{11}c_{11} + a_{11}c_{21} & a_{11}c_{12} + a_{12}c_{22} \\ a_{21}c_{11} + a_{22}c_{21} & a_{21}c_{12} + a_{22}c_{22} \end{bmatrix}$

Therefore,

$$A(B + C) = AB + AC.$$

The statement is true.

79. $c(A + B)$

$$= c\left(\begin{bmatrix} a_{11} & a_{12} \\ a_{21} & a_{22} \end{bmatrix} + \begin{bmatrix} b_{11} & b_{12} \\ b_{21} & b_{22} \end{bmatrix}\right)$$

$$= c\begin{bmatrix} a_{11} + b_{11} & a_{12} + b_{12} \\ a_{21} + b_{21} & a_{22} + b_{22} \end{bmatrix}$$

$$= \begin{bmatrix} c(a_{11} + b_{11}) & c(a_{12} + b_{12}) \\ c(a_{21} + b_{21}) & c(a_{22} + b_{22}) \end{bmatrix}$$

$$= \begin{bmatrix} ca_{11} + cb_{11} & ca_{12} + cb_{12} \\ ca_{21} + cb_{21} & ca_{22} + cb_{22} \end{bmatrix}$$

$$= \begin{bmatrix} ca_{11} & ca_{12} \\ ca_{21} & ca_{22} \end{bmatrix} + \begin{bmatrix} cb_{11} & cb_{12} \\ cb_{21} & cb_{22} \end{bmatrix}$$

$$= c\begin{bmatrix} a_{11} & a_{12} \\ a_{21} & a_{22} \end{bmatrix} = c\begin{bmatrix} b_{11} & b_{12} \\ b_{21} & b_{22} \end{bmatrix}$$

Therefore,

$$c(A + B) = cA + cB.$$

The statement is true.

81. $c(A)d = c\begin{bmatrix} a_{11} & a_{12} \\ a_{21} & a_{22} \end{bmatrix}d$

$$= \begin{bmatrix} ca_{11} & ca_{12} \\ ca_{21} & ca_{22} \end{bmatrix}d$$

$$= \begin{bmatrix} (ca_{11})d & (ca_{12})d \\ (ca_{21})d & c(a_{22})d \end{bmatrix}$$

$$= \begin{bmatrix} cd(a_{11}) & cd(a_{12}) \\ cd(a_{21}) & cd(a_{22}) \end{bmatrix}$$

$$= cd\begin{bmatrix} a_{11} & a_{12} \\ a_{21} & a_{22} \end{bmatrix}$$

Therefore,

$$c(A)d = cdA.$$

The statement is true.

83. The statement

$$(A + B)(A - B) = A^2 - B^2$$

is false.

Let $A = \begin{bmatrix} 1 & 3 \\ 4 & 5 \end{bmatrix}$ and $B = \begin{bmatrix} 2 & 3 \\ 1 & 5 \end{bmatrix}$.

$$(A + B)(A - B) = \left(\begin{bmatrix} 1 & 3 \\ 4 & 5 \end{bmatrix} + \begin{bmatrix} 2 & 3 \\ 1 & 5 \end{bmatrix}\right)$$

$$\cdot \left(\begin{bmatrix} 1 & 3 \\ 4 & 5 \end{bmatrix} - \begin{bmatrix} 2 & 3 \\ 1 & 5 \end{bmatrix}\right)$$

$$= \begin{bmatrix} 3 & 6 \\ 5 & 10 \end{bmatrix}\begin{bmatrix} -1 & 0 \\ 3 & 0 \end{bmatrix}$$

$$= \begin{bmatrix} 15 & 0 \\ 25 & 0 \end{bmatrix}$$

$$A^2 - B^2 = \begin{bmatrix} 1 & 3 \\ 4 & 5 \end{bmatrix}\begin{bmatrix} 1 & 3 \\ 4 & 5 \end{bmatrix} - \begin{bmatrix} 2 & 3 \\ 1 & 5 \end{bmatrix}\begin{bmatrix} 2 & 3 \\ 1 & 5 \end{bmatrix}$$

$$= \begin{bmatrix} 13 & 18 \\ 24 & 37 \end{bmatrix} - \begin{bmatrix} 7 & 21 \\ 7 & 28 \end{bmatrix}$$

$$= \begin{bmatrix} 6 & -3 \\ 17 & 9 \end{bmatrix}$$

We see that

$$(A + B)(A - B) \neq A^2 - B^2.$$

Section 6.5

1. $\begin{vmatrix} 2 & 5 \\ 4 & -7 \end{vmatrix} = 2(-7) - 4(5) = -34$

3. $\begin{vmatrix} -9 & 7 \\ 2 & 6 \end{vmatrix} = -9(6) - 2(7) = -68$

5. $\begin{vmatrix} y & 3 \\ -2 & x \end{vmatrix} = yx - (-2)(3) = yx + 6$

7. $\begin{vmatrix} -2 & 0 & 1 \\ 1 & 2 & 0 \\ 4 & 2 & 1 \end{vmatrix}$

Cofactor of 1: $(-1)\begin{vmatrix} 0 & 1 \\ 2 & 1 \end{vmatrix} = -1(-2)$
$$= 2$$

Cofactor of 2: $1\begin{vmatrix} -2 & 1 \\ 4 & 1 \end{vmatrix} = 1(-6)$
$$= -6$$

Cofactor of 0: $(-1)\begin{vmatrix} -2 & 0 \\ 4 & 2 \end{vmatrix} = -1(-4)$

$\qquad\qquad\qquad\qquad = 4$

9. $\begin{vmatrix} 1 & 2 & -1 \\ 2 & 3 & -2 \\ -1 & 4 & 1 \end{vmatrix}$

Cofactor of 2: $(-1)\begin{vmatrix} 2 & -1 \\ 4 & 1 \end{vmatrix} = -1(6)$

$\qquad\qquad\qquad\qquad = -6$

Cofactor of 3: $1\begin{vmatrix} 1 & -1 \\ -1 & 1 \end{vmatrix} = 1(0)$

$\qquad\qquad\qquad\qquad = 0$

Cofactor of -2: $-1\begin{vmatrix} 1 & 2 \\ -1 & 4 \end{vmatrix} = -1(6)$

$\qquad\qquad\qquad\qquad = -6$

11. $\begin{vmatrix} 1 & 0 & 0 \\ 0 & 1 & 0 \\ 0 & 0 & 1 \end{vmatrix}$

Expand by minors about row 1.

$\begin{vmatrix} 1 & 0 & 0 \\ 0 & 1 & 0 \\ 0 & 0 & 1 \end{vmatrix} = 1\begin{vmatrix} 1 & 0 \\ 0 & 1 \end{vmatrix} - 0\begin{vmatrix} 0 & 0 \\ 0 & 1 \end{vmatrix} + 0\begin{vmatrix} 0 & 1 \\ 0 & 0 \end{vmatrix}$

$\qquad\qquad = 1(1) - 0 + 0$

$\qquad\qquad = 1$

13. $\begin{vmatrix} -2 & 0 & 1 \\ 0 & 1 & 0 \\ 0 & 0 & -1 \end{vmatrix}$

Expand by minors about row 3 since row 3 has two zeros.

$\begin{vmatrix} -2 & 0 & 1 \\ 0 & 1 & 0 \\ 0 & 0 & -1 \end{vmatrix}$

$= 0\begin{vmatrix} 0 & 1 \\ 1 & 0 \end{vmatrix} - 0\begin{vmatrix} -2 & 1 \\ 0 & 0 \end{vmatrix} + (-1)\begin{vmatrix} -2 & 0 \\ 0 & 1 \end{vmatrix}$

$= 0 - 0 - 1(-2 - 0)$

$= 2$

15. $\begin{vmatrix} 0 & 5 & 2 \\ 0 & 3 & -1 \\ 0 & -4 & 7 \end{vmatrix}$

Expand by minors about column 1.

$\begin{vmatrix} 0 & 5 & 2 \\ 0 & 3 & -1 \\ 0 & -4 & 7 \end{vmatrix}$

$= 0\begin{vmatrix} 3 & -1 \\ -4 & 7 \end{vmatrix} - 0\begin{vmatrix} 5 & 2 \\ -4 & 7 \end{vmatrix} + 0\begin{vmatrix} 5 & 2 \\ 3 & -1 \end{vmatrix}$

$= 0 + 0 + 0$

$= 0$

17. $\begin{vmatrix} 0 & 3 & y \\ 0 & 4 & 2 \\ 1 & 0 & 1 \end{vmatrix}$

Expand by minors about column 1 since it has two zeros.

$\begin{vmatrix} 0 & 3 & y \\ 0 & 4 & 2 \\ 1 & 0 & 1 \end{vmatrix}$

$= 0\begin{vmatrix} 4 & 2 \\ 0 & 1 \end{vmatrix} - 0\begin{vmatrix} 3 & y \\ 0 & 1 \end{vmatrix} + 1\begin{vmatrix} 3 & y \\ 4 & 2 \end{vmatrix}$

$= 0 - 0 + 1(6 - 4y)$

$= 6 - 4y$

19. $\begin{vmatrix} .4 & -.8 & .6 \\ .3 & .9 & .7 \\ 3.1 & 4.1 & -2.8 \end{vmatrix}$

Expand by minors about the first row.

$= .4\begin{vmatrix} .9 & .7 \\ 4.1 & -2.8 \end{vmatrix} - (-.8)\begin{vmatrix} .3 & .7 \\ 3.1 & -2.8 \end{vmatrix}$

$\qquad + .6\begin{vmatrix} .3 & .9 \\ 3.1 & 4.1 \end{vmatrix}$

$= (.4)(-5.39) + (.8)(-3.01)$

$\qquad + (.6)(-1.56)$

$= -5.5$

21.
$$\begin{vmatrix} 2 & 0 & 0 & 1 \\ -2 & 0 & 6 & 0 \\ 2 & 4 & 0 & 1 \\ 2 & 4 & 1 & 2 \end{vmatrix}$$

Expand by minors about row 1.

$$\begin{vmatrix} 2 & 0 & 0 & 1 \\ -2 & 0 & 6 & 0 \\ 2 & 4 & 0 & 1 \\ 2 & 4 & 1 & 2 \end{vmatrix}$$

$$= 2\begin{vmatrix} 0 & 6 & 0 \\ 4 & 0 & 1 \\ 4 & 1 & 2 \end{vmatrix} - 0\begin{vmatrix} -2 & 6 & 0 \\ 2 & 0 & 1 \\ 2 & 1 & 2 \end{vmatrix}$$

$$+ 0\begin{vmatrix} -2 & 0 & 0 \\ 2 & 4 & 1 \\ 2 & 4 & 2 \end{vmatrix} - 1\begin{vmatrix} -2 & 0 & 6 \\ 2 & 4 & 0 \\ 2 & 4 & 1 \end{vmatrix}$$

$$= 2\left(0\begin{vmatrix} 0 & 1 \\ 1 & 2 \end{vmatrix} - 6\begin{vmatrix} 4 & 1 \\ 4 & 2 \end{vmatrix} + 0\begin{vmatrix} 4 & 0 \\ 4 & 1 \end{vmatrix} \right) - 0$$

$$+ 0 - 1\left(-2\begin{vmatrix} 4 & 0 \\ 4 & 1 \end{vmatrix} - 0\begin{vmatrix} 2 & 0 \\ 2 & 1 \end{vmatrix} + 6\begin{vmatrix} 2 & 4 \\ 2 & 4 \end{vmatrix} \right)$$

$$= 2[0 - 6(4) + 0]$$

$$- 1[-2(4) - 0 + 6(0)]$$

$$= 2(-24) - (-8)$$

$$= -40$$

23.
$$\begin{vmatrix} x & 2 & 1 \\ -1 & x & 4 \\ -2 & 0 & 5 \end{vmatrix}$$

Expand by minors about row 3.

$$\begin{vmatrix} x & 2 & 1 \\ -1 & x & 4 \\ -2 & 0 & 5 \end{vmatrix}$$

$$= -2\begin{vmatrix} 2 & 1 \\ x & 4 \end{vmatrix} - 0\begin{vmatrix} x & 1 \\ -1 & 4 \end{vmatrix} + 5\begin{vmatrix} x & 2 \\ -1 & x \end{vmatrix}$$

$$= -2(8 - x) - 0 + 5(x^2 + 2)$$

$$= -16 + 2x + 5x^2 + 10$$

$$= 5x^2 + 2x - 6$$

24. The equation is

$$5x^2 + 2x - 6 = 45,$$

which is a quadratic equation.

25. $5x^2 + 2x - 6 = 45$

Rewrite the equation in standard form.

$$5x^2 + 2x - 51 = 0$$

Solve this equation by factoring.

$$(5x + 17)(x - 3) = 0$$

$$5x + 17 = 0 \quad \text{or} \quad x - 3 = 0$$

$$5x = -17$$

$$x = -\frac{17}{5}$$

$$x = -3.4 \quad \text{or} \quad x = 3$$

Solution set: $\{-3.4, 3\}$

26. If $x = 3$, the determinant becomes

$$\begin{vmatrix} 3 & 2 & 1 \\ -1 & 3 & 4 \\ -2 & 0 & 5 \end{vmatrix}$$

$$= -2\begin{vmatrix} 2 & 1 \\ 3 & 4 \end{vmatrix} - 0\begin{vmatrix} 3 & 1 \\ -1 & 4 \end{vmatrix} + 5\begin{vmatrix} 3 & 2 \\ -1 & 3 \end{vmatrix}$$

$$= -2(5) - 0 + 5(11)$$

$$= -10 + 55$$

$$= 45.$$

If $x = 3.4$, the determinant becomes

$$\begin{vmatrix} -3.4 & 2 & 1 \\ -1 & -3.4 & 4 \\ -2 & 0 & 5 \end{vmatrix}$$

$$= -2\begin{vmatrix} 2 & 1 \\ -3.4 & 4 \end{vmatrix} - 0\begin{vmatrix} -3.4 & 1 \\ -1 & 4 \end{vmatrix}$$

$$+ 5\begin{vmatrix} -3.4 & 2 \\ -1 & -3.4 \end{vmatrix}$$

$$= -2(11.4) - 0 + 5(13.56)$$

$$= -22.8 + 67.8$$

$$= 45.$$

27. To solve the equation

$$\begin{vmatrix} -2 & 0 & 1 \\ -1 & 3 & x \\ 5 & -2 & 0 \end{vmatrix} = 3,$$

expand by minors about the first row.

$$-2 \begin{vmatrix} 3 & x \\ -2 & 0 \end{vmatrix} - 0 \begin{vmatrix} -1 & x \\ 5 & 0 \end{vmatrix} + 1 \begin{vmatrix} -1 & 3 \\ 5 & -2 \end{vmatrix} = 3$$

$$-2(0 + 2x) - 0 + 1(2 - 15) = 3$$

$$-4x - 13 = 3$$

$$-4x = 16$$

$$x = -4$$

Solution set: $\{-4\}$

29. $\begin{vmatrix} 5 & 3x & -3 \\ 0 & 2 & -1 \\ 4 & -1 & x \end{vmatrix} = -7$

Expand about the second row.

$$-0 \begin{vmatrix} 3x & -3 \\ -1 & x \end{vmatrix} + 2 \begin{vmatrix} 5 & -3 \\ 4 & x \end{vmatrix} - (-1) \begin{vmatrix} 5 & 3x \\ 4 & -1 \end{vmatrix} = -7$$

$$2(5x + 12) + (-5 - 12x) = -7$$

$$10x + 24 - 5 - 12x = -7$$

$$-2x + 19 = 2x$$

$$26 = 2x$$

$$13 = x$$

Solution set: $\{13\}$

31. $\begin{vmatrix} 2 & 3 \\ 2 & 3 \end{vmatrix} = 0$

This statement is true by Theorem 5, since the two rows of the matrix $\begin{bmatrix} 2 & 3 \\ 2 & 3 \end{bmatrix}$ are identical.

33. $\begin{vmatrix} 2 & 0 \\ 3 & 0 \end{vmatrix} = 0$

This statement is true by Theorem 1, since every element of column 2 of the matrix $\begin{bmatrix} 2 & 0 \\ 3 & 0 \end{bmatrix}$ is 0.

35. $\begin{vmatrix} 1 & 0 & 0 \\ 1 & 0 & 1 \\ 3 & 0 & 0 \end{vmatrix} = 0$

This statement is true by Theorem 1, since every element in column 2 of the corresponding matrix is 0.

37. $\begin{vmatrix} 7z & 8x & 2y \\ z & x & y \\ 7z & 7x & 7y \end{vmatrix} = 0$

Row 3 of the matrix for this determinant equals row 2 multiplied by 7, so, by Theorem 4, we have

$$\begin{vmatrix} 7z & 8x & 2y \\ z & x & y \\ 7z & 7x & 7y \end{vmatrix} = 7 \begin{vmatrix} 7z & 8x & 2y \\ z & x & y \\ z & x & y \end{vmatrix}.$$

By Theorem 5, the determinant of a matrix with two identical rows equals 0, so the value of the original determinant is

$$7(0) = 0.$$

39. $\begin{vmatrix} 4 & -2 \\ 3 & 8 \end{vmatrix} = \begin{vmatrix} 4 & 3 \\ -2 & 8 \end{vmatrix}$

Since rows and columns of the corresponding matrix are interchanged, Theorem 2 says that the determinant is not changed.

41.
$$\begin{vmatrix} -1 & 8 & 9 \\ 0 & 2 & 1 \\ 3 & 2 & 0 \end{vmatrix} = -\begin{vmatrix} 8 & -1 & 9 \\ 2 & 0 & 1 \\ 2 & 3 & 0 \end{vmatrix}$$

By Theorem 3, interchanging two columns of a matrix reverses the sign of the determinant. Note that columns 1 and 2 have been interchanged here.

43. $-\dfrac{1}{2}\begin{vmatrix} 5 & -8 & 2 \\ 3 & -6 & 9 \\ 2 & 4 & 4 \end{vmatrix} = \begin{vmatrix} 5 & 4 & 2 \\ 3 & 3 & 9 \\ 2 & -2 & 4 \end{vmatrix}$

Theorem 4 says that if every element of column 2 of the matrix

$\begin{bmatrix} 5 & -8 & 2 \\ 3 & -6 & 9 \\ 2 & 4 & 4 \end{bmatrix}$ is multiplied by $-1/2$,

then the determinant of the new

matrix, $\begin{bmatrix} 5 & 4 & 2 \\ 3 & 3 & 9 \\ 2 & -2 & 4 \end{bmatrix}$, is $-\dfrac{1}{2}\begin{vmatrix} 5 & -8 & 2 \\ 3 & -6 & 9 \\ 2 & 4 & 4 \end{vmatrix}$.

45. $\begin{vmatrix} 3 & -4 \\ 2 & 5 \end{vmatrix} = \begin{vmatrix} 3 & -4 \\ 5 & 1 \end{vmatrix}$

By Theorem 6, if a multiple of a row of a matrix is added to the corresponding elements of another row, the determinant is unchanged. In this case, row 1 of the matrix $\begin{bmatrix} 3 & -4 \\ 2 & 5 \end{bmatrix}$

was added to row 2 to obtain the

matrix $\begin{bmatrix} 3 & -4 \\ 5 & 1 \end{bmatrix}$.

47. $\begin{vmatrix} -4 & 2 & 1 \\ 3 & 0 & 5 \\ -1 & 4 & -2 \end{vmatrix} = \begin{vmatrix} -4 & 2 & 1+(-4)k \\ 3 & 0 & 5+3k \\ -1 & 4 & -2+(-1)k \end{vmatrix}$

Multiply column 1 of $\begin{bmatrix} -4 & 2 & 1 \\ 3 & 0 & 5 \\ -1 & 4 & -2 \end{bmatrix}$

by k and add it to column 3. By Theorem 6, these determinants are equal.

49. $\begin{vmatrix} -5 & 10 \\ 6 & -12 \end{vmatrix}$

Multiply column 1 of the corresponding matrix by 2 and add the result to column 2.

$$\begin{bmatrix} -5 & 10 \\ 6 & -12 \end{bmatrix} = \begin{bmatrix} -5 & 0 \\ 6 & 0 \end{bmatrix} = 0$$

51. $\begin{vmatrix} 6 & 8 & -12 \\ -1 & 0 & 2 \\ 4 & 0 & -8 \end{vmatrix}$

Multiply column 1 of the matrix for this determinant by 2 and add the result to column 3.
Thus,

$$\begin{vmatrix} 6 & 8 & -12 \\ -1 & 0 & 2 \\ 4 & 0 & -8 \end{vmatrix} = \begin{vmatrix} 6 & 8 & 0 \\ -1 & 16 & 0 \\ 4 & 0 & 0 \end{vmatrix} = 0.$$

53. $\begin{vmatrix} -4 & 1 & 4 \\ 2 & 0 & 1 \\ 0 & 2 & 4 \end{vmatrix} = \begin{vmatrix} 0 & 1 & 0 \\ 2 & 0 & 1 \\ 8 & 4 & -4 \end{vmatrix}$ $\begin{array}{l} 4C2 + C1 \\ -4C2 + C3 \end{array}$

$$= 0\begin{vmatrix} 0 & 1 \\ 4 & -4 \end{vmatrix} - 1\begin{vmatrix} 2 & 1 \\ 8 & -4 \end{vmatrix}$$

$$+ 0\begin{vmatrix} 2 & 0 \\ 8 & 4 \end{vmatrix}$$

$$= 0 - 1(-16) + 0$$

$$= 16$$

55.
$$\begin{vmatrix} 2 & -1 & 1 & 0 \\ 1 & 1 & 0 & 1 \\ 0 & -1 & 1 & 1 \\ 1 & 2 & 1 & 2 \end{vmatrix} = \begin{vmatrix} 2 & 0 & 0 & -1 \\ 1 & 1 & 0 & 1 \\ 0 & -1 & 1 & 1 \\ 1 & 2 & 1 & 2 \end{vmatrix}$$

Multiply row 3 of the corresponding matrix by −1 and add to row 1.

$$= \begin{vmatrix} 2 & 0 & 0 & -1 \\ 1 & 1 & 0 & 1 \\ 0 & -1 & 1 & 1 \\ 1 & 3 & 0 & 1 \end{vmatrix}$$

Multiply row 3 of the corresponding matrix by −1 and add to row 4.
Expand the determinant about column 3.

$$= 0\begin{vmatrix} 1 & 1 & 1 \\ 0 & -1 & 1 \\ 1 & 3 & 1 \end{vmatrix} - 0\begin{vmatrix} 2 & 0 & -1 \\ 0 & -1 & 1 \\ 1 & 3 & 1 \end{vmatrix}$$

$$+ 1\begin{vmatrix} 2 & 0 & -1 \\ 1 & 1 & 1 \\ 1 & 3 & 1 \end{vmatrix} - 0\begin{vmatrix} 2 & 0 & -1 \\ 1 & 1 & 1 \\ 0 & -1 & 1 \end{vmatrix}$$

$$= \begin{vmatrix} 2 & 0 & -1 \\ 1 & 1 & 1 \\ 1 & 3 & 1 \end{vmatrix}$$

Multiply row 2 by −2 and add to row 1.

$$= \begin{vmatrix} 0 & -2 & -3 \\ 1 & 1 & 1 \\ 1 & 3 & 1 \end{vmatrix}$$

Multiply row 2 by −1 and add to row 3.

$$= \begin{vmatrix} 0 & -2 & -3 \\ 1 & 1 & 1 \\ 0 & 2 & 0 \end{vmatrix}$$

Expand about column 1.

$$= 0\begin{vmatrix} 1 & 1 \\ 2 & 0 \end{vmatrix} - 1\begin{vmatrix} -2 & -3 \\ 2 & 0 \end{vmatrix} + 0\begin{vmatrix} -2 & -3 \\ 1 & 1 \end{vmatrix}$$

$$= -1(6) = -6$$

57. P(0, 0), Q(0, 2), R(1, 4)

Find

$$D = \frac{1}{2}\begin{vmatrix} x_1 & y_1 & 1 \\ x_2 & y_2 & 1 \\ x_3 & y_3 & 1 \end{vmatrix},$$

where $P = (x_1, y_1) = (0, 0)$, $Q(x_2, y_2) = (0, 2)$, and $R = (x_3, y_3) = (1, 4)$.

$$\frac{1}{2}\begin{vmatrix} 0 & 0 & 1 \\ 0 & 2 & 1 \\ 1 & 4 & 1 \end{vmatrix}$$

$$= \frac{1}{2}\left[0\begin{vmatrix} 2 & 1 \\ 4 & 1 \end{vmatrix} - 0\begin{vmatrix} 0 & 1 \\ 1 & 1 \end{vmatrix} + 1\begin{vmatrix} 0 & 2 \\ 1 & 4 \end{vmatrix}\right]$$

$$= \frac{1}{2}[0 - 0 + 1(-2)]$$

$$= -1$$

Area of triangle of $|D| = |-1| = 1$

59. P(2, 5), Q(−1, 3), R(4, 0)

$$\frac{1}{2}\begin{vmatrix} 2 & 5 & 1 \\ -1 & 3 & 1 \\ 4 & 0 & 1 \end{vmatrix}$$

$$= \frac{1}{2}\left[4\begin{vmatrix} 5 & 1 \\ 3 & 1 \end{vmatrix} - 0\begin{vmatrix} 2 & 1 \\ -1 & 1 \end{vmatrix} + 1\begin{vmatrix} 2 & 5 \\ -1 & 3 \end{vmatrix}\right]$$

$$= \frac{1}{2}[4(2) - 0 + 1(11)]$$

$$= \frac{1}{2}(8 + 11)$$

$$= \frac{19}{2} = 9.5$$

Area of triangle = $|9.5| = 9.5$

61. P(4, 7), Q(5, −2), R(1, 1)

$$\frac{1}{2}\begin{vmatrix} 4 & 7 & 1 \\ 1 & 1 & 1 \\ 5 & -2 & 1 \end{vmatrix}$$

$$= \frac{1}{2}\left[4\begin{vmatrix} 1 & 1 \\ -2 & 1 \end{vmatrix} - 7\begin{vmatrix} 1 & 1 \\ 5 & 1 \end{vmatrix} + 1\begin{vmatrix} 1 & 1 \\ 5 & -2 \end{vmatrix}\right]$$

$$= \frac{1}{2}[4(3) - 7(-4) + 1(-7)]$$

$$= \frac{33}{2} = 16.5$$

Area of triangle = $|16.5|$ = 16.5

63. Using a graphing calculator with matrix capabilities, we obtain

$$\begin{vmatrix} \pi & \sqrt{2} \\ e & \sqrt{3} \end{vmatrix} = 1.597167065.$$

65. Using a graphing calculator with matrix capabilities, we obtain

$$\begin{vmatrix} .29 & .36 & -.51 \\ -.16 & 1.24 & 3.26 \\ 2.43 & 3.84 & -6.15 \end{vmatrix} = -1.494192.$$

67. $\begin{vmatrix} x & y & 1 \\ x_1 & y_1 & 1 \\ x_2 & y_2 & 1 \end{vmatrix}$

$$= \begin{vmatrix} x & y & 1 \\ x_1 - x & y_1 - y & 0 \\ x_2 - x & y_2 - y & 0 \end{vmatrix} \begin{array}{l} -1R1 + R2 \\ -1R1 + R3 \end{array}$$

$$= 1 \begin{vmatrix} x_1 - x & y_1 - y \\ x_2 - x & y_2 - y \end{vmatrix} - 0 \begin{vmatrix} x & y \\ x_2 - x & y_2 - y \end{vmatrix}$$

$$+ 0 \begin{vmatrix} x & y \\ x_1 - x & y_1 - y \end{vmatrix}$$

Expand about column 3

$$= (x_1 - x)(y_2 - y)$$
$$\quad - (y_1 - y)(x_2 - x)$$

If $\begin{vmatrix} x & y & 1 \\ x_1 & y_1 & 1 \\ x_2 & y_2 & 1 \end{vmatrix} = 0$, then

$(x_1 - x)(y_2 - y) - (y_1 - y)(x_2 - x) = 0$

$(x_1 - x)(y_2 - y) = (y_1 - y)(x_2 - x)$

$$\frac{(x_1 - x)(y_2 - y)}{(x_1 - x)(x_2 - x)} = \frac{(y_1 - y)(x_2 - x)}{(x_1 - x)(x_2 - x)}$$

$$\frac{y_2 - y}{x_2 - x} = \frac{y_1 - y}{x_1 - x}$$

$$\frac{y - y_2}{x - x_2} = \frac{y - y_1}{x - x_1}.$$

If (x, y) is a point on the line through (x_1, y_1) and (x_2, y_2), then the slope of the line is

$$m = \frac{y - y_2}{x - x_2}.$$

Therefore,

$$m = \frac{y - y_1}{x - x_1}$$

or

$$y - y_1 = m(x - x_1).$$

This is the point-slope form of the equation of the line through (x_1, y_1) and (x_2, y_2).

69. Find the slope of each line.

$$a_1x + b_1y = c_1$$
$$b_1y = -a_1x + c_1$$
$$y = -\frac{a_1}{b_1}x + \frac{c_1}{b_1}$$

$$a_2x + b_2y = c_2$$
$$b_2y = -a_2x + c_2$$
$$y = -\frac{a_2}{b_2}x + \frac{c_2}{b_2}$$

Notice that c_1 and c_2 represent the y-intercepts of the two lines. The restriction $c_1 \neq c_2$ guarantees that the two lines cannot be the same line.

Therefore, if their slopes are equal, the lines must be parallel. Set the two slopes equal to each other.

$$-\frac{a_1}{b_1} = -\frac{a_2}{b_2}$$

$$\frac{a_1}{b_1} = \frac{a_2}{b_2}$$

$$a_1b_2 = a_2b_1$$

$$a_1b_2 - a_2b_1 = 0$$

By the definition of the determinant of a 2 × 2 matrix, the last equation is

$$\begin{vmatrix} a_1 & b_1 \\ a_2 & b_2 \end{vmatrix} = 0.$$

71. Since rows 1 and 3 are identical, by Determinant Theorem 5,

$$\begin{vmatrix} 1 & 1 & 1 \\ 1+x & 1+y & 1 \\ 1 & 1 & 1 \end{vmatrix} = 0.$$

73.

$$|A| = \begin{vmatrix} a_{11} & a_{12} & a_{13} \\ a_{21} & a_{22} & a_{23} \\ a_{31} & a_{32} & a_{33} \end{vmatrix}$$

$$= a_{31}\begin{vmatrix} a_{12} & a_{13} \\ a_{22} & a_{23} \end{vmatrix} - a_{32}\begin{vmatrix} a_{11} & a_{13} \\ a_{21} & a_{23} \end{vmatrix}$$

$$+ a_{33}\begin{vmatrix} a_{11} & a_{12} \\ a_{21} & a_{22} \end{vmatrix}$$

$$= a_{31}(a_{12}a_{23} - a_{13}a_{22})$$
$$- a_{32}(a_{11}a_{23} - a_{13}a_{21})$$
$$+ a_{33}(a_{11}a_{22} - a_{12}a_{21})$$

$$= a_{31}a_{12}a_{23} - a_{31}a_{13}a_{22} - a_{32}a_{11}a_{23}$$
$$+ a_{32}a_{13}a_{21} + a_{33}a_{11}a_{22}$$
$$- a_{33}a_{12}a_{21}$$

$$= (a_{11}a_{22}a_{33} + a_{12}a_{23}a_{31} + a_{13}a_{21}a_{32})$$
$$- (a_{31}a_{22}a_{13} + a_{32}a_{23}a_{11} + a_{33}a_{21}a_{12})$$

This is $|A|$ as given in the definition.

75.

$$B = \begin{vmatrix} a_{13} & a_{12} & a_{11} \\ a_{23} & a_{22} & a_{21} \\ a_{33} & a_{32} & a_{31} \end{vmatrix}$$

We find $|B|$ by expanding about row 1.

$$|B| = a_{13}\begin{vmatrix} a_{22} & a_{21} \\ a_{32} & a_{31} \end{vmatrix} - a_{12}\begin{vmatrix} a_{23} & a_{21} \\ a_{33} & a_{31} \end{vmatrix}$$

$$+ a_{11}\begin{vmatrix} a_{23} & a_{22} \\ a_{33} & a_{32} \end{vmatrix}$$

$$= a_{13}(a_{22}a_{31} - a_{21}a_{32})$$
$$- a_{12}(a_{23}a_{31} - a_{21}a_{33})$$
$$+ a_{11}(a_{23}a_{32} - a_{22}a_{33})$$

$$= a_{13}a_{22}a_{31} - a_{13}a_{21}a_{32} - a_{12}a_{23}a_{31}$$
$$+ a_{12}a_{21}a_{33} + a_{11}a_{23}a_{32} - a_{11}a_{22}a_{33}$$

$$= (-a_{11}a_{22}a_{33} - a_{12}a_{23}a_{31} - a_{13}a_{21}a_{32})$$
$$+ (a_{31}a_{22}a_{13} + a_{32}a_{23}a_{11} + a_{33}a_{21}a_{12}$$

$$= -1[(a_{11}a_{22}a_{33} + a_{12}a_{23}a_{31} + a_{13}a_{21}a_{32}$$
$$- (a_{31}a_{22}a_{13} + a_{32}a_{23}a_{11} + a_{33}a_{21}a_{12})]$$

$$= -1|A| = -|A|$$

77.

$$B = \begin{vmatrix} a_{11}+ka_{12} & a_{12} & a_{13} \\ a_{21}+ka_{22} & a_{22} & a_{23} \\ a_{31}+ka_{32} & a_{32} & a_{33} \end{vmatrix}$$

We find $|B|$ by expanding about column 3.

$$|B| = a_{13}\begin{vmatrix} a_{21}+ka_{22} & a_{22} \\ a_{31}+ka_{32} & a_{32} \end{vmatrix}$$

$$- a_{23}\begin{vmatrix} a_{11}+ka_{12} & a_{12} \\ a_{31}+ka_{32} & a_{32} \end{vmatrix}$$

$$+ a_{33}\begin{vmatrix} a_{11}+ka_{12} & a_{12} \\ a_{21}+ka_{22} & a_{22} \end{vmatrix}$$

$$= a_{13}(a_{21}a_{32}+ka_{22}a_{32}-a_{22}a_{31}-ka_{22}a_{32})$$
$$- a_{23}(a_{11}a_{32}+ka_{12}a_{32}-a_{12}a_{31}-ka_{12}a_{32})$$
$$+ a_{33}(a_{11}a_{22}+ka_{12}a_{22}-a_{12}a_{21}-ka_{12}a_{22})$$

$$= a_{13}a_{21}a_{32} + ka_{13}a_{22}a_{32} - a_{13}a_{22}a_{31}$$
$$- ka_{13}a_{22}a_{32} - a_{23}a_{11}a_{32} + ka_{23}a_{12}a_{32}$$
$$+ a_{23}a_{12}a_{31} + ka_{23}a_{12}a_{32} + a_{33}a_{11}a_{22}$$
$$+ ka_{33}a_{12}a_{22} - a_{33}a_{12}a_{21} + ka_{33}a_{12}a_{22}$$

$$= (a_{11}a_{22}a_{33} + a_{12}a_{23}a_{31} + a_{13}a_{21}a_{32})$$
$$- (a_{31}a_{22}a_{13} + a_{32}a_{23}a_{11} + a_{33}a_{21}a_{12})$$

$$= |A| \quad \textit{Definition of determinant of 3 × 3 matrix}$$

79. Let $A = \begin{bmatrix} a & b \\ c & d \end{bmatrix}$ and $B = \begin{bmatrix} r & s \\ t & u \end{bmatrix}$.

Then, $AB = \begin{bmatrix} ar+bt & as+bu \\ cr+dt & cs+du \end{bmatrix}$.

$$|A| \cdot |B| = \begin{vmatrix} a & b \\ c & d \end{vmatrix} \cdot \begin{vmatrix} r & s \\ t & u \end{vmatrix}$$
$$= (ad - bc)(ru - st)$$
$$= adru - adst - bcru + bcst$$

$$|AB| = \begin{vmatrix} ar+bt & as+bu \\ cr+dt & cs+du \end{vmatrix}$$
$$= (ar + bt)(cs + du)$$
$$\quad - (as + bu)(cr + dt)$$
$$= arcs + ardu + btcs + btdu$$
$$\quad - ascr - asdt - bucr$$
$$\quad - budt$$
$$= adru - adst - bcru + bcst$$

Therefore, $|AB| = |A| \cdot |B|$.

Section 6.6

1. $x + y = 4$
$2x - y = 2$

$$D = \begin{vmatrix} 1 & 1 \\ 2 & -1 \end{vmatrix} = -3$$

To find D_x, replace the first column of $\begin{vmatrix} 1 & 1 \\ 2 & -1 \end{vmatrix}$ with $\begin{matrix} 4 \\ 2 \end{matrix}$.

$$D_x = \begin{vmatrix} 4 & 1 \\ 2 & -1 \end{vmatrix} = -6$$

To form D_y, replace the second column of D with $\begin{matrix} 4 \\ 2 \end{matrix}$.

$$D_y = \begin{vmatrix} 1 & 4 \\ 2 & 2 \end{vmatrix} = -6$$

$$x = \frac{D_x}{D} = \frac{-6}{-3} = 2$$

$$y = \frac{D_y}{D} = \frac{-6}{-3} = 2$$

Solution set: $\{(2, 2)\}$

3. $4x + 3y = -7$
$2x + 3y = -11$

$$D = \begin{vmatrix} 4 & 3 \\ 2 & 3 \end{vmatrix} = 6$$

To form D_x, replace the first column of $\begin{vmatrix} 4 & 3 \\ 2 & 3 \end{vmatrix}$ with $\begin{matrix} -7 \\ -11 \end{matrix}$.

$$D_x = \begin{vmatrix} -7 & 3 \\ -11 & 3 \end{vmatrix} = 12$$

To form D_y, replace the second column of D with $\begin{matrix} -7 \\ -11 \end{matrix}$.

$$D_y = \begin{vmatrix} 4 & -7 \\ 2 & -11 \end{vmatrix} = -30$$

$$x = \frac{D_x}{D} = \frac{12}{6} = 2$$

$$y = \frac{D_y}{D} = \frac{-30}{6} = -5$$

Solution set: $\{(2, -5)\}$

5. $5x + 4y = 10$
$3x - 7y = 6$

$$D = \begin{vmatrix} 5 & 4 \\ 3 & -7 \end{vmatrix} = -35 - 12 = -47$$

$$D_x = \begin{vmatrix} 10 & 4 \\ 6 & -7 \end{vmatrix} = -70 - 24 = -94$$

$$D_y = \begin{vmatrix} 5 & 10 \\ 3 & 6 \end{vmatrix} = 30 - 30 = 0$$

$$x = \frac{D_x}{D} = \frac{-94}{-47} = 2$$

$$y = \frac{D_y}{D} = \frac{0}{-47} = 0$$

Solution set: $\{(2, 0)\}$

7. $2x - 3y = -5$

  $x + 5y = 17$

$$D = \begin{vmatrix} 2 & -3 \\ 1 & 5 \end{vmatrix} = 13$$

$$D_x = \begin{vmatrix} -5 & -3 \\ 17 & 5 \end{vmatrix} = 26$$

$$D_y = \begin{vmatrix} 2 & -5 \\ 1 & 17 \end{vmatrix} = 39$$

$$x = \frac{D_x}{D} = \frac{26}{13} = 2$$

$$y = \frac{D_y}{D} = \frac{39}{13} = 3$$

Solution set: $\{(2, 3)\}$

9. $1.5x + 3y = 5$ (1)

  $2x + 4y = 3$ (2)

$$D = \begin{vmatrix} 1.5 & 3 \\ 2 & 4 \end{vmatrix} = 6 - 6 = 0$$

Since D = 0, Cramer's rule does not apply. To determine whether the system is inconsistent or contains dependent equations, use the elimination method.

  $6x + 12y = 20$ *Multiply (1) by 4*

  $\underline{-6x - 12y = -9}$ *Multiply (2) by -3*

    $0 = 11$ *False*

The system is inconsistent.

Solution set: ∅

11. $3x + 2y = 4$ (1)

  $6x + 4y = 8$ (2)

$$D = \begin{vmatrix} 3 & 2 \\ 6 & 4 \end{vmatrix} = 0$$

Since D = 0, Cramer's rule does not apply. To determine whether the system is inconsistent or contains dependent equations, use the elimination method. Multiply equation (1) by -2 and add the result to equation (2).

  $-6x - 4y = -8$

  $\underline{6x + 4y = 8}$

    $0 = 0$ *True*

This shows that equations (1) and (2) are dependent.

To write the solution set with y as the arbitrary variable, solve equation (1) for x in terms of y.

$$3x + 2y = 4$$

$$3x = 4 - 2y$$

$$x = \frac{4 - 2y}{3}$$

Solution set: $\left\{\left(\frac{4 - 2y}{3}, y\right)\right\}$

13. $4x - y + 3z = -3$

  $3x + y + z = 0$

  $2x - y + 4z = 0$

$$D = \begin{vmatrix} 4 & -1 & 3 \\ 3 & 1 & 1 \\ 2 & -1 & 4 \end{vmatrix}$$

Add row 1 to row 2; then add -1 times row 1 to row 3.

$$D = \begin{vmatrix} 4 & -1 & 3 \\ 7 & 0 & 4 \\ -2 & 0 & 1 \end{vmatrix}$$

Expand about column 2 to get

$$D = -(-1)\begin{vmatrix} 7 & 4 \\ -2 & 1 \end{vmatrix}$$

$$= 1(15)$$

$$= 15.$$

Replace the first column of D with $\begin{array}{c} -3 \\ 0 \\ 0 \end{array}$ to find D_x.

$$D_x = \begin{vmatrix} -3 & -1 & 3 \\ 0 & 1 & 1 \\ 0 & -1 & 4 \end{vmatrix}$$

Expand about column 1 to get

$$D_x = -3\begin{vmatrix} 1 & 1 \\ -1 & 4 \end{vmatrix} = -3(5)$$

$$= -15.$$

Replace the second column of D with $\begin{array}{c} -3 \\ 0 \\ 0 \end{array}$ to get

$$D_y = \begin{vmatrix} 4 & -3 & 3 \\ 3 & 0 & 1 \\ 2 & 0 & 4 \end{vmatrix} = -(-3)\begin{vmatrix} 3 & 1 \\ 2 & 4 \end{vmatrix}$$

$$= 3(10)$$

$$= 30.$$

To find D_z, replace the third column of D with $\begin{array}{c} -3 \\ 0 \\ 0 \end{array}$.

$$D_z = \begin{vmatrix} 4 & -1 & -3 \\ 3 & 1 & 0 \\ 2 & -1 & 0 \end{vmatrix}$$

Expand about column 3 to get

$$D_z = -3\begin{vmatrix} 3 & 1 \\ 2 & -1 \end{vmatrix}$$

$$= -3(-5) = 15.$$

$$x = \frac{D_x}{D} = \frac{-15}{15} = -1$$

$$y = \frac{D_y}{D} = \frac{30}{15} = 2$$

$$z = \frac{D_z}{D} = \frac{15}{15} = 1$$

Solution set: $\{(-1, 2, 1)\}$

15. $2x - y + 4z = -2$
     $3x + 2y - z = -3$
     $x + 4y + 2z = 17$

$$D = \begin{vmatrix} 2 & -1 & 4 \\ 3 & 2 & -1 \\ 1 & 4 & 2 \end{vmatrix}$$

Multiply row 1 by 2 and add to row 2; then multiply row 1 by 4 and add to row 3.

$$D = \begin{vmatrix} 2 & -1 & 4 \\ 7 & 0 & 7 \\ 9 & 0 & 18 \end{vmatrix}$$

Expand about column 2.

$$D = -(-1)\begin{vmatrix} 7 & 7 \\ 9 & 18 \end{vmatrix} = 63$$

$$D_x = \begin{vmatrix} -2 & -1 & 4 \\ -3 & 2 & -1 \\ 17 & 4 & 2 \end{vmatrix}$$

Multiply row 1 by 2 and add to row 2; then multiply row 1 by 4 and add to row 3.

$$D_x = \begin{vmatrix} -2 & -1 & 4 \\ -7 & 0 & 7 \\ 9 & 0 & 18 \end{vmatrix}$$

Expand about column 2.

$$D_x = -(-1)\begin{vmatrix} -7 & 7 \\ 9 & 18 \end{vmatrix} = -189$$

$$D_y = \begin{vmatrix} 2 & -2 & 4 \\ 3 & -3 & -1 \\ 1 & 17 & 2 \end{vmatrix}$$

Add column 2 to column 1.

$$D_y = \begin{vmatrix} 0 & -2 & 4 \\ 0 & -3 & -1 \\ 18 & 17 & 2 \end{vmatrix}$$

Expand about column 1.

$$D_y = 18 \begin{vmatrix} -2 & 4 \\ -3 & -1 \end{vmatrix} = 18(14)$$

$$= 252$$

$$D_z = \begin{vmatrix} 2 & -1 & -2 \\ 3 & 2 & -3 \\ 1 & 4 & 17 \end{vmatrix}$$

Add column 3 to column 1.

$$D_z = \begin{vmatrix} 0 & -1 & -2 \\ 0 & 2 & -3 \\ 18 & 4 & 17 \end{vmatrix}$$

Expand about column 1.

$$D_z = 18 \begin{vmatrix} -1 & -2 \\ 2 & -3 \end{vmatrix}$$

$$= 18(7) = 126$$

$$x = \frac{D_x}{D} = \frac{-189}{63} = -3$$

$$y = \frac{D_y}{D} = \frac{252}{63} = 4$$

$$z = \frac{D_z}{D} = \frac{126}{63} = 2$$

Solution set: $\{(-3, 4, 2)\}$

17. $4x - 3y + z = -1$

$5x + 7y + 2z = -2$

$3x - 5y - z = 1$

$$D = \begin{vmatrix} 4 & -3 & 1 \\ 5 & 7 & 2 \\ 3 & -5 & -1 \end{vmatrix}$$

Multiply row 3 by 2 and add to row 2. Add row 1 to row 3.

$$D = \begin{vmatrix} 4 & -3 & 1 \\ 11 & -3 & 0 \\ 7 & -8 & 0 \end{vmatrix}$$

Expand about column 3 to get

$$D = 1 \begin{vmatrix} 11 & -3 \\ 7 & -8 \end{vmatrix} = -67.$$

$$D_x = \begin{vmatrix} -1 & -3 & 1 \\ -2 & 7 & 2 \\ 1 & -5 & -1 \end{vmatrix}$$

Add column 1 to get column 3 to get

$$D_x = \begin{vmatrix} -1 & 3 & 0 \\ -2 & 7 & 0 \\ 1 & -5 & 0 \end{vmatrix} = 0,$$

since it has a column of zeros.

$$D_y = \begin{vmatrix} 4 & -1 & 1 \\ 5 & -2 & 2 \\ 3 & 1 & -1 \end{vmatrix}$$

Add column 2 to column 3 to get

$$D_y = \begin{vmatrix} 4 & -1 & 0 \\ 5 & -2 & 0 \\ 3 & 1 & 0 \end{vmatrix} = 0,$$

since it has a column of zeros.

$$D_z = \begin{vmatrix} 4 & -3 & -1 \\ 5 & 7 & -2 \\ 3 & -5 & 1 \end{vmatrix}$$

Add row 3 to row 1. Add twice row 3 to row 2.

$$D_z = \begin{vmatrix} 7 & -8 & 0 \\ 11 & -3 & 0 \\ 3 & -5 & 1 \end{vmatrix}$$

Expand about column 3 to get

$$D_z = 1 \begin{vmatrix} 7 & -8 \\ 11 & -3 \end{vmatrix} = 67.$$

$$x = \frac{D_x}{D} = \frac{0}{-67} = 0$$

$$y = \frac{D_y}{D} = \frac{0}{-67} = 0$$

$$z = \frac{D_z}{D} = \frac{67}{-67} = -1$$

Solution set: $\{(0, 0, -1)\}$

19. $x + 2y + 3z - 4 = 0$

$4x + 3y + 2z - 1 = 0$

$-x - 2y - 3z = 0$

Rewrite the system.

$x + 2y + 3z = 4$ (1)

$4x + 3y + 2z = 1$ (2)

$-x - 2y - 3z = 0$ (3)

$$D = \begin{vmatrix} 1 & 2 & 3 \\ 4 & 3 & 2 \\ -1 & -2 & -3 \end{vmatrix}$$

Add row 1 to row 3.

$$D = \begin{vmatrix} 1 & 2 & 3 \\ 4 & 3 & 2 \\ 0 & 0 & 0 \end{vmatrix} = 0$$

Since $D = 0$, Cramer's rule does not apply. Use the elimination method. Add equations (1) and (3).

$x + 2y + 3z = 4$

$\underline{-x - 2y - 3z = 0}$

$\qquad\qquad 0 = 4$ *False*

The system is inconsistent.
Solution set: $\varnothing$

21. $-2x - 2y + 3z = 4$ (1)

$5x + 7y - z = 2$ (2)

$2x + 2y - 3z = -4$ (3)

$$D = \begin{vmatrix} -2 & -2 & 3 \\ 5 & 7 & -1 \\ 2 & 2 & -3 \end{vmatrix}$$

Add row 1 to row 3.

$$D = \begin{vmatrix} -2 & -2 & 3 \\ 5 & 7 & -1 \\ 0 & 0 & 0 \end{vmatrix} = 0,$$

since there is a row of zeros.
Since $D = 0$, Cramer's rule does not apply. Use the elimination method.

Add equations (1) and (3).

$-2x - 2y + 3z = 4$

$\underline{2x + 2y - 3z = -4}$

$\qquad\qquad 0 = 0$

Equations (1) and (3) are dependent. Solve the system made up of equations (2) and (3) in terms of the arbitrary variable z.

To eliminate x, multiply equation (2) by -2 and equation (3) by 5 and add the results.

$-10x - 14y + 2z = -4$

$\underline{10x + 10y - 15z = -20}$

$\qquad -4y - 13z = -24$

Solve for y in terms of z.

$$-4y = -24 + 13z$$

$$y = \frac{24 - 13z}{4}$$

Now, express x also in terms of z by solving equation (3) for x and substituting $\frac{24 - 13z}{4}$ for y in the result.

$2x + 2y - 3z = -4$

$\qquad 2x = -2y + 3z - 4$

$$x = \frac{-2y + 13z - 4}{2}$$

$$= \frac{-2\left(\frac{24 - 13z}{2}\right) + 3z - 4}{2}$$

$$= \frac{\frac{-24 + 13z}{2} + 3z - 4}{2}$$

$$= \frac{-24 + 13z + 6z - 8}{4}$$

$$x = \frac{-32 + 19z}{4}$$

Solution set (with z arbitrary):

$$\left\{ \left(\frac{-32 + 19z}{4}, \frac{24 - 13z}{4}, z \right) \right\}$$

23. $2x + 3y = 13$

$2y - z = 5$

$x + 2z = 4$

$$D = \begin{vmatrix} 2 & 3 & 0 \\ 0 & 2 & -1 \\ 1 & 0 & 2 \end{vmatrix}$$

Add twice row 2 to row 3.

$$D = \begin{vmatrix} 2 & 3 & 0 \\ 0 & 2 & -1 \\ 1 & 4 & 0 \end{vmatrix}$$

Expand about column 3.

$$D = -(-1) \begin{vmatrix} 2 & 3 \\ 1 & 4 \end{vmatrix} = 5$$

$$D_x = \begin{vmatrix} 13 & 3 & 0 \\ 5 & 2 & -1 \\ 4 & 0 & 2 \end{vmatrix}$$

Add 1/2 row 3 to row 2.

$$D_x = \begin{vmatrix} 13 & 3 & 0 \\ 7 & 2 & 0 \\ 4 & 0 & 2 \end{vmatrix}$$

Expand about column 3.

$$D_x = 2 \begin{vmatrix} 13 & 3 \\ 7 & 2 \end{vmatrix} = 10$$

$$D_y = \begin{vmatrix} 2 & 13 & 0 \\ 0 & 5 & -1 \\ 1 & 4 & 2 \end{vmatrix}$$

Add 5 times column 3 to column 2.

$$D_y = \begin{vmatrix} 2 & 13 & 0 \\ 0 & 0 & -1 \\ 1 & 14 & 2 \end{vmatrix}$$

Expand about row 2.

$$D_y = -(-1) \begin{vmatrix} 2 & 13 \\ 1 & 14 \end{vmatrix} = 15$$

$$D_z = \begin{vmatrix} 2 & 3 & 13 \\ 0 & 2 & 5 \\ 1 & 0 & 4 \end{vmatrix}$$

Multiply row 3 by -2 and add to row 1.

$$D_z = \begin{vmatrix} 0 & 3 & 5 \\ 0 & 2 & 5 \\ 1 & 0 & 4 \end{vmatrix}$$

Expand about column 1.

$$D_z = 1 \begin{vmatrix} 3 & 5 \\ 2 & 5 \end{vmatrix} = 5$$

$$x = \frac{D_x}{D} = \frac{10}{5} = 2$$

$$y = \frac{D_y}{D} = \frac{15}{5} = 3$$

$$z = \frac{D_z}{D} = \frac{5}{5} = 1$$

Solution set: $\{(2, 3, 1)\}$

25. $5x - y = -4$

$3x + 2z = 4$

$4y + 3z = 22$

$$D = \begin{vmatrix} 5 & -1 & 0 \\ 3 & 0 & 2 \\ 0 & 4 & 3 \end{vmatrix}$$

Add 4 times row 1 to row 3.

$$D = \begin{vmatrix} 5 & -1 & 0 \\ 3 & 0 & 2 \\ 20 & 0 & 3 \end{vmatrix}$$

Expand about column 2.

$$D = -(-1) \begin{vmatrix} 3 & 2 \\ 20 & 3 \end{vmatrix} = -31$$

$$D_x = \begin{vmatrix} -4 & -1 & 0 \\ 4 & 0 & 2 \\ 22 & 4 & 3 \end{vmatrix}$$

Add 4 times row 1 to row 3.

$$D_x = \begin{vmatrix} -4 & -1 & 0 \\ 4 & 0 & 2 \\ 6 & 0 & 3 \end{vmatrix}$$

Expand about column 2.

$$D_x = -(-1)\begin{vmatrix} 4 & 2 \\ 6 & 3 \end{vmatrix} = 0$$

$$D_y = \begin{vmatrix} 5 & -4 & 0 \\ 3 & 4 & 2 \\ 0 & 22 & 3 \end{vmatrix}$$

Add column 2 to column 1.

$$D_y = \begin{vmatrix} 1 & -4 & 0 \\ 7 & 4 & 2 \\ 22 & 22 & 3 \end{vmatrix}$$

Add 4 times column 1 to column 2.

$$D_y = \begin{vmatrix} 1 & 0 & 0 \\ 7 & 32 & 2 \\ 22 & 110 & 3 \end{vmatrix}$$

Expand about row 1.

$$D_y = 1\begin{vmatrix} 32 & 2 \\ 110 & 3 \end{vmatrix}$$

$$= 96 - 220 = -124$$

$$D_z = \begin{vmatrix} 5 & -1 & -4 \\ 3 & 0 & 4 \\ 0 & 4 & 22 \end{vmatrix}$$

Add 4 times row 1 to row 3.

$$D_z = \begin{vmatrix} 5 & -1 & -4 \\ 3 & 0 & 4 \\ 20 & 0 & 6 \end{vmatrix}$$

Expand about column 2.

$$D_z = -(-1)\begin{vmatrix} 3 & 4 \\ 20 & 6 \end{vmatrix}$$

$$= 1(18 - 80) = -62$$

$$x = \frac{D_x}{D} = \frac{0}{-31} = 0$$

$$y = \frac{D_y}{D} = \frac{-124}{-31} = 4$$

$$z = \frac{D_z}{D} = \frac{-62}{-31} = 2$$

Solution set: $\{(0, 4, 2)\}$

27. $x + 2y = 10$

$3x + 4z = 7$

$-y - z = 1$

$$D = \begin{vmatrix} 1 & 2 & 0 \\ 3 & 0 & 4 \\ 0 & -1 & -1 \end{vmatrix}$$

Multiply row 1 by -3 and add to row 2.

$$D = \begin{vmatrix} 1 & 2 & 0 \\ 0 & -6 & 4 \\ 0 & -1 & -1 \end{vmatrix}$$

Expand about column 1.

$$D = 1\begin{vmatrix} -6 & 4 \\ -1 & -1 \end{vmatrix} = 10$$

$$D_x = \begin{vmatrix} 10 & 2 & 0 \\ 7 & 0 & 4 \\ 1 & -1 & -1 \end{vmatrix}$$

Add column 1 to column 2 and to column 3.

$$D_x = \begin{vmatrix} 10 & 12 & 10 \\ 7 & 7 & 11 \\ 1 & 0 & 0 \end{vmatrix}$$

Expand about row 3.

$$D_x = 1\begin{vmatrix} 12 & 10 \\ 7 & 11 \end{vmatrix} = 62$$

$$D_y = \begin{vmatrix} 1 & 10 & 0 \\ 3 & 7 & 4 \\ 0 & 1 & -1 \end{vmatrix}$$

Add column 2 to column 3.

$$D_y = \begin{vmatrix} 1 & 10 & 10 \\ 3 & 7 & 11 \\ 0 & 1 & 0 \end{vmatrix}$$

Expand about row 3.

$D_y = -1 \begin{vmatrix} 1 & 10 \\ 3 & 11 \end{vmatrix} = 19$

$D_z = \begin{vmatrix} 1 & 2 & 10 \\ 3 & 0 & 7 \\ 0 & -1 & 1 \end{vmatrix}$

Add column 3 to column 2.

$D_z = \begin{vmatrix} 1 & 12 & 10 \\ 3 & 7 & 7 \\ 0 & 0 & 1 \end{vmatrix}$

Expand about row 3.

$D_z = 1 \begin{vmatrix} 1 & 12 \\ 3 & 7 \end{vmatrix} = -29$

$x = \dfrac{D_x}{D} = \dfrac{62}{10} = \dfrac{31}{5}$

$y = \dfrac{D_y}{D} = \dfrac{19}{10}$

$z = \dfrac{D_z}{D} = \dfrac{-29}{10} = -\dfrac{29}{10}$

Solution set: $\left\{ \left(\dfrac{31}{5}, \dfrac{19}{10}, -\dfrac{29}{10} \right) \right\}$

31.
$\begin{aligned} x + 3y - 2z - w &= 9 \quad (1) \\ 4x + y + z + 2w &= 2 \quad (2) \\ -3x - y + z - w &= -5 \quad (3) \\ x - y - 3z - 2w &= 2 \quad (4) \end{aligned}$

$D = \begin{vmatrix} 1 & 3 & -2 & -1 \\ 4 & 1 & 1 & 2 \\ -3 & -1 & 1 & -1 \\ 1 & -1 & -3 & -2 \end{vmatrix}$

Expand about column 1.

$= 1 \begin{vmatrix} -1 & 1 & 2 \\ -1 & 1 & -1 \\ -1 & -3 & -2 \end{vmatrix} - 4 \begin{vmatrix} 3 & -2 & -1 \\ -1 & 1 & -1 \\ -1 & -3 & -2 \end{vmatrix}$

$\quad - 3 \begin{vmatrix} 3 & -2 & -1 \\ 1 & 1 & 2 \\ -1 & -3 & -2 \end{vmatrix} - 1 \begin{vmatrix} 3 & -2 & -1 \\ 1 & 1 & 2 \\ -1 & 1 & 1 \end{vmatrix}$

$= 1[-1(-2 - 3) + (-2 + 6) - 1(-1 - 2)]$

$\quad - 4[3(-2 - 3) + 1(4 - 3) - 1(2 + 1)]$

$\quad - 3[3(-2 + 6) - 1(4 - 3) - 1(-4 + 1)]$

$\quad - 1[3(1 - 2) - 1(-2 + 1) - 1(-4 + 1)]$

$= 12 + 68 - 42 - 1$

$= 37$

$D_x = \begin{vmatrix} 9 & 3 & -2 & -1 \\ 2 & 1 & 1 & 2 \\ -5 & -1 & 1 & -1 \\ 2 & -1 & -3 & -2 \end{vmatrix}$

$= -3 \begin{vmatrix} 2 & 1 & 2 \\ -5 & 1 & -1 \\ 2 & -3 & -2 \end{vmatrix} + 1 \begin{vmatrix} 9 & -2 & -1 \\ -5 & 1 & -1 \\ 2 & -3 & -2 \end{vmatrix}$

$\quad + 1 \begin{vmatrix} 9 & -2 & -1 \\ 2 & 1 & 2 \\ 2 & -3 & -2 \end{vmatrix} - 1 \begin{vmatrix} 9 & -2 & -1 \\ -2 & 1 & 2 \\ -5 & 1 & -1 \end{vmatrix}$

$= -3[2(-2 - 3) - 1(10 + 2) + 2(15 - 2)]$

$\quad + 1[-1(15 - 2) + 1(-27 + 4) - 2(9 - 10)]$

$\quad + 1[-1(6 - 2) - 2(-27 + 4) - 2(9 + 4)]$

$\quad - 1[-1(2 + 5) - 2(9 - 10) - 1(9 + 4)]$

$\quad = -12 - 34 + 28 + 18 = 0$

$x = \dfrac{D_x}{D} = \dfrac{0}{37} = 0$

Substitute 0 for x in equations (1), (2), and (3).

The resulting system of 3 equations in 3 variables may be solved by Cramer's rule.

$\begin{aligned} 3y - 2z - w &= 9 \quad (4) \\ y + z + 2w &= 2 \quad (5) \\ -y + z - w &= -5 \quad (6) \end{aligned}$

Find D, D_x, D_y, and D_z for this system.

$D = \begin{vmatrix} 3 & -2 & -1 \\ 1 & 1 & 2 \\ -1 & -1 & -1 \end{vmatrix} = -9$

$D_y = \begin{vmatrix} 9 & -2 & -1 \\ 2 & 1 & 2 \\ -5 & 1 & -1 \end{vmatrix} = -18$

$$D_z = \begin{vmatrix} 3 & 9 & -1 \\ 1 & 2 & 2 \\ -1 & -5 & -1 \end{vmatrix} = 18$$

$$D_w = \begin{vmatrix} 3 & -2 & 9 \\ 1 & 1 & 2 \\ -1 & 1 & -5 \end{vmatrix} = -9$$

$$y = \frac{D_y}{D} = \frac{-18}{-9} = 2$$

$$z = \frac{D_z}{D} = \frac{18}{-9} = -2$$

$$w = \frac{D_w}{D} = \frac{-9}{-9} = 1$$

Solution set: $\{(0, 2, -2, 1)\}$

33.
$$5x + 3y - 2z + w = 9$$
$$-3x + y - 6z + 2w = -33$$
$$2x + 2y - z + 3w = 5$$
$$4x + 3y - z + 8w = 12$$

$$D = \begin{vmatrix} 5 & 3 & -2 & 1 \\ -3 & 1 & -6 & 2 \\ 2 & 2 & -1 & 3 \\ 4 & 3 & -1 & 8 \end{vmatrix} = 117$$

$$D_x = \begin{vmatrix} 9 & 3 & -2 & 1 \\ -33 & 1 & -6 & 2 \\ 5 & 2 & -1 & 3 \\ 12 & 3 & -1 & 8 \end{vmatrix} = 234$$

$$D_y = \begin{vmatrix} 5 & 9 & -2 & 1 \\ -3 & -33 & -6 & 2 \\ 2 & 5 & -1 & 3 \\ 4 & 12 & -1 & 8 \end{vmatrix} = 351$$

$$D_z = \begin{vmatrix} 5 & 3 & 9 & 1 \\ -3 & 1 & -33 & 2 \\ 2 & 2 & 5 & 3 \\ 4 & 3 & 12 & 8 \end{vmatrix} = 585$$

$$D_w = \begin{vmatrix} 5 & 3 & -2 & 9 \\ -3 & 1 & -6 & -33 \\ 2 & 2 & -1 & 5 \\ 4 & 3 & -1 & 12 \end{vmatrix} = 0$$

$$x = \frac{D_x}{D} = \frac{234}{117} = 2$$

$$y = \frac{D_y}{D} = \frac{351}{117} = 3$$

$$z = \frac{D_z}{D} = \frac{585}{117} = 5$$

$$w = \frac{D_w}{D} = \frac{0}{117} = 0$$

Solution set: $\{(2, 3, 5, 0)\}$

35.
$$2x + 3y = 4$$
$$5x + 6y = 7$$

Solve this system by the Gauss-Jordan method.

$$\begin{bmatrix} 2 & 3 & | & 4 \\ 5 & 6 & | & 7 \end{bmatrix}$$

$$\begin{bmatrix} 1 & \frac{3}{2} & | & 2 \\ 5 & 6 & | & 7 \end{bmatrix} \quad \frac{1}{2}R1$$

$$\begin{bmatrix} 1 & \frac{3}{2} & | & 2 \\ 0 & -\frac{3}{2} & | & -3 \end{bmatrix} \quad -5R1 + R2$$

$$\begin{bmatrix} 1 & \frac{3}{2} & | & 2 \\ 0 & 1 & | & 2 \end{bmatrix} \quad -\frac{2}{3}R2$$

$$\begin{bmatrix} 1 & 0 & | & -1 \\ 0 & 1 & | & 2 \end{bmatrix} \quad -\frac{3}{2}R2 + R1$$

Solution set: $\{(-1, 2)\}$

36. Choose the integers 8, 9, 10, 11, 12, 13. This gives the system

$$8x + 9y = 10$$
$$11x + 12y = 13.$$

Solve this system by the Gauss-Jordan method.

$$\begin{bmatrix} 8 & 9 & | & 10 \\ 11 & 12 & | & 13 \end{bmatrix}$$

$$\begin{bmatrix} 1 & \frac{9}{8} & \Big| & \frac{5}{4} \\ 11 & 12 & \Big| & 13 \end{bmatrix} \quad \frac{1}{8}R1$$

$$\begin{bmatrix} 1 & \frac{9}{8} & \Big| & \frac{5}{4} \\ 0 & -\frac{3}{8} & \Big| & -\frac{3}{4} \end{bmatrix} \quad -11R1 + R2$$

$$\begin{bmatrix} 1 & \frac{9}{8} & \Big| & \frac{5}{4} \\ 0 & 1 & \Big| & 2 \end{bmatrix} \quad -\frac{8}{3}R2$$

$$\begin{bmatrix} 1 & 0 & \Big| & -1 \\ 0 & 1 & \Big| & 2 \end{bmatrix} \quad -\frac{9}{8}R2 + R1$$

Solution set: $\{(-1, 2)\}$

37. The solutions are the same.

38.
$$nx + (n + 1)y = n + 2$$
$$(n + 3)x + (n + 4)y = n + 5$$

$$D = \begin{vmatrix} n & n + 1 \\ n + 3 & n + 4 \end{vmatrix}$$

$$= n(n + 4) - (n + 1)(n + 3)$$
$$= (n^2 + 4n) - (n^2 + 4n + 3)$$
$$= n^2 + 4n - n^2 - 4n - 3$$
$$= -3$$

39. $D_x = \begin{vmatrix} n + 2 & n + 1 \\ n + 5 & n + 4 \end{vmatrix}$

$$= (n + 2)(n + 4) - (n + 1)(n + 5)$$
$$= (n^2 + 6n + 8) - (n^2 + 6n + 5)$$
$$= n^2 + 6n + 8 - n^2 - 6n - 5$$
$$= 3$$

$$D_y = \begin{vmatrix} n & n + 2 \\ n + 3 & n + 5 \end{vmatrix}$$

$$= n(n + 5) - (n + 2)(n + 3)$$
$$= (n^2 + 5n) - (n^2 + 5n + 6)$$
$$= n^2 + 5n - n^2 - 5n - 6$$
$$= -6$$

40. $x = \dfrac{D_x}{D} = \dfrac{3}{-3} = -1$

$y = \dfrac{D_y}{D} = \dfrac{-6}{-3} = 2$

The solution set for every such system is $\{(-1, 2)\}$.

41. $\dfrac{\sqrt{3}}{2}(W_1 + W_2) = 100$

$$W_1 - W_2 = 0$$

Use the distributive property to rewrite the first equation; then use Cramer's rule.

$$\frac{\sqrt{3}}{2}W_1 + \frac{\sqrt{3}}{2}W_2 = 100$$

$$W_1 - W_2 = 0$$

$$D = \begin{vmatrix} \dfrac{\sqrt{3}}{2} & \dfrac{\sqrt{3}}{2} \\ 1 & -1 \end{vmatrix} = -\sqrt{3}$$

$$D_{W_1} = \begin{vmatrix} 100 & \dfrac{\sqrt{3}}{2} \\ 0 & -1 \end{vmatrix} = -100$$

$$D_{W_2} = \begin{vmatrix} \dfrac{\sqrt{3}}{2} & 100 \\ 1 & 0 \end{vmatrix} = -100$$

$$W_1 = \frac{D_{W_1}}{D} = \frac{-100}{-\sqrt{3}} = \frac{100}{\sqrt{3}} = \frac{100\sqrt{3}}{3} \approx 58$$

$$W_2 = \frac{D_{W_2}}{D} = \frac{-100}{-\sqrt{3}} \approx 58$$

Both W_1 and W_2 are approximately 58 lb.

43. $bx + y = a^2$

$ax + y = b^2$

$$D = \begin{vmatrix} b & 1 \\ a & 1 \end{vmatrix} = b - a$$

$D_x = \begin{vmatrix} a^2 & 1 \\ b^2 & 1 \end{vmatrix} = a^2 - b^2$

$D_y = \begin{vmatrix} b & a^2 \\ a & b^2 \end{vmatrix} = b^3 - a^3$

$x = \dfrac{D_x}{D} = \dfrac{a^2 - b^2}{b - a}$

$ = \dfrac{(a + b)(a - b)}{(b - a)}$

Factor numerator as difference of two squares

$ = -(a + b) = -a - b$

$y = \dfrac{D_y}{D} = \dfrac{a^3 - a^3}{b - a}$

$ = \dfrac{(b - a)(b^2 + ab + a^2)}{b - a}$

Factor numerator as difference of two cubes

$ = b^2 + ab + a^2$

Solution set:

$\{(-a - b,\ a^2 + ab + b^2)\}$

45. $b^2x + a^2y = b^2$

$\ ax + by = a$

$D = \begin{vmatrix} b^2 & a^2 \\ a & b \end{vmatrix} = b^3 - a^3$

Note that for Cramer's rule to apply, $b^3 \neq a^3$, which is equivalent to $b \neq a$ or $b - a \neq 0$.

$D_x = \begin{vmatrix} b^2 & a^2 \\ a & b \end{vmatrix} = b^3 - a^3$

$D_y = \begin{vmatrix} b^2 & b^2 \\ a & a \end{vmatrix} = ab^2 - ab^2 = 0$

$x = \dfrac{D_x}{D} = \dfrac{b^3 - a^3}{b^3 - a^3} = 1$

$y = \dfrac{D_y}{D} = \dfrac{0}{b^3 - a^3} = 0$

Solution set: $\{(1,\ 0)\}$

Section 6.7

1. $\begin{bmatrix} 5 & 7 \\ 2 & 3 \end{bmatrix}\begin{bmatrix} 3 & -7 \\ -2 & 5 \end{bmatrix}$

$= \begin{bmatrix} 5(3) + 7(-2) & 5(-7) + 7(5) \\ 2(3) + 3(-2) & 2(-7) + 3(5) \end{bmatrix}$

$= \begin{bmatrix} 1 & 0 \\ 0 & 1 \end{bmatrix}$

$\begin{bmatrix} 3 & -7 \\ -2 & 5 \end{bmatrix}\begin{bmatrix} 5 & 7 \\ 2 & 3 \end{bmatrix}$

$= \begin{bmatrix} 3(5) + (-7)(2) & 3(7) + (-7)(3) \\ (-2)(5) + 5(2) & (-2)(7) + 5(3) \end{bmatrix}$

$= \begin{bmatrix} 1 & 0 \\ 0 & 1 \end{bmatrix}$

Since the products obtained by multiplying the matrices in either order are both the 2×2 identity matrix, the given matrices are inverses of each other.

3. $\begin{bmatrix} -1 & 2 \\ 3 & -5 \end{bmatrix}\begin{bmatrix} -5 & -2 \\ -3 & -1 \end{bmatrix} = \begin{bmatrix} -1 & 0 \\ 0 & -1 \end{bmatrix}$

Since this product is not the 2×2 identity matrix, the given matrices are not inverses of each other.

5.
$$\begin{bmatrix} 0 & 1 & 0 \\ 0 & 0 & -2 \\ 1 & -1 & 0 \end{bmatrix}\begin{bmatrix} 1 & 0 & 1 \\ 1 & 0 & 0 \\ 0 & -1 & 0 \end{bmatrix} = \begin{bmatrix} 1 & 0 & 0 \\ 0 & 2 & 0 \\ 0 & 0 & 1 \end{bmatrix}$$

Since this product is not the 2 × 2 identity matrix, the given matrices are not inverses of each other.

7.
$$\begin{bmatrix} -1 & -1 & -1 \\ 4 & 5 & 0 \\ 0 & 1 & -3 \end{bmatrix}\begin{bmatrix} 15 & 4 & -5 \\ -12 & -3 & 4 \\ -4 & -1 & 1 \end{bmatrix}$$

$$= \begin{bmatrix} 1 & 0 & 0 \\ 0 & 1 & 0 \\ 0 & 0 & 1 \end{bmatrix} = I_3$$

$$\begin{bmatrix} 15 & 4 & -5 \\ -12 & -3 & 4 \\ -4 & -1 & 1 \end{bmatrix}\begin{bmatrix} -1 & -1 & -1 \\ 4 & 5 & 0 \\ 0 & 1 & -3 \end{bmatrix}$$

$$= \begin{bmatrix} 1 & 0 & 0 \\ 0 & 1 & 0 \\ 0 & 0 & 1 \end{bmatrix} = I_3$$

The given matrices are inverses of each other.

9. Find the inverse of $A = \begin{bmatrix} -1 & 2 \\ -2 & -1 \end{bmatrix}$, if it exists.

$$[A|I_2] = \begin{bmatrix} -1 & 2 & | & 1 & 0 \\ -2 & -1 & | & 0 & 1 \end{bmatrix}$$

$$\begin{bmatrix} 1 & -2 & | & -1 & 0 \\ -2 & -1 & | & 0 & 1 \end{bmatrix} \quad -1R1$$

$$\begin{bmatrix} 1 & -2 & | & -1 & 0 \\ 0 & -5 & | & -2 & 1 \end{bmatrix} \quad 2R1 + R2$$

$$\begin{bmatrix} 1 & -2 & | & -1 & 0 \\ 0 & 1 & | & \frac{2}{5} & -\frac{1}{5} \end{bmatrix} \quad -\frac{1}{5}R2$$

$$\begin{bmatrix} 1 & 0 & | & -\frac{1}{5} & -\frac{2}{5} \\ 0 & 1 & | & \frac{2}{5} & -\frac{1}{5} \end{bmatrix} \quad 2R2 + R1$$

$$A^{-1} = \begin{bmatrix} -\frac{1}{5} & -\frac{2}{5} \\ \frac{2}{5} & -\frac{1}{5} \end{bmatrix}$$

11. Find the inverse of $A = \begin{bmatrix} -1 & -2 \\ 3 & 4 \end{bmatrix}$, if it exists.

$$[A|I_2] = \begin{bmatrix} -1 & -2 & | & 1 & 0 \\ 3 & 4 & | & 0 & 1 \end{bmatrix}$$

$$\begin{bmatrix} -1 & -2 & | & 1 & 0 \\ 0 & -2 & | & 3 & 1 \end{bmatrix} \quad 3R1 + R2$$

$$\begin{bmatrix} 1 & 2 & | & -1 & 0 \\ 0 & -2 & | & 3 & 1 \end{bmatrix} \quad -1R1$$

$$\begin{bmatrix} 1 & 0 & | & 2 & 1 \\ 0 & -2 & | & 3 & 1 \end{bmatrix} \quad R2 + R1$$

$$\begin{bmatrix} 1 & 0 & | & 2 & 1 \\ 0 & 1 & | & -\frac{3}{2} & -\frac{1}{2} \end{bmatrix} \quad -\frac{1}{2}R2$$

$$A^{-1} = \begin{bmatrix} 2 & 1 \\ -\frac{3}{2} & -\frac{1}{2} \end{bmatrix}$$

13. Find the inverse of $\begin{bmatrix} 5 & 10 \\ -3 & -6 \end{bmatrix}$, if it exists.

$$[A|I_2] = \begin{bmatrix} 5 & 10 & | & 1 & 0 \\ -3 & -6 & | & 0 & 1 \end{bmatrix}$$

$$\begin{bmatrix} 1 & 2 & | & \frac{1}{5} & 0 \\ -3 & -6 & | & 0 & 1 \end{bmatrix} \quad \frac{1}{5}R1$$

$$\begin{bmatrix} 1 & 2 & | & \frac{1}{5} & 0 \\ 0 & 0 & | & \frac{3}{5} & 1 \end{bmatrix} \quad 3R1 + R2$$

At this point, the matrix should be changed so that the second-row, second-column element will be 1. Since that element is now 0, the desired transformation cannot be completed. Therefore, the inverse of the given matrix does not exist.

15. $A = \begin{bmatrix} 1 & 0 & 1 \\ 0 & -1 & 0 \\ 2 & 1 & 1 \end{bmatrix}$

$[A|I_3] = \begin{bmatrix} 1 & 0 & 1 & | & 1 & 0 & 0 \\ 0 & -1 & 0 & | & 0 & 1 & 0 \\ 2 & 1 & 1 & | & 0 & 0 & 1 \end{bmatrix}$

$\begin{bmatrix} 1 & 0 & 1 & | & 1 & 0 & 0 \\ 0 & -1 & 0 & | & 0 & 1 & 0 \\ 0 & 1 & -1 & | & -2 & 0 & 1 \end{bmatrix}$ $-2R1 + R3$

$\begin{bmatrix} 1 & 0 & 1 & | & 1 & 0 & 0 \\ 0 & 1 & 0 & | & 0 & -1 & 0 \\ 0 & 1 & -1 & | & -2 & 0 & 1 \end{bmatrix}$ $-1R2$

$\begin{bmatrix} 1 & 0 & 1 & | & 1 & 0 & 0 \\ 0 & 1 & 0 & | & 0 & -1 & 0 \\ 0 & 0 & -1 & | & -2 & 1 & 1 \end{bmatrix}$ $-1R2 + R3$

$\begin{bmatrix} 1 & 0 & 1 & | & 1 & 0 & 0 \\ 0 & 1 & 0 & | & 0 & -1 & 0 \\ 0 & 0 & 1 & | & 2 & -1 & -1 \end{bmatrix}$ $-1R3$

$\begin{bmatrix} 1 & 0 & 0 & | & -1 & 1 & 1 \\ 0 & 1 & 0 & | & 0 & -1 & 0 \\ 0 & 0 & 1 & | & 2 & -1 & -1 \end{bmatrix}$ $-1R3 + R1$

$A^{-1} = \begin{bmatrix} -1 & 1 & 1 \\ 0 & -1 & 0 \\ 2 & -1 & -1 \end{bmatrix}$

17. $A = \begin{bmatrix} 1 & 3 & 3 \\ 1 & 4 & 3 \\ 1 & 3 & 4 \end{bmatrix}$

$\begin{bmatrix} 1 & 3 & 3 & | & 1 & 0 & 0 \\ 1 & 4 & 3 & | & 0 & 1 & 0 \\ 1 & 3 & 4 & | & 0 & 0 & 1 \end{bmatrix}$

$\begin{bmatrix} 1 & 3 & 3 & | & 1 & 0 & 0 \\ 0 & 1 & 0 & | & -1 & 1 & 0 \\ 0 & 0 & 1 & | & -1 & 0 & 1 \end{bmatrix}$ $-1R1 + R2$ $-1R1 + R3$

$\begin{bmatrix} 1 & 0 & 3 & | & 4 & -3 & 0 \\ 0 & 1 & 0 & | & -1 & 1 & 0 \\ 0 & 0 & 1 & | & -1 & 0 & 1 \end{bmatrix}$ $-3R2 + R1$

$\begin{bmatrix} 1 & 0 & 0 & | & 7 & -3 & -3 \\ 0 & 1 & 0 & | & -1 & 1 & 0 \\ 0 & 0 & 1 & | & -1 & 0 & 1 \end{bmatrix}$ $-3R3 + R1$

$A^{-1} = \begin{bmatrix} 7 & -3 & -3 \\ -1 & 1 & 0 \\ -1 & 0 & 1 \end{bmatrix}$

19. $A = \begin{bmatrix} 2 & 2 & -4 \\ 2 & 6 & 0 \\ -3 & -3 & 5 \end{bmatrix}$

$\begin{bmatrix} 2 & 2 & -4 & | & 1 & 0 & 0 \\ 2 & 6 & 0 & | & 0 & 1 & 0 \\ -3 & -3 & 5 & | & 0 & 0 & 1 \end{bmatrix}$

$\begin{bmatrix} 1 & 1 & -2 & | & \frac{1}{2} & 0 & 0 \\ 2 & 6 & 0 & | & 0 & 1 & 0 \\ -3 & -3 & 5 & | & 0 & 0 & 1 \end{bmatrix}$ $\frac{1}{2}R1$

$\begin{bmatrix} 1 & 1 & -2 & | & \frac{1}{2} & 0 & 0 \\ 0 & 4 & 4 & | & -1 & 1 & 0 \\ 0 & 0 & -1 & | & \frac{3}{2} & 0 & 1 \end{bmatrix}$ $-2R1 + R2$ $3R1 + R3$

$\begin{bmatrix} 1 & 1 & -2 & | & \frac{1}{2} & 0 & 0 \\ 0 & 1 & 1 & | & -\frac{1}{4} & \frac{1}{4} & 0 \\ 0 & 0 & -1 & | & \frac{3}{2} & 0 & 1 \end{bmatrix}$ $\frac{1}{4}R2$

$\begin{bmatrix} 1 & 0 & -3 & | & \frac{3}{4} & -\frac{1}{4} & 0 \\ 0 & 1 & 1 & | & -\frac{1}{4} & \frac{1}{4} & 0 \\ 0 & 0 & -1 & | & \frac{3}{2} & 0 & 1 \end{bmatrix}$ $-1R2 + R1$

$\begin{bmatrix} 1 & 0 & -3 & | & \frac{3}{4} & -\frac{1}{4} & 0 \\ 0 & 1 & 1 & | & -\frac{1}{4} & \frac{1}{4} & 0 \\ 0 & 0 & 1 & | & -\frac{3}{2} & 0 & -1 \end{bmatrix}$ $-1R3$

$\begin{bmatrix} 1 & 0 & 0 & | & -\frac{15}{4} & -\frac{1}{4} & -3 \\ 0 & 1 & 0 & | & \frac{5}{4} & \frac{1}{4} & 1 \\ 0 & 0 & 1 & | & -\frac{3}{2} & 0 & -1 \end{bmatrix}$ $3R3 + R1$ $-1R3 + R2$

$A^{-1} = \begin{bmatrix} -\frac{15}{4} & -\frac{1}{4} & -3 \\ \frac{5}{4} & \frac{1}{4} & 1 \\ -\frac{3}{2} & 0 & -1 \end{bmatrix}$

21. $A = \begin{bmatrix} 1 & 1 & 0 & 2 \\ 2 & -1 & 1 & -1 \\ 3 & 3 & 2 & -2 \\ 1 & 2 & 1 & 0 \end{bmatrix}$

$\left[\begin{array}{cccc|cccc} 1 & 1 & 0 & 2 & 1 & 0 & 0 & 0 \\ 2 & -1 & 1 & -1 & 0 & 1 & 0 & 0 \\ 3 & 3 & 2 & -2 & 0 & 0 & 1 & 0 \\ 1 & 2 & 1 & 0 & 0 & 0 & 0 & 1 \end{array}\right]$

$\left[\begin{array}{cccc|cccc} 1 & 1 & 0 & 2 & 1 & 0 & 0 & 0 \\ 0 & -3 & 1 & -5 & -2 & 1 & 0 & 0 \\ 0 & 0 & 2 & -8 & -3 & 0 & 1 & 0 \\ 0 & 1 & 1 & -2 & -1 & 0 & 0 & 1 \end{array}\right]$ $\begin{array}{l} -2R1 + R2 \\ -3R1 + R3 \\ -1R1 + R4 \end{array}$

$\left[\begin{array}{cccc|cccc} 1 & 1 & 0 & 2 & 1 & 0 & 0 & 0 \\ 0 & 1 & -\frac{1}{3} & \frac{5}{3} & \frac{2}{3} & -\frac{1}{3} & 0 & 0 \\ 0 & 0 & 2 & -8 & -3 & 0 & 1 & 0 \\ 0 & 1 & 1 & -2 & -1 & 0 & 0 & 1 \end{array}\right]$ $-\frac{1}{3}R2$

$\left[\begin{array}{cccc|cccc} 1 & 0 & \frac{1}{3} & \frac{1}{3} & \frac{1}{3} & \frac{1}{3} & 0 & 0 \\ 0 & 1 & -\frac{1}{3} & \frac{5}{3} & \frac{2}{3} & -\frac{1}{3} & 0 & 0 \\ 0 & 0 & 2 & -8 & -3 & 0 & 1 & 0 \\ 0 & 0 & \frac{4}{3} & -\frac{11}{3} & -\frac{5}{3} & \frac{1}{3} & 0 & 1 \end{array}\right]$ $\begin{array}{l} -1R2 + R1 \\ \\ \\ -1R2 + R4 \end{array}$

$\left[\begin{array}{cccc|cccc} 1 & 0 & \frac{1}{3} & \frac{1}{3} & \frac{1}{3} & \frac{1}{3} & 0 & 0 \\ 0 & 1 & -\frac{1}{3} & \frac{5}{3} & \frac{2}{3} & -\frac{1}{3} & 0 & 0 \\ 0 & 0 & 1 & -4 & -\frac{3}{2} & 0 & \frac{1}{2} & 0 \\ 0 & 0 & \frac{4}{3} & -\frac{11}{3} & -\frac{5}{3} & \frac{1}{3} & 0 & 1 \end{array}\right]$ $\frac{1}{2}R3$

$\left[\begin{array}{cccc|cccc} 1 & 0 & 0 & \frac{5}{3} & \frac{5}{6} & \frac{1}{3} & -\frac{1}{6} & 0 \\ 0 & 1 & 0 & \frac{1}{3} & \frac{1}{6} & -\frac{1}{3} & \frac{1}{6} & 0 \\ 0 & 0 & 1 & -4 & -\frac{3}{2} & 0 & \frac{1}{2} & 0 \\ 0 & 0 & 0 & \frac{5}{3} & \frac{1}{3} & \frac{1}{3} & -\frac{2}{3} & 1 \end{array}\right]$ $\begin{array}{l} \frac{1}{3}R3 + R1 \\ \frac{1}{3}R3 + R2 \\ \\ -\frac{4}{3}R3 + R4 \end{array}$

$\left[\begin{array}{cccc|cccc} 1 & 0 & 0 & \frac{5}{3} & \frac{5}{6} & \frac{1}{3} & -\frac{1}{6} & 0 \\ 0 & 1 & 0 & \frac{1}{3} & \frac{1}{6} & -\frac{1}{3} & \frac{1}{6} & 0 \\ 0 & 0 & 1 & -4 & -\frac{3}{2} & 0 & \frac{1}{2} & 0 \\ 0 & 0 & 0 & 1 & \frac{1}{5} & \frac{1}{5} & -\frac{2}{5} & \frac{3}{5} \end{array}\right]$ $\frac{3}{5}R4$

$\left[\begin{array}{cccc|cccc} 1 & 0 & 0 & 0 & \frac{1}{2} & 0 & \frac{1}{2} & -1 \\ 0 & 1 & 0 & 0 & \frac{1}{10} & -\frac{2}{5} & \frac{3}{10} & -\frac{1}{5} \\ 0 & 0 & 1 & 0 & -\frac{7}{10} & \frac{4}{5} & -\frac{11}{10} & \frac{12}{5} \\ 0 & 0 & 0 & 1 & \frac{1}{5} & \frac{1}{5} & -\frac{2}{5} & \frac{3}{5} \end{array}\right]$ $\begin{array}{l} -\frac{5}{3}R4 + R1 \\ -\frac{1}{3}R4 + R2 \\ 4R4 + R3 \end{array}$

$A^{-1} = \begin{bmatrix} \frac{1}{2} & 0 & \frac{1}{2} & -1 \\ \frac{1}{10} & -\frac{2}{5} & \frac{3}{10} & -\frac{1}{5} \\ -\frac{7}{10} & \frac{4}{5} & -\frac{11}{10} & \frac{12}{5} \\ \frac{1}{5} & \frac{1}{5} & -\frac{2}{5} & \frac{3}{5} \end{bmatrix}$

23. $A^{-1} = \begin{bmatrix} 5 & -9 \\ -1 & 2 \end{bmatrix}$

Find $A = (A^{-1})^{-1}$.

$\left[\begin{array}{cc|cc} 5 & -9 & 1 & 0 \\ -1 & 2 & 0 & 1 \end{array}\right]$

$\left[\begin{array}{cc|cc} 1 & -\frac{9}{5} & \frac{1}{5} & 0 \\ -1 & 2 & 0 & 1 \end{array}\right]$ $\frac{1}{5}R1$

$\left[\begin{array}{cc|cc} 1 & -\frac{9}{5} & \frac{1}{5} & 0 \\ 0 & \frac{1}{5} & \frac{1}{5} & 1 \end{array}\right]$ $1R1 + R2$

$\left[\begin{array}{cc|cc} 1 & -\frac{9}{5} & \frac{1}{5} & 0 \\ 0 & 1 & 1 & 5 \end{array}\right]$ $5R2$

$\begin{bmatrix} 1 & 0 & | & 2 & 9 \\ 0 & 1 & | & 1 & 5 \end{bmatrix}$ $\frac{9}{5}R2 + R1$

$A = (A^{-1})^{-1} = \begin{bmatrix} 2 & 9 \\ 1 & 5 \end{bmatrix}$

25. $A^{-1} = \begin{bmatrix} \frac{2}{3} & -\frac{1}{3} & 0 \\ \frac{1}{3} & -\frac{5}{3} & 1 \\ \frac{1}{3} & \frac{1}{3} & 0 \end{bmatrix}$

Find $A = (A^{-1})^{-1}$.

$\begin{bmatrix} \frac{2}{3} & -\frac{1}{3} & 0 & | & 1 & 0 & 0 \\ \frac{1}{3} & -\frac{5}{3} & 1 & | & 0 & 1 & 0 \\ \frac{1}{3} & \frac{1}{3} & 0 & | & 0 & 0 & 1 \end{bmatrix}$

$\begin{bmatrix} 1 & -\frac{1}{2} & 0 & | & \frac{3}{2} & 0 & 0 \\ \frac{1}{3} & -\frac{5}{3} & 1 & | & 0 & 1 & 0 \\ \frac{1}{3} & \frac{1}{3} & 0 & | & 0 & 0 & 1 \end{bmatrix}$ $\frac{3}{2}R1$

$\begin{bmatrix} 1 & -\frac{1}{2} & 0 & | & \frac{3}{2} & 0 & 0 \\ 0 & -\frac{3}{2} & 1 & | & -\frac{1}{2} & 1 & 0 \\ 0 & \frac{1}{2} & 0 & | & -\frac{1}{2} & 0 & 1 \end{bmatrix}$ $-\frac{1}{3}R1 + R2$ $-\frac{1}{3}R1 + R3$

$\begin{bmatrix} 1 & -\frac{1}{2} & 0 & | & \frac{3}{2} & 0 & 0 \\ 0 & 1 & -\frac{2}{3} & | & \frac{1}{3} & -\frac{2}{3} & 0 \\ 0 & \frac{1}{2} & 0 & | & -\frac{1}{2} & 0 & 1 \end{bmatrix}$ $-\frac{2}{3}R2$

$\begin{bmatrix} 1 & 0 & -\frac{1}{3} & | & \frac{5}{3} & -\frac{1}{3} & 0 \\ 0 & 1 & -\frac{2}{3} & | & \frac{1}{3} & -\frac{2}{3} & 0 \\ 0 & 0 & \frac{1}{3} & | & -\frac{2}{3} & \frac{1}{3} & 1 \end{bmatrix}$ $\frac{1}{2}R2 + R1$ $-\frac{1}{2}R2 + R3$

$\begin{bmatrix} 1 & 0 & -\frac{1}{3} & | & \frac{5}{3} & -\frac{1}{3} & 0 \\ 0 & 1 & -\frac{2}{3} & | & \frac{1}{3} & -\frac{2}{3} & 0 \\ 0 & 0 & 1 & | & -2 & 1 & 3 \end{bmatrix}$ $3R3$

$\begin{bmatrix} 1 & 0 & 0 & | & 1 & 0 & 1 \\ 0 & 1 & 0 & | & -1 & 0 & 2 \\ 0 & 0 & 1 & | & -2 & 1 & 3 \end{bmatrix}$ $\frac{1}{3}R3 + R1$ $\frac{2}{3}R3 + R2$

$A = (A^{-1})^{-1} = \begin{bmatrix} 1 & 0 & 1 \\ -1 & 0 & 2 \\ -2 & 1 & 3 \end{bmatrix}$

27. If $A = \begin{bmatrix} a & b \\ c & d \end{bmatrix}$, then $ad - bc$ is the determinant of matrix A.

28. If

$A^{-1} = \begin{bmatrix} \dfrac{d}{ad - bc} & \dfrac{-b}{ad - bc} \\ \dfrac{-c}{ad - bc} & \dfrac{a}{ad - bc} \end{bmatrix}$,

then

$A^1 = \begin{bmatrix} \dfrac{d}{|A|} & \dfrac{-b}{|A|} \\ \dfrac{-c}{|A|} & \dfrac{a}{|A|} \end{bmatrix}$.

29. $A^{-1} = \begin{bmatrix} \dfrac{d}{|A|} & \dfrac{-b}{|A|} \\ \dfrac{-c}{|A|} & \dfrac{a}{|A|} \end{bmatrix}$

$= \begin{bmatrix} \dfrac{1}{|A|}d & \dfrac{1}{|A|}(-b) \\ \dfrac{1}{|A|}(-c) & \dfrac{1}{|A|}a \end{bmatrix}$

$= \dfrac{1}{|A|} \begin{bmatrix} d & -b \\ -c & a \end{bmatrix}$

31. $A = \begin{bmatrix} 4 & 2 \\ 7 & 3 \end{bmatrix}$

$|A| = 12 - 14 = -2$

$A^{-1} = \dfrac{1}{-2}\begin{bmatrix} 3 & -2 \\ -7 & 4 \end{bmatrix} = \begin{bmatrix} -\dfrac{3}{2} & 1 \\ \dfrac{7}{2} & -2 \end{bmatrix}$

32.   The inverse of a 2 × 2 matrix A does not exist if the determinant of A has the value *zero*.

33. $-x + y = 1$

$2x - y = 1$

$A = \begin{bmatrix} -1 & 1 \\ 2 & -1 \end{bmatrix}, \quad X = \begin{bmatrix} x \\ y \end{bmatrix}, \quad B = \begin{bmatrix} 1 \\ 1 \end{bmatrix}$

Find A^{-1}.

$[A|I_2] = \begin{bmatrix} -1 & 1 & | & 1 & 0 \\ 2 & -1 & | & 0 & 1 \end{bmatrix}$

$\begin{bmatrix} -1 & 1 & | & 1 & 0 \\ 0 & 1 & | & 2 & 1 \end{bmatrix}$ $2R1 + R2$

$\begin{bmatrix} -1 & 0 & | & -1 & -1 \\ 0 & 1 & | & 2 & 1 \end{bmatrix}$ $-1R2 + R1$

$\begin{bmatrix} 1 & 0 & | & 1 & 1 \\ 0 & 1 & | & 2 & 1 \end{bmatrix}$ $-1R1$

$A^{-1} = \begin{bmatrix} 1 & 1 \\ 2 & 1 \end{bmatrix}$

$X = A^{-1}B = \begin{bmatrix} 1 & 1 \\ 2 & 1 \end{bmatrix}\begin{bmatrix} 1 \\ 1 \end{bmatrix} = \begin{bmatrix} 2 \\ 3 \end{bmatrix}$

Solution set: $\{(2, 3)\}$

35. $2x - y = -8$

$3x + y = -2$

$A = \begin{bmatrix} 2 & -1 \\ 3 & 1 \end{bmatrix}, \quad X = \begin{bmatrix} x \\ y \end{bmatrix}, \quad B = \begin{bmatrix} -8 \\ -2 \end{bmatrix}$

Find A^{-1}.

$[A|I_2] = \begin{bmatrix} 2 & -1 & | & 1 & 0 \\ 3 & 1 & | & 0 & 1 \end{bmatrix}$

$\begin{bmatrix} 1 & -\dfrac{1}{2} & | & \dfrac{1}{2} & 0 \\ 3 & 1 & | & 0 & 1 \end{bmatrix}$ $\dfrac{1}{2}R1$

$\begin{bmatrix} 1 & -\dfrac{1}{2} & | & \dfrac{1}{2} & 0 \\ 0 & \dfrac{5}{2} & | & -\dfrac{3}{2} & 1 \end{bmatrix}$ $-3R1 + R2$

$\begin{bmatrix} 1 & -\dfrac{1}{2} & | & \dfrac{1}{2} & 0 \\ 0 & 1 & | & -\dfrac{3}{5} & \dfrac{2}{5} \end{bmatrix}$ $\dfrac{2}{5}R2$

$\begin{bmatrix} 1 & 0 & | & \dfrac{1}{5} & \dfrac{1}{5} \\ 0 & 1 & | & -\dfrac{3}{5} & \dfrac{2}{5} \end{bmatrix}$ $\dfrac{1}{2}R2 + R1$

$A^{-1} = \begin{bmatrix} \dfrac{1}{5} & \dfrac{1}{5} \\ -\dfrac{3}{5} & \dfrac{2}{5} \end{bmatrix}$

$X = A^{-1}B = \begin{bmatrix} \dfrac{1}{5} & \dfrac{1}{5} \\ -\dfrac{3}{5} & \dfrac{2}{5} \end{bmatrix}\begin{bmatrix} -8 \\ -2 \end{bmatrix}$

$= \begin{bmatrix} -2 \\ 4 \end{bmatrix}$

Solution set: $\{(-2, 4)\}$

37. $2x + 3y = -10$

$3x + 4y = -12$

$A = \begin{bmatrix} 2 & 3 \\ 3 & 4 \end{bmatrix}, \quad X = \begin{bmatrix} x \\ y \end{bmatrix}, \quad B = \begin{bmatrix} -10 \\ -12 \end{bmatrix}$

Find A^{-1}.

$[A|I_2] = \begin{bmatrix} 2 & 3 & | & 1 & 0 \\ 3 & 4 & | & 0 & 1 \end{bmatrix}$

$\begin{bmatrix} 1 & \dfrac{3}{2} & | & \dfrac{1}{2} & 0 \\ 3 & 4 & | & 0 & 1 \end{bmatrix}$ $\dfrac{1}{2}R1$

$$\begin{bmatrix} 1 & \frac{3}{2} & \bigm| & \frac{1}{2} & 0 \\ 0 & -\frac{1}{2} & \bigm| & -\frac{3}{2} & 1 \end{bmatrix} \quad -3R1 + R2$$

$$\begin{bmatrix} 1 & \frac{3}{2} & \bigm| & \frac{1}{2} & 0 \\ 0 & 1 & \bigm| & 3 & -2 \end{bmatrix} \quad -2R2$$

$$\begin{bmatrix} 1 & 0 & \bigm| & -4 & 3 \\ 0 & 1 & \bigm| & 3 & -2 \end{bmatrix} \quad -\frac{3}{2}R2 + R1$$

$$A^{-1} = \begin{bmatrix} -4 & 3 \\ 3 & -2 \end{bmatrix}$$

$$X = A^{-1}B = \begin{bmatrix} -4 & 3 \\ 3 & -2 \end{bmatrix}\begin{bmatrix} 10 \\ 15 \end{bmatrix}$$

$$= \begin{bmatrix} 4 \\ -6 \end{bmatrix}$$

Solution set: $\{(4, -6)\}$

39. $x + 3y + 3z = 1$
$x + 4y + 3z = 0$
$x + 3y + 4z = -1$

$$A = \begin{bmatrix} 1 & 3 & 3 \\ 1 & 4 & 3 \\ 1 & 3 & 4 \end{bmatrix}, \quad X = \begin{bmatrix} x \\ y \\ z \end{bmatrix}, \quad B = \begin{bmatrix} 1 \\ 0 \\ -1 \end{bmatrix}$$

From Exercise 17,

$$A^{-1} = \begin{bmatrix} 7 & -3 & -3 \\ -1 & 1 & 0 \\ -1 & 0 & 1 \end{bmatrix}.$$

$$X = A^{-1}B = \begin{bmatrix} 7 & -3 & -3 \\ -1 & 1 & 0 \\ -1 & 0 & 1 \end{bmatrix}\begin{bmatrix} 1 \\ 0 \\ -1 \end{bmatrix}\begin{bmatrix} 10 \\ -1 \\ -2 \end{bmatrix}$$

Solution set: $\{(10, -1, -2)\}$

41. $2x + 2y - 4z = 12$
$2x + 6y = 16$
$-3x - 3y + 5z = -20$

$$A = \begin{bmatrix} 2 & 2 & -4 \\ 2 & 6 & 0 \\ -3 & -3 & 5 \end{bmatrix}, \quad X = \begin{bmatrix} x \\ y \\ z \end{bmatrix},$$

$$B = \begin{bmatrix} 12 \\ 16 \\ -20 \end{bmatrix}$$

From Exercise 19,

$$A^{-1} = \begin{bmatrix} -\frac{15}{4} & -\frac{1}{4} & -3 \\ \frac{5}{4} & \frac{1}{4} & 1 \\ -\frac{3}{2} & 0 & -1 \end{bmatrix}.$$

$$X = A^{-1}B = \begin{bmatrix} -\frac{15}{4} & -\frac{1}{4} & -3 \\ \frac{5}{4} & \frac{1}{4} & 1 \\ -\frac{3}{2} & 0 & -1 \end{bmatrix}\begin{bmatrix} 12 \\ 16 \\ -20 \end{bmatrix}$$

$$= \begin{bmatrix} 11 \\ -1 \\ 2 \end{bmatrix}$$

Solution set: $\{(11, -1, 2)\}$

43. $x + y + 2w = 3$
$2x - y + z - w = 3$
$3x + 3y + 2z - 2w = 5$
$x + 2y + z = 3$

$$A = \begin{bmatrix} 1 & 1 & 0 & 2 \\ 2 & -1 & 1 & -1 \\ 3 & 3 & 2 & -2 \\ 1 & 2 & 1 & 0 \end{bmatrix}, \quad X = \begin{bmatrix} x \\ y \\ z \\ w \end{bmatrix},$$

$$B = \begin{bmatrix} 3 \\ 3 \\ 5 \\ 3 \end{bmatrix}$$

From Exercise 21,

$$A^{-1} = \begin{bmatrix} \frac{1}{2} & 0 & \frac{1}{2} & -1 \\ \frac{1}{10} & -\frac{2}{5} & \frac{3}{10} & -\frac{1}{5} \\ -\frac{7}{10} & \frac{4}{5} & -\frac{11}{10} & \frac{12}{5} \\ \frac{1}{5} & \frac{1}{5} & -\frac{2}{5} & \frac{3}{5} \end{bmatrix}.$$

$$X = A^{-1}B = \begin{bmatrix} \frac{1}{2} & 0 & \frac{1}{2} & -1 \\ \frac{1}{10} & -\frac{2}{5} & \frac{3}{10} & -\frac{1}{5} \\ -\frac{7}{10} & \frac{4}{5} & -\frac{11}{10} & \frac{12}{5} \\ \frac{1}{5} & \frac{1}{5} & -\frac{2}{5} & \frac{3}{5} \end{bmatrix} \begin{bmatrix} 3 \\ 3 \\ 5 \\ 3 \end{bmatrix}$$

$$= \begin{bmatrix} 1 \\ 0 \\ 2 \\ 1 \end{bmatrix}$$

Solution set: $\{(1, 0, 2, 1)\}$

45. **(a)** $602.7 = a + 5.543b + 37.14c$
$656.7 = a + 6.933b + 41.30c$
$778.5 = a + 7.638b + 45.62c$

(b) $A = \begin{bmatrix} 1 & 5.543 & 37.14 \\ 1 & 6.933 & 41.30 \\ 1 & 7.638 & 45.62 \end{bmatrix}$, $X = \begin{bmatrix} a \\ b \\ c \end{bmatrix}$

$B = \begin{bmatrix} 602.7 \\ 656.7 \\ 778.5 \end{bmatrix}$

$X = A^{-1}B$

Using a graphing calculator with matrix capabilities, we obtain

$$\begin{bmatrix} a \\ b \\ c \end{bmatrix} = \begin{bmatrix} -490.547375 \\ -89 \\ 42.71875 \end{bmatrix}.$$

Thus, $a \approx -490.547$, $b \approx -89$, $c \approx 42.71875$.

(c) $S = -490.547 - 89A + 42.71875B$

(d) If $A = 7.752$ and $B = 47.38$, the predicted value of S is given by

$S = -490.547 - 89(7.752)$
$+ 42.71875(47.38)$
$= 843.539375 \approx 843.5.$

The predicted value is approximately 843.5.

(e) If $A = 8.9$ and $B = 66.25$, the predicted value of S is given by

$S = -490.547 - 89(8.9)$
$+ 42.71875(66.25)$
$= 1547.470188 \approx 1547.5.$

The predicted value is approximately 1547.5.

Using only three consecutive years t forecast six years into the future i probably not very accurate.

47. **(a)** The message UBNL can be repre-sented by the 21 2 4 12. Writing these numbers in a matrix with two rows produces the matrix $C = \begin{bmatrix} 21 & 14 \\ 2 & 12 \end{bmatrix}$

To reverse the process, we must multiply C by $A^{-1} = \begin{bmatrix} -2 & -1 \\ 5 & 2 \end{bmatrix}$.

$$A^{-1}C = \begin{bmatrix} -2 & -1 \\ 5 & 2 \end{bmatrix} \begin{bmatrix} 21 & 14 \\ 2 & 12 \end{bmatrix}$$

$$= \begin{bmatrix} -44 & -40 \\ 109 & 94 \end{bmatrix}$$

Scaling the matrix elements between 1 and 26 results in the matrix B.

$-44 + 2(26) = 8$
$-40 + 2(26) = 12$
$109 - 4(26) = 5$
$94 - 3(26) = 16$

$B = \begin{bmatrix} 8 & 12 \\ 5 & 16 \end{bmatrix}$

From B the numbers are 8 5 12 16 which decode into the word HELP.

(b) The message QNABMV can be represented by the 17 14 1 2 13 22. Writing these numbers in a matrix with two rows produces the matrix $C = \begin{bmatrix} 17 & 1 & 13 \\ 14 & 2 & 22 \end{bmatrix}$. To reverse the process we must multiply by $A^{-1} = \begin{bmatrix} -2 & -1 \\ 5 & 2 \end{bmatrix}$.

$$A^{-1}C = \begin{bmatrix} -2 & -1 \\ 5 & 2 \end{bmatrix} \begin{bmatrix} 17 & 1 & 13 \\ 14 & 2 & 22 \end{bmatrix}$$

$$= \begin{bmatrix} -48 & -4 & -48 \\ 113 & 9 & 109 \end{bmatrix}$$

Scaling the matrix elements between 1 and 26 results in the matrix B.

$$-48 + 2(26) = 4$$
$$-4 + 1(26) = 22$$
$$-48 + 2(26) = 4$$
$$113 - 4(26) = 9$$
$$9 - 0(26) = 9$$
$$109 - 4(26) = 5$$

$$B = \begin{bmatrix} 4 & 22 & 4 \\ 9 & 9 & 5 \end{bmatrix}$$

From B the numbers are

$$4 \quad 9 \quad 22 \quad 9 \quad 4 \quad 5,$$

which decode into the word DIVIDE.

49. $A = \begin{bmatrix} \sqrt{2} & .5 \\ -17 & \frac{1}{2} \end{bmatrix}$

Using a graphing calculator with matrix capabilities, we obtain

$$A^{-1} = \begin{bmatrix} .0543058761 & -.0543058761 \\ 1.846399787 & .153600213 \end{bmatrix}.$$

51. $\quad x - \sqrt{2}y = 2.6$
$\quad .75x + y = -7$

$$A = \begin{bmatrix} 1 & -\sqrt{2} \\ .75 & 1 \end{bmatrix}, \quad X = \begin{bmatrix} x \\ y \end{bmatrix}, \quad B = \begin{bmatrix} 2.6 \\ -7 \end{bmatrix}$$

$$X = A^{-1}B$$

Using a graphing calculator with matrix capabilities, we obtain

$$X = \begin{bmatrix} -3.542308934 \\ -4.343268299 \end{bmatrix}.$$

Solution set:
$\{(-3.542308934, \ -4.343268299)\}$

53. $\quad \pi x + ey + \sqrt{2}z = 1$
$\quad ex + \pi y + \sqrt{2}z = 2$
$\quad \sqrt{2}x + ey + \pi z = 3$

$$A = \begin{bmatrix} \pi & e & \sqrt{2} \\ e & \pi & \sqrt{2} \\ \sqrt{2} & e & \pi \end{bmatrix}, \quad X = \begin{bmatrix} x \\ y \\ z \end{bmatrix}, \text{ and}$$

$$B = \begin{bmatrix} 1 \\ 2 \\ 3 \end{bmatrix}$$

$$X = A^{-1}B$$

Using a graphing calculator with matrix capabilities, we obtain

$$X = \begin{bmatrix} -.9704156959 \\ 1.391914631 \\ .1874077432 \end{bmatrix}.$$

Solution set:
$\{(-.9704156959, \ 1.391914631, \\ \qquad\quad .1874077432)\}$

55. $A = \begin{bmatrix} a & b \\ c & d \end{bmatrix}$ and $O = \begin{bmatrix} 0 & 0 \\ 0 & 0 \end{bmatrix}$

$$A \cdot O = \begin{bmatrix} a & b \\ c & d \end{bmatrix} \begin{bmatrix} 0 & 0 \\ 0 & 0 \end{bmatrix} = \begin{bmatrix} 0 & 0 \\ 0 & 0 \end{bmatrix} = O$$

$$O \cdot A = \begin{bmatrix} 0 & 0 \\ 0 & 0 \end{bmatrix} \begin{bmatrix} a & b \\ c & d \end{bmatrix} = \begin{bmatrix} 0 & 0 \\ 0 & 0 \end{bmatrix} = O$$

Therefore, $A \cdot O = O \cdot A = O$

57. Suppose A is a square matrix with two inverses, call them R and S. Then, AR = I and AS = I with R ≠ S. Therefore,

$$AR = AS.$$

Multiply both sides on the left by R.

R(AR) = R(AS)

(RA)R = (RA)S   *Associative property*

   IR = IS   *Inverse property*

    R = S   *Identity property*

This contradicts our assumption that R ≠ S.

Therefore, A has no more than one inverse.

59. $(B^{-1}A^{-1})(AB)$

= $B^{-1}[A^{-1}(AB)]$ *Associative property*

= $B^{-1}[(A^{-1}A)B]$ *Associative property*

= $B^{-1}(IB)$ *Inverse property*

= $B^{-1}B$ *Identity property*

= I *Inverse prorerty*

Since

$$(B^{-1}A^{-1})(AB) = I,$$
$$B^{-1}A^{-1} = (AB)^{-1}.$$

61. $A = \begin{bmatrix} 1 & 0 & 0 \\ 0 & 0 & -1 \\ 0 & 1 & -1 \end{bmatrix}$

$A^2 = AA = \begin{bmatrix} 1 & 0 & 0 \\ 0 & 0 & -1 \\ 0 & 1 & -1 \end{bmatrix}\begin{bmatrix} 1 & 0 & 0 \\ 0 & 0 & -1 \\ 0 & 1 & -1 \end{bmatrix}$

$= \begin{bmatrix} 1 & 0 & 0 \\ 0 & -1 & 1 \\ 0 & -1 & 0 \end{bmatrix}$

$A^3 = AA^2 = \begin{bmatrix} 1 & 0 & 0 \\ 0 & 0 & -1 \\ 0 & 1 & -1 \end{bmatrix}\begin{bmatrix} 1 & 0 & 0 \\ 0 & -1 & 1 \\ 0 & -1 & 0 \end{bmatrix}$

$= \begin{bmatrix} 1 & 0 & 0 \\ 0 & 1 & 0 \\ 0 & 0 & 1 \end{bmatrix}$

Since $AA^2 = I$, $A^2 = A^{-1}$. Therefore,

$$A^{-1} = \begin{bmatrix} 1 & 0 & 0 \\ 0 & -1 & 1 \\ 0 & -1 & 0 \end{bmatrix}.$$

Section 6.8

For Exercises 1–15, see the answer graphs in the back of the textbook.

1. x ≤ 3

The boundary is the vertical line x = 3. Because of the = portion of ≤, the boundary is included in the graph, so draw a solid line. Select any test point not on the line, such as (0, 0). Since 0 ≤ 3 is a true statement, shade the side of the line containing (0, 0).

3. x + 2y ≤ 6

The boundary is the line x + 2y = 6, which can be graphed using the x-intercept 6 and y-intercept 3. The boundary is included in the graph, so draw a solid line. Use (0, 0) as a test point. Since 0 + 2(0) ≤ 6 is a true statement, shade the line of the graph containing (0, 0).

5. $2x + 3y \geq 4$

The boundary is the line $2x + 3y = 4$. The boundary is included in the graph, so draw a solid line. Use $(0, 0)$ as a test point. Since $2(0) + 3(0) \geq 4$ is false, shade the side of the line that does not contain $(0, 0)$.

7. $3x - 5y > 6$

The boundary is the line $3x - 5y = 6$. Since the inequality symbol is $>$, not $\geq$, the boundary is not included in the graph, so draw a dashed line. Use $(0, 0)$ as a test point. Since $3(0) - 5(0) > 6$ is false, shade the side of the line that does not include $(0, 0)$.

9. $5x \leq 4y - 2$

The boundary is the line $5x = 4y - 2$. Draw a solid line. Use $(0, 0)$ as a test point. Since $5(0) \leq 4(0) - 2$ is false, shade the side of the line that does not include $(0, 0)$.

11. $y < 3x^2 + 2$

The boundary is the parabola $y = 3x^2 + 2$. Since the inequality symbol is $<$, draw a dashed curve. Use $(0, 0)$ as a test point. Since $0 < 3(0)^2 + 2$ is true, shade the region that includes $(0, 0)$.

13. $y > (x - 1)^2 + 2$

The boundary is the parabola $y = (x - 1)^2 + 2$, with vertex $(1, 2)$. Since the inequality symbol is $>$, draw a dashed curve. Use $(0, 0)$ as a test point. Since $0 > (0 - 1)^2 + 2$ is false, shade the region that does not include $(0, 0)$.

15. $x^2 + (y + 3)^2 \leq 16$

The boundary is a circle with center $(0, -3)$ and radius 4. Draw a solid circle to show that the boundary is included in the graph. Use $(0, 0)$ as a test point. Since $0^2 + (0 + 3)^2 \leq 16$ is true, shade the region that includes $(0, 0)$, that is, the interior of the circle.

19. $Ax + By \geq C$, $B > 0$

Solve this inequality for y.

$$By \geq -Ax + C$$

Since $B > 0$, the inequality symbol is not reversed when both sides are divided by B.

$$y \geq -\frac{A}{B}x + \frac{C}{B}$$

You would shade above the line.

21. The graph of

$$(x - 5)^2 + (y - 2)^2 = 4$$

is a circle with center (5, 2) and radius $r = \sqrt{4} = 2$. The graph of

$$(x - 5)^2 + (y - 2)^2 < 4$$

is the region inside this circle. The correct response is (b).

23. The graph of $y \leq 3x - 6$ is the region below the line with slope 3 and y-intercept −6. This is graph C.

25. The graph of $y \leq -3x - 6$ is the region below the line with slope −3 and y-intercept −6. This is graph A.

For Exercises 27–53, see the answer graphs in the textbook.

27. $x + y \geq 0$
$2x - y \geq 3$

Graph $x + y = 0$ as a solid line through the origin with a slope of −1. Shade the region above this line.
Graph $2x - y = 3$ as a solid line with x-intercept 3/2 and y-intercept −3. Shade the region below this line.
The solution set is the common region, which is shaded in the final graph.

29. $2x + y > 2$
$x - 3y < 6$

Graph $2x + y = 2$ as a dashed line with y-intercept 2 and x-intercept 1. Shade the region above this line.
Graph $x - 3y = 6$ as a dashed line with y-intercept −2 and x-intercept 6. Shade the region above this line. The solution set is the common region, which is shaded in the final graph.

31. $3x + 5y \leq 15$
$x - 3y \geq 9$

Graph $3x + 5y = 15$ as a solid line with y-intercept 3 and x-intercept 5. Shade the region below this line.
Graph $x - 3y = 9$ as a solid line with y-intercept −3 and x-intercept 9. Shade the region below this line.
The solution set is the common region, which is shaded in the final graph.

33. $4x - 3y \leq 12$
$y \leq x^2$

Graph $4x - 3y = 12$ as a solid line with y-intercept −4 and x-intercept 3. Shade the region above this line.
Graph the solid parabola $y = x^2$. Shade the region outside of this parabola.

The solution set is the intersection of these two regions, which is shaded in the final graph.

35. $x + y \leq 9$

$x \leq -y^2$

Graph $x + y = 9$ as a solid line with y-intercept 9 and x-intercept 9. Shade the region below this line. Graph the solid horizontal parabola $x = -y^2$. Shade the region inside of this parabola.

The solution set is the intersection of these two regions, which is shaded in the final graph.

37. $y \leq (x + 2)^2$

$y \geq -2x^2$

Graph $y = (x + 2)^2$ as a solid parabola opening up with a vertex at $(-2, 0)$. Shade the region below the parabola.

Graph $y = -2x^2$ as a solid parabola opening down with a vertex at the origin. Shade the region above the parbola.

The solution set is the intersection of these two regions, which is shaded in the final graph.

39. $x + y \leq 36$

$-4 \leq x \leq 4$

Graph $x + y = 36$ as solid line with y-intercept 36 and x-intercept 36. Shade the region below this line.

Graph the vertical lines $x = -4$ and $x = 4$ as solid lines. Shade the region between these lines.

The solution set is the intersection of these two regions, which is shaded in the final graph.

41. $y \geq (x - 2)^2 + 3$

$y \leq -(x - 1)^2 + 6$

Graph $y = (x - 2)^2 + 3$ as a solid parabola opening up with a vertex at $(2, 3)$. Shade the region above the parabola.

Graph $y = -(x - 1)^2 + 6$ as a solid parabola opening down with a vertex at $(1, 6)$. Shade the region below the parbola.

The solution set is the intersection of these two regions, which is shaded in the final graph.

43. $3x - 2y \geq 6$

$x + y \leq -5$

$y \leq 4$

Graph $3x - 2y = 6$ as a solid line and shade the region below it.

Graph $x + y = -5$ as a solid line and shade the region below it.

Graph $y = 4$ as a solid horizontal line and shade the region below it.

The solution set is the intersection of these three regions, which is shaded in the final graph.

45. $-2 < x < 2$

$y > 1$

$x - y > 0$

Graph the vertical lines $x = -2$ and $x = 2$ as a dashed line. Shade the region between the two lines.
Graph the horizontal line $y = 1$ as a dashed line. Shade the region above the line.
Graph the line $x - y = 0$ as a dashed line through the origin with a slope of 1. Shade the region below this line.
The solution set is the intersection of these three regions, which is shaded in the final graph.

47. $x \leq 4$

$x \geq 0$

$y \geq 0$

$x + 2y \geq 2$

Graph $x = 4$ as a solid vertical line. Shade the region to the left of this line.
Graph $x = 0$ as a solid vertical line. (This is the y-axis.) Shade the region to the right of this line.
Graph $y = 0$ as a solid horizontal line. (This is the x-axis.) Shade the region above the line.
Graph $x + 2y = 2$ as a solid line with x-intercept 2 and y-intercept 1. Shade the region above the line.

The solution set is the intersection of these four regions, which is shaded in the final graph.

49. $2x + 3y \leq 12$

$2x + 3y > -6$

$3x + y < 4$

$x \geq 0$

$y \geq 0$

Graph $2x + 3y = 12$ as a solid line and shade the region below it.
Graph $2x + 3y = 6$ as a dashed line and shade the region above it.
Graph $3x + y = 4$ as a dashed line and shade the region below it.
$x = 0$ is the y-axis. Shade the region to the right of it.
$y = 0$ is the x-axis. Shade the region above it. The solution set is the intersection of these five regions, which is shaded in the final graph. The open circles at $(0, 4)$ and $\left(\frac{4}{3}, 0\right)$ indicate that those points are not included in the solution (due to the fact that the boundary line on which they lie, $3x + y = 4$, is not included).

51. $y \leq \left(\frac{1}{2}\right)^x$

$y \geq 4$

Graph $y = \left(\frac{1}{2}\right)^x$ using a solid curve passing through the points $(-2, 4)$, $(-1, 2)$, $(0, 1)$, $\left(1, \frac{1}{2}\right)$, and $\left(2, \frac{1}{4}\right)$.
Shade the region below this curve.
Graph the solid horizontal line $y = 4$ and shade the region above it.

The solution set consists of the intersection of these two regions, which is shaded in the final graph.

53. $y \leq \log x$

    $y \geq |x - 2|$

    Graph $y = \log x$ using a solid curve passing through the points $\left(\frac{1}{10}, -1\right)$, $(1, 0)$, and $(10, 1)$.

    Shade the region below the curve.

    Graph $y = |x - 2|$ using a solid curve. This is the same as the graph of $y = |x|$, but translated 2 units to the right. Shade the region above.

    The solution set is the intersection of the two regions, which is shaded in the final graph.

55. $y \geq x$

    $y \leq 2x - 3$

    The graph is the region above the line $y = x$ and below the line $y = 2x - 2$. This is graph A.

57. $x^2 + y^2 \leq 16$

    $y \geq 0$

    The graph is the region inside the circle $x^2 + y^2 = 16$ and above the horizontal line $y = 0$. This is graph B.

59. $3x + 2y \geq 6$

    Solve the inequality for y.

    $$2y \leq -3x + 6$$

    $$y \leq -\frac{3}{2}x + 3$$

Enter $y_1 = (-3/2)x + 3$ and use a graphing calculator to shade the region above the line.

61. $x + y \geq 2$

    $x + y \leq 6$

    Solve each inequality for y.

    Enter $y_1 = -x + 2$ and $y_2 = -x + 6$. Use a graphing calculator to shade the region above the graph of $y_1 = -x + 2$ and below the graph of $y_2 = -x + 6$.

63. $y \geq 2^x$

    $y \leq 8$

    Enter $y_1 = 2^x$ and $y_2 = 8$. Use a graphing calculator to shade the region above the graph of $y_1 = 2^x$ and below the graph of $y_2 = 8$.

65. Since we are in the first quadrant, $x \geq 0$ and $y \geq 0$.

    The lines $x + 2y - 8 = 0$ and $x + 2y = 12$ are parallel, with $x + 2y = 12$ having the greater y-intercept. Therefore, we must shade below $x + 2y = 12$ and above $x + 2y - 8 = 0$

    The system is

    $$x + 2y - 8 \geq 0$$

    $$x + 2y \leq 12$$

    $$x \geq 0, \ y \geq 0.$$

67.

Point	Value of 3x + 5y
(1, 1)	3(1) + 5(1) = 8 ← Minimum
(2, 7)	3(2) + 5(7) = 41
(5, 10)	3(5) + 5(10) = 65 ← Maximum
(6, 3)	3(6) + 5(3) = 33

The maximum value is 65 at (5, 10).
The minimum value is 8 at (1, 1).

69.

Point	Value of 3x + 5y
(1, 0)	3(1) + 5(0) = 3 ← Minimum
(1, 10)	3(1) + 5(10) = 53
(7, 9)	3(7) + 5(9) = 66 ← Maximum
(7, 6)	3(7) + 5(6) = 51

The maximum value is 66 at (7, 9).
The minimum value is 3 at (1, 0).

71.

Point	Value of 10y
(1, 0)	10(0) = 0 ← Minimum
(1, 10)	10(10) = 100 ← Maximum
(7, 9)	10(9) = 90
(7, 6)	10(6) = 60

The maximum value is 100 at (1, 10).
The minimum value is 0 at (1, 0).

73. Let x = the number of Brand X pills;
y = the number of Brand Y pills.

Then

$$3000x + 1000y \geq 6000$$
$$45x + 50y \geq 195$$
$$75x + 200y \geq 600$$
$$x \geq 0, \; y \geq 0.$$

Graph 3000x + 1000y = 6000 as a solid line with x-intercept 2 and y-intercept 6. Shade the region above the line.

Graph 45x + 50y = 195 as a solid line with x-intercept 4.$\overline{3}$ and y-intercept 3.9. Shade the region above the line.

Graph 75x + 200y = 600 as a solid line with x-intercept 8 and y-intercept 3. Shade the region above the line.

Graph x = 0 (the y-axis) as a solid line and shade the region to the right of it.

Graph y = 0 (the x-axis) as a solid line and shade the region above it. The region of feasible solutions is the intersection of these five regions. See the answer graph in the textbook.

75. Let P = number of pigs;
G = number of geese.

P + G ≤ 16	*Total number of animals*
G ≤ 12	*No more than 12 geese*
50P + 20G ≤ 500	*$500 available to spend*

Maximize the profit 80G + 40P. Find the region of feasible solutions by graphing on a horizontal P-axis and a vertical G-axis.

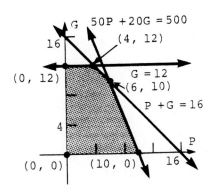

The boundaries of the graph are

$$P + G = 16$$
$$G = 12$$
$$50P + 20G = 500.$$

Also, $P \geq 0$ and $G \geq 0$, so the P- and G-axes are boundaries. The vertices are $(0, 0)$, $(0, 12)$, $(4, 12)$, $(10, 0)$, and $(6, 10)$, the intersection of $P + G = 16$ and $50P + 20G = 500$. Check the value of $80G + 40P$ at each vertex to find the maximum profit.

Point (P, G)	Profit = 80G + 40P
(0, 0)	0
(0, 12)	960
(4, 12)	1120 ← Maximum
(10, 0)	400
(6, 10)	1040

The maximum profit is $1120 with 4 pigs and 12 geese.

77. Let x = number of cabinet #1;
     y = number of cabinet #2.

The cost constraint is

$$10x + 20y \leq 140.$$

The space constraint is

$$6x + 8y \leq 72.$$

Since the numbers of cabinets cannot be negative, we also have

$$x \geq 0$$
$$y \geq 0.$$

We want to maximize the volume of files, given by $8x + 12y$.
Find the region of feasible solutions by graphing.

$$20x + 40y = 280 \text{ (shade below)}$$
$$6x + 8y = 72 \text{ (shade below)}$$
$$x = 0 \text{ (shade right)}$$
$$y = 0 \text{ (shade above)}$$

The vertices are at $(0, 7)$, $(0, 0)$, $(12, 0)$, and the intersection of $20x + 40y = 280$ and $6x + 8y = 72$, which is the point $(8, 3)$.

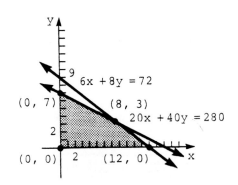

Find the value of 8x + 12y at each vertex.

Point	Value = 8x + 12y
(0, 7)	84
(0, 0)	0
(12, 0)	96
(8, 3)	100 ← Maximum

8x + 12y is maximized at (8, 3). She should get 8 #1 cabinets and 3 #2 cabinets. This will correspond to maximum storage capacity of 100 cu ft.

79. Let x = number of gallons of gasoline;

   y = number of gallons of fuel oil.

The constraints are

$$x \geq 0, \ y \geq 0$$
$$x \geq 2y$$
$$y \geq 3{,}000{,}000$$
$$x \leq 6{,}400{,}000.$$

Maximize revenue, given by 1.9x + 1.5y.

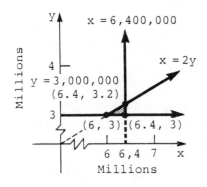

The boundaries are

$$x = 0$$
$$y = 0$$
$$x = 2y$$
$$y = 3{,}000{,}000$$
$$x = 6{,}400{,}000.$$

The vertices of the region of feasible solutions are

   (6.4, 3), (6, 3) and (6.4, 3.2).

Testing these points in the expression to be maximized will show that (6.4, 3.2) will maximize revenue. 6.4 million gallons of gasoline and 3.2 million gallons of fuel oil should be produced for maximum revenue of

$$1.9(6.4) + 1.5(3.2) = 16.96$$

or $16,960,000.

Chapter 6   Review Exercises

1. 3x − 5y = 7   (1)
   2x + 3y = 30   (2)

Multiply equation (1) by 3, and multiply equation (2) by 5; then add the resulting equations.

$$9x - 15y = \ 21$$
$$\underline{10x + 15y = 150}$$
$$19x \qquad = 171$$
$$x = \frac{171}{19}$$
$$x = 9$$

Substitute this value into equation (2).

$$2(9) + 3y = 30$$
$$18 + 3y = 30$$
$$3y = 12$$
$$y = 4$$

Solution set: $\{(9, 4)\}$

3. $6x - 2y = 4$ (1)
    $4x + 5y = 9$ (2)

Multiply equation (1) by 5 and equation (2) by 2; then add the resulting equations.

$$
\begin{array}{r}
30x - 10y = 20 \\
\underline{8x + 10y = 18} \\
38x = 38
\end{array}
$$

$$x = 1$$

Substitute this value into equation (2).

$$4(1) + 5y = 9$$
$$5y = 5$$
$$y = 1$$

Solution set: $\{(1, 1)\}$

5. $.2x + .5y = 6$ (1)
    $.4x + y = 9$ (2)

Multiply equation (1) by -2 and add the result to equation (2).

$$
\begin{array}{r}
-.4x - y = -12 \\
\underline{.4x + y = 9} \\
0 = -3
\end{array}
$$

Since this is a false statement, the system is inconsistent.

Solution set: ∅

7. $2x - 5y + 3z = -1$ (1)
    $x + 4y - 2z = 9$ (2)
    $-x + 2y + 4z = 5$ (3)

First, we eliminate x. Multiply equation (2) by -2 and add the result to equation (1).

$$
\begin{array}{r}
2x - 5y + 3z = -1 \\
\underline{-2x - 8y + 4z = -18} \\
-13y + 7z = -19 \quad (4)
\end{array}
$$

Next, add equations (2) and (3).

$$
\begin{array}{r}
x + 4y - 2z = 9 \quad (2) \\
\underline{-x + 2y + 4z = 5} \quad (3) \\
6y + 2z = 14 \quad (5)
\end{array}
$$

Now, we solve the system

$$-13y + 7z = -19 \quad (4)$$
$$6y + 2z = 14. \quad (5)$$

Multiply equation (4) by 2, multiply equation (5) by -7, and add the resulting equations.

$$
\begin{array}{r}
-26y + 14z = -38 \\
\underline{-42y - 14z = -98} \\
-68y = -136
\end{array}
$$

$$y = 2$$

Substitute this value into equation (4).

$$-13(2) + 7z = -19$$
$$-26 + 7z = -19$$
$$7z = 7$$
$$z = 1$$

Substitute these values into equation (2).

$$x + 4(2) - 2(1) = 9$$
$$x + 8 - 2 = 9$$
$$x = 3$$

Solution set: $\{(3, 2, 1)\}$

9. Let x = amount of rice
and y = amount of soybeans.

$$15x + 22.5y = 9.5 \quad (1)$$
$$810x + 270y = 324 \quad (2)$$

Multiply equation (1) by −12 and add
the result to equation (2).

$$-180x - 270y = -114$$
$$\underline{810x + 270y = 324}$$
$$630x = 210$$
$$x = \frac{1}{3}$$

Substitute 1/3 for x in equation (1)
and solve for y.

$$15\left(\frac{1}{3}\right) + 22.5y = 9.5$$
$$5 + 22.5y = 9.5$$
$$22.5y = 4.5$$
$$y = .20 = \frac{1}{5}$$

1/3 cup of rice and 1/5 cup of soybeans should be used.

11. Let x = the number of blankets,
y = the number of rugs,
and z = the number of skirts.

$$24x + 30y + 12z = 306 \quad (1)$$
$$4x + 5y + 3z = 59 \quad (2)$$
$$15x + 18y + 9z = 201 \quad (3)$$

Simplify equations (1) and (3).

$$4x + 5y + 2z = 51 \quad (4)$$
$$4x + 5y + 3z = 59 \quad (5)$$
$$5x + 6y + 3z = 67 \quad (6)$$

Multiply equation (4) by −1 and add
the result to equation (5).

$$-4x - 5y - 2z = -51$$
$$\underline{4x + 5y + 3z = 59}$$
$$z = 8$$

Substitute 8 for z in equations (5)
and (6) and simplify.

$$4x + 5y = 35 \quad (7)$$
$$5x + 6y = 43 \quad (8)$$

Multiply equation (7) by 5 and equation (8) by −4 and add the results.

$$20x + 25y = 175$$
$$\underline{-20x - 24y = -172}$$
$$y = 3$$

Substitute 3 for y in equation (7)
and solve for x.

$$4x + 5(3) = 35$$
$$4x = 20$$
$$x = 5$$

5 blankets, 3 rugs, and 8 skirts can
be made.

13. Since $y = ax^2 + bx + c$ and the
points (1, −2.3), (2, −1.3), and
(3, 4.5) are on the parabola, we
have the following equations:

$$-2.3 = a(1)^2 + b(1) + c$$
$$-1.3 = a(2)^2 + b(2) + c$$
$$4.5 = a(3)^2 + b(3) + c.$$

This becomes the following system:

$$a + b + c = -2.3 \quad (1)$$
$$4a + 2b + c = -1.3 \quad (2)$$
$$9a + 3b + c = 4.5. \quad (3)$$

First, we eliminate c. Multiply equation (1) by -1 and add the result to equation (2).

$$
\begin{array}{rcl}
-a - b - c & = & 2.3 \quad (1) \\
\underline{4a + 2b + c} & = & \underline{-1.3} \quad (2) \\
3a + b & = & 1 \quad (4)
\end{array}
$$

Next, multiply equation (2) by -1 and add to equation (3).

$$
\begin{array}{rcl}
-4a - 2b - c & = & 1.3 \quad (2) \\
\underline{9a + 3b + c} & = & \underline{4.5} \quad (3) \\
5a + b & = & 5.8 \quad (5)
\end{array}
$$

We now solve the system

$$
\begin{array}{rcll}
3a + b & = & 1 & \quad (4) \\
5a + b & = & 5.8. & \quad (5)
\end{array}
$$

Next, eliminate b. Multiply equation (4) by -1 and add the result to equation (5).

$$
\begin{array}{rcl}
-3a - b & = & -1 \\
\underline{5a + b} & = & \underline{5.8} \\
2a & = & 4.8 \\
a & = & 2.4
\end{array}
$$

Substitute this value into equation (4).

$$
\begin{array}{rcl}
3(2.4) + b & = & 1 \\
7.2 + b & = & 1 \\
b & = & -6.2
\end{array}
$$

Substitute these values into equation (1).

$$
\begin{array}{rcl}
(2.4) + (-6.2) + c & = & -2.3 \\
-3.8 + c & = & -2.3 \\
c & = & 1.5
\end{array}
$$

The equation of the parabola is

$$y = 2.4x^2 - 6.2x + 1.5.$$

15.
$$
\begin{array}{rcll}
3x - 4y + z & = & 2 & \quad (1) \\
2x + y & = & 1 & \quad (2)
\end{array}
$$

Solve equation (2) for y.

$$y = 1 - 2x$$

Substitute $1 - 2x$ for y in equation (1) and solve for z.

$$
\begin{array}{rcl}
3x - 4(1 - 2x) + z & = & 2 \\
3x - 4 + 8x + z & = & 2 \\
11x + z & = & 6 \\
z & = & 6 - 11x
\end{array}
$$

Solution set: $\{(x, \ 1 - 2x, \ 6 - 11x)\}$

17.
$$
\begin{array}{rcll}
2x^2 + 3y^2 & = & 30 & \quad (1) \\
x^2 + y^2 & = & 13 & \quad (2)
\end{array}
$$

Multiply equation (2) by -2 and add the result to equation (1).

$$
\begin{array}{rcl}
2x^2 + 3y^2 & = & 30 \\
\underline{-2x^2 - 2y^2} & = & \underline{-26} \\
y^2 & = & 4 \\
y & = & \pm 2
\end{array}
$$

To find the corresponding values of x, substitute back into equation (2).

If $y = 2$, then

$$
\begin{array}{rcl}
x^2 + 4 & = & 13 \\
x^2 & = & 9 \\
x & = & \pm 3.
\end{array}
$$

If $y = -2$, we also obtain $x = \pm 3$.

Solutions set:
$$\{(3, 2), \ (3, -2), \ (-3, 2), \ (-3, -2)\}$$

19. $x^2 + 2xy + y^2 = 4$ (1)

$x - 3y = -2$ (2)

Factor the left side of equation (1) and solve equation (2) for x.

$$(x + y)^2 = 4 \quad (3)$$
$$x = 3y - 2 \quad (4)$$

Substitute $3y - 2$ for x in equation (3).

$$(3y - 2 + y)^2 = 4$$
$$(4y - 2)^2 = 4$$

Solve this quadratic equation by the square root property.

$4y - 2 = 2$ or $4y - 2 = -2$

$\qquad 4y = 4 \qquad\qquad 4y = 0$

$\qquad y = 1$ or $\qquad y = 0$

Substitute back into equation (4) to find the corresponding values of x.

If $y = 1$, $x = 3(1) - 2 = 1$.

If $y = 0$, $x = 3(0) - 2 = -2$.

Solutions set: $\{(1, 1), (-2, 0)\}$

21. $x^2 + y^2 = 144$ (1)

$x + 2y = 8$ (2)

Solve equation (2) for x.

$$x - 2y + 8 \quad (3)$$

Substitute $-2y + 8$ for x in equation (1).

$$(-2y + 8)^2 + y^2 = 144$$
$$4y^2 - 32y + 64 + y^2 = 144$$
$$5y^2 - 32y - 80 = 0$$

Solve this equation by the quadratic formula.

$$y = \frac{32 \pm \sqrt{1,024 + 1,600}}{10}$$

$$= \frac{32 \pm \sqrt{2,624}}{10}$$

$$= \frac{32 \pm 8\sqrt{41}}{10}$$

$$= \frac{16 \pm 4\sqrt{41}}{5}$$

Use equation (3) to solve for x.

$$x = -2\left(\frac{16 \pm 4\sqrt{41}}{5}\right) + 8$$

$$= \frac{-32 \mp 8\sqrt{41}}{5} + \frac{40}{5}$$

$$= \frac{8 \mp 8\sqrt{41}}{5}$$

The circle and the line have two points in common,

$$\left(\frac{8 - 8\sqrt{41}}{5}, \frac{16 + 4\sqrt{41}}{5}\right)$$

and

$$\left(\frac{8 + 8\sqrt{41}}{5}, \frac{16 - 4\sqrt{41}}{5}\right).$$

23. $5x + 2y = -10$

$3x - 5y = -6$

$$\begin{bmatrix} 5 & 2 & \bigm| & -10 \\ 3 & -5 & \bigm| & -6 \end{bmatrix}$$

$$\begin{bmatrix} 1 & \frac{2}{5} & \bigm| & -2 \\ 3 & -5 & \bigm| & -6 \end{bmatrix} \quad \frac{1}{5}R1$$

$$\begin{bmatrix} 1 & \frac{2}{5} & \bigm| & -2 \\ 0 & -\frac{31}{5} & \bigm| & 0 \end{bmatrix} \quad -3R1 + R2$$

$$\begin{bmatrix} 1 & \frac{2}{5} & \bigm| & -2 \\ 0 & 1 & \bigm| & 0 \end{bmatrix} \quad -\frac{5}{31}R2$$

$$\begin{bmatrix} 1 & 0 & \bigm| & -2 \\ 0 & 1 & \bigm| & 0 \end{bmatrix} \quad -\frac{2}{5}R2 + R1$$

Solution set: $\{(-2, 0)\}$

25.

$$x - z = -3$$
$$y + z = 6$$
$$2x - 3z = -9$$

$$\begin{bmatrix} 1 & 0 & -1 & | & -3 \\ 0 & 1 & 1 & | & 6 \\ 2 & 0 & -3 & | & -9 \end{bmatrix}$$

$$\begin{bmatrix} 1 & 0 & -1 & | & -3 \\ 0 & 1 & 1 & | & 6 \\ 0 & 0 & -1 & | & -3 \end{bmatrix} \quad -2R1 + R3$$

$$\begin{bmatrix} 1 & 0 & -1 & | & -3 \\ 0 & 1 & 0 & | & 3 \\ 0 & 0 & 1 & | & 3 \end{bmatrix} \quad \begin{array}{l} R2 + R1 \\ \\ -1R3 \end{array}$$

$$\begin{bmatrix} 1 & 0 & 0 & | & 0 \\ 0 & 1 & 0 & | & 3 \\ 0 & 0 & 1 & | & 3 \end{bmatrix} \quad R3 + R1$$

Solution set: $\{(0, 3, 3)\}$

27. Let x = the number of pounds of
$4.60 tea;

y = the number of pounds of
$5.75 tea;

z = the number of pounds of
$6.50 tea.

$$x + y + z = 20$$
(1) Total pounds

$$4.6x + 5.75y + 6.5z = 20(5.25)$$
(2) Total value

$$x = y + z$$
(3) Amount of $4.60
tea equals sum of
other two

Rewrite the system so that each equation is in standard form.

$$x + y + z = 20$$
$$4.6x + 5.75y + 6.5z = 105$$
$$x - y - x = 0$$

Write the augmented matrix; then solve by the Gauss–Jordan method.

$$\begin{bmatrix} 1 & 1 & 1 & | & 20 \\ 4.6 & 5.75 & 6.5 & | & 105 \\ 1 & -1 & -1 & | & 0 \end{bmatrix}$$

$$\begin{bmatrix} 1 & 1 & 1 & | & 20 \\ 0 & 1.15 & 1.9 & | & 13 \\ 0 & -2 & -2 & | & -20 \end{bmatrix} \quad \begin{array}{l} -4.6R1 + R2 \\ -1R1 + R3 \end{array}$$

$$\begin{bmatrix} 1 & 1 & 1 & | & 20 \\ 0 & 1 & \dfrac{1.9}{1.15} & | & \dfrac{13}{1.15} \\ 0 & -2 & -2 & | & -20 \end{bmatrix} \quad \dfrac{1}{1.15}R2$$

$$\begin{bmatrix} 1 & 1 & 1 & | & 20 \\ 0 & 1 & \dfrac{1.9}{1.15} & | & \dfrac{13}{1.15} \\ 0 & 0 & \dfrac{1.5}{1.15} & | & \dfrac{3}{1.15} \end{bmatrix} \quad 2R2 + R3$$

$$\begin{bmatrix} 1 & 1 & 1 & | & 20 \\ 0 & 1 & \dfrac{1.9}{1.15} & | & \dfrac{13}{1.15} \\ 0 & 0 & 1 & | & 2 \end{bmatrix} \quad \dfrac{1.15}{1.5}R3$$

$$\begin{bmatrix} 1 & 1 & 0 & | & 18 \\ 0 & 1 & 0 & | & 8 \\ 0 & 0 & 1 & | & 2 \end{bmatrix} \quad \begin{array}{l} -R3 + R1 \\ -\dfrac{1.9}{1.15}R3 + R2 \end{array}$$

$$\begin{bmatrix} 1 & 0 & 0 & | & 10 \\ 0 & 1 & 0 & | & 8 \\ 0 & 0 & 1 & | & 2 \end{bmatrix} \quad -1R2 + R1$$

From the final matrix, we have x = 10, y = 8, and z = 2. Therefore, 10 lb of $4.60 tea, 8 lb of $5.75 tea, and 2 lb of $6.50 tea should be used.

29. Let x = the number of fives;

y = the number of tens;

z = the number of twenties.

$$5x + 10y + 20z = 2480 \qquad (1)$$
$$x + y + z = 290 \qquad (2)$$
$$10y = 20z + 60 \qquad (3)$$

Rewrite the system so that each equation is in standard form.

$$5x + 10y + 20z = 2480$$
$$x + y + z = 290$$
$$10y - 20z = 60$$

Write the augmented matrix; then solve by the Gauss–Jordan method.

$$\begin{bmatrix} 5 & 10 & 20 & 2480 \\ 1 & 1 & 1 & 290 \\ 0 & 10 & -20 & 60 \end{bmatrix}$$

$$\begin{bmatrix} 1 & 2 & 4 & 496 \\ 1 & 1 & 1 & 290 \\ 0 & 10 & -20 & 60 \end{bmatrix} \quad \frac{1}{5}R1$$

$$\begin{bmatrix} 1 & 2 & 4 & 496 \\ 0 & -1 & -3 & -206 \\ 0 & 10 & -20 & 60 \end{bmatrix} \quad -R1 + R2$$

$$\begin{bmatrix} 1 & 2 & 4 & 496 \\ 0 & -1 & -3 & -206 \\ 0 & 0 & -50 & -2000 \end{bmatrix} \quad 10R2 + R3$$

$$\begin{bmatrix} 1 & 2 & 4 & 496 \\ 0 & 1 & 3 & 206 \\ 0 & 0 & 1 & 40 \end{bmatrix} \quad \begin{array}{l} -1R2 \\ -\frac{1}{50}R3 \end{array}$$

$$\begin{bmatrix} 1 & 0 & -2 & 84 \\ 0 & 1 & 3 & 206 \\ 0 & 0 & 1 & 40 \end{bmatrix} \quad -2R2 + R1$$

$$\begin{bmatrix} 1 & 0 & 0 & 164 \\ 0 & 1 & 0 & 86 \\ 0 & 0 & 1 & 40 \end{bmatrix} \quad \begin{array}{l} 2R3 + R1 \\ -3R3 + R2 \end{array}$$

From the final matrix, we have x = 164, y = 86, and z = 40. Thus, the cashier has 164 fives, 86 tens, and 40 twenties.

31. $\begin{bmatrix} 5 & x + 2 \\ -6y & z \end{bmatrix} = \begin{bmatrix} a & 3x - 1 \\ 5y & 9 \end{bmatrix}$

$a = 5 \quad x + 2 = 3x - 1$
$$3 = 2x$$
$$\frac{3}{2} = x$$

$-6y = 5y \qquad z = 9$
$$0 = 11y$$
$$0 = y$$

Thus, $a = 5$, $x = 3/2$, $y = 0$, and $z = 9$.

33. $\begin{bmatrix} 3 \\ 2 \\ 5 \end{bmatrix} - \begin{bmatrix} 8 \\ -4 \\ 6 \end{bmatrix} + \begin{bmatrix} 1 \\ 0 \\ 2 \end{bmatrix}$

$$= \begin{bmatrix} -5 \\ 6 \\ 1 \end{bmatrix} + \begin{bmatrix} 1 \\ 0 \\ 2 \end{bmatrix} = \begin{bmatrix} -4 \\ 6 \\ 1 \end{bmatrix}$$

35. $\begin{bmatrix} -3 & 4 \\ 2 & 8 \end{bmatrix} \begin{bmatrix} -1 & 0 \\ 2 & 5 \end{bmatrix}$

$$= \begin{bmatrix} -3(-1) + 4(2) & -3(0) + 4(5) \\ 2(-1) + 8(2) & 2(0) + 8(5) \end{bmatrix}$$

$$= \begin{bmatrix} 11 & 20 \\ 14 & 40 \end{bmatrix}$$

37. $\begin{bmatrix} 1 & -2 & 4 & 2 \\ 0 & 1 & -1 & 8 \end{bmatrix} \begin{bmatrix} -1 \\ 2 \\ 0 \\ 1 \end{bmatrix}$

$$= \begin{bmatrix} 1(-1) + (-2)(2) + 4(0) + 2(1) \\ 0(-1) + 1(2) + (-1)(0) + 8(1) \end{bmatrix}$$

$$= \begin{bmatrix} -3 \\ 10 \end{bmatrix}$$

39. $A = (A^T)^T$

$$= \begin{bmatrix} 5 & 1 & 3 \\ -2 & 4 & 8 \end{bmatrix}$$

41. $\begin{vmatrix} -2 & 4 \\ 0 & 3 \end{vmatrix} = -2(3) - (0)(4) = -6$

43. $\begin{vmatrix} -2 & 4 & 1 \\ 3 & 0 & 2 \\ -1 & 0 & 3 \end{vmatrix}$

Expand by minors about the second column.

$$= -4 \begin{vmatrix} 3 & 2 \\ -1 & 3 \end{vmatrix} + 0 \begin{vmatrix} -2 & 1 \\ -1 & 3 \end{vmatrix} - 0 \begin{vmatrix} -2 & 1 \\ 3 & 2 \end{vmatrix}$$

$$= -4(9 + 2) + 0 - 0$$

$$= -4(11) = -44$$

45. $\begin{vmatrix} -1 & 0 & 2 & -3 \\ 0 & 4 & 4 & -1 \\ -6 & 0 & 3 & -5 \\ 0 & -2 & 1 & 0 \end{vmatrix}$

Expand about column 1.

$$= (-1) \begin{vmatrix} 4 & 4 & -1 \\ 0 & 3 & -5 \\ -2 & 1 & 0 \end{vmatrix} - 0$$

$$+ (-6) \begin{vmatrix} 0 & 2 & -3 \\ 4 & 4 & -1 \\ -2 & 1 & 0 \end{vmatrix} - 0$$

$$= (-1)\left\{ 4 \begin{vmatrix} 3 & -5 \\ 1 & 0 \end{vmatrix} - 0 + (-2) \begin{vmatrix} 4 & -1 \\ 3 & -5 \end{vmatrix} \right\}$$

$$+ (-6)\left\{ 0 - 4 \begin{vmatrix} 2 & -3 \\ 1 & 0 \end{vmatrix} + (-2) \begin{vmatrix} 2 & -3 \\ 4 & -1 \end{vmatrix} \right\}$$

$$= (-4)(5) + (2)(-17) + (24)(3)$$
$$+ (12)(10)$$

$$= -20 - 34 + 72 + 120$$

$$= 138$$

47. $\begin{vmatrix} 8 & 9 & 2 \\ 0 & 0 & 0 \\ 3 & 1 & 4 \end{vmatrix} = 0$

By Determinant Theorem 1, since every element of row 2 is 0, the value of the determinant is 0.

49. $\begin{vmatrix} 8 & 2 \\ 4 & 3 \end{vmatrix} = 2 \begin{vmatrix} 4 & 1 \\ 4 & 3 \end{vmatrix}$

By Theorem 4, since every element of the first row of $\begin{bmatrix} 4 & 1 \\ 4 & 3 \end{bmatrix}$ is multiplied by 2 to obtain $\begin{bmatrix} 8 & 2 \\ 4 & 3 \end{bmatrix}$, the determinant is multiplied by 2.

51. $\begin{vmatrix} 5 & -1 & 2 \\ 3 & -2 & 0 \\ -4 & 1 & 2 \end{vmatrix} = \begin{vmatrix} 5 & -1 & 2 \\ 8 & -3 & 2 \\ -4 & 1 & 2 \end{vmatrix}$

Row 2 of the matrix corresponding to the determinant on the right is the sum of rows 1 and 2 of the determinant corresponding to the determinant on the left. By Theorem 6, the determinants are equal.

53. $\begin{vmatrix} 6t & 2 & 0 \\ 1 & 5 & 3 \\ t & 2 & -1 \end{vmatrix} = 2t$

Expand about row 1.

$$6t \begin{vmatrix} 5 & 3 \\ 2 & -1 \end{vmatrix} - 2 \begin{vmatrix} 1 & 3 \\ t & -1 \end{vmatrix} + 0 \begin{vmatrix} 1 & 5 \\ t & 2 \end{vmatrix} = 2t$$

$$6t(-5 - 6) - 2(-1 - 3t) + 0 = 2t$$

$$-66t + 2 + 6t = 2t$$

$$-60t + 2 = 2t$$

$$2 = 62t$$

$$\frac{2}{62} = t$$

$$t = \frac{1}{31}$$

Solution set: $\left\{\left(\frac{1}{31}\right)\right\}$

55. $3x + y = -1$

$5x + 4y = 10$

$$D = \begin{vmatrix} 3 & 1 \\ 5 & 4 \end{vmatrix} = 7$$

$$D_x = \begin{vmatrix} -1 & 1 \\ 10 & 4 \end{vmatrix} = -14$$

$$D_y = \begin{vmatrix} 3 & -1 \\ 5 & 10 \end{vmatrix} = 35$$

$$x = \frac{D_x}{D} = \frac{-14}{7} = -2$$

$$y = \frac{D_y}{D} = \frac{35}{7} = 5$$

Solution set: $\{(-2, 5)\}$

57. $3x + 2y + z = 2$

$4x - y + 3z = -16$

$x + 3y - z = 12$

$$D = \begin{vmatrix} 3 & 2 & 1 \\ 4 & -1 & 3 \\ 1 & 3 & -2 \end{vmatrix}$$

Add 3 times column 3 to column 2.

$$D = \begin{vmatrix} 3 & 5 & 1 \\ 4 & 8 & 3 \\ 1 & 0 & -1 \end{vmatrix}$$

Add row 1 row to 3.

$$D = \begin{vmatrix} 3 & 5 & 4 \\ 4 & 8 & 0 \\ 1 & 0 & 0 \end{vmatrix}$$

Expand about row 3.

$$D = 1\begin{vmatrix} 5 & 4 \\ 8 & 7 \end{vmatrix} = 35 - 32 = 3$$

$$D_x = \begin{vmatrix} 2 & 2 & 1 \\ -16 & -1 & 3 \\ 12 & 3 & -1 \end{vmatrix}$$

Add -2 times column 3 to column 2.

$$D_x = \begin{vmatrix} 2 & 0 & 1 \\ -16 & -7 & 3 \\ 12 & 5 & -1 \end{vmatrix}$$

Add -2 times column 3 to column 1.

$$D_x = \begin{vmatrix} 0 & 0 & 1 \\ -22 & -7 & 3 \\ 14 & 5 & -1 \end{vmatrix}$$

Expand about row 1.

$$D_x = 1\begin{vmatrix} -22 & -7 \\ 14 & 5 \end{vmatrix} = -110 + 98 = -12$$

$$D_y = \begin{vmatrix} 3 & 2 & 1 \\ 4 & -16 & 3 \\ 1 & 12 & -1 \end{vmatrix}$$

Add column 1 to column 3.

$$D_y = \begin{vmatrix} 3 & 2 & 4 \\ 4 & -16 & 7 \\ 1 & 12 & 0 \end{vmatrix}$$

Add 12 times column 1 to column 2.

$$D_y = \begin{vmatrix} 3 & -34 & 4 \\ 4 & -64 & 7 \\ 1 & 0 & 0 \end{vmatrix}$$

Expand about row 3.

$$D_y = 1\begin{vmatrix} -34 & 4 \\ -64 & 7 \end{vmatrix} = -238 - (-256) = 18$$

$$D_z = \begin{vmatrix} 3 & 2 & 2 \\ 4 & -1 & -16 \\ 1 & 3 & 12 \end{vmatrix}$$

Add 4 times column 1 to column 3.

$$D_z = \begin{vmatrix} 3 & 2 & 14 \\ 4 & -1 & 0 \\ 1 & 3 & 16 \end{vmatrix}$$

Add 4 times column 2 to column 1.

$$D_z = \begin{vmatrix} 11 & 2 & 14 \\ 0 & -1 & 0 \\ 13 & 3 & 16 \end{vmatrix}$$

Expand about row 2.

$$D_z = -1 \begin{vmatrix} 11 & 14 \\ 13 & 16 \end{vmatrix} = -(176 - 182) = 6$$

$$x = \frac{D_x}{D} = \frac{-12}{3} = -4$$

$$y = \frac{D_y}{D} = \frac{18}{3} = 6$$

$$z = \frac{D_z}{D} = \frac{6}{3} = 2$$

Solution set: $\{(-4, 6, 2)\}$

59. Find the inverse of $A = \begin{bmatrix} 2 & 1 \\ 5 & 3 \end{bmatrix}$, if it exists.

$$[A \mid I_2] = \begin{bmatrix} 2 & 1 & | & 1 & 0 \\ 5 & 3 & | & 0 & 1 \end{bmatrix}$$

$$\begin{bmatrix} 1 & \frac{1}{2} & | & \frac{1}{2} & 0 \\ 5 & 3 & | & 0 & 1 \end{bmatrix} \quad \frac{1}{2}R1$$

$$\begin{bmatrix} 1 & \frac{1}{2} & | & \frac{1}{2} & 0 \\ 0 & \frac{1}{2} & | & -\frac{5}{2} & 1 \end{bmatrix} \quad -5R1 + R2$$

$$\begin{bmatrix} 1 & \frac{1}{2} & | & \frac{1}{2} & 0 \\ 0 & 1 & | & -5 & 2 \end{bmatrix} \quad 2R2$$

$$\begin{bmatrix} 1 & 0 & | & 3 & -1 \\ 0 & 1 & | & -5 & 2 \end{bmatrix} \quad -\frac{1}{2}R2 + R1$$

$$A^{-1} = \begin{bmatrix} 3 & -1 \\ -5 & 2 \end{bmatrix}$$

61. Find the inverse of $A = \begin{bmatrix} 2 & -1 & 0 \\ 1 & 0 & 1 \\ 1 & -2 & 0 \end{bmatrix}$ if it exists.

$$[A \mid I_3] = \begin{bmatrix} 2 & -1 & 0 & | & 1 & 0 & 0 \\ 1 & 0 & 1 & | & 0 & 1 & 0 \\ 1 & -2 & 0 & | & 0 & 0 & 1 \end{bmatrix}$$

$$\begin{bmatrix} 1 & -1 & -1 & | & 1 & -1 & 0 \\ 1 & 0 & 1 & | & 0 & 1 & 0 \\ 1 & -2 & 0 & | & 0 & 0 & 1 \end{bmatrix} \quad -1R2 + R1$$

$$\begin{bmatrix} 1 & -1 & -1 & | & 1 & -1 & 0 \\ 0 & 1 & 2 & | & -1 & 2 & 0 \\ 1 & -2 & 0 & | & 0 & 0 & 1 \end{bmatrix} \quad -1R1 + R2$$

$$\begin{bmatrix} 1 & -1 & -1 & | & 1 & -1 & 0 \\ 0 & 1 & 2 & | & -1 & 2 & 0 \\ 0 & -1 & 1 & | & -1 & 1 & 1 \end{bmatrix} \quad -1R1 + R3$$

$$\begin{bmatrix} 1 & -1 & -1 & | & 1 & -1 & 0 \\ 0 & 1 & 2 & | & -1 & 2 & 0 \\ 0 & 0 & 3 & | & -2 & 3 & 1 \end{bmatrix} \quad R2 + R3$$

$$\begin{bmatrix} 1 & -1 & -1 & | & 1 & -1 & 0 \\ 0 & 1 & 2 & | & -1 & 2 & 0 \\ 0 & 0 & 1 & | & -\frac{2}{3} & 1 & \frac{1}{3} \end{bmatrix} \quad \frac{1}{3}R3$$

$$\begin{bmatrix} 1 & -1 & -1 & | & 1 & -1 & 0 \\ 0 & 1 & 0 & | & \frac{1}{3} & 0 & -\frac{2}{3} \\ 0 & 0 & 1 & | & -\frac{2}{3} & 1 & \frac{1}{3} \end{bmatrix} \quad -2R3 + R2$$

$$\begin{bmatrix} 1 & 0 & -1 & | & \frac{4}{3} & -1 & -\frac{2}{3} \\ 0 & 1 & 0 & | & \frac{1}{3} & 0 & -\frac{2}{3} \\ 0 & 0 & 1 & | & -\frac{2}{3} & 1 & \frac{1}{3} \end{bmatrix} \quad R2 + R1$$

$$\begin{bmatrix} 1 & 0 & 0 & | & \frac{2}{3} & 0 & -\frac{1}{3} \\ 0 & 1 & 0 & | & \frac{1}{3} & 0 & -\frac{2}{3} \\ 0 & 0 & 1 & | & -\frac{2}{3} & 1 & \frac{1}{3} \end{bmatrix} \quad R3 + R1$$

$$A^{-1} = \begin{bmatrix} \frac{2}{3} & 0 & -\frac{1}{3} \\ \frac{1}{3} & 0 & -\frac{2}{3} \\ -\frac{2}{3} & 1 & \frac{1}{3} \end{bmatrix}$$

63. $x + y + z = 1$
$2x - y = -2$
$3y + z = 2$

$$A = \begin{bmatrix} 1 & 1 & 1 \\ 2 & -1 & 0 \\ 0 & 3 & 1 \end{bmatrix}, \quad X = \begin{bmatrix} x \\ y \\ z \end{bmatrix}, \quad B = \begin{bmatrix} 1 \\ -2 \\ 2 \end{bmatrix}$$

Find A^{-1}.

$$[A \,|\, I_3] = \begin{bmatrix} 1 & 1 & 1 & | & 1 & 0 & 0 \\ 2 & -1 & 0 & | & 0 & 1 & 0 \\ 0 & 3 & 1 & | & 0 & 0 & 1 \end{bmatrix}$$

$$\begin{bmatrix} 1 & 1 & 1 & | & 1 & 0 & 0 \\ 0 & -3 & -2 & | & -2 & 1 & 0 \\ 0 & 3 & 1 & | & 0 & 0 & 1 \end{bmatrix} \quad -2R1 + R2$$

$$\begin{bmatrix} 1 & 1 & 1 & | & 1 & 0 & 0 \\ 0 & -3 & -2 & | & -2 & 1 & 0 \\ 0 & 0 & -1 & | & -2 & 1 & 1 \end{bmatrix} \quad R2 + R3$$

$$\begin{bmatrix} 1 & 1 & 1 & | & 1 & 0 & 0 \\ 0 & 1 & \frac{2}{3} & | & \frac{2}{3} & -\frac{1}{3} & 0 \\ 0 & 0 & -1 & | & -2 & 1 & 1 \end{bmatrix} \quad -\frac{1}{3}R2$$

$$\begin{bmatrix} 1 & 0 & \frac{1}{3} & | & \frac{1}{3} & \frac{1}{3} & 0 \\ 0 & 1 & \frac{2}{3} & | & \frac{2}{3} & -\frac{1}{3} & 0 \\ 0 & 0 & -1 & | & -2 & 1 & 1 \end{bmatrix} \quad -R2 + R1$$

$$\begin{bmatrix} 1 & 0 & \frac{1}{3} & | & \frac{1}{3} & \frac{1}{3} & 0 \\ 0 & 1 & \frac{2}{3} & | & \frac{2}{3} & -\frac{1}{3} & 0 \\ 0 & 0 & 1 & | & 2 & -1 & -1 \end{bmatrix} \quad -1R3$$

$$\begin{bmatrix} 1 & 0 & 0 & | & -\frac{1}{3} & \frac{2}{3} & \frac{1}{3} \\ 0 & 1 & \frac{2}{3} & | & \frac{2}{3} & -\frac{1}{3} & 0 \\ 0 & 0 & 1 & | & 2 & -1 & -1 \end{bmatrix} \quad -\frac{1}{3}R3 + R1$$

$$\begin{bmatrix} 1 & 0 & 0 & | & -\frac{1}{3} & \frac{2}{3} & \frac{1}{3} \\ 0 & 1 & 0 & | & -\frac{2}{3} & \frac{1}{3} & \frac{2}{3} \\ 0 & 0 & 1 & | & 2 & -1 & -1 \end{bmatrix} \quad -\frac{2}{3}R3 + R2$$

$$A^{-1} = \begin{bmatrix} -\frac{1}{3} & \frac{2}{3} & \frac{1}{3} \\ -\frac{2}{3} & \frac{1}{3} & \frac{2}{3} \\ 2 & -1 & -1 \end{bmatrix}$$

$X = A^{-1}B$

$$= \begin{bmatrix} -\frac{1}{3} & \frac{2}{3} & \frac{1}{3} \\ -\frac{2}{3} & \frac{1}{3} & \frac{2}{3} \\ 2 & -1 & -1 \end{bmatrix} \begin{bmatrix} 1 \\ -2 \\ 2 \end{bmatrix}$$

$$= \begin{bmatrix} -1 \\ 0 \\ 2 \end{bmatrix}$$

Solution set: $\{(-1, 0, 2)\}$

For Exercises 65 and 67, see the answer graphs in the textbook.

65. $x + y \leq 6$
$2x - y \geq 3$

Graph the solid line $x + y = 6$, which has x-intercept 6 and y-intercept 6. Shade the region below this line.
Graph the solid line $2x - y = 3$, which has x-intercept $\frac{3}{2}$ and y-intercept -3. Shade the region below this line.
The solution set is the intersection of these two regions, which is shaded in the final graph.

67. $x^2 + y^2 \leq 144$
$x^2 + y^2 \geq 16$

Graph $x^2 + y^2 = 144$ as a solid circle with a center at the origin and radius 12. Shade the region that is the interior of this circle.

Graph $x^2 + y^2 = 16$ as a solid circle with a center at the origin and radius 4. Shade the region that is the exterior of this circle.

The solution set is the intersection of these two regions, which is the region between the two circles.

This region is shaded in the final graph.

69.   Find $x \geq 0$ and $y \geq 0$ such that

$$y \leq -3x + 5$$
$$y \geq 4x - 3$$

and $7x + 14y$ is maximized.

Graph the solid lines $y = -3x + 5$, $y = 4x - 3$, $x = 0$ (the y-axis) and $y = 0$ (the x-axis). Shade the region satisfying all of these inequalities. This is the region below the line $y = -3x + 5$, above the line $y = 4x - 3$, and in Quadrant I.

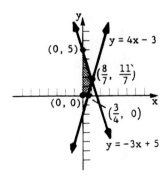

The vertices of the region of feasible solutions are $(0, 0)$, $(0, 5)$, $(3/4, 0)$, and $(8/7, 11/7)$.

The last of these points is the intersection of the lines $y = -3x + 5$ and $y = 4x - 3$. Its coordinates can be found by solving the system

$$y = -3x + 5$$
$$y = 4x - 3$$

by substitution or elimination.

Evaluate $7x + 14y$ at each vertex.

Point	Value of $7x + 14y$
$(0, 0)$	$7(0) + 14(0) = 0$
$(0, 5)$	$7(0) + 14(5) = 70$
$\left(\frac{3}{4}, 0\right)$	$7\left(\frac{3}{4}\right) + 14(0) = \frac{21}{4} = 5.25$
$\left(\frac{8}{7}, \frac{11}{7}\right)$	$7\left(\frac{8}{7}\right) + 14\left(\frac{11}{7}\right) = 30$

The maximum value is 70 at $(0, 5)$.

71.   Let x = the number of kilograms of the half and half mixture;

y = the number of kilograms of the other mixture.

Find $x \geq 0$ and $y \geq 0$ such that

$$\frac{1}{2}x + \frac{1}{3}y \leq 100$$

$$\frac{1}{2}x + \frac{2}{3}y \leq 125$$

and $6.00x + 4.80y$ is maximized.

Graph the lines $\frac{1}{2}x + \frac{1}{3}y = 100$ and $\frac{1}{2}x + \frac{2}{3}y = 125$ in quadrant I.

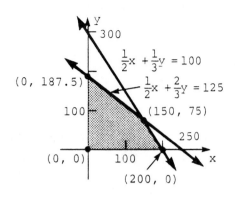

The vertices are (0, 0), (200, 0), (150, 75), and (0, 187.5).

Point	Value = 6.00x + 4.80y
(0, 0)	6.00(0) + 4.80(0) = 0
(200, 0)	6.00(200) + 4.80(0) = 1200
(150, 75)	6.00(150) + 4.80(75) = 1260
(0, 187.5)	6.00(0) + 4.80(187.5) = 900

Make 150 kg of the half and half mixture and 75 kg of the other mixture for a maximum revenue of $1260.

Chapter 6   Test

1. 3x − y = 9   (1)
    x + 2y = 10   (2)

Solve equation (2) for x.

$$x = 10 - 2y \quad (3)$$

Substitute this result into equation (1) and solve for y.

$$3(10 - 2y) - y = 9$$
$$30 - 6y - y = 9$$
$$-7y = -21$$
$$y = 3$$

Substitute y = 3 back into equation (3) to find x.

$$x = 10 - 2(3)$$
$$x = 4$$

Solution set: $\{(4, 3)\}$

2. 6x + 9y = −21   (1)
    4x + 6y = −14   (2)

Solve equation (1) for x.

$$6x = -9y - 21$$
$$x = \frac{-9y - 21}{6}$$
$$x = \frac{-3y - 7}{2} \quad (3)$$

Substitute this result into equation (2).

$$4\left(\frac{-3y - 7}{2}\right) + 6y = -14$$
$$2(-3y - 7) + 6y = -14$$
$$-6y - 14 + 6y = -14$$
$$-14 = -14$$

The equations are dependent. We express the solution set with y as the arbitrary variable.

Solution set: $\left\{\left(\frac{-3y - 7}{2}, y\right)\right\}$

3. $\frac{1}{4}x - \frac{1}{3}y = -\frac{5}{12}$ (1)

    $\frac{1}{10}x + \frac{1}{5}y = \frac{1}{2}$ (2)

To eliminate fractions, multiply equation (1) by 12 and equation (2) by 10.

$$3x + 4y = -5 \quad (3)$$
$$x + 2y = 5 \quad (4)$$

Multiply equation (4) by 2 and add the result to equation (3).

$$3x - 4y = -5$$
$$\underline{2x + 4y = 10}$$
$$5x \qquad = 5$$
$$x = 1$$

Substitute $x = 1$ in equation (4) to find y.

$$1 + 2y = 5$$
$$2y = 4$$
$$y = 2$$

Solution set: $\{(1, 2)\}$

4. $\quad x - 2y = 4 \quad (1)$
$\quad -2x + 4y = 6 \quad (2)$

Multiply equation (1) by 2 and add the result to equation (2).

$$2x - 4y = 8$$
$$\underline{-2x + 4y = 6}$$
$$0 = 14$$

The system is inconsistent.
Solution set: $\emptyset$

5. $\quad 2x + y + z = 3 \quad (1)$
$\quad x + 2y - z = 3 \quad (2)$
$\quad 3x - y + z = 5 \quad (3)$

Eliminate z first. Add equations (1) and (2).

$$3x + 3y = 6 \quad (4)$$

Add equations (2) and (3).

$$4x + y = 8 \quad (5)$$

Multiply equation (5) by -3 and add the result to equation (4).

$$3x + 3y = 6$$
$$\underline{-12x - 3y = -24}$$
$$-9x \qquad = -18$$
$$x = 2$$

Substitute $x = 2$ and in equation (5) to find y.

$$4(2) + y = 8$$
$$y = 0$$

Substitute $x = 2$ and $y = 0$ in equation (1) to find z.

$$2(2) + 0 + z = 3$$
$$z = -1$$

Solution set: $\{(2, 0, -1)\}$

6. $\quad 2x^2 + y^2 = 6 \quad (1)$
$\quad x^2 - 4y^2 = -15 \quad (2)$

$$8x^2 + 4y^2 = 24 \quad \textit{Multiply (1) by 4}$$
$$\underline{x^2 - 4y^2 = -15} \quad (2)$$
$$9x^2 \qquad = 9$$
$$x^2 = 1$$
$$x = \pm 1$$

Substitute these values into equation (1) and solve for y.

If $x = 1$, $\quad 2(1)^2 + y^2 = 6$
$$2 + y^2 = 6$$
$$y^2 = 4$$
$$y = \pm 2.$$

Thus, (1, 2) and (1, -2) are solutions.

If $x = -1$, $2(-1)^2 + y^2 = 6$

$$2 + y^2 = 6$$
$$y^2 = 4$$
$$y = \pm 2.$$

Thus, $(-1, 2)$ and $(-1, -2)$ are solutions.

Solution set:
$\{(1, 2), (-1, 2), (1, -2), (-1, -2)\}$

7. $x^2 + y^2 = 25$ (1)
    $x + y = 7$ (2)

Solve equation (2) for x.

$$x = 7 - y$$

Substitute this result into equation (1).

$$(7 - y)^2 + y^2 = 25$$
$$49 - 14y + y^2 + y^2 = 25$$
$$2y^2 - 14y + 24 = 0$$
$$y^2 - 7y + 12 = 0$$
$$(y - 3)(y - 4) = 0$$
$$y = 3 \quad \text{or} \quad y = 4$$

If $y = 3$, $x = 7 - 3 = 4$.
If $y = 4$, $x = 7 - 4 = 3$.

Solution set: $\{(3, 4), (4, 3)\}$

8. The system will have exactly one solution if the line is tangent to the circle. See the answer sketch in the textbook.

9. Let x and y represent the numbers.

    $x + y = -1$ (1)
    $x^2 + y^2 = 61$ (2)

Rewrite equation (1) as $y = -x - 1$ and substitute $-x - 1$ for y in (2).

$$x^2 + (-x - 1)^2 = 61$$
$$x^2 + x^2 + 2x + 1 = 61$$
$$2x^2 + 2x - 60 = 0$$
$$x^2 + x - 30 = 0$$
$$(x + 6)(x - 5) = 0$$
$$x = -6 \quad \text{or} \quad x = 5$$

Substitute these values in equation (1) to find the corresponding values of y.

If $x = -6$, $-6 + y = -1$ or $y = 5$.
If $x = 5$, $5 + y = -1$ or $y = -6$.
The same pair of numbers results from both cases.
The numbers are 5 and -6.

10. $3a - 2b = 13$
     $4a - b = 19$

Write the augmented matrix.

$$\begin{bmatrix} 3 & -2 & | & 13 \\ 4 & -1 & | & 19 \end{bmatrix}$$

$$\begin{bmatrix} 1 & -\frac{2}{3} & | & \frac{13}{3} \\ 4 & -1 & | & 19 \end{bmatrix} \quad \frac{1}{3}R1$$

$$\begin{bmatrix} 1 & -\frac{2}{3} & | & \frac{13}{3} \\ 0 & \frac{5}{3} & | & \frac{5}{3} \end{bmatrix} \quad -4R1 + R2$$

$$\begin{bmatrix} 1 & -\frac{2}{3} & | & \frac{13}{3} \\ 0 & 1 & | & 1 \end{bmatrix} \quad \frac{3}{5}R2$$

$$\begin{bmatrix} 1 & 0 & | & 5 \\ 0 & 1 & | & 1 \end{bmatrix} \quad \frac{2}{3}R2 + R1$$

Solution set: $\{(5, 1)\}$

11. $3a - 4b + 2c = 15$ (1)

$2a - b + c = 13$ (2)

$a + 2b - c = 5$ (3)

Write the augmented matrix.

$$\begin{bmatrix} 3 & -4 & 2 & | & 15 \\ 2 & -1 & 1 & | & 13 \\ 1 & 2 & -1 & | & 5 \end{bmatrix}$$

$$\begin{bmatrix} 1 & 2 & -1 & | & 5 \\ 2 & -1 & 1 & | & 13 \\ 3 & -4 & 2 & | & 15 \end{bmatrix} \quad R1 \leftrightarrow R3$$

$$\begin{bmatrix} 1 & 2 & -1 & | & 5 \\ 0 & -5 & 3 & | & 3 \\ 0 & -10 & 5 & | & 0 \end{bmatrix} \quad \begin{matrix} -2R1 + R2 \\ -3R1 + R3 \end{matrix}$$

$$\begin{bmatrix} 1 & 2 & -1 & | & 5 \\ 0 & 1 & -\frac{3}{5} & | & -\frac{3}{5} \\ 0 & -10 & 5 & | & 0 \end{bmatrix} \quad -\frac{1}{5}R2$$

$$\begin{bmatrix} 1 & 2 & -1 & | & 5 \\ 0 & 1 & -\frac{3}{5} & | & -\frac{3}{5} \\ 0 & 0 & -1 & | & -6 \end{bmatrix} \quad 10R2 + R3$$

$$\begin{bmatrix} 1 & 2 & -1 & | & 5 \\ 0 & 1 & -\frac{3}{5} & | & -\frac{3}{5} \\ 0 & 0 & 1 & | & 6 \end{bmatrix} \quad -1R3$$

$$\begin{bmatrix} 1 & 2 & 0 & | & 11 \\ 0 & 1 & 0 & | & 3 \\ 0 & 0 & 1 & | & 6 \end{bmatrix} \quad \begin{matrix} R3 + R1 \\ \frac{3}{5}R3 + R2 \end{matrix}$$

$$\begin{bmatrix} 1 & 0 & 0 & | & 5 \\ 0 & 1 & 0 & | & 3 \\ 0 & 0 & 1 & | & 6 \end{bmatrix} \quad -2R2 + R1$$

Solution set: $\{(5, 3, 6)\}$

12. Since $y = ax^2 + bx + c$, and the points $(-1, -.95)$, $(1, -.35)$, and $(2, -.8)$ are on the graph, we have the following equations:

$$-.95 = a(-1)^2 + b(-1) + c$$

$$-.35 = a(1)^2 + b(1) + c$$

$$-.8 = a(2)^2 + b(2) + c.$$

This becomes the following system

$a - b + c = -.95$ (1)

$a + b + c = -.35$ (2)

$4a + 2b + c = -.8.$ (3)

First, eliminate b by adding equations (1) and (2).

$a - b + c = -.95$ (1)

$\underline{a + b + c = -.35}$ (2)

$2a + 2c = -1.3$ (4)

Next, multiply equation (2) by -2 and add the result to equation (3).

$-2a - 2b - 2c = .7$

$\underline{4a + 2b + c = -.8}$

$2a - c = -.1$ (5)

We now solve the system

$2a + 2c = -1.3$ (4)

$2a - c = -.1.$ (5)

Multiply equation (5) by -1 and add the result to equation (1).

$2a + 2c = -1.3$

$\underline{-2a + c = .1}$

$3c = -1.2$

$c = -.4$

Substitute this value into equation (5).

$2a - (-.4) = -.1$

$2a + .4 = -.1$

$2a = -.5$

$a = -.25$

Substitute these values into equation (2).

$(-.25) + b + (-.4) = -.35$

$b - .65 = -.35$

$b = .3$

The equation of the parabola is

$$y = -.25x^2 + .3x - .4.$$

13. Let x, y, and z be the three numbers.

The sum of the three numbers is 2, so

$$x + y + z = 2. \qquad (1)$$

The first number is equal to the sum of the other two, so

$$x = y + z. \qquad (2)$$

The third number is the result of subtracting the first from the second, so

$$z = y - x. \qquad (3)$$

These three equations make up the following system:

$$x + y + z = 2 \qquad (1)$$
$$x - y - z = 0 \qquad (4)$$
$$x - y + z = 0. \qquad (5)$$

Add equations (1) and (4).

$$2x = 2$$
$$x = 1$$

Add equations (4) and (5).

$$2x - 2y = 0 \qquad (6)$$

Substitute x = 1 in equation (6) to find y.

$$2(1) - 2y = 0$$
$$y = 1$$

Substitute x = 1 and y = 1 in equation (1) to find z.

$$1 + 1 + z = 2$$
$$z = 0$$

The first number is 1, the second number is 1, the third number is 0.

14. $\begin{bmatrix} 5 & x + 6 \\ 0 & 4 \end{bmatrix} = \begin{bmatrix} y - 2 & 4 - x \\ 0 & w + 7 \end{bmatrix}$

All corresponding elements, position by position, of the two matrices must be equal.

$$5 = y - 2 \quad x + 6 = 4 - x \quad 4 = w + 7$$
$$7 = y \qquad\qquad 2x = -2 \qquad -3 = w$$
$$\qquad\qquad\qquad x = -1$$

Thus, x = -1, y = 7, and w = -3.

15. $3\begin{bmatrix} 2 & 3 \\ 1 & -4 \\ 5 & 9 \end{bmatrix} - \begin{bmatrix} -2 & 6 \\ 3 & -1 \\ 0 & 8 \end{bmatrix}$

$$= \begin{bmatrix} 6 & 9 \\ 3 & -12 \\ 15 & 27 \end{bmatrix} + \begin{bmatrix} 2 & -6 \\ -3 & 1 \\ 0 & -8 \end{bmatrix}$$

$$= \begin{bmatrix} 8 & 3 \\ 0 & -11 \\ 5 & 19 \end{bmatrix}$$

16. $\begin{bmatrix} 1 \\ 2 \end{bmatrix} + \begin{bmatrix} 4 \\ -6 \end{bmatrix} + \begin{bmatrix} 2 & 8 \\ -7 & 5 \end{bmatrix}$

The first two matrices are 2 × 1 and the third is 2 × 2. Only matrices of the same size can be added, so it is not possible to find this sum.

17. $\begin{bmatrix} 2 & 1 & -3 \\ 4 & 0 & 5 \end{bmatrix}\begin{bmatrix} 1 & 3 \\ 2 & 4 \\ 3 & -2 \end{bmatrix}$

$$= \begin{bmatrix} 2(1)+1(2)+(-3)(3) & 2(3)+1(4)+(-3)(-2) \\ 4(1)+0(2)+5(3) & 4(3)+0(4)+5(-2) \end{bmatrix}$$

$$= \begin{bmatrix} -5 & 16 \\ 19 & 2 \end{bmatrix}$$

18. $\begin{bmatrix} 2 & -4 \\ 3 & 5 \end{bmatrix} \begin{bmatrix} 4 \\ 2 \\ 7 \end{bmatrix}$

The first matrix is 2 × 2 and the second is 1 × 3. The product of two matrices can be found only if the number of columns of the first matrix is the same as the number of rows of the second matrix. The first matrix has two columns and the second has one row, so it is not possible to find this product.

19. There are associative, distributive, and identity properties that apply to multiplication of matrices, but matrix multiplication is not commutative. The correct choice is (a).

20. $\begin{vmatrix} 6 & 8 \\ 2 & -7 \end{vmatrix} = 6(-7) - 2(8)$
$= -58$

21. $\begin{vmatrix} 2 & 0 & 8 \\ -1 & 7 & 9 \\ 12 & 5 & -3 \end{vmatrix}$

This determinant may be evaluated by expanding about any row or any column. Choose the first row or second column because they contain a 0. We will expand by minors about the first row.

$\begin{vmatrix} 2 & 0 & 8 \\ -1 & 7 & 9 \\ 12 & 5 & -3 \end{vmatrix}$

$= 2\begin{vmatrix} 7 & 9 \\ 5 & -3 \end{vmatrix} - 0\begin{vmatrix} -1 & 9 \\ 12 & -3 \end{vmatrix} + 8\begin{vmatrix} -1 & 7 \\ 12 & 5 \end{vmatrix}$

$= 2[7(-3) - 5(9) - 0$
$\qquad + 8[(-1)(5) - 12(7)]$

$= 2(-21 - 45) + 8(-5 - 84)$

$= 2(-66) + 8(-89)$

$= -132 - 712$

$= -844$

22. $\begin{vmatrix} 6 & 7 \\ -5 & 2 \end{vmatrix} = -\begin{vmatrix} -5 & 2 \\ 6 & 7 \end{vmatrix}$

Rows 1 and 2 have been interchanged. Interchanging two rows of a matrix reverses the sign of the determinant (Theorem 3).

23. $\begin{vmatrix} 7 & 2 & -1 \\ 5 & -4 & 3 \\ 6 & -2 & 1 \end{vmatrix} = \begin{vmatrix} 7 & 2 & -1 \\ -7 & 0 & 1 \\ 13 & 0 & 0 \end{vmatrix}$

−2 times row 3 has been added to row 2, and row 1 has been added to row 3. The determinant of a matrix is unchanged if a multiple of a row of the matrix is added to the corresponding elements of another row (Theorem 6).

24. $2x - 3y = -33$
$4x + 5y = 11$

$D = \begin{vmatrix} 2 & -3 \\ 4 & 5 \end{vmatrix} = 2(5) - 4(-3) = 22$

$D_x = \begin{vmatrix} -33 & -3 \\ 11 & 5 \end{vmatrix} = -33(5) - 11(-3)$
$\qquad\qquad\qquad = -132$

$D_y = \begin{vmatrix} 2 & -33 \\ 4 & 11 \end{vmatrix} = 2(11) - 4(33) = 154$

$$x = \frac{D_x}{D} = \frac{-132}{22} = -6$$

$$y = \frac{D_y}{D} = \frac{154}{22} = 7$$

Solution set: $\{(-6, 7)\}$

25. $x + y - z = -4$
$2x - 3y - z = 5$
$x + 2y + 2z = 3$

Expand about row 1.

$$D = \begin{vmatrix} 1 & 1 & -1 \\ 2 & -3 & -1 \\ 1 & 2 & 2 \end{vmatrix}$$

$$= 1\begin{vmatrix} -3 & -1 \\ 2 & 2 \end{vmatrix} - 1\begin{vmatrix} 2 & -1 \\ 1 & 2 \end{vmatrix}$$

$$+ (-1)\begin{vmatrix} 2 & -3 \\ 1 & 2 \end{vmatrix}$$

$$= -4 - 1(5) - 1(7)$$

$$= -16$$

Expand about row 1.

$$D_x = \begin{vmatrix} -4 & 1 & -1 \\ 5 & -3 & -1 \\ 3 & 2 & 2 \end{vmatrix}$$

$$= 4\begin{vmatrix} -3 & -1 \\ 2 & 2 \end{vmatrix} - 1\begin{vmatrix} 5 & -1 \\ 3 & 2 \end{vmatrix}$$

$$+ (-1)\begin{vmatrix} 5 & -3 \\ 3 & 2 \end{vmatrix}$$

$$= -4(-4) - 1(13) - 1(19)$$

$$= -16$$

Expand about row 1.

$$D_y = \begin{vmatrix} 1 & -4 & -1 \\ 2 & 5 & -1 \\ 1 & 3 & 2 \end{vmatrix}$$

$$= 1\begin{vmatrix} 5 & -1 \\ 3 & 2 \end{vmatrix} - (-4)\begin{vmatrix} 2 & -1 \\ 1 & 2 \end{vmatrix}$$

$$+ (-1)\begin{vmatrix} 2 & 5 \\ 1 & 3 \end{vmatrix}$$

$$= 13 + 4(5) - 1(1)$$

$$= 32$$

Expand about row 1.

$$D_z = \begin{vmatrix} 1 & 1 & -4 \\ 2 & -3 & 5 \\ 1 & 2 & 3 \end{vmatrix}$$

$$= 1\begin{vmatrix} -3 & 5 \\ 2 & 3 \end{vmatrix} - 1\begin{vmatrix} 2 & 5 \\ 1 & 3 \end{vmatrix}$$

$$+ (-4)\begin{vmatrix} 2 & -3 \\ 1 & 2 \end{vmatrix}$$

$$= 1(-19) - 1(1) - 4(7)$$

$$= -48$$

$$x = \frac{D_x}{D} = \frac{-16}{-16} = 1$$

$$x = \frac{D_y}{D} = \frac{32}{-16} = -2$$

$$z = \frac{D_z}{D} = \frac{-48}{-16} = 3$$

Solution set: $\{(1, -2, 3)\}$

26. Find the inverse of $A = \begin{bmatrix} -8 & 5 \\ 3 & -2 \end{bmatrix}$, if it exists.

Form the augmented matrix $[A|I_2]$.

$$[A|I_2] = \begin{bmatrix} -8 & 5 & | & 1 & 0 \\ 3 & -2 & | & 0 & 1 \end{bmatrix}$$

Perform row transformations on $[A|I_2]$ until a matrix of the form $[I_2|B]$ is obtained.

$$[A|I_2] = \begin{bmatrix} 1 & -\frac{5}{8} & | & -\frac{1}{8} & 0 \\ 3 & -2 & | & 0 & 1 \end{bmatrix} \quad -\frac{1}{8}R1$$

$$\begin{bmatrix} 1 & -\frac{5}{8} & | & -\frac{1}{8} & 0 \\ 0 & -\frac{1}{8} & | & \frac{3}{8} & 1 \end{bmatrix} \quad -3R1 + R2$$

$$\begin{bmatrix} 1 & 0 & | & -2 & -5 \\ 0 & -\frac{1}{8} & | & \frac{3}{8} & 1 \end{bmatrix} \quad -5R2 + R1$$

$$[I_2 | B] = \begin{bmatrix} 1 & 0 & | & -2 & -5 \\ 0 & 1 & | & -3 & -8 \end{bmatrix} \quad -8R2$$

$$A^{-1} = B = \begin{bmatrix} -2 & -5 \\ -3 & -8 \end{bmatrix}$$

27. Find the inverse of $A = \begin{bmatrix} 4 & 12 \\ 2 & 6 \end{bmatrix}$, if it exists.

$$[A | I_2] = \begin{bmatrix} 4 & 12 & | & 1 & 0 \\ 2 & 6 & | & 0 & 1 \end{bmatrix}$$

$$\begin{bmatrix} 1 & 3 & | & \frac{1}{4} & 0 \\ 2 & 6 & | & 0 & 1 \end{bmatrix} \quad \frac{1}{4}R1$$

$$\begin{bmatrix} 1 & 3 & | & \frac{1}{4} & 0 \\ 0 & 0 & | & -\frac{1}{2} & 1 \end{bmatrix} \quad -2R1 + R2$$

The second row, second column element is now 0, so the desired transformation cannot be completed. Therefore, the inverse of the given matrix does not exist.

28. Find the inverse of

$$A = \begin{bmatrix} 1 & 3 & 4 \\ 2 & 7 & 8 \\ -2 & -5 & -7 \end{bmatrix}, \text{ if it exists.}$$

$$[A | I_3] = \begin{bmatrix} 1 & 3 & 4 & | & 1 & 0 & 0 \\ 2 & 7 & 8 & | & 0 & 1 & 0 \\ -2 & -5 & -7 & | & 0 & 0 & 1 \end{bmatrix}$$

$$\begin{bmatrix} 1 & 3 & 4 & | & 1 & 0 & 0 \\ 0 & 1 & 0 & | & -2 & 1 & 0 \\ -2 & -5 & -7 & | & 0 & 0 & 1 \end{bmatrix} \quad -2R1 + R2$$

$$\begin{bmatrix} 1 & 3 & 4 & | & 1 & 0 & 0 \\ 0 & 1 & 0 & | & -2 & 1 & 0 \\ 0 & 1 & 1 & | & 2 & 0 & 1 \end{bmatrix} \quad 2R1 + R3$$

$$\begin{bmatrix} 1 & 0 & 1 & | & -5 & 0 & -3 \\ 0 & 1 & 0 & | & -2 & 1 & 0 \\ 2 & 1 & 1 & | & 2 & 0 & 1 \end{bmatrix} \quad -3R3 + R1$$

$$\begin{bmatrix} 1 & 0 & 1 & | & -5 & 0 & -3 \\ 0 & 1 & 0 & | & -2 & 1 & 0 \\ 0 & 0 & 1 & | & -4 & -1 & 1 \end{bmatrix} \quad -1R2 + R3$$

$$\begin{bmatrix} 1 & 0 & 0 & | & -9 & 1 & -4 \\ 0 & 1 & 0 & | & -2 & 1 & 0 \\ 0 & 0 & 1 & | & 4 & -1 & 1 \end{bmatrix} \quad -1R3 + R1$$

$$A^{-1} = \begin{bmatrix} -9 & 1 & -4 \\ -2 & 1 & 0 \\ 4 & -1 & 1 \end{bmatrix}$$

29. $2x + y = -6$

$3x - y = -29$

Represent the system as a matrix equation as follows.

Let $A = \begin{bmatrix} 2 & 1 \\ 3 & -1 \end{bmatrix}$, $X = \begin{bmatrix} x \\ y \end{bmatrix}$,

$B = \begin{bmatrix} -6 \\ -29 \end{bmatrix}$.

Then $AX = B$.

Find A^{-1}.

$$\begin{bmatrix} 2 & 1 & | & 1 & 0 \\ 3 & -1 & | & 0 & 1 \end{bmatrix}$$

$$\begin{bmatrix} 1 & \frac{1}{2} & | & \frac{1}{2} & 0 \\ 3 & -1 & | & 0 & 1 \end{bmatrix} \quad \frac{1}{2}R1$$

$$\begin{bmatrix} 1 & \frac{1}{2} & | & \frac{1}{2} & 0 \\ 0 & -\frac{5}{2} & | & -\frac{3}{2} & 1 \end{bmatrix} \quad -3R1 + R2$$

$$\begin{bmatrix} 1 & 0 & | & \frac{1}{5} & \frac{1}{5} \\ 0 & -\frac{5}{2} & | & -\frac{3}{2} & 1 \end{bmatrix} \quad \frac{1}{5}R2 + R1$$

$$\begin{bmatrix} 1 & 0 & | & \frac{1}{5} & \frac{1}{5} \\ 0 & 1 & | & \frac{3}{5} & -\frac{2}{5} \end{bmatrix} \quad -\frac{2}{5}R2$$

Thus,

$$A^{-1} = \begin{bmatrix} \frac{1}{5} & \frac{1}{5} \\ \frac{3}{5} & -\frac{2}{5} \end{bmatrix}.$$

$$A^{-1}B = \begin{bmatrix} \frac{1}{5} & \frac{1}{5} \\ \frac{3}{5} & -\frac{2}{5} \end{bmatrix}\begin{bmatrix} -6 \\ -29 \end{bmatrix}$$

$$= \begin{bmatrix} -\frac{6}{5} + \left(-\frac{29}{5}\right) \\ -\frac{18}{5} + \frac{58}{5} \end{bmatrix} = \begin{bmatrix} -\frac{35}{5} \\ \frac{40}{5} \end{bmatrix}$$

$$= \begin{bmatrix} -7 \\ 8 \end{bmatrix}$$

Since $X = A^{-1}B$,

$$X = \begin{bmatrix} -7 \\ 8 \end{bmatrix}.$$

Solution set: $\{(-7, 8)\}$

30. $x + y = 5$
     $y - 2z = 23$
     $x + 3z = -27$

Let $A = \begin{bmatrix} 1 & 1 & 0 \\ 0 & 1 & -2 \\ 1 & 0 & 3 \end{bmatrix}$,

$$X = \begin{bmatrix} x \\ y \\ z \end{bmatrix},$$

$$B = \begin{bmatrix} 5 \\ 23 \\ -27 \end{bmatrix}.$$

Then $AX = B$.

Find A^{-1}.

$$\begin{bmatrix} 1 & 1 & 0 & | & 1 & 0 & 0 \\ 0 & 1 & -2 & | & 0 & 1 & 0 \\ 1 & 0 & 3 & | & 0 & 0 & 1 \end{bmatrix}$$

$$\begin{bmatrix} 1 & 1 & 0 & | & 1 & 0 & 0 \\ 0 & 1 & -2 & | & 0 & 1 & 0 \\ 0 & -1 & 3 & | & -1 & 0 & 1 \end{bmatrix} \quad -R1 + R3$$

$$\begin{bmatrix} 1 & 0 & 3 & | & 0 & 0 & 1 \\ 0 & 1 & -2 & | & 0 & 1 & 0 \\ 0 & -1 & 3 & | & -1 & 0 & 1 \end{bmatrix} \quad R3 + R1$$

$$\begin{bmatrix} 1 & 0 & 3 & | & 0 & 0 & 1 \\ 0 & 1 & -2 & | & 0 & 1 & 0 \\ 0 & 0 & 1 & | & -1 & 1 & 1 \end{bmatrix} \quad R2 + R3$$

$$\begin{bmatrix} 1 & 0 & 0 & | & 3 & -3 & -2 \\ 0 & 1 & -2 & | & 0 & 1 & 0 \\ 0 & 0 & 1 & | & -1 & 1 & 1 \end{bmatrix} \quad -3R1 + R1$$

$$\begin{bmatrix} 1 & 0 & 0 & | & 3 & -3 & -2 \\ 0 & 1 & 0 & | & -2 & 3 & 2 \\ 0 & 0 & 1 & | & -1 & 1 & 1 \end{bmatrix} \quad 2R3 + R2$$

Thus,

$$A^{-1} = \begin{bmatrix} 3 & -3 & -2 \\ -2 & 3 & 2 \\ -1 & 1 & 1 \end{bmatrix}.$$

$$A^{-1}B = \begin{bmatrix} 3 & -3 & -2 \\ -2 & 3 & 2 \\ -1 & 1 & 1 \end{bmatrix}\begin{bmatrix} 5 \\ 23 \\ -27 \end{bmatrix}$$

$$= \begin{bmatrix} 0 \\ 5 \\ -9 \end{bmatrix} = X$$

Solution set: $\{(0, 5, -9)\}$

31. $x - 3y \geq 6$
     $y^2 \leq 16 - x^2$

Graph $x - 3y = 6$ as a solid line with x-intercept 6 and y-intercept of -2. Shade the region below the line.

Graph $y^2 = 16 - x^2$ or $x^2 + y^2 = 16$ as a solid circle with a center at the origin and radius 4. Shade the region which is the interior of the circle.

The solution set is the intersection at these two regions, which is the region shaded in the final graph. See the answer graph in the back of the textbook.

32. Find $x \geq 0$ and $y \geq 0$ such that

$$x + 2y \leq 24$$
$$3x + 4y \leq 60$$

and $2x + 3y$ is maximized.

Graph the equation $x + 2y = 24$ as a solid line with x-intercept 24 and y-intercept 12.
Graph the equation $3x + 4y = 60$ as a solid line with x-intercept 20 and y-intercept 15.
Graph the equation $x = 0$ (the y-axis).
Graph the equation $y = 0$ (the x-axis).
The region of feasible solutions is the intersection of the four regions.

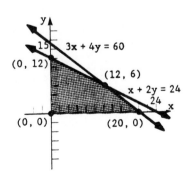

Three vertices are $(0, 0)$, $(0, 12)$, and $(20, 0)$.
To find the fourth vertex, we must solve the system

$$x + 2y = 24 \quad (1)$$
$$3x + 4y = 60. \quad (2)$$

Eliminate x. Multiply equation (1) by -3 and add the result to equation (2).

$$
\begin{array}{r}
-3x - 6y = -72 \\
3x + 4y = 60 \\
\hline
-2y = -12 \\
y = 6
\end{array}
$$

Substitute this value into equation (1).

$$x + 2(6) = 24$$
$$x + 12 = 24$$
$$x = 24$$

The fourth vertex is $(12, 6)$.
Find the value of $2x + 3y$ at each vertex.

Point	Value of $2x + 3y$
$(0, 0)$	$2(0) + 3(0) = 0$
$(0, 12)$	$2(0) + 3(12) = 36$
$(12, 6)$	$2(12) + 3(6) = 42$
$(20, 0)$	$2(20) + 3(0) = 40$

The maximum value is 42 at $(12, 6)$.

33. Let x = number of VIP rings;
y = number of SST rings.

We translate the given information into the following linear programming problem.
Maximize $30x + 40y$ if

$$x + y \leq 24$$
$$3x + 2y \leq 60$$
$$x \geq 0, y \geq 0.$$

Graph $x + y = 24$ as a solid line with x-intercept 24 and y-intercept 24.

Graph $3x + 2y \leq 60$ as a solid line with x-intercept 20 and y-intercept 30.

Graph $x = 0$ (y-axis) as a solid line.

Graph $y = 0$ (x-axis) as a solid line.

The region of feasible solutions is the intersection of the four regions.

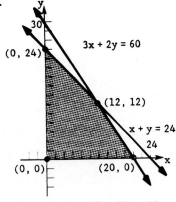

Three vertices are (0, 0), (0, 24), and (20, 0).

To find the fourth vertex, we must solve the system

$$x + y = 24 \quad (1)$$
$$3x + 2y = 60. \quad (2)$$

Eliminate x. Multiply equation (1) by -3 and add the result to equation (2).

$$\begin{array}{r} -3x - 3y = -72 \\ \underline{3x + 2y = 60} \\ -y = -12 \\ y = 12 \end{array}$$

Substitute this value into equation (1).

$$x + 12 = 24$$
$$x = 12$$

The fourth vertex is (12, 12).

Point	Value of $30x + 40y$
(0, 0)	$30(0) + 40(0) = 0$
(0, 24)	$30(0) + 40(24) = 960$
(12, 12)	$30(12) + 40(12) = 840$
(20, 0)	$30(20) + 40(0) = 600$

The maximum profit is $960 when no VIP rings are made and 24 SST rings are made.

CHAPTER 7 ANALYTIC GEOMETRY

Section 7.1

For Exercises 3–17, see the answer graphs in the textbook.

3. $x = -y^2$

$= -1(y - 0)^2 + 0$

The vertex is $(0, 0)$. The graph opens to the left and has the same shape as $y = x^2$. The axis is the horizontal line $y = 0$ (the x-axis). Use the vertex and axis and plot a few additional points.

x	-4	-1	0	-1	-4
y	-2	-1	0	1	2

5. $x = (y - 3)^2$

The vertex is $(0, 3)$. The graph opens to the right and has the same shape as $x = y^2$. It is a translation of the graph of $y = y^2$ 3 units up. The axis is the horizontal line $y = 3$.

x	4	1	0	1	4
y	1	2	3	4	5

7. $x = (y - 4)^2 + 2$

The vertex is $(2, 4)$. The graph opens to the right and has the same shape as $x = y^2$. It is a translation of the graph of $x = y^2$ 2 units to the right and 4 units up. The axis is the horizontal line $y = 4$.

x	6	3	2	3	6
y	2	3	4	5	6

9. $x = -3(y - 1)^2 + 2$

The vertex is $(2, 1)$. The graph opens to the left and has the same shape as $x = -3y^2$. It is a translation of the graph of $x = -3y^2$ 1 unit up and 2 units to the right. The axis is the horizontal line $y = 1$.

x	-10	-1	2	-1	-10
y	-1	0	1	2	3

11. $x = \frac{1}{2}(y - 1)^2 + 4$

The vertex is $(4, 1)$. The graph opens to the right and has the same shape as $x = (1/2)y^2$. It is a translation of the graph of $x = (1/2)y^2$ 4 units to the right and 1 unit up. The axis is the horizontal line $y = 1$.

x	6	4.5	4	4.5	6
y	-1	0	1	2	3

13. $x = y^2 + 4y + 2$

Complete the square on y to find the vertex and axis.

$$x = (y^2 + 4y \quad) + 2$$
$$= (y^2 + 4y + 4 - 4) + 2$$
$$= (y^2 + 4y + 4) - 4 + 2$$
$$= (y + 2)^2 - 2$$

The vertex is $(-2, -2)$ and the axis is the horizontal line $y = -2$. The graph opens to the right and has the same shape as $x = y^2$. It is a translation of the graph of $x = y^2$ 2 units to the left and 2 units down.

x	2	-1	-2	-1	2
y	-4	-3	-2	-1	0

15. $x = -4y^2 - 4y + 3$

Complete the square on y to find the vertex and axis.

$$x = -4(y^2 + y \quad) + 3$$
$$= -4\left(y^2 + y + \frac{1}{4} - \frac{1}{4}\right) + 3$$
$$= -4\left(y^2 + y + \frac{1}{4}\right) - 4\left(-\frac{1}{4}\right) + 3$$
$$= -4\left(y^2 + y + \frac{1}{4}\right) + 1 + 3$$
$$= -4\left(y + \frac{1}{2}\right)^2 + 4$$

The vertex is $(4, -1/2)$ and the axis is the horizontal line $y = -1/2$. The graph opens to the left and has the same shape as $x = -4y^2$. It is a translation of the graph of $x = -4y^2$ 4 units to the right and 1/2 unit down.

x	-5	0	4	0	-5
y	$-2\frac{1}{2}$	$-1\frac{1}{2}$	$-\frac{1}{2}$	$\frac{1}{2}$	1

17. $2x = y^2 - 4y + 6$

$$x = \frac{1}{2}y^2 - 2y + 3$$

Complete the square on y to find the vertex and axis.

$$x = \frac{1}{2}(y^2 - 4y + 4 - 4) + 3$$
$$= \frac{1}{2}(y^2 - 4y + 4) - 2 + 3$$
$$= \frac{1}{2}(y - 2)^2 + 1$$

The vertex is $(1, 2)$, and the axis is the horizontal line $y = 2$. The graph opens to the right and has the same shape as $x = (1/2)y^2$. It is a translation of the graph of $x = (1/2)y^2$ 1 unit to the right and 2 units up.

x	3	1.5	1	1.5	3
y	0	1	2	3	4

19. $x^2 = 24y$

The equation has the form $x^2 = 4py$, with $4p = 24$, so $p = 6$. The parabola is vertical, with focus $(0, 6)$, directrix $y = -6$, and the y-axis as axis of the parabola.

21.
$$y = -4x^2$$
$$-\frac{1}{4}y = x^2$$
$$-\frac{1}{4} = 4p$$
$$-\frac{1}{16} = p$$

Focus: $\left(0, -\frac{1}{16}\right)$

Directrix: $y = \frac{1}{16}$

Axis: y-axis

23. $x = -32y^2$

$-\dfrac{1}{32}x = y^2$

This equation has the form $y^2 = 4px$, with $4p = -1/32$, so $p = -1/128$. The parabola is horizontal, with focus $(-1/128, 0)$, directrix $x = 1/128$, and the x–axis as axis of the parabola.

25. $x = -\dfrac{1}{4}y^2$

$-4x = y^2$

$4p = -4$

$p = -1$

Focus: $(-1, 0)$
Directrix: $x = 1$
Axis: x–axis

27. $(y - 3)^2 = 12(x - 1)$

$4p = 12$

$p = 3$

The vertex is $(1, 3)$.
The parabola opens to the right, so the focus is 3 units to the right of the vertex.
Focus: $(4, 3)$
Directrix: $x = -2$
Axis: $y = 3$

29. $(x - 7)^2 = 16(y + 5)$

$4p = 16$

$p = 4$

The vertex is $(7, -5)$.
The parabola opens upward, so the focus is 4 units above the vertex and the directrix is 5 units below the vertex.

Focus: $(7, -1)$
Directrix: $y = -9$
Axis: $x = 7$

31. Focus $(5, 0)$, vertex at the origin

Since the focus $(5, 0)$ is on the x–axis, the parabola is horizontal. It opens to the right because $p = 5$ is positive. The equation has the form

$$y^2 = 4px.$$

Substituting 5 for p, we find that an equation for this parabola is

$$y^2 = 4(5)x$$
$$y^2 = 20x.$$

33. Focus $(0, 1/4)$, vertex at the origin

Since the focus $(0, 1/4)$ is on the y–axis, the parabola is vertical. It opens up because p is positive. The equation has the form

$$x^2 = 4py.$$

Substituting $1/4$ for p, we find that an equation for this parabola is

$$x^2 = 4\left(\dfrac{1}{4}\right)y$$
$$x^2 = y.$$

35. Through $(\sqrt{3}, 3)$, opending upward, vertex at the origin

Since the parabola opens upward, it is vertical, so the equation is of the form

$$x^2 = 4py.$$

Substitute $\sqrt{3}$ for x and 3 for y and solve for p.

$$(\sqrt{3})^2 = 4p(3)$$
$$3 = 12p$$
$$\frac{1}{4} = p$$

Thus, an equation of the parabola is

$$x^2 = 4\left(\frac{1}{4}\right)y$$
$$x^2 = y.$$

37. Through (3, 2), symmetric with respect to the x-axis, vertex at the origin

Since the parabola is symmetric with respect to the x-axis, it is horizontal, so the equation is of the form

$$y^2 = 4px.$$

Substitute 3 for x and 2 for y and solve for p.

$$2^2 = 4p(3)$$
$$4 = 12p$$
$$\frac{1}{3} = p$$

Thus, an equation for the parabola is

$$y^2 = 4\left(\frac{1}{3}\right)x$$
$$y^2 = \frac{4}{3}x.$$

39. Vertex (4, 3), focus (4, 5)

Since the focus is above the vertex, the axis is vertical and the parabola opens upward. The distance between the vertex and the focus is

5 − 3 = 2. Since the parabola opens upward, choose p = 2. The equation will have the form

$$(x - h)^2 = 4p(y - k).$$

Substitute p = 2, h = 4, and k = 3 to find the required equation.

$$(x - 4)^2 = 4(2)(y - 3)$$
$$(x - 4)^2 = 8(y - 3)$$

41. Vertex (−5, 6), focus (2, 6)

Since the focus is to the right of the vertex, the axis is horizontal and the parabola opens to the right. The distance between the vertex and the focus is 2 − (−5) = 7. Since the parabola opens to the right, choose p = 7. The equation will have the form

$$(y - k)^2 = 4p(x - h).$$

Substitute p = 7, h = −5, and k = 6.

$$(y - 6)^2 = 4(7)[x - (-5)]$$
$$(y - 6)^2 = 28(x + 5)$$

For Exercises 43 and 45, see the answer graphs in the textbook.

43. $x = 3y^2 + 6y - 4$

Complete the square on y.

$$x = 3(y^2 + 2y + 1 - 1) - 4$$
$$= 3(y^2 + 2y + 1) + 3(-1) - 4$$
$$= 3(y + 1)^2 - 7$$

Here, a = 3, h = −7, and k = −1.

Rewrite the equation in the form of two functions:

$$Y_1 = k + \sqrt{\frac{x - h}{a}} \quad \text{and} \quad Y_2 = k - \sqrt{\frac{x - h}{a}}$$

$$Y_1 = -1 + \sqrt{\frac{x + 7}{3}} \quad \text{and} \quad Y_2 = -1 - \sqrt{\frac{x + 7}{3}}.$$

Graph both functions on the same screen.

45. $x + 2 = -(y + 1)^2$

$$x = -(y + 1)^2 - 2$$

Here, $a = -1$, $h = -2$, and $k = -1$.
Rewrite the equation in the form of two functions:

$$Y_1 = k + \sqrt{\frac{x - h}{a}} \quad \text{and} \quad Y_2 = k - \sqrt{\frac{x - h}{a}}$$

$$Y_1 = -1 + \sqrt{\frac{x + 2}{-1}} \quad \text{and} \quad Y_2 = -1 - \sqrt{\frac{x + 2}{-1}}$$

$$Y_1 = -1 + \sqrt{-x - 2} \quad \text{and} \quad Y_2 = -1 - \sqrt{-x - 2}.$$

Graph both functions on the same screen.

47. (a) $y = x - \dfrac{g}{1922}x^2$

For Earth, $g = 32.2$, so the equation is

$$y = x - \frac{32.2}{1922}x^2 = x - \frac{16.1}{961}x^2.$$

For Mars, $g = 12.6$, so the equation is

$$y = x - \frac{12.6}{1922}x^2 = x - \frac{6.3}{961}x^2.$$

Graph both equations on the same screen using the simultaneous mode. See the answer graph in the textbook.

(b) From the graph, we see that the ball hits the ground (the y-coordinate returns to 0) at $x \approx 153$ ft on Mars and $x \approx 60$ ft on Earth. Therefore, the difference between the horizontal distances traveled between the two balls is approximately

$$153 - 60 = 93 \text{ ft.}$$

49. Sketch a cross-section of the dish. Place this parabola on a coordinate system with the vertex at the origin. (This solution will show that the focus lies outside of the dish.)

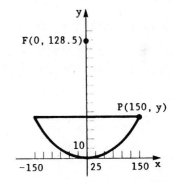

Since the parabola has vertex $(0, 0)$ and a vertical axis (the y-axis), it has an equation of the form

$$x^2 = 4py.$$

Since the focus is $(0, 128.5)$, we have $p = 128.5$, and the equation is

$$x^2 = 4(128.5)y$$
$$x^2 = 514y$$
$$\text{or} \quad y = \frac{1}{514}x^2.$$

Since the diameter (distance across the top) of the dish is 300 ft, the radius (distance halfway across the top) is 150 ft. To find the maximum depth of the dish, find the y-coordinate of point P, which lies on the parabola and has an x-coordinate of 150.

$$y = \frac{1}{514}(150)^2$$
$$\approx 43.8$$

The maximum depth of the parabolic dish is approximately 43.8 ft.

51. Place the parabola on a coordinate system and label points. (Other locations for the parabola are possible besides the one shown below.)

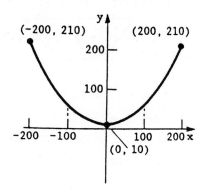

The vertex of the parabola is (0, 10).
The equation has the form

$$4p(y - 10) = x^2.$$

Substitute $x = 200$ and $y = 210$ to solve for p.

$$4p(210 - 10) = 200^2$$
$$800p = 40{,}000$$
$$p = 50$$

The equation of the parabola is

$$4(50)(y - 10) = x^2$$
$$200(y - 10) = x^2$$
$$y - 10 = \frac{1}{200}x^2$$
$$y = \frac{1}{200}x^2 + 10.$$

The vertical cables are located at $x = \pm100$.
Their heights are

$$y = \frac{1}{200}(\pm100)^2 + 10$$
$$= 60 \text{ ft.}$$

53. The equation is of the form

$$x = ay^2 + by + c.$$

Substituting $x = -5$, $y = 1$, we get

$$-5 = a(1)^2 + b(1) + c$$
$$-5 = a + b + c. \qquad (1)$$

Substituting $x = -14$, $y = -2$, we get

$$-14 = a(-2)^2 + b(-2) + c$$
$$-14 = 4a - 2b + c. \qquad (2)$$

Substituting $x = -10$, $y = 2$, we get

$$-10 = a(2)^2 + b(2) + c$$
$$-10 = 4a + 2b + c. \qquad (3)$$

54. The system of three equations is

$$a + b + c = -5 \qquad (1)$$
$$4a - 2b + c = -14 \qquad (2)$$
$$4a + 2b + c = -10. \qquad (3)$$

Add equations (2) and (3).

$$4a - 2b + c = -14$$
$$\underline{4a + 2b + c = -10}$$
$$8a \qquad + 2c = -24 \qquad (4)$$

Add 2 times equation (1) to equation (2).

$$2a + 2b + 2c = -10$$
$$\underline{4a - 2b + c = -14}$$
$$6a + 3c = -24 \quad (5)$$

Add -3 times equation (4) to 2 times equation (5).

$$-24a - 6c = 72$$
$$\underline{12a + 6c = -48}$$
$$-12a = 24$$
$$a = -2$$

Substitute $a = -2$ into equation (4).

$$8(-2) + 2c = -24$$
$$2c = -8$$
$$c = -4$$

Substitute $a = -2$ and $c = -4$ into equation (1).

$$-2 + b - 4 = -5$$
$$b = 1$$

Solution set: $\{(-2, 1, -4)\}$

55. Since $a = -2 < 0$, the parabola opens to the left.

56. Substituting $a = -2$, $b = 1$, and $c = -4$, the equation of the parabola is

$$x = -2y^2 + y - 4.$$

57. Since the axis of the parabola is parallel to the x-axis, its equation has the form

$$x - h = a(y - k)^2$$

where $(h, k) = (1, 2)$, or

$$x - 1 = a(y - 2)^2.$$

The point (13, 4) satisfies this equation, so

$$13 - 1 = a(4 - 2)^2$$
$$12 = 4a$$
$$a = 3.$$

The equation is

$$x - 1 = 3(y - 2)^2$$

or

$$(y - 2)^2 = \frac{1}{3}(x - 1).$$

Section 7.2

1. A circle is a "special case" of an ellipse because if the two foci of an ellipse are allowed to coincide, the resulting figure is a circle.

For Exercises 3–13, see the answer graphs in the textbook.

3. $\dfrac{x^2}{25} + \dfrac{y^2}{9} = 1$

The graph is an ellipse with center (0, 0). Rewrite the given equation as

$$\frac{x^2}{5^2} + \frac{y^2}{3^2} = 1.$$

Since $5 > 3$, we have $a = 5$ and $b = 3$, and the major axis is horizontal. Thus, the vertices are (-5, 0) and (5, 0). The endpoints of the minor axis are (0, -3) and (0, 3). Find the foci.

$$c^2 = a^2 - b^2$$
$$= 25 - 9 = 16$$
$$c = 4$$

Since the major axis lies on the x-axis, the foci are (-4, 0) and (4, 0). Plot the points (-5, 0), (5, 0), (0, -3), and (0, 3). Draw a smooth curve through these points.

5. $\dfrac{x^2}{9} + y^2 = 1$

Rewrite the equation as

$$\frac{x^2}{9} + \frac{y^2}{1} = 1.$$

The center is (0, 0). The vertices are (-3, 0) and (3, 0). The endpoints of the minor axis are (0, -1) and (0, 1).

$$c^2 = a^2 - b^2$$
$$= 9 - 1 = 8$$
$$c = \sqrt{8} = 2\sqrt{2}$$

The foci are $(-2\sqrt{2},\ 0)$ and $(2\sqrt{2},\ 0)$.

7. $9x^2 + y^2 = 81$

Divide both sides by 81.

$$\frac{x^2}{9} + \frac{y^2}{81} = 1$$

The center is (0, 0).
The vertices are (0, -9) and (0, 9).
The endpoints of the minor axis are (-3, 0) and (3, 0).

$$c^2 = a^2 - b^2$$
$$= 81 - 9 = 72$$
$$c = \sqrt{72} = 6\sqrt{2}$$

The foci are $(0,\ -6\sqrt{2})$ and $(0,\ 6\sqrt{2})$.

9. $4x^2 + 25y^2 = 100$

Divide both sides by 100.

$$\frac{x^2}{25} + \frac{y^2}{4} = 1$$

The center is (0, 0).
The vertices are (-5, 0) and (5, 0).
The endpoints of the minor axis are (0, -2) and (0, 2).

$$c^2 = a^2 - b^2$$
$$= 25 - 4 = 21$$
$$c = \sqrt{21}$$

The foci are $(-\sqrt{21},\ 0)$ and $(\sqrt{21},\ 0)$.

11. $\dfrac{(x-2)^2}{25} + \dfrac{(y-1)^2}{4} = 1$

The center is (2, 1).
We have a = 5 and b = 2.
Since a = 5 is associated with x^2, the major axis of the ellipse is horizontal. The vertices are on a horizontal line through (2, 1), while the endpoints of the minor axis are on the vertical line through (2, 1). The vertices are 5 units to the left and 5 units to the right of the vertex at (-3, 1) and (7, 1).
The endpoints of the minor axis are 2 units below and 2 units above the vertex at (2, -1) and (2, 3).

$$c^2 = a^2 - b^2$$
$$= 25 - 4 = 21$$
$$c = \sqrt{21}$$

The foci are $(2 - \sqrt{21},\ 1)$ and $(2 + \sqrt{21},\ 1)$.

13. $\dfrac{(x + 3)^2}{16} + \dfrac{(y - 2)^2}{36} = 1$

The center is (-3, 2).

We have a = 6 and b = 4.

Since a = 6 is associated with y^2, the major axis of the ellipse is vertical. The vertices are on the vertical line through (-3, 2), and the endpoints of the minor axis are on the horizontal line through (-3, 2). The vertices are 6 units above and 6 units below the vertex at (-3, -4) and (-3, 8). The endpoints of the minor axis are 4 units to the left and 4 units to the right of the vertex at (-7, 2) and (1, 2).

$$c^2 = a^2 - b^2$$
$$= 36 - 16 = 20$$
$$c = \sqrt{20} = 2\sqrt{5}$$

The foci are $(-3, 2 - 2\sqrt{5})$ and $(-3, 2 + 2\sqrt{5})$.

15. x-intercepts ±5; foci at (-3, 0), (3, 0)

From the given informtion, a = 5 and c = 3. Find b^2.

$$c^2 = a^2 - b^2$$
$$9 = 25 - b^2$$
$$b^2 = 16$$

Since the foci are on the x-axis and the ellipse is centered at the origin, the equation has the form

$$\dfrac{x^2}{a^2} + \dfrac{y^2}{b^2} = 1.$$

Since $a^2 = 25$ and $b^2 = 16$, the equation of the ellipse is

$$\dfrac{x^2}{25} + \dfrac{y^2}{16} = 1.$$

17. Major axis with length 6; foci at (0, 2), (0, -2)

The length of the major axis is 2a, so we have

$$2a = 6$$
$$a = 3.$$

From the foci, we have c = 2. Solve for b^2.

$$c^2 = a^2 - b^2$$
$$4 = 9 - b^2$$
$$b^2 = 5$$

Since the foci are on the y-axis and the ellipse is centered at the origin, the eqution has the form

$$\dfrac{x^2}{b^2} + \dfrac{y^2}{a^2} = 1.$$

Thus, the equation is

$$\dfrac{x^2}{5} + \dfrac{y^2}{9} = 1.$$

19. Center at (5, 2); minor axis vertical, with length 8; c = 3

The length of the minor axis is 2b, so b = 4. Find a^2.

$$c^2 = a^2 - b^2$$
$$9 = a^2 - 16$$
$$a^2 = 25$$

Since the center is (5, 2) and the minor axis is vertical, the equation has the form

$$\frac{(x - 5)^2}{a^2} + \frac{(y - 2)^2}{b^2} = 1.$$

Thus, the equation is

$$\frac{(x - 5)^2}{25} + \frac{(y - 2)^2}{16} = 1.$$

21. Vertices at (4, 9), (4, 1); minor axis with length 6

The length of the minor axis is 2b, so b = 3.

The distance between the vertices is 9 − 1 = 8, so 2a = 8 and thus a = 4. The center is halfway between the vertices, so the center is (4, 5). Since the vertices lie on the vertical line x = 4, the major axis is vertical, so the equation is of the form

$$\frac{(x - 4)^2}{b^2} + \frac{(y - 5)^2}{a^2} = 1.$$

Thus, the equation is

$$\frac{(x - 4)^2}{9} + \frac{(y - 5)^2}{16} = 1.$$

23. Foci at (0, −3), (0, 3); (8, 3) on ellipse

The distance between the foci is 3 − (−3) = 6, so 2c = 6 and thus c = 3. The center is halfway between the foci, so the center is (0, 0).

Since the foci lie on the y−axis, the major axis is vertical, so the equation is of the form

$$\frac{x^2}{b^2} + \frac{y^2}{a^2} = 1.$$

Let P(8, 3), F′(0, −3), and F(0, 3) represent the point on the ellipse and the foci, respectively. Recall that for any point P on an ellipse,

$$PF' + PF = 2a.$$

Then

$$\begin{aligned} PF' + PF &= \sqrt{(8 - 0)^2 + (3 + 3)^2} \\ &\quad + \sqrt{(8 - 0)^2 + (3 - 3)^2} \\ &= 10 + 8 \\ &= 18 \\ 2a &= 18 \\ a &= 9. \end{aligned}$$

Solve for b^2.

$$\begin{aligned} c^2 &= a^2 - b^2 \\ b^2 &= 72. \end{aligned}$$

Thus the equation is

$$\frac{x^2}{72} + \frac{y^2}{81} = 1.$$

25. Foci at (0, 4), (0, −4); sum of distances from foci to point on ellipse is 10

The distance between the foci is 4 − (−4) = 8, so 2c = 8 and thus c = 4. The center is halfway between the foci, so the center is (0, 0). Since the foci lie on the y−axis, the equation is of the form

$$\frac{x^2}{b^2} + \frac{y^2}{a^2} = 1.$$

The sum of the distances from the foci to any point on the ellipse is 10, so

$$2a = 10$$
$$a = 5.$$

Solve for b^2.

$$c^2 = a^2 - b^2$$
$$16 = 25 - b^2$$
$$b^2 = 9$$

Thus, the equation is

$$\frac{x^2}{9} + \frac{y^2}{25} = 1.$$

27. Eccentricity $\frac{3}{4}$, foci at $(0, -2)$, $(0, 2)$

The foci are on the y-axis, so the equation has the form

$$\frac{x^2}{b^2} + \frac{y^2}{a^2} = 1.$$

From the foci, we have $c = 2$.
Use the eccentricity to find a.

$$e = \frac{c}{a}$$
$$\frac{3}{4} = \frac{2}{a}$$
$$a = \frac{8}{3}$$

Now, solve for b^2.

$$c^2 = a^2 - b^2$$
$$4 = \frac{64}{9} - b^2$$
$$b^2 = \frac{28}{9}$$

Thus, the eqution is

$$\frac{x^2}{28/9} + \frac{y^2}{64/9} = 1$$

or $\frac{9x^2}{28} + \frac{9y^2}{64} = 1.$

For Exercises 29–35, see the answer graphs in the textbook.

29. $\frac{y}{2} = \sqrt{1 - \frac{x^2}{25}}$

Square both sides to get

$$\frac{y^2}{4} = 1 - \frac{x^2}{25}$$
$$\frac{x^2}{25} + \frac{y^2}{4} = 1,$$

which is the equation of an ellipse with x-intercepts ± 5 and y-intercepts ± 2. Since

$$\sqrt{1 - \frac{x^2}{25}} \geq 0,$$

the only possible values of y are those making

$$\frac{y}{2} \geq 0$$

or $y \geq 0.$

The graph of the original equation is the upper half of the ellipse. By applying the vertical line test, we see that this is the graph of a function.

31. $x = -\sqrt{1 - \frac{y^2}{64}}$

Square both sides to get

$$x^2 = 1 - \frac{y^2}{64}$$
$$x^2 + \frac{y^2}{64} = 1,$$

the equation of an ellipse centered at (0, 0), with vertices (0, 8) and (0, -8) and minor axis endpoints at (-1, 0) and (1, 0).

Since

$$-\sqrt{1 - \frac{y^2}{64}} \le 0,$$

we must have x ≤ 0, so the graph of the original equation is the left half of the ellipse.

The vertical line test shows that this is not the graph of a function.

33. $\dfrac{x^2}{16} + \dfrac{y^2}{4} = 1$

Solve the equation for y.

$$\frac{y^2}{4} = 1 - \frac{x^2}{16}$$

$$\frac{y}{2} = \pm\sqrt{1 - \frac{x^2}{16}}$$

$$y = \pm 2\sqrt{1 - \frac{x^2}{16}}$$

Graph

$$Y_1 = 2\sqrt{1 - \frac{x^2}{16}}$$

and

$$Y_2 = -2\sqrt{1 - \frac{x^2}{16}}$$

on the same screen.

35. $\dfrac{(x - 3)^2}{25} + \dfrac{y^2}{9} = 1$

Solve the equation for y.

$$\frac{(x - 3)^2}{25} + \frac{y^2}{9} = 1$$

$$\frac{y^2}{9} = 1 - \frac{(x - 3)^2}{25}$$

$$\frac{y}{3} = \pm\sqrt{1 - \frac{(x - 3)^2}{25}}$$

$$y = \pm 3\sqrt{1 - \frac{(x - 3)^2}{25}}$$

Graph

$$Y_1 = 3\sqrt{1 - \frac{(x - 3)^2}{25}}$$

and

$$Y_2 = -3\sqrt{1 - \frac{(x - 3)^2}{25}}$$

on the same screen.

37. $\dfrac{(x - 2)^2}{9} + \dfrac{(y + 1)^2}{16} = 1$

Solve this equation for y.

$$\frac{(y + 1)^2}{16} = 1 - \frac{(x - 2)^2}{9}$$

$$(y + 1)^2 = 16\left(1 - \frac{(x - 2)^2}{9}\right)$$

$$(y + 1)^2 = 16 - \frac{16(x - 2)^2}{9}$$

$$y + 1 = \pm\sqrt{16 - \frac{16(x - 2)^2}{9}}$$

$$y = -1 \pm\sqrt{16 - \frac{16(x - 2)^2}{9}}$$

We obtain the two functions

$$Y_1 = -1 + \sqrt{16 - \frac{16(x - 2)^2}{9}}$$

and

$$Y_2 = -1 - \sqrt{16 - \frac{16(x - 2)^2}{9}}.$$

Graph these two functions on the same screen. Use a "square" viewing window to see the true shape of the ellipse.

See the answer graph in the textbook.

38. From Figure 14, we see that the domain of the ellipse is $[-1, 5]$.

39. To find the domain analytically, the radicand must be positive. Thus, we would need to solve the inequality

$$16 - \frac{16(x-2)^2}{9} \geq 0.$$

40. Let $y = 16 - \frac{16(x-2)^2}{9}$.

The graph of this equation is a parabola.

41. Graph $Y = 16 - \frac{16(x-2)^2}{9}$ with a graphing calculator in the window $[-10, 10]$ by $[-10, 20]$. See the answer graph in the textbook.

42. The graph of

$$y = 16 - \frac{16(x-2)^2}{9}$$

lies above or on the x-axis in the interval $[-1, 5]$.

43. In Figure 14, we see that the domain is $[-1, 5]$. This corresponds to the solution set found graphically in Exercise 42.

44.
$$16 - \frac{16(x-2)^2}{9} \geq 0$$

Multiply both sides by 9.

$$144 - 16(x-2)^2 \geq 0$$

Divide both sides by 16.

$$9 - (x-2)^2 \geq 0$$
$$9 - (x^2 - 4x + 4) \geq 0$$
$$9 - x^2 + 4x - 4 \geq 0$$
$$-x^2 + 4x + 5 \geq 0$$

Divide both sides by -1; reverse the inequality symbol.

$$x^2 - 4x - 5 \leq 0$$

Solve the corresponding equation.

$$(x - 5)(x + 1) = 0$$

Use a sign graph.

Sign of $(x - 5)(x + 1)$

Thus,

$$(x - 5)(x + 1) \leq 0$$

in the interval $[-1, 5]$.
Since

$$x^2 - 4x - 5 \leq 0$$

in the interval $[-1, 5]$, it follows that

$$-x^2 - 4x + 5 \geq 0$$

and also

$$16 - \frac{16(x-2)^2}{9} \geq 0$$

in the interval $[-1, 5]$.

47. **(a)** The graph of the orbit of the satellite is determined by the equation

$$\frac{x^2}{a^2} + \frac{y^2}{b^2} = 1$$

with a = 4465 and b = 4462. In order to graph this equation, we must solve the equation for y.

$$\frac{x^2}{4465^2} + \frac{y^2}{4462^2} = 1$$

$$\frac{y^2}{4462^2} = 1 - \frac{x^2}{4465^2}$$

$$y^2 = 4462^2\left(1 - \frac{x^2}{4465^2}\right)$$

$$y = \pm\sqrt{4462^2\left(1 - \frac{x^2}{4465^2}\right)}$$

$$y = \pm 4462\sqrt{1 - \frac{x^2}{4465^2}}$$

From the last equation, we obtain the two functions

$$Y_1 = 4462\sqrt{1 - \frac{x^2}{4465^2}}$$

$$\text{and} \quad Y_2 = -4462\sqrt{1 - \frac{x^2}{4465^2}}$$

$$= -Y_1.$$

The graph of Earth can be represented by a circle of radius 3960 centered at one focus. To determine the foci of the orbit we must determine c.

$$c^2 = a^2 - b^2$$

$$= 4465^2 - 4462^2$$

$$= 26{,}781$$

$$c \approx 163.6$$

If the center of Earth is located at (163.6, 0), then the equation of the circle will be

$$(x - 163.6)^2 + y^2 = 3960^2.$$

Solving for y, we find that

$$y = \pm\sqrt{3960^2 - (x - 163.6)^2}.$$

From the last equation, we obtain the two functions

$$Y_3 = \sqrt{3960^2 - (x - 163.6)^2}$$

$$\text{and} \quad Y_4 = -\sqrt{3960^2 - (x - 163.6)^2}$$

$$= -Y^3.$$

Graph all four functions on the same screen, using the window

[−6750, 6750] by [−4500, 4500].

See the answer graph in the textbook.

(b) From the graph we can see that the distance is maximum and minimum when the orbits intersect the x-axis.

The x-intercepts of the satellite's orbit are ±4465. The x-intercepts of Earth's surface occur when

$$(x - 163.6)^2 + 0^2 = 3960^2$$

$$(x - 163.6)^2 = 3960^2$$

$$x - 163.6 = \pm 3960$$

$$x = 163.6 \pm 3960$$

$$x = 4123.6 \quad \text{or} \quad x = -3796.4.$$

The minimum distance is

$$4465 - 4123.6 = 341.4$$

$$\approx 341 \text{ miles}$$

and the maximum distance is

$$-3796.4 - (-4465) = 668.6$$

$$\approx 669 \text{ miles}.$$

49. (a) v_{max}

$$= \frac{2\pi a}{P}\sqrt{\frac{1 + e}{1 - e}}$$

$$= \frac{2\pi \times 1.496 \times 10^8}{60 \times 60 \times 24 \times 365.25}$$

$$\times \sqrt{\frac{1 + .0167}{1 - .0167}}$$

$$\approx 30.3 \text{ km/sec}$$

v_{min}

$$= \frac{2\pi a}{P}\sqrt{\frac{1 - e}{1 + e}}$$

$$= \frac{2\pi \times 1.496 \times 10^8}{60 \times 60 \times 24 \times 365.25}$$

$$\times \sqrt{\frac{1 - .0167}{1 + .0167}}$$

$$\approx 29.3 \text{ km/sec}$$

(b) For a circle $e = 0$, so

$$v_{max} = v_{min} = \frac{2\pi}{P}.$$

The minimum and maximum velocities are equal. Therefore, the planet's velocity is constant.

(c) A planet is at its maximum and minimum distance from a focus when it is located at the vertices of the ellipse. Thus, the minimum and maximum velocities of a planet will occur at the vertices of the elliptical orbit.

51. An ellipse with major axis 620 ft and minor axis 513 ft has $2a = 620$ and $2b = 513$, or $a = 310$ and $b = 256.5$.

The distance between the center and a focus is c, where

$$c^2 = a^2 - b^2$$
$$c^2 = 310^2 - (256.5)^2$$
$$c^2 = 96,100 - 65,792.25$$
$$c^2 = 30,307.75$$
$$c \approx 174.1.$$

(The negative value of c is rejected.) The distance between the two foci of the ellipse is

$$2c = 2(174.1) = 348.2.$$

There are 348.2 ft between the foci of the Roman Coliseum.

53. The shortest distance between Halley's comet and the sun is $a - c$ and the greatest distance is $a + c$. Since the eccentricity is .9673,

$$\frac{c}{a} = .9673$$
$$c = .9673a.$$

Since the greatest distance is 3281,

$$a + c = 3281$$
$$a + .9673a = 3281$$
$$1.9673a = 3281$$
$$a \approx 1667.768$$
$$c \approx 1613.232$$
$$a - c \approx 1667.768 - 1613.232$$
$$\approx 55.$$

Thus, the shortest distance between Halley's comet and the sun is approximately 55 million miles.

55. To graph $\frac{x^2}{16} + \frac{y^2}{12} = 1$ with a graphing calculator, solve for y.

$$\frac{y^2}{12} = 1 - \frac{x^2}{16}$$

$$y^2 = 12\left(1 - \frac{x^2}{16}\right)$$

$$y = \pm\sqrt{12\left(1 - \frac{x^2}{16}\right)}$$

Graph

$$Y_1 = \sqrt{12\left(1 - \frac{x^2}{16}\right)}$$

and

$$Y_2 = -\sqrt{12\left(1 - \frac{x^2}{16}\right)}$$

on the same screen.

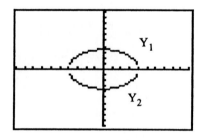

Use TRACE to find the coordinates of several points on the ellipse. The required calculations are shown here for one sample point,

$$P(-2.258065, 2.8593458).$$

Using the distance formula, we find

Distance of P from (-2, 0)

$$= \sqrt{[-2.258065 - (-2)]^2 + (2.8593458 - 0)^2}$$

$$\approx 2.8709678$$

and

Distance of P from (2, 0)

$$= \sqrt{(-2.258065 - 2)^2 + (2.8593458 - 0)^2}$$

$$\approx 5.1290327.$$

Thus,

[Distance of P from (-2, 0)]

+ [Distance of P from (2, 0)]

$$\approx 2.8709678 + 5.1290327$$

$$= 8.0000005$$

$$\approx 8.$$

Section 7.3

1. $\frac{x^2}{25} + \frac{y^2}{9} = 1$

This is an equation of an ellipse with x-intercepts ±5 and y-intercepts ±3. The correct graph is C.

3. $\frac{x^2}{9} - \frac{y^2}{25} = 1$

This is the graph of a hyperbola with x-intercepts ±3 and no y-intercepts. The correct graph is D.

For Exercises 5-25, see the answer graphs in the textbook.

5. $\frac{x^2}{16} - \frac{y^2}{9} = 1$

This equation may be written as

$$\frac{x^2}{4^2} - \frac{y^2}{3^2} = 1,$$

which has the form

$$\frac{x^2}{a^2} - \frac{y^2}{b^2} = 1.$$

The hyperbola is centered at $(0, 0)$ with branches opening to the left and right. The graph has x-intercepts ± 4, so the vertices are $(-4, 0)$ and $(4, 0)$. There are no y-intercepts. The foci are on the x-axis.

$$c^2 = a^2 + b^2$$
$$= 16 + 9 = 25$$
$$c = 5$$

The foci are $(-5, 0)$ and $(5, 0)$. The asymptotes are

$$y = \pm \frac{b}{a}x = \pm \frac{3}{4}x.$$

7. $\dfrac{y^2}{25} - \dfrac{x^2}{49} = 1$

This equation may be written as

$$\frac{y^2}{5^2} - \frac{x^2}{7^2} = 1,$$

which has the form

$$\frac{y^2}{a^2} - \frac{x^2}{b^2} = 1.$$

The hyperbola is centered at $(0, 0)$ with branches opening upward and downward. The graph has y-intercepts ± 5, so the vertices are $(0, -5)$ and $(0, 5)$. There are no x-intercepts. The foci are on the y-axis.

$$c^2 = a^2 + b^2$$
$$= 25 + 49 = 74$$
$$c = \sqrt{74}$$

The foci are $(0, -\sqrt{74})$ and $(0, \sqrt{74})$. The asymptotes are

$$y = \pm \frac{a}{b}x = \pm \frac{5}{7}x.$$

9. $x^2 - y^2 = 9$

Divide both sides by 9.

$$\frac{x^2}{9} - \frac{y^2}{9} = 1$$

Thus, $a = 3$ and $b = 3$.
The center is $(0, 0)$, and the vertices are $(-3, 0)$ and $(3, 0)$.

$$c^2 = a^2 + b^2$$
$$= 9 + 9 = 18$$
$$c = \sqrt{18} = 3\sqrt{2}$$

The foci are $(-3\sqrt{2}, 0)$ and $(3\sqrt{2}, 0)$. The asymptotes are

$$y = \pm \frac{b}{a}x = \pm \frac{3}{3}x = \pm x.$$

11. $9x^2 - 25y^2 = 225$

Divide both sides by 225.

$$\frac{9x^2}{225} - \frac{25y^2}{225} = \frac{225}{225}$$
$$\frac{x^2}{25} - \frac{y^2}{9} = 1$$

Thus, $a = 5$ and $b = 3$.
The center is $(0, 0)$, and the vertices are $(-5, 0)$ and $(5, 0)$.

$$c^2 = a^2 + b^2$$
$$= 25 + 9 = 34$$
$$c = \sqrt{34}$$

The foci are $(-\sqrt{34}, 0)$ and $(\sqrt{34}, 0)$. The asymptotes are

$$y = \pm \frac{b}{a}x = \pm \frac{3}{5}x.$$

13. $4x^2 - y^2 = -16$

Divide both sides by -16 and rearrange.

$$\frac{4x^2}{-16} - \frac{y^2}{-16} = \frac{-16}{-16}$$

$$-\frac{x^2}{4} + \frac{y^2}{16} = 1$$

$$\frac{y^2}{16} - \frac{x^2}{4} = 1$$

Thus, $a = 4$ and $b = 2$.
The center is $(0, 0)$, and the vertices are $(0, -4)$ and $(0, 4)$.

$$c^2 = a^2 + b^2$$
$$= 16 + 4 = 20$$
$$c = \sqrt{20} = 2\sqrt{5}$$

The foci are $(0, -2\sqrt{5})$ and $(0, 2\sqrt{5})$.
The asymptotes are

$$y = \pm\frac{a}{b}x$$

$$= \pm\frac{4}{2}x = \pm 2x.$$

15. $9x^2 - 4y^2 = 1$

Rewrite in standard form.

$$\frac{x^2}{1/9} - \frac{y^2}{1/4} = 1$$

Thus, $a = 1/3$ and $b = 1/2$.
The center is $(0, 0)$, and the vertices are $(-1/3, 0)$ and $(1/3, 0)$.

$$c^2 = a^2 + b^2$$
$$= \frac{1}{9} + \frac{1}{4} = \frac{13}{36}$$
$$c = \sqrt{\frac{13}{36}} = \frac{\sqrt{13}}{6}$$

The foci are $(-\sqrt{13}/6, 0)$ and $(\sqrt{13}/6, 0)$. The asymptotes are

$$y = \pm\frac{b}{a}x$$

$$= \pm\frac{1/2}{1/3}x = \pm\frac{3}{2}x.$$

17. $\frac{(y - 7)^2}{36} - \frac{(x - 4)^2}{64} = 1$

Here, $a = 6$ and $b = 8$.
The center is $(4, 7)$. The vertices are 6 units above and below the center $(4, 7)$. These points are $(4, 1)$ and $(4, 13)$.

$$c^2 = a^2 + b^2$$
$$= 36 + 64 = 100$$
$$c = \sqrt{100} = 10$$

The foci are 10 units above and below the center $(4, 7)$. The foci are $(4, -3)$ and $(4, 17)$. The asymptotes are

$$y - k = \pm\frac{a}{b}(x - h)$$

$$y - 7 = \pm\frac{6}{8}(x - 4)$$

$$y = 7 \pm \frac{3}{4}(x - 4).$$

19. $\frac{(x + 3)^2}{16} - \frac{(y - 2)^2}{9} = 1$

Here, $a = 4$ and $b = 3$.
The center is $(-3, 2)$. The vertices are 4 units to the left and right of the center $(-3, 2)$. These points are $(-7, 2)$ and $(1, 2)$.

$$c^2 = a^2 + b^2$$
$$= 16 + 9 = 25$$
$$c = \sqrt{25} = 5$$

The foci are 5 units to the left and right of the center $(-3, 2)$. These points are $(-8, 2)$ and $(2, 2)$. The asymptotes are

$$y - k = \pm\frac{b}{a}(x - h)$$

$$y - 2 = \pm\frac{3}{4}(x + 3)$$

$$y = 2 \pm \frac{3}{4}(x + 3).$$

21. $16(x + 5)^2 - (y - 3)^2 = 1$

Rewrite in standard form.

$$\frac{(x + 5)^2}{1/16} - \frac{(y - 3)^2}{1} = 1$$

Thus, $a = 1/4$ and $b = 1$.
The center is $(-5, 3)$. The vertices are $1/4$ unit to the left and right of the center $(-5, 3)$. These points are $(-21/4, 3)$ and $(-19/4, 3)$.

$$c^2 = a^2 + b^2$$

$$= \frac{1}{16} + 1 = \frac{17}{16}$$

$$c = \sqrt{\frac{17}{16}} = \frac{\sqrt{17}}{4}$$

The foci are $\sqrt{17}/4$ units to the left and right of the center $(-5, 3)$.
These points are $\left(-5 - \frac{\sqrt{17}}{4}, 3\right)$ and

$\left(-5 + \frac{\sqrt{17}}{4}, 3\right)$. The asymptotes are

$$y - k = \pm\frac{b}{a}(x - h)$$

$$y - 3 = \pm\frac{1}{1/4}(x + 5)$$

$$y = 3 \pm 4(x + 5).$$

23. $\frac{y}{3} = \sqrt{1 + \frac{x^2}{16}}$

Square both sides and write in standard form.

$$\frac{y^2}{9} = 1 + \frac{x^2}{16}$$

$$\frac{y^2}{9} - \frac{x^2}{16} = 1$$

This is the equation of a hyperbola centered at $(0, 0)$, with vertices $(0, \pm 3)$ and asymptotes $y = \pm\frac{3}{4}x$. The original equation is the top half of the hyperbola. The vertical line test shows this is the graph of a function.

25. $5x = -\sqrt{1 + 4y^2}$

Square both sides and write in standard form.

$$25x^2 = 1 + 4y^2$$

$$25x^2 - 4y^2 = 1$$

$$\frac{x^2}{1/25} - \frac{y^2}{1/4} = 1$$

This is the equation of a hyperbola centered at $(0, 0)$, with vertices $(\pm 1/5, 0)$ and asymptotes $y = \pm(25/4)x$. The original equation is the left half of the hyperbola. The vertical line test shows this is not the graph of a function.

27. x-intercepts ± 4; foci at $(-5, 0)$, $(5, 0)$

Since the x-intercepts are ± 4, the equation has the form

$$\frac{x^2}{a^2} - \frac{y^2}{b^2} = 1.$$

$a = 4$, so $a^2 = 16$.

The foci are at $(\pm5, 0)$, so $c = 5$.

$$c^2 = a^2 + b^2$$
$$25 = 16 + b^2$$
$$9 = b^2$$

The equation is

$$\frac{x^2}{16} - \frac{y^2}{9} = 1.$$

29. Vertices at $(0, 6)$, $(0, -6)$; asymptotes $y = \pm(1/2)x$

Since the vertices are at $(0, \pm6)$, the equation has the form

$$\frac{y^2}{a^2} - \frac{x^2}{b^2} = 1.$$

$a = 6$, so $a^2 = 36$.
The slopes of the asymptotes are $\pm1/2$, so

$$\frac{a}{b} = \pm\frac{1}{2}$$
$$\frac{6}{b} = \pm\frac{1}{2}$$
$$b = \pm12$$
$$b^2 = 144.$$

The equation is

$$\frac{y^2}{36} - \frac{x^2}{144} = 1.$$

31. Vertices at $(-3, 0)$, $(3, 0)$; passing through $(6, 1)$

Since the vertices are at $(\pm3, 0)$, the equation has the form

$$\frac{x^2}{9} - \frac{y^2}{b^2} = 1.$$

The hyperbola goes through the point $(6, 1)$, so substitute $x = 6$ and $y = 1$ into the equation and solve for b^2.

$$\frac{36}{9} - \frac{1}{b^2} = 1$$
$$4 - \frac{1}{b^2} = 1$$
$$-\frac{1}{b^2} = -3$$
$$b^2 = \frac{1}{3}$$

The equation is

$$\frac{x^2}{9} - \frac{y^2}{1/3} = 1$$

or $\qquad \dfrac{x^2}{9} - 3y^2 = 1.$

33. Foci at $(0, \sqrt{13})$, $(0, -\sqrt{13})$; asymptotes $y = \pm5x$

Since the foci are on the y-axis, the equation has the form

$$\frac{y^2}{a^2} - \frac{x^2}{b^2} = 1.$$

$c = \sqrt{13}$, so

$$c^2 = a^2 + b^2$$
$$13 = a^2 + b^2. \quad (1)$$

The slopes of the asymptotes are ±5. Use one of the slopes to find a in terms of b and a^2 in terms of b^2.

$$\frac{a}{b} = 5$$
$$a = 5b$$
$$a^2 = 25b^2$$

Substitute $a^2 = 25b^2$ into equation (1).

$$13 = 25b^2 + b^2$$
$$13 = 26b^2$$
$$b^2 = \frac{1}{2}$$
$$a^2 = 25\left(\frac{1}{2}\right) = \frac{25}{2}$$

The equation is

$$\frac{y^2}{25/2} - \frac{x^2}{1/2} = 1$$

or $\quad \dfrac{2y^2}{25} - 2x^2 = 1.$

35. Vertices at $(4, 5)$, $(4, 1)$; asymptotes $y - 3 = \pm 7(x - 4)$

The center is halfway between the vertices at $(4, 3)$. Since the distance between the vertices is 4, we have $2a = 4$, and thus $a = 2$.
The equation has the form

$$\frac{(y - 3)^2}{4} - \frac{(x - 4)^2}{b^2} = 1$$

The slopes of the asymptotes are ± 7. Use one of the slopes to find b and b^2.

$$\frac{a}{b} = 7$$

$$\frac{2}{b} = 7$$

$$b = \frac{2}{7}$$

$$b^2 = \frac{4}{49}.$$

The equation is

$$\frac{(y - 3)^2}{4} - \frac{(x - 4)^2}{4/49} = 1$$

$$\frac{(y - 3)^2}{4} - \frac{49(x - 4)^2}{4} = 1.$$

37. Center at $(1, -2)$; focus at $(4, -2)$; vertex at $(3, -2)$

The center is halfway between the vertices. Since one vertex is $(3, -2)$, the other vertex must be $(-1, -2)$, and $a = 2$.

The equation has the form

$$\frac{(x - 1)^2}{4} - \frac{(y + 2)^2}{b^2} = 1.$$

One focus is at $(4, -2)$, 3 units from the center, so $c = 3$.

$$c^2 = a^2 + b^2$$

$$9 = 4 + b^2$$

$$b^2 = 5$$

The equation is

$$\frac{(x - 1)^2}{4} - \frac{(y + 2)^2}{5} = 1.$$

39. Eccentricity 3; center at $(0, 0)$; vertex at $(0, 7)$

Since the center and vertex lie on a vertical line, the equation is of the form

$$\frac{y^2}{a^2} - \frac{x^2}{b^2} = 1.$$

The distance between the center and a vertex is 7 so $a = 7$. Use the eccentricity to find c.

$$e = \frac{c}{a}$$

$$3 = \frac{c}{7}$$

$$c = 21$$

Now solve for b^2.

$$c^2 = a^2 + b^2$$

$$441 = 49 + b^2$$

$$392 = b^2$$

The equation is

$$\frac{y^2}{49} - \frac{x^2}{392} = 1.$$

41. Vertices at $(-2, 10)$, $(-2, 2)$; eccentricity $5/4$

The center is halfway between the vertices at $(-2, 6)$. The distance from the center to each vertex is 4, so $a = 4$.

Use the eccentricity to find c.

$$e = \frac{c}{a}$$

$$\frac{5}{4} = \frac{c}{4}$$

$$c = 5$$

Now solve for b^2.

$$c^2 = a^2 + b^2$$

$$25 = 16 + b^2$$

$$b^2 = 9$$

The equation is

$$\frac{(y - 6)^2}{16} - \frac{(x + 2)^2}{9} = 1.$$

For Exercises 43 and 45, see the answer graphs in the textbook.

43. $\dfrac{x^2}{4} - \dfrac{y^2}{16} = 1$

Solve the equation for y.

$$\frac{x^2}{4} - 1 = \frac{y^2}{16}$$

$$\pm\sqrt{\frac{x^2}{4} - 1} = \frac{y}{4}$$

$$y = \pm 4\sqrt{\frac{x^2}{4} - 1}$$

Graph

$$Y_1 = 4\sqrt{\frac{x^2}{4} - 1}$$

and

$$Y_2 = -4\sqrt{\frac{x^2}{4} - 1}$$

on the same screen.

45. $4y^2 - 36x^2 = 144$

Divide by 144.

$$\frac{y^2}{36} - \frac{x^2}{4} = 1$$

Solve this equation for y.

$$\frac{y^2}{36} = \frac{x^2}{4} + 1$$

$$y^2 = 36\left(\frac{x^2}{4} + 1\right)$$

$$y = \pm\sqrt{36\left(\frac{x^2}{4} + 1\right)}$$

$$= \pm 6\sqrt{\frac{x^2}{4} + 1}$$

Graph

$$Y_1 = 6\sqrt{\frac{x^2}{4} + 1}$$

and

$$Y_2 = -6\sqrt{\frac{x^2}{4} + 1}$$

on the same screen.

47. $\dfrac{x^2}{4} - y^2 = 1$

Solve the equation for y.

$$\frac{x^2}{4} - 1 = y^2$$

$$y = \pm\sqrt{\frac{x^2}{4} - 1}$$

The positive square root is

$$y = \sqrt{\frac{x^2}{4} - 1}.$$

48. Write the equation in standard form.

$$\frac{x^2}{4} - y^2 = 1$$

$$\frac{x^2}{4} - \frac{y^2}{1} = 1$$

Here, a = 2 and b = 1. The slopes of the asymptotes are ±b/a = ±1/2. The equation of the asymptote with positive slope is

$$y = \frac{1}{2}x.$$

49. $y = \sqrt{\frac{x^2}{4} - 1}$

At x = 50,

$$y = \sqrt{\frac{50^2}{4} - 1} \approx 24.98.$$

50. On the asymptote $y = \frac{1}{2}x$,

when x = 50,

$$y = \frac{1}{2} \cdot 50 = 25.$$

51. Because 24.98 < 25, the graph of $y = \sqrt{\frac{x^2}{4} - 1}$ lies below the graph of $y = \frac{1}{2}x$ when x = 50.

52. If we choose x-values larger than 50, the y-values on the hyperbola will approach the y-values on the asymptote.

53. (a) We must determine a and b in the equation

$$\frac{x^2}{a^2} - \frac{y^2}{b^2} = 1.$$

The asymptotes are y = x and y = -x, which have slopes of 1 and -1, respectively, so a = b. Look at the small right triangle that is shown in quadrant III.

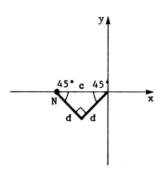

The line y = x intersects quadrants I and III at a 45° angle. Since the right angle vertex of the triangle lies on the line y = x, we know that this triangle is a 45°-45°-90° triangle (an isosceles right triangle). Thus, both legs of the triangle have length d, and, by the Pythagorean theorem,

$$c^2 = d^2 + d^2$$
$$c^2 = 2d^2$$
$$c = d\sqrt{2}. \quad (1)$$

Thus, the coordinates of N are $(-d\sqrt{2}, 0)$. Since N is a focus of the hyperbola, c represents the distance between the center of the hyperbola, which is (0, 0), and either focus.

Sine a = b, we have

$$c^2 = a^2 + b^2$$
$$c^2 = 2a^2$$
$$c = a\sqrt{2}. \quad (2)$$

From equations (1) and (2), we have

$$d\sqrt{2} = a\sqrt{2}$$
$$d = a.$$

Thus,

$$a = b = d = 5 \times 10^{-14}.$$

The equation of the trajectory of A is given by

$$\frac{x^2}{a^2} - \frac{y^2}{b^2} = 1$$

$$\frac{x^2}{(5 \times 10^{-14})^2} - \frac{y^2}{(5 \times 10^{-14})^2} = 1$$

$$x^2 - y^2 = (5 \times 10^{-14})^2$$

$$= 25 \times 10^{-28}$$

$$= 2.5 \times 10^{-27}$$

$$x^2 = y^2 + (2.5 \times 10^{-27})$$

$$x = \sqrt{y^2 + (2.5 \times 10^{-27})}.$$

(We choose the positive square root since the trajectory occurs only where x > 0. This equation represents the right half of the hyperbola, as shown in the figure in the textbook.)

(b) The minimum distance between their centers is

$$c + a = d\sqrt{2} + d$$

$$= (5 \times 10^{-14})\sqrt{2} + (5 \times 10^{-14})$$

$$\approx 12.07 \times 10^{-14}$$

$$\approx 1.2 \times 10^{-13} \text{ m.}$$

55. A sketch illustrates this exercise.

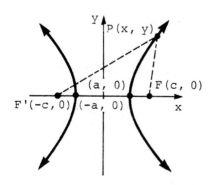

$$d(P, F') - d(P, F) = 2a$$

$$\sqrt{(x + c)^2 + (y - 0)^2} - \sqrt{(x - c)^2 + (y - 0)^2}$$

$$= 2a$$

$$\sqrt{(x + c)^2 + (y - 0)^2} = 2a + \sqrt{(x + c)^2 + (y - 0)^2}$$

$$(x + c)^2 + y^2 = 4a^2 + 4a\sqrt{(x - c)^2 + y^2}$$
$$+ (x - c)^2 + y^2$$
Square both sides

$$x^2 + 2xc + c^2 + y^2 = 4a^2 + 4a\sqrt{(x - c)^2 + y^2}$$
$$+ x^2 - 2xc + c^2 + y^2$$
Square binomials

$$4xc = 4a^2 + 4a\sqrt{(x - c)^2 + y^2}$$
Simplify

$$xc = a^2 + a\sqrt{(x - c)^2 + y^2}$$
Divide by 4

$$cx - a^2 = a\sqrt{(x - c)^2 + y^2}$$

$$c^2x^2 - 2cxa^2 + a^4 = a^2(x^2 - 2xc + c^2 + y^2)$$
Square both sides again

$$c^2x^2 - 2cxa^2 + a^4 = a^2x^2 - 2a^2xc + a^2c^2$$
$$+ a^2y^2$$
Distributive property

$$c^2x^2 - a^2x^2 - a^2y^2 = a^2c^2 - a^4 \quad \textit{Simplify}$$

$$x^2(c^2 - a^2) - a^2y^2 = a^2(c^2 - a^2)$$
Greatest common factors

$$\frac{x^2}{a^2} - \frac{y^2}{c^2 - a^2} = 1 \quad \begin{array}{l} \textit{Divide by} \\ a^2(c^2 - a^2) \end{array}$$

$$\frac{x^2}{a^2} - \frac{y^2}{b^2} = 1 \quad b^2 = c^2 - a^2$$

57. The graph in Figure 21 has the equation

$$\frac{(y + 2)^2}{9} - \frac{(x + 3)^2}{4} = 1.$$

Solve this equation for y.

$$\frac{(y + 2)^2}{9} = 1 + \frac{(x + 3)^2}{4}$$

$$\frac{y + 2}{3} = \pm\sqrt{1 + \frac{(x + 3)^2}{4}}$$

$$y + 2 = \pm 3\sqrt{1 + \frac{(x + 3)^2}{4}}$$

$$y = -2 \pm 3\sqrt{1 + \frac{(x + 3)^2}{4}}$$

Graph

$$Y_1 = -2 + 3\sqrt{1 + \frac{(x + 3)^2}{4}}$$

and

$$Y_2 = -2 - 3\sqrt{1 + \frac{(x + 3)^2}{4}}$$

on the same screen.

See the answer graph in the text-book.

Section 7.4

1.
$$x^2 + y^2 = 144$$
$$(x - 0)^2 + (y - 0)^2 = 12^2$$

The graph of this equation is a circle.

3. $y = 2x^2 + 3x - 4$

The graph of this equation is a parabola.

5. $x - 1 = -3(y - 4)^2$

The graph of this equation is a parabola.

7. $\frac{x^2}{49} + \frac{y^2}{100} = 1$

$$\frac{x^2}{7^2} + \frac{y^2}{10^2} = 1$$

The graph of this equation is an ellipse.

9. $\frac{x^2}{4} - \frac{y^2}{16} = 1$

$$\frac{x^2}{2^2} - \frac{y^2}{4^2} = 1$$

The graph of this equation is a hyperbola.

11. $\frac{x^2}{25} - \frac{y^2}{25} = 1$

$$\frac{x^2}{5^2} - \frac{y^2}{5^2} = 1$$

The graph of this equation is a hyperbola.

13. $\frac{x^2}{4} = 1 - \frac{y^2}{9}$

$$\frac{x^2}{4} + \frac{y^2}{9} = 1$$

The graph of this equation is an ellipse.

15. $\frac{x^2}{4} + \frac{y^2}{4} = 1$

$$x^2 + y^2 = 4$$

The graph of this equation is a circle.

17.
$$x^2 = 25 + y^2$$
$$x^2 - y^2 = 25$$
$$\frac{x^2}{25} - \frac{y^2}{25} = 1$$

The graph is a hyperbola.

19. $9x^2 + 36y^2 = 36$

$$\frac{x^2}{4} + \frac{y^2}{1} = 1$$

The graph is an ellipse.

21. $\frac{(x + 3)^2}{16} + \frac{(y - 2)^2}{16} = 1$

$$(x + 3)^2 + (y - 2)^2 = 16$$

The graph is a circle.

23. $y^2 - 4y = x + 4$

$y^2 - 4y - 4 = x$

The graph is a parabola.

25. $(x + 7)^2 + (y - 5)^2 + 4 = 0$

$(x + y)^2 + (y - 5)^2 = -4$

This is the form of the equation of a circle, but r^2 cannot be negative, so there is no graph.

27. $3x^2 + 6x + 3y^2 - 12y = 12$

$x^2 + 2x + y^2 - 4y = 4$

$(x^2 + 2x + 1) + (y^2 - 4y + 4) = 4 + 1 + 4$

$(x + 1)^2 + (y - 2)^2 = 9$

The graph is a circle.

29. $x^2 - 6x + y = 0$

$y = -x^2 + 6x$

The graph is a parabola.

31. $4x^2 - 8x - y^2 - 6y = 6$

$4(x^2 - 2x + 1) - 1(y^2 + 6y + 9) = 6 + 4 - 9$

$4(x - 1)^2 - 1(y + 3)^2 = 1$

$\dfrac{(x - 1)^2}{1/4} - \dfrac{(y + 3)^2}{1} = 1$

The graph is a hyperbola.

33. $4x^2 - 8x + 9y^2 + 54y = -84$

$4(x^2 - 2x + 1) + 9(y^2 + 6y + 9) = -84 + 4 + 81$

$4(x - 1)^2 + 9(y + 3)^2 = 1$

$\dfrac{(x - 1)^2}{1/4} + \dfrac{(y + 3)^2}{1/9} = 1$

The graph is an ellipse.

35. $6x^2 - 12x + 6y^2 - 18y + 25 = 0$

$6\left(x^2 - 2x + 1\right) + 6\left(y^2 - 3y + \dfrac{9}{4}\right) = -25 + 6 + \dfrac{27}{2}$

$6(x - 1)^2 + 6\left(y - \dfrac{3}{2}\right)^2 = -\dfrac{50}{2} + \dfrac{12}{2} + \dfrac{27}{2}$

$6(x - 1)^2 + 6\left(y - \dfrac{3}{2}\right)^2 = -\dfrac{11}{2}$

$(x - 1)^2 + \left(y - \dfrac{3}{2}\right)^2 = -\dfrac{11}{12}$

This is the form of the equation of a circle, but r^2 cannot be negative, so there is no graph.

37.   The definition of an ellipse states that "an ellipse is the set of all points in a plane the sum of whose distances from two fixed points is constant."

Therefore, the set of all points in a plane for which the sum of the distances from the points (5, 0) and (−5, 0) is 14 is an ellipse with foci (5, 0) and (−5, 0).

39.   Refer to the "Geometric Character-ization of Conic Sections" box in the textbook. We see that

[Distance of P from F]

   = e • [Distance of P from L],

so this conic section has eccen-tricity 1 1/2 = 3/2. Since e > 1, this is a hyperbola.

41.   From the graph, we see that P = (−3, 8), F = (3, 0) and L is the vertical line x = 27.

By the distance formula, we have

$$\text{Distance of P from F} = \sqrt{(-6)^2 + 8^2}$$
$$= \sqrt{36 + 64}$$
$$= \sqrt{100} = 10.$$

The distance from a point to a line is defined as the perpendicular distance, so

$$\text{Distance of P from L} = |27 - (-3)|$$
$$= 30.$$

$$e = \frac{\text{Distance of P from F}}{\text{Distance of P from L}}$$
$$= \frac{10}{30} = \frac{1}{3}$$

43. From the graph, we see that F = $(\sqrt{2}, 0)$ and L is the vertical line $x = -\sqrt{2}$.

Choose (0, 0), the vertex of the parabola, as P.

$$\text{Distance of P from F} = \sqrt{2}$$
$$\text{Distance of P from L} = \sqrt{2}$$

$$e = \frac{\text{Distance of P from F}}{\text{Distance of P from L}}$$

$$= \frac{\sqrt{2}}{\sqrt{2}} = 1$$

45. From the graph, we see that P = (9, −7.5), F = (9, 0) and L is the vertical line x = 4.

$$\text{Distance of P from F} = 7.5$$
$$\text{Distance of P from L} = 5$$

$$e = \frac{\text{Distance of P from F}}{\text{Distance of P from L}}$$

$$= \frac{7.5}{5} = \frac{3}{2}$$

47. **(a)** $y = x^2$

This is a parabola with vertex (0, 0) opening upward.
The correct graph is C.

(b) $x = y^2$

This is a parabola with vertex (0, 0) opening to the right.
The correct graph is D.

(c) $x = 2(y + 3)^2 - 4$

This is a parabola with vertex (−4, −3) opening to the right.
The correct graph is F.

(d) $y = 2(x + 3)^2 - 4$

This is a parabola with vertex (−3, −4) opening upward.
The correct graph is I.

(e) $y = -\frac{1}{3}x^2$

This is a parabola with vertex (0, 0) opening downward.
The correct graph is G.

(f) $x = -\frac{1}{3}y^2$

This is a parabola with vertex (0, 0) opening to the left.
The correct graph is A.

(g) $x^2 + y^2 = 25$

This is a circle centered at (0, 0) with a radius of 5.
The correct graph is J.

(h) $(x - 3)^2 + (y + 4)^2 = 25$

This is a circle centered at (3, −4) with a radius of 5.
The correct graph is H.

(i) $(x + 3)^2 + (y - 4)^2 = 25$

This is a circle centered at $(-3, 4)$ with a radius of 5.

The correct graph is B.

(j) $x^2 + y^2 = -4$

This is the form of the equation of a circle centered at the origin, but i^2 cannot be negative, so there is no graph.

The correct "graph" is E.

49.
$$\frac{k}{\sqrt{D}} = \frac{2.82 \times 10^7}{\sqrt{42.5 \times 10^6}}$$

$$= \frac{2.82 \times 10^7}{\sqrt{42.5 \times 10^3}}$$

$$\approx .432568 \times 10^4$$

$$\approx 4326$$

Since $V = 2090$, we have

$$V < \frac{k}{\sqrt{D}},$$

so the shape of the satellite's trajectory was elliptic.

53. $\dfrac{x^2}{16} + \dfrac{y^2}{12} = 1$

Solve for y.

$$\frac{y^2}{12} = 1 - \frac{x^2}{16}$$

$$y^2 = 12\left(1 - \frac{x^2}{16}\right)$$

$$y = \pm\sqrt{12\left(1 - \frac{x^2}{16}\right)}$$

Using a graphing calculator, graph

$$Y_1 = \sqrt{12\left(1 - \frac{x^2}{16}\right)}$$

and $Y_2 = -\sqrt{12\left(1 - \frac{x^2}{16}\right)}$

on the same screen.

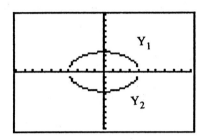

Use TRACE to find the coordinates of several points on the ellipse. The required calculations are shown here for one sample point,

$$P(1.9148936, 3.0413627).$$

Using the distance formula, we find

Distance of P from $(2, 0)$

$= \sqrt{(1.9148936 - 2)^2 + (3.0413627 - 0)^2}$

$\approx 3.0425532.$

The distance between a point and a line is defined as the perpendicular distance, so

Distance of P from the line $x = 8$

= distance of P from $(8, 3.0413627)$

$= |1.9148936 - 8|$

$= 6.0851064.$

Since

$$\frac{1}{2}(6.0851064) = 3.0425532,$$

we have shown that

[Distance of P from $(2, 0)$]

$= \dfrac{1}{2}[\text{Distance of P from the line } x = 8]$.

Chapter 7 Review Exercises

For Exercises 1–7, see the answer graphs in the textbook.

1. $x = 4(y - 5)^2 + 2$

   The graph is a parabola opening to the right. The vertex is $(2, 5)$, and the axis is the horizontal line $y = 5$.

x	18	6	2	6	18
y	3	4	5	6	7

3. $x = 5y^2 - 5y + 3$

   Complete the square on y to find the vertex and axis of this parabola.

   $$x = 5\left(y^2 - y + \frac{1}{4} - \frac{1}{4}\right) + 3$$

   $$= 5\left(y - \frac{1}{2}\right)^2 + 5\left(-\frac{1}{4}\right) + 3$$

   $$= 5\left(y - \frac{1}{2}\right)^2 + \frac{7}{4}$$

   The parabola opens to the right. The vertex is $(7/4, 1/2)$ and the axis is the horizontal line $y = 1/2$.

x	$\frac{27}{4}$	3	$\frac{7}{4}$	3	$\frac{27}{4}$
y	$-\frac{1}{2}$	0	$\frac{1}{2}$	1	$\frac{3}{2}$

5. $y^2 = -\frac{2}{3}x$

   The equation has the form $y^2 = 4px$, with $4p = -2/3$, so $p = -1/6$. The parabola is horizontal, with focus $(-1/6, 0)$, directrix $x = 1/6$, and the line $y = 0$ (the x-axis) as axis of the parabola.

x	-6	$-\frac{3}{2}$	0
y	±2	±1	0

7. $3x^2 = y$

   $$x^2 = \frac{1}{3}y$$

   The equation has the form $x^2 = 4py$, with $4p = 1/3$, so $p = 1/12$. The parabola is vertical, with focus $(0, 1/12)$, directrix $y = -1/12$, and the y-axis as axis of the parabola.

x	±2	±1	0
y	12	3	0

9. Focus $(4, 0)$, vertex at the origin

   Since the focus is on the x-axis, the parabola is horizontal. It opens to the right since $p = 4$ is positive. The equation has the form

   $$y^2 = 4px.$$

   Substituting 4 for p, we get

   $$y^2 = 16x$$

   or $\qquad x = \frac{1}{16}y^2.$

11. Through $(-3, 4)$, opening upward, vertex at the origin

    The form of the equation is

    $$x^2 = 4py.$$

Substitute x = -3 and y = 4, and solve for p.

$$(-3)^2 = 4p(4)$$
$$9 = 16p$$
$$p = \frac{9}{16}$$

The equation is

$$x^2 = 4\left(\frac{9}{16}\right)y = \frac{9}{4}y$$

or $y = \frac{4}{9}x^2$.

13. $y^2 + 9x^2 = 9$

$$x^2 + \frac{y^2}{9} = 1$$

The graph of this equation is an ellipse.

15. $3y^2 - 5x^2 = 30$

$$\frac{y^2}{10} - \frac{x^2}{6} = 1$$

The graph of this equation is a hyperbola.

17. $4x^2 - y = 0$

$$y = 4x^2$$

The graph of this equation is a parabola.

19. $4x^2 - 8x + 9y^2 + 36 = -4$
$$4(x^2 - 2x) + 9(y^2 + 4y) = -4$$
$$4(x^2 - 2x + 1) + 9(y^2 + 4y + 4) = -4 + 4 + 36$$
$$4(x - 1)^2 + 9(y + 2)^2 = 36$$
$$\frac{(x - 1)^2}{9} + \frac{(y + 2)^2}{4} = 1$$

The graph of this equation is an ellipse.

21. $4x^2 + y^2 = 36$

$$\frac{4x^2}{36} + \frac{y^2}{36} = \frac{36}{36}$$
$$\frac{x^2}{9} + \frac{y^2}{36} = 1$$

This is an ellipse centered at (0, 0), with vertices at (0, ±6) and endpoints of the minor axis at (±3, 0). The correct graph is F.

23. $(x - 2)^2 + (y + 3)^2 = 36$

This is a circle centered at (2, -3), with a radius of 6. The correct graph is A.

25. $(y - 1)^2 - (x - 2)^2 = 36$

$$\frac{(y - 1)^2}{36} - \frac{(x - 2)^2}{36} = 1$$

This is a hyperbola centered at (2, 1), opening upward and downward, with vertices at (2, -5) and (2, 7). The correct graph is B.

For Exercises 27–41, see the answer graphs in the textbook.

27. $\frac{x^2}{4} + \frac{y^2}{9} = 1$

The graph is an ellipse with vertices at (0, ±3). The endpoints of the minor axis are (±2, 0).

29. $\frac{x^2}{64} - \frac{y^2}{36} = 1$

The graph is a hyperbola with vertices (±8, 0). The asymptotes are the lines $y = \pm\frac{3}{4}x$.

31. $\dfrac{(x + 1)^2}{16} + \dfrac{(y - 1)^2}{16} = 1$

Rewrite in standard form

$$(x + 1)^2 + (y - 1)^2 = 16$$

The graph is a circle with center $(-1, 1)$ and radius 4.

33. $4x^2 + 9y^2 = 36$

$$\dfrac{x^2}{9} + \dfrac{y^2}{4} = 1$$

The graph is an ellipse with vertices at $(\pm 3, 0)$ and endpoints of the minor axis at $(0, \pm 2)$.

35. $\dfrac{(x - 3)^2}{4} + (y + 1)^2 = 1$

The graph is an ellipse centered at $(3, -1)$. Since $a = 2$, the vertices are $(3 - 2, -1) = (1, -1)$ and $(3 + 2, -1) = (5, -1)$. Since $b = 1$, the endpoints of the minor axis are

$$(3, -1 - 1) = (3, -2)$$

and $(3, -1 + 1) = (3, 0)$.

37. $\dfrac{(y + 2)^2}{4} - \dfrac{(x + 3)^2}{9} = 1$

The graph is a hyperbola centered at $(-3, -2)$. Since $a = 2$, the vertices are at $(-3, -2 - 2) = (-3, -4)$ and $(-3, -2 + 2) = (-3, 0)$. The asymptotes are the lines

$$y + 2 = \pm\dfrac{2}{3}(x + 3)$$

$$y = \pm\dfrac{2}{3}(x + 3) - 2.$$

For Exercises 39 and 41, see the answer graphs in the textbook.

39. $\dfrac{x}{3} = -\sqrt{1 - \dfrac{y^2}{16}}$ (1)

Square both sides.

$$\dfrac{x^2}{9} = 1 - \dfrac{y^2}{16}$$

$$\dfrac{x^2}{9} + \dfrac{y^2}{16} = 1 \quad (2)$$

Equation (2) has a graph which is an ellipse with vertices at $(0, -4)$ and $(0, 4)$.

The graph of equation (1) is the left half of this ellipse. The vertical line test shows that this relation is not a function.

41. $y = -\sqrt{1 + x^2}$ (1)

Square both sides.

$$y^2 = 1 + x^2$$

$$y^2 - x^2 = 1$$

$$\dfrac{y^2}{1} - \dfrac{x^2}{1} = 1 \quad (2)$$

The graph of equation (2) is a hyperbola with vertices at $(0, \pm 1)$. The graph of equation (1) is the bottom half of the hyperbola. The vertical line test shows that the relation is a function.

43. Ellipse; vertex at $(0, 4)$, focus at $(0, 2)$, center at the origin

Since the vertex is at $(0, 4)$, we have $a = 4$, so the equation is of the form

$$\dfrac{x^2}{b^2} + \dfrac{y^2}{16} = 1.$$

Since there is a focus at $(0, 2)$, we have $c = 2$.

$$c^2 = a^2 - b^2$$
$$4 = 16 - b^2$$
$$b^2 = 12$$

The equation is

$$\frac{x^2}{12} + \frac{y^2}{16} = 1.$$

45. Hyperbola; focus at $(0, -5)$, transverse axis of length 8, center at the origin.

Since the focus is at $(0, -5)$, $c = 5$, the hyperbola opens up and down, and the equation has the form

$$\frac{y^2}{a^2} - \frac{x^2}{b^2} = 1.$$

Since the transverse axis has length 8, $2a = 8$, so $a = 4$.

$$c^2 = a^2 + b^2$$
$$25 = 16 + b^2$$
$$9 = b^2$$

The equation is

$$\frac{y^2}{16} - \frac{x^2}{9} = 1.$$

47. Parabola with focus at $(3, 2)$ and directrix $x = -3$

Since the directrix is $x = -3$ and the focus is $(3, 2)$, the vertex must be $(0, 2)$, and $p = 3$. The parabola opens to the right and the equation is

$$4px = (y - 2)^2$$
$$12x = (y - 2)^2$$
$$x = \frac{1}{12}(y - 2)^2.$$

49. Ellipse with foci at $(-2, 0)$ and $(2, 0)$ and major axis of length 10

Since the foci are at $(\pm 2, 0)$, $c = 2$, and the equation is of the form

$$\frac{x^2}{a^2} + \frac{y^2}{b^2} = 1.$$

Since the major axis is of length 10, $2a = 10$, so $a = 5$.

$$c^2 = a^2 - b^2$$
$$4 = 25 - b^2$$
$$b^2 = 21$$

The equation is

$$\frac{x^2}{25} + \frac{y^2}{21} = 1.$$

51. Hyperbola with x-intercepts ± 3; foci at $(-5, 0)$, $(5, 0)$

The foci are $(\pm 5, 0)$, so $c = 5$. The x-intercepts are ± 3, so $a = 3$.

$$c^2 = a^2 + b^2$$
$$25 = 9 + b^2$$
$$16 = b^2$$

The equation is

$$\frac{x^2}{9} - \frac{y^2}{16} = 1.$$

53. The points $F'(0, 0)$ and $F(4, 0)$ are the foci, so the center of the ellipse is $(2, 0)$. For any point P on the ellipse,

$$PF' + PF = 2a = 8,$$

so $a = 4$.
Solve for b^2.

$$c^2 = a^2 - b^2$$
$$4 = 16 - b^2$$
$$b^2 = 12$$

The equation is

$$\frac{(x - 2)^2}{16} + \frac{y^2}{12} = 1.$$

55. Graph A is an ellipse, so $e < 1$.
Graph B is a parabola, so $e = 1$.
Graph C is a circle, so $e = 0$.
Graph D is a hyperbola, so $e > 1$.

In increasing order of eccentricity, the graphs are C, A, B, D.

57. Since the eccentricity is .964,

$$e = \frac{c}{a} = .964$$
$$c = .964a.$$

The closest distance to the sun is 89, so

$$a - c = 89$$
$$a - .964a = 89$$
$$.036a = 89$$
$$a = 2472.\overline{2}$$
$$c = 2383.\overline{2}$$

Solve for b^2.

$$c^2 = a^2 - b^2$$
$$5,679,748 = 6,111,883 - b^2$$
$$b^2 = 432,135$$

The equation is

$$\frac{x^2}{6,111,883} + \frac{y^2}{432,135} = 1.$$

Chapter 7 Test

For Exercises 1 and 2, see the answer graphs in the textbook.

1. $y = -x^2 + 6x$

Complete the square on x to find the vertex and axis.

$$y = -(x^2 - 6x + 9 - 9)$$
$$= -(x - 3)^2 + 9$$

The parabola open downward. The vertex is $(3, 9)$ and the axis is the vertical line $x = 3$.

x	1	2	3	4	5
y	5	8	9	8	5

2. $x = 4y^2 + 8y$

Complete the square on y to find the vertex and axis.

$$x = 4(y^2 + 2y + 1 - 1)$$
$$= 4(y + 1)^2 - 4$$

The parabola opens to the right. The vertex is $(-4, -1)$ and the axis is the horizontal line $y = -1$.

x	12	0	-4	0	12
y	-3	-2	-1	0	1

3. $x = 8y^2$

$\frac{1}{8}x = y^2$

The parabola is of the form

$$4px = y^2,$$

so $4p = 1/8$, $p = 1/32$.
The focus is $(1/32, 0)$ and the
directrix is the line $x = -1/32$.

4.   Parabola; vertex at $(2, 3)$, passing
through $(-18, 1)$, opening to the
left

The equation is of the form

$$(y - 3)^2 = 4p(x - 2).$$

Substitute $x = -18$ and $y = 1$, and
solve for p.

$$(1 - 3)^2 = 4p(-18 - 2)$$
$$-80p = 4$$
$$p = -\frac{1}{20}$$

The equation is

$$(y - 3)^2 = -\frac{1}{5}(x - 2).$$

For Exercises 6–8, see the answer graphs
in the textbook.

6. $\frac{(x - 8)^2}{100} + \frac{(y - 5)^2}{49} = 1$

The graph is an ellipse centered at
$(8, 5)$. Since $a = 10$, the vertices
are $(8 - 10, 5) = (-2, 5)$ and
$(8 + 10, 5) = (18, 5)$. Since $b = 7$,
the endpoints of the minor axis are

$$(8, 5 - 7) = (8, -2)$$
and $(8, 5 + 7) = (8, 12).$

7. $16x^2 + 4y^2 = 64$

$\frac{x^2}{4} + \frac{y^2}{16} = 1$

The graph is an ellipse with vertices $(0, \pm 4)$. The endpoints of the
minor axis are $(\pm 2, 0)$.

8. $y = -\sqrt{1 - \frac{x^2}{36}}$

Square both sides.

$$y^2 = 1 - \frac{x^2}{36}$$
$$\frac{x^2}{36} + y^2 = 1$$

This is the equation of an ellipse
with vertices $(\pm 6, 0)$ and endpoints
of the minor axis at $(0, \pm 1)$.
The graph of the original equation
is the bottom half of the ellipse.
The vertical line test shows that
this relation is a function.

9.   Ellipse; centered at the origin,
horizontal major axis with length 6,
minor axis with length 4

Since the major axis is horizontal
and has length 6, $2a = 6$, so $a = 3$.
Since the minor axis has length 4,
$2b = 4$, so $b = 2$.
The equation is

$$\frac{x^2}{9} + \frac{y^2}{4} = 1.$$

10.   Place the arch on a coordinate system with the center of the ellipse
at the origin.

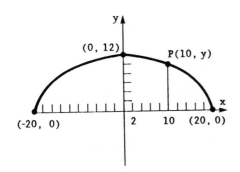

Since the arch is 40 ft wide,
2a = 40, so a = 20.
Since the arch is 12 ft high at the
center, b = 12.
The equation is

$$\frac{x^2}{400} + \frac{y^2}{144} = 1.$$

At a distance of 10 ft from the center of the bottom, x = 10.
Find the y-coordinate of the point P
on the hyperbola whose x-coordinate
is 10.

$$\frac{x^2}{400} + \frac{y^2}{144} = 1$$

$$\frac{100}{400} + \frac{y^2}{144} = 1 \qquad Let\ x = 10$$

$$\frac{y^2}{144} = \frac{3}{4}$$

$$y^2 = 108$$

$$y = \sqrt{108} \quad y \geq 0$$

$$y \approx 10.39$$

The arch is approximately 10.39 ft
high 10 ft from the center of the
bottom.

11. $\frac{x^2}{4} - \frac{y^2}{4} = 1$

The graph is a hyperbola with vertices (±2, 0). The asymptotes are
the lines y = ±(4/4)x = ±x.

12. $9x^2 - 4y^2 = 36$

$$\frac{x^2}{4} - \frac{y^2}{9} = 1$$

The graph is a hyperbola with
vertices (±2, 0). The asymptotes
are the lines $y = ±\frac{3}{2}x$.

13. Hyperbola; x-intercepts ±5, foci at
(−6, 0) and (6, 0)

The x-intercepts are ±5, so a = 5.
The foci are (±6, 0), so c = 6.
Solve for b^2.

$$c^2 = a^2 + b^2$$

$$36 = 25 + b^2$$

$$11 = b^2$$

The equation is

$$\frac{x^2}{25} - \frac{y^2}{11} = 1.$$

14. $x^2 + 8x + y^2 - 4y + 2 = 0$

Since the x^2 and y^2 terms have the
same positive coefficient, the graph
of this equation is a circle.

15. $5x^2 + 10x - 2y^2 - 12y - 23 = 0$

Since the x^2 and y^2 terms have coefficients of different sign, the
graph of this equation is a hyperbola.

16. $3x^2 + 10y^2 - 30 = 0$

Since the x^2 and y^2 terms have
different positive coefficients,
the graph of this equation is an
ellipse.

17. $x^2 - 4y = 0$

   Since only the x term is squared, the graph of this equation is a parabola.

18. $(x + 9)^2 + (y - 3)^2 = 0$

   This is the equation of a "circle" with radius 0. The graph of this equation is a point.

19. $x^2 + 4x + y^2 - 6y + 30 = 0$

   $(x^2 + 4x + 4) + (y^2 - 6y + 9) = -30 + 4 + 9$

   $(x + 2)^2 + (y - 3)^2 = -17$

   This equation has the form of the equation of a circle. However, since r^2 cannot be negative, there is no graph of this equation.

20. $\dfrac{x^2}{25} - \dfrac{y^2}{49} = 1$

   Solve the equation for y.

   $$\frac{x^2}{25} - 1 = \frac{y^2}{49}$$

   $$\pm\sqrt{\frac{x^2}{25} - 1} = \frac{y}{7}$$

   $$\pm 7\sqrt{\frac{x^2}{25} - 1} = y$$

   Thus, the functions

   $$Y_1 = 7\sqrt{\frac{x^2}{25} - 1}$$

   $$\text{and} \quad Y_2 = -7\sqrt{\frac{x^2}{25} - 1}$$

   were used to obtain the graph.

CHAPTER 8   FURTHER TOPICS IN ALGEBRA

Section 8.1

1. $a_n = 4n + 10$

   Replace n with 1, 2, 3, 4, and 5.

   $n = 1: a_1 = 4(1) + 10 = 14$
   $n = 2: a_2 = 4(2) + 10 = 18$
   $n = 3: a_3 = 4(3) + 10 = 22$
   $n = 4: a_4 = 4(4) + 10 = 26$
   $n = 5: a_5 = 4(5) + 10 = 30$

   The first five terms are

   14, 18, 22, 26, and 30.

3. $a_n = 2^{n-1}$

   Replace n with 1, 2, 3, 4, and 5.

   $n = 1: a_1 = 2^{1-1} = 2^0 = 1$
   $n = 2: a_2 = 2^{2-1} = 2^1 = 2$
   $n = 3: a_3 = 2^{3-1} = 2^2 = 4$
   $n = 4: a_4 = 2^{4-1} = 2^3 = 8$
   $n = 5: a_5 = 2^{5-1} = 2^4 = 16$

   The first five terms are

   1, 2, 4, 8, and 16.

5. $a_n = \left(\frac{1}{3}\right)^n (n - 1)$

   Replace n with 1, 2, 3, 4, and 5.

   $n = 1: a_1 = \left(\frac{1}{3}\right)^1 (1 - 1) = \frac{1}{3}(0) = 0$
   $n = 2: a_2 = \left(\frac{1}{3}\right)^2 (2 - 1) = \frac{1}{9}(1) = \frac{1}{9}$
   $n = 3: a_3 = \left(\frac{1}{3}\right)^3 (3 - 1) = \frac{1}{27}(2) = \frac{2}{27}$
   $n = 4: a_4 = \left(\frac{1}{3}\right)^4 (4 - 1) = \frac{1}{81}(3) = \frac{1}{27}$
   $n = 5: a_5 = \left(\frac{1}{3}\right)^5 (5 - 1) = \frac{1}{243}(4)$

   $= \frac{4}{243}$

The first five terms are

$0, \frac{1}{9}, \frac{2}{27}, \frac{1}{27},$ and $\frac{4}{243}.$

7. $a_n = (-1)^n (2n)$

   Replace n with 1, 2, 3, 4, and 5.

   $n = 1: a_1 = (-1)^1 [2(1)] = -1(2)$
   $= -2$
   $n = 2: a_2 = (-1)^2 [2(2)] = 1(4) = 4$
   $n = 3: a_3 = (-1)^3 [2(3)] = -1(6) = -6$
   $n = 4: a_4 = (-1)^4 [2(4)] = 1(8) = 8$
   $n = 5: a_5 = (-1)^5 [2(5)] = -1(10)$
   $= -10$

   The first five terms are

   -2, 4, -6, 8, and -10.

9. $a_n = \frac{4n - 1}{n^2 + 2}$

   Replace n with 1, 2, 3, 4, and 5.

   $n = 1: a_1 = \frac{4(1) - 1}{(1)^2 + 2} = \frac{4 - 1}{1 + 2} = \frac{3}{3} = 1$
   $n = 2: a_2 = \frac{4(2) - 1}{(2)^2 + 2} = \frac{8 - 1}{4 + 2} = \frac{7}{6}$
   $n = 3: a_3 = \frac{4(3) - 1}{(3)^2 + 2} = \frac{12 - 1}{9 + 2}$
   $= \frac{11}{11} = 1$
   $n = 4: a_4 = \frac{4(4) - 1}{(4)^2 + 2} = \frac{16 - 1}{16 + 2}$
   $= \frac{15}{18} = \frac{5}{6}$
   $n = 5: a_5 = \frac{4(5) - 1}{(5)^2 + 2} = \frac{20 - 1}{25 + 2} = \frac{19}{27}$

   The first five terms are

   $1, \frac{7}{6}, 1, \frac{5}{6},$ and $\frac{19}{27}.$

13. The sequence of the days of the week has as its domain

$$\{1, 2, 3, 4, 5, 6, 7\}.$$

Therefore, it is a finite sequence.

15. The sequence 1, 2, 3, 4 has as its domain $\{1, 2, 3, 4\}$. Therefore, it is a finite sequence.

17. The sequence 1, 2, 3, 4, ... has as its domain $\{1, 2, 3, 4, ...\}$. Therefore, the sequence is infinite.

19. The sequence $a_1 = 3$ and for $2 \leq n \leq 10$, $a_n = 3 \cdot a_{n-1}$ has as its domain $\{1, 2, 3, ..., 10\}$. Therefore, the sequence is finite.

21. $a_1 = -2$, $a_n = a_{n-1} + 3$, for $n > 1$

$n = 2$: $a_2 = a_1 + 3 = (-2) + 3 = 1$

$n = 3$: $a_3 = a_2 + 3 = (1) + 3 = 4$

$n = 4$: $a_4 = a_3 + 3 = (4) + 3 = 7$

The first four terms are

$$-2, 1, 4, \text{ and } 7.$$

23. $a_1 = 1$, $a_2 = 1$, $a_n = a_{n-1} + a_{n-2}$, for $n \geq 3$

$n = 3$: $a_3 = a_2 + a_1 = 1 + 1 = 2$

$n = 4$: $a_4 = a_3 + a_2 = 2 + 1 = 3$

The first four terms are

$$1, 1, 2, \text{ and } 3.$$

25. $\displaystyle\sum_{i=1}^{5} (2i + 1)$

Start with $i = 1$; end with $i = 5$.

$\displaystyle\sum_{i=1}^{5} (2i + 1)$

$= (2 + 1) + (4 + 1) + (6 + 1)$
$\quad + (8 + 1) + (10 + 1)$

$= 3 + 5 + 7 + 9 + 11$

$= 35$

27. $\displaystyle\sum_{i=1}^{4} \frac{1}{j}$

$= \dfrac{1}{1} + \dfrac{1}{2} + \dfrac{1}{3} + \dfrac{1}{4} = \dfrac{25}{12}$

29. $\displaystyle\sum_{i=1}^{4} i^i$

$= 1^1 + 2^2 + 3^3 + 4^4$

$= 1 + 4 + 27 + 256$

$= 288$

31. $\displaystyle\sum_{i=1}^{6} (-1)^k \cdot k$

$= (-1)^1 \cdot 1 + (-1)^2 \cdot 2 + (-1)^3 \cdot 3$
$\quad + (-1)^4 \cdot 4 + (-1)^5 \cdot 5 + (-1)^6 \cdot 6$

$= -1 + 2 - 3 + 4 - 5 + 6$

$= 3$

33. $\displaystyle\sum_{i=1}^{5} (2x_i + 3)$

$= (2x_1 + 3) + (2x_2 + 3) + (2x_3 + 3)$
$\quad + (2x_4 + 3) + (2x_5 + 3)$

$= [2(-2) + 3] + [2(-1) + 3]$
$\quad + [2(0) + 3] + [2(1) + 3]$
$\quad + [2(2) + 3]$

$= -1 + 1 + 3 + 5 + 7$

35. $\displaystyle\sum_{i=1}^{3} (3x_i - x_i^2)$

$= (3x_1 - x_1^2) + (3x_2 - x_2^2)$
$\quad + (3x_3 - x_3^2)$

$= [3(-2) - (-2)^2] + [3(-1) - (-1)^2]$
$\quad + [3(0) - (0)^2]$

$= (-6 - 4) + (-3 - 1) + 0$

$= -10 - 4 + 0$

37. $\displaystyle\sum_{i=2}^{5} \frac{x_i + 1}{x_i + 2}$

$= \dfrac{x_2 + 1}{x_2 + 2} + \dfrac{x_3 + 1}{x_3 + 2} + \dfrac{x_4 + 1}{x_4 + 2} + \dfrac{x_5 + 1}{x_5 + 2}$

$= \dfrac{(-1) + 1}{(-1) + 2} + \dfrac{(0) + 1}{(0) + 2} + \dfrac{(1) + 1}{(1) + 2}$
$\quad + \dfrac{(2) + 1}{(2) + 2}$

$= 0 + \dfrac{1}{2} + \dfrac{2}{3} + \dfrac{3}{4}$

39. $f(x) = 4x - 7$, $\Delta x = .5$

$\displaystyle\sum_{i=1}^{4} f(x_i)\Delta x$

$= f(x_1)\Delta x + f(x_2)\Delta x + f(x_3)\Delta x$
$\quad + f(x_4)\Delta x$

$= (4x_1 - 7)(.5) + (4x_2 - 7)(.5)$
$\quad + (4x_3 - 7)(.5) + (4x_4 - 7)(.5)$

$= [4(0) - 7](.5) + [4(2) - 7](.5)$
$\quad + [4(4) - 7](.5) + [4(6) - 7](.5)$

$= (-7)(.5) + (1)(.5) + (9)(.5)$
$\quad + (17)(.5)$

$= -3.5 + .5 + 4.5 + 8.5$

41. $f(x) = 2x^2$, $\Delta x = .5$

$\displaystyle\sum_{i=1}^{4} f(x_1)\Delta x$

$= 2(x_1)^2(.5) + 2(x_2)^2(.5)$
$\quad + 2(x_3)^2(.5) + 2(x_4)^2(.5)$

$= 2(0)^2(.5) + 2(2)^2(.5) + 2(4)^2(.5)$
$\quad + 2(6)^2(.5)$

$= 0 + (8)(.5) + (32)(.5) + (72)(.5)$

$= 0 + 4 + 16 + 36$

43. $f(x) = \dfrac{-2}{x + 1}$

$\displaystyle\sum_{i=1}^{4} f(x_i)\Delta x$

$= f(x_1)\Delta x + f(x_2)\Delta x + f(x_3)\Delta x$
$\quad + f(x_4)\Delta x$

$= \dfrac{-2}{x_1 + 1}(.5) + \dfrac{-2}{x_2 + 1}(.5)$
$\quad + \dfrac{-2}{x_3 + 1}(.5) + \dfrac{-2}{x_4 + 1}(.5)$

$= \dfrac{-2}{(0) + 1}(.5) + \dfrac{-2}{(2) + 1}(.5)$
$\quad + \dfrac{-2}{(4) + 1}(.5) + \dfrac{-2}{(6) + 1}(5)$

$= (-2)(.5) + \left(-\dfrac{2}{3}\right)(.5) + \left(-\dfrac{2}{5}\right)(.5)$
$\quad + \left(-\dfrac{2}{7}\right)(.5)$

$= -1 - \dfrac{1}{3} - \dfrac{1}{5} - \dfrac{1}{7}$

45. $\displaystyle\sum_{i=1}^{5} (6 - 3i);\ 3$

Let the new index of summation be j,
with $j = i + 2$ or $i = j - 2$.

Thus,

$6 - 3i = 6 - 3(j - 2) = 12 - 3j.$

If $i = 1$, $j = 3$.

If $i = 5$, $j = 7$.

Therefore,

$$\sum_{i=1}^{5} (6 - 3i) = \sum_{i=1}^{7} (12 - 3j).$$

47. $\displaystyle\sum_{i=1}^{10} 2(3)^i;\ \ 0$

Let the new index of summation be j,
with $j = i - 1$ or $i = j + 1$.

Thus,

$$2(3)^i = 2(3)^{j+1}.$$

If $i = 1$, $j = 0$.

If $i = 10$, $j = 9$.

Therefore,

$$\sum_{i=1}^{10} 2(3)^i = \sum_{j=0}^{9} 2(3)^{j+1}.$$

49. $\displaystyle\sum_{i=-1}^{9} (i^2 - 2i);\ \ 0$

Let the new index of summation be j,
with $j = i + 1$ or $i = j - 1$.

Thus,

$$\begin{aligned}
i^2 - 2i &= [(j - 1)^2 - 2(j - 1)] \\
&= j^2 - 2j + 1 - 2j + 2 \\
&= j^2 - 4j + 3.
\end{aligned}$$

If $i = -1$, $j = 0$.

If $i = 9$, $j = 8$.

Therefore,

$$\sum_{i=-1}^{9} (i^2 - 2i)$$

$$= \sum_{j=0}^{8} [(j - 1)^2 - 2(j - 1)]$$

or

$$\sum_{j=0}^{8} (j^2 - 4j + 3).$$

51. $\displaystyle\sum_{i=1}^{5} (5i + 3) = \sum_{i=1}^{5} 5i + \sum_{i=1}^{5} 3$

$$= 5 \sum_{i=1}^{5} i + 5(3)$$

$$= 5\left[\frac{5(5 + 1)}{2}\right] + 15$$

$$= 5(15) + 15$$

$$= 90$$

53. $\displaystyle\sum_{i=1}^{5} (4i^2 - 2i + 6)$

$$= \sum_{i=1}^{5} 4i^2 - \sum_{i=1}^{5} 2i + \sum_{i=1}^{5} 6$$

$$= 4 \sum_{i=1}^{5} i^2 - 2 \sum_{i=1}^{5} i + 5(6)$$

$$= 4\left[\frac{5(5 + 1)(10 + 1)}{6}\right]$$

$$- 2\left[\frac{5(5 + 1)}{2}\right] + 30$$

$$= 220 - 30 + 30$$

$$= 220$$

55. $\displaystyle\sum_{i=1}^{4} (3i^3 + 2i - 4)$

$$= \sum_{i=1}^{4} 3i^3 + \sum_{i=1}^{4} 2i - \sum_{i=1}^{4} 4$$

$$= 3 \sum_{i=1}^{4} i^3 + 2 \sum_{i=1}^{4} i - 4(4)$$

$$= 3\left[\frac{4^2(4 + 1)^2}{4}\right] + 2\left[\frac{4(4 + 1)}{2}\right] - 16$$

$$= 300 + 20 - 16$$

$$= 304$$

57. $a_n = \dfrac{n + 4}{2n}$

Using the sequence graphing capability of a graphing calculator, the given sequence appears to converge to 1/2.

59. $a_n = 2e^n$

Using the sequence graphing capability of a graphing calculator, the given sequence appears to diverge.

61. $a_n = \left(p + \dfrac{1}{n}\right)^n$

Using the sequence graphing capability of a graphing calculator, the given sequence appears to converge to $e \approx 2.71828$.

63. **(a)** Since the number of bacteria doubles every 40 minutes, it follows that

$$N_{j+1} = 2N_j \text{ for } j \geq 1.$$

(b) Two hours is 120 minutes. If

$$120 = 40(j - 1),$$

then $\qquad 3 = j - 1$

$$j = 4.$$

$N_1 = 230$, $N_2 = 460$, $N_3 = 920$, and $N_4 = 1840$.

If there are initially 230 bacteria, then there will be 1840 bacteria after two hours.

(c) We must graph the sequence $N_{j+1} = 2N_j$ for $j = 1, 2, 3, \ldots, 7$ if $N_1 = 230$.

See the answer graph in the textbook.

(d) The growth is very rapid. Since there is a doubling of the bacteria at equal intervals, their growth is exponential.

Section 8.2

1. 2, 5, 8, 11, ...

$$d = a_2 - a_1 = 5 - 2 = 3$$

3. 3, -2, -7, -12, ...

$$d = -2 - 3$$
$$d = -5$$

5. x + 3y, 2x + 5y, 3x + 7y, ...

$$\begin{aligned} d &= (2x + 5y) - (x + 3y) \\ &= 2x + 5y - x - 3y \\ &= x + 2y \end{aligned}$$

7. $a_1 = 8$ and $d = 6$

Starting with $a_1 = 8$, add $d = 6$ to each term to get the next term.

$$\begin{aligned} a_2 &= 8 + 6 = 14 \\ a_3 &= 14 + 6 = 20 \\ a_4 &= 20 + 6 = 26 \\ a_5 &= 26 + 6 = 32 \end{aligned}$$

The first five terms are

8, 14, 20, 26, and 32.

9. $a_1 = 5$, $d = -2$

$$a_2 = 5 + (-2) = 3$$
$$a_3 = 3 + (-2) = 1$$
$$a_4 = 1 + (-2) = -1$$
$$a^5 = -1 + (-2) = -3$$

The first five terms are

$$5, \ 3, \ 1, \ -1, \ \text{and} \ -3.$$

11. $a_3 = 10$, $d = -2$

$$a_4 = 10 + d = 10 + (-2) = 8$$
$$a_5 = 8 + d = 8 + (-2) = 6$$

Subtract the common difference -2 to find the earlier terms.

$$a_2 = a_3 - d = 10 - (-2) = 12$$
$$a_1 = a_2 - d = 12 - (-2) = 14$$

The first five terms are

$$14, \ 12, \ 10, \ 8, \ \text{and} \ 6.$$

13. $a_1 = 5$, $d = 2$

$$a_8 = a_1 + 7d$$
$$= 5 + 7(2)$$
$$a_8 = 19$$

$$a_n = a_1 + (n - 1)d$$
$$= 5 + (n - 1)(2)$$
$$a_n = 3 + 2n$$

15. $a_3 = 2$, $d = 1$

First, find a_1.

$$a_3 = a_1 + 2d$$
$$2 = a_1 + 2(1)$$
$$a_1 = 0$$

$$a_8 = a_1 + 7d$$
$$a_8 = 0 + 7(1) = 7$$

$$a_n = a_1 + (n - 1)d$$
$$= 0 + (n - 1)(1)$$
$$a_n = n - 1$$

17. $a_1 = 8$, $a_2 = 6$

First, find d.

$$a_2 = a_1 + d$$
$$6 = 8 + d$$
$$d = -2$$

$$a_8 = a_1 + 7d$$
$$a_8 = 8 + 7(-2) = -6$$

$$a_n = a_1 + (n - 1)d$$
$$= 8 + (n - 1)(-2)$$
$$= 10 - 2n$$

19. $a_{10} = 6$, $a_{12} = 15$

First, find a_1 and d.

$$a_n = a_1 + (n - 1)d$$
$$a_{10} = a_1 + 9d$$
$$6 = a_1 + 9d \quad (1)$$
$$a_{12} = a_1 + 11d$$
$$15 = a_1 + 11d \quad (2)$$

Solve the system formed by equations (1) and (2) by the substitution method. Solve equation (1) for a_1.

$$a_1 = 6 - 9d \quad (3)$$

Substitute $6 - 9d$ for a_1 in equation (2).

$$15 = (6 - 9d) + 11d$$
$$15 = 6 + 2d$$
$$9 = 2d$$
$$\frac{9}{2} = d$$

Now substitute $d = \frac{9}{2}$ into equation (3) to find a_1.

$$a_1 = 6 - 9\left(\frac{9}{2}\right)$$

$$= \frac{12}{2} - \frac{81}{2}$$

$$a_1 = -\frac{69}{2}$$

Now find a_8 and a_n.

$$a_8 = a_1 + 7d$$

$$= -\frac{69}{2} + 7\left(\frac{9}{2}\right)$$

$$= -\frac{69}{2} + \frac{63}{2}$$

$$a_8 = -\frac{6}{2} = -3$$

$$a_n = a_1 + (n - 1)(d)$$

$$= -\frac{69}{2} + (n - 1)\left(\frac{9}{2}\right)$$

$$= -\frac{69}{2} + \frac{9}{2}n - \frac{9}{2}$$

$$= -\frac{78}{2} + \frac{9}{2}n$$

$$a_n = -39 + \frac{9}{2}n$$

21. $a_1 = x$, $a_2 = x + 3$

$$d = (x + 3) - x = 3$$

$$a_8 = a_1 + 7d$$

$$= x + 7(3)$$

$$a_8 = x + 21$$

$$a_n = a_1 + (n - 1)d$$

$$= x + (n - 1)(3)$$

$$a_n = x + 3n - 3$$

23. $a_5 = 27$ and $a_{15} = 87$

$$a_5 = a_1 + (5 - 1)d$$

$$a_5 = a_1 + 4d$$

$$27 = a_1 + 4d \quad (1)$$

$$a_{15} = a_1 + (15 - 1)d$$

$$a_{15} = a_1 + 14d$$

$$87 = a_1 + 14d \quad (2)$$

Form a system of equations using equations (1) and (2).

$$a_1 + 4d = 27 \quad (1)$$

$$a_1 + 14d = 87 \quad (2)$$

Multiply equation (1) by 7 and equation (2) by -2; then add the results.

$$7a_1 + 28d = 189$$

$$\underline{-2a_1 - 28d = 174}$$

$$5a_1 \qquad = 15$$

$$a_1 = 3$$

25. $S_{16} = -160$ and $a_{16} = -25$

$$S_n = \frac{n}{2}(a_1 + a_n)$$

$$S_{16} = \frac{16}{2}(a_1 + a_{16})$$

$$-160 = 8(a_1 - 25)$$

$$-160 = 8a_1 - 200$$

$$40 = 8a_1$$

$$a_1 = 5$$

27. Find S_{10} if $a_1 = 8$ and $d = 3$.

S_{10} represents the sum of the first ten terms of a sequence.

To find S_{10}, substitute $n = 10$, $a_1 = 8$, and $d = 3$ into the formula

$$S_n = \frac{n}{2}[2a_1 + (n - 1)d].$$

$$S_{10} = \frac{10}{2}[2(8) + (9)(3)]$$

$$= 5(16 + 27) = 5(43)$$

$$= 215$$

29. Find S_{10} if $a_3 = 5$ and $a_4 = 8$.

Find a_1 and d first.

$$d = a_4 - a_3 = 8 - 5 = 3$$

$$a_3 = a_1 + 2d$$

$$5 = a_1 + 2 \cdot 3$$

$$-1 = a_1$$

$$S_n = \frac{n}{2}[2a_1 + (n - 1)d$$

$$S_{10} = \frac{10}{2}[2(-1) + (9)(3)]$$

$$= 5(-2 + 27) = 5(25)$$

$$= 125$$

31. To find S_{10} for the sequence 5, 9, 13, ..., substitute $n = 10$, $a_1 = 5$, and $d = 4$ into the formula

$$S_n = \frac{n}{2}[2a_1 + (n - 1)d].$$

$$S_{10} = \frac{10}{2}[2(5) + (9)(4)]$$

$$= 5(10 + 36) = 230$$

33. $a_1 = 10$, $a_{10} = 5.5$

$$S_n = \frac{n}{2}(a_1 + a_n)$$

$$S_{10} = \frac{10}{2}(10 + 5.5)$$

$$= 5(15.5)$$

$$= 77.5$$

35. $S_{20} = 1090$, $a_{20} = 102$

The formula for the sum is

$$S_n = \frac{n}{2}(a_1 + a_n), \text{ so}$$

$$S_{20} = \frac{20}{2}(a_1 + a_{20}).$$

$$1090 = 10(a_1 + 102)$$

Solve for a_1.

$$109 = a_1 + 102$$

$$7 = a_1$$

Solve for d.

$$a_n = a_1 + (n - 1)d$$

$$a_{20} = a_1 + (20 - 1)d$$

$$102 = 7 + 19d$$

$$95 = 19d$$

$$5 = d$$

37. $S_{12} = -108$, $a_{12} = -19$

The formula for the sum is

$$S_n = \frac{n}{2}(a_1 + a_n), \text{ so}$$

$$S_{12} = \frac{12}{2}(a_1 + a_{12}).$$

$$-108 = 6(a_1 - 19)$$

$$-108 = 6a_1 - 114$$

$$6 = 6a_1$$

$$1 = a_1$$

Solve for d.

$$a_n = a_1 + (n - 1)d$$

$$a_{12} = a_1 + (12 - 1)d$$

$$-19 = 1 + 11d$$

$$-20 = 11d$$

$$-\frac{20}{11} = d$$

39. $\sum\limits_{i=1}^{3} (i + 4)$

This is a sum of three terms having a common difference of 1, so it is the sum of the first three terms of the arithmetic sequence having

$$a_1 = 1 + 4 = 5,$$
$$n = 3,$$

and

$$a_n = a_3 = 3 + 4 = 7.$$

Thus,

$$\sum\limits_{i=1}^{3} (i + 4) = S_3 = \frac{3}{2}(a_1 + a_3)$$
$$= \frac{3}{2}(5 + 7)$$
$$= 18.$$

41. $\sum\limits_{j=1}^{10} (2j + 3)$

There is a common difference of $2 \cdot 1$ or 2, so this is the sum of an arithmetic sequence with

$$a_1 = 2(1) + 3 = 5,$$
$$n = 10,$$

and

$$a_n = a_{10} = 2(10) + 3 = 23.$$

Thus,

$$\sum\limits_{j=1}^{10} (2j + 3) = S_{10} = \frac{10}{2}(a_1 + a_{10})$$
$$= 5(5 + 23)$$
$$= 140.$$

43. $\sum\limits_{i=1}^{12} (-5 - 8i)$

This is the sum of an arithmetic sum sequence with

$$d = -8$$
$$a_1 = -5 - 8(1) = -13$$
$$n = 12$$
$$a_n = a_{12} = -5 - 8(12) = -101.$$

$$\sum\limits_{i=1}^{12} (-5 - 8i)$$
$$= S_{12} = \frac{12}{2}[(-13) + (-101)]$$
$$= 6(-114)$$
$$= -684$$

45. $\sum\limits_{i=1}^{1000} i$

This is the sum of an arithmetic sequence with

$$d = 1$$
$$a_1 = 1$$
$$n = 1000$$
$$a_n = a_{1000} = 1000.$$

$$\sum\limits_{i=1}^{1000} i = S_{1000} = \frac{1000}{2}(1 + 1000)$$
$$= 500(1001)$$
$$= 500,500$$

47. $f(x) = mx + b$
$f(1) = m(1) + b = m + b$
$f(2) = m(2) + b = 2m + b$
$f(3) = m(3) + b = 3m + b$

48. The sequence $f(1)$, $f(2)$, $f(3)$ is an arithmetic sequence since the difference between any two adjacent terms is m.

49. The common difference is the difference between any two adjacent terms.

$$d = (3m + b) - (2m + b) = m$$

50. From Exercise 47, we know that $a_1 = m + b$. From Exercise 49, we know that $d = m$. Therefore,

$$\begin{aligned}
a_n &= a_1 + (n - 1)d \\
&= (m + b) + (n - 1)m \\
&= m + b + nm - m \\
a_n &= mn + b.
\end{aligned}$$

51. $a_n = 4.2n + 9.73$

Using the sequence feature of a graphing calculator, we obtain

$$S_{10} = 328.3.$$

53. $a_n = \sqrt{8}n + \sqrt{3}$

Using the sequence feature of a graphing calculator, we obtain

$$S_{10} \approx 172.884.$$

55. Find the sum of all the integers from 51 to 71.

$$\sum_{i=51}^{71} i = \sum_{i=1}^{71} i - \sum_{i=1}^{50} i$$

$$= S_{71} - S_{50}$$

We know $a_1 = 1$, $d = 1$, $a_{50} = 50$, and $a_{71} = 71$. Thus,

$$S_{71} = \frac{71}{2}(1 + 71) = 71(36) = 2556,$$

$$S_{50} = \frac{50}{2}(1 + 50) = 25(51) = 1275.$$

Thus, the sum is

$$S_{71} - S_{50} = 2556 - 1275 = 1281.$$

57. In every 12-hour cycle, the clock will chime $1 + 2 + 3 + \ldots + 12$ times.

$a_1 = 1$, $n = 12$, $a_{12} = 12$

$$S_{12} = \frac{12}{2}(1 + 12) = 6(13) = 78$$

Since there are two 12-hour cycles in 1 day, every day the clock will chime $2(78) = 156$ times.
Since there are 30 days in this month, the clock will chime $156 \cdot 30 = 4680$ times.

59. $a_1 = 49{,}000$
    $d = 580$
    $n = 11$

Find a_{10}.

$$\begin{aligned}
a_{10} &= a_1 + (11 - 1)580 \\
&= 49{,}000 + 5800 \\
&= 54{,}800
\end{aligned}$$

Five years from now, the population will be 54,800.

61. $a_1 = 18$
    $a_{31} = 28$
    $n = 31$

Find S_{31}.

$$\begin{aligned}
S_{31} &= \frac{31}{2}(18 + 28) \\
&= \frac{31}{2}(46) \\
&= 713
\end{aligned}$$

713 inches of material would be needed.

63. Assume that a_1, a_2, a_3, ... is an arithmetic sequence. Consider the sequence $a_1{}^2$, $a_2{}^2$, $a_3{}^2$, ... Find differences of successive terms.

$$\begin{aligned}
a_2{}^2 - a_1{}^2 &= (a_1 + d)^2 - a_1{}^2 \\
&= a_1{}^2 + 2a_1 d + d^2 - a_1{}^2 \\
&= 2a_1 d + d^2
\end{aligned}$$

$$\begin{aligned}
a_3{}^2 - a_2{}^2 &= (a_1 + 2d)^2 - (a_1 + d)^2 \\
&= a_1{}^2 + 4a_1 d + 4d^2 \\
&\quad - (a_1{}^2 + 2a_1 d + d^2) \\
&= 2a_1 d + 3d^2
\end{aligned}$$

For the sequence of squared terms to be arithmetic, these two differences must have the same value. That is,

$$\begin{aligned}
2a_1 d + d^2 &= 2a_1 d + 3d^2 \\
d^2 &= 3d^2 \\
0 &= 2d^2 \\
0 &= d^2 \\
0 &= d.
\end{aligned}$$

For $d = 0$ to be true in the sequence a_1, a_2, a_3, ..., all terms of the sequence must be the same constant.

65. Consider the arithmetic sequence a_1, a_2, a_3, a_4, a_5, ... Then, consider the sequence a_1, a_3, a_5, ...

$$\begin{aligned}
a_3 - a_1 &= (a_1 + 2d) - a_1 = 2d \\
a_5 - a_3 &= (a_1 + 4d) - (a_1 + 2d) \\
&= 2d \\
a_n - a_{n-2} &= [a_1 + (n - 1)d] \\
&\quad - [a_1 + (n - 2 - 1)d] \\
&= a_1 + nd - d - (a_1 + nd - 3d) \\
&= a_1 + nd - d - a_1 - nd + 3d \\
&= 2d
\end{aligned}$$

Since there is a common difference, $2d$, between these terms, a_1, a_3, a_5, ... is an arithmetic sequence.

Section 8.3

1. $a_1 = 5/3$, $r = 3$, $n = 4$

$$\begin{aligned}
a_2 &= a_1 r = \left(\frac{5}{3}\right)(3) = 5 \\
a_3 &= a_2 r = 5(3) = 15 \\
a_4 &= a_3 r = 15(3) = 45
\end{aligned}$$

The first four terms of the sequence are

$$\frac{5}{3}, \; 5, \; 15, \; \text{and } 45.$$

3. $a_4 = 5$, $a_5 = 10$, $n = 5$

First find r.

$$r = \frac{a_5}{a_4} = \frac{10}{5} = 2$$

$$a_3 = \frac{a_4}{r} = \frac{5}{2}$$

$$a_2 = \frac{a_3}{r} = \frac{\frac{5}{2}}{2} = \frac{5}{4}$$

$$a_1 = \frac{a_2}{r} = \frac{\frac{5}{4}}{2} = \frac{5}{8}$$

The first five terms are

$$\frac{5}{8}, \; \frac{5}{4}, \; \frac{5}{2}, \; 5, \text{ and } 10.$$

5. $a_1 = 5$, $r = -2$

$$\begin{aligned}
a_5 &= a_1 r^{5-1} \\
a_5 &= 5(-2)^4 = 80 \\
a_n &= a_1 r^{n-1} \\
a_n &= 5(-2)^{n-1}
\end{aligned}$$

7. $a_2 = -4$, $r = 3$

First find a_1.

$$a_1 = \frac{a_2}{r} = \frac{-4}{3}$$

Now find a_5 and a_n.

$$a_5 = a_1 r^{5-1}$$

$$a_5 = \left(-\frac{4}{3}\right)(3)^4 = -108$$

$$a_n = a_1 r^{n-1}$$

$$a_n = \left(-\frac{4}{3}\right)(3)^{n-1}$$

Note that

$$\left(-\frac{4}{3}\right)(3)^{n-1} = (-4)(3^{-1})(3)^{n-1}$$

$$= (-4)(3)^{n-2},$$

so $a_n = (-4)(3)^{n-2}$ is an equivalent formula for the nth term of this sequence.

9. $a_4 = 243$, $r = -3$

First find a_1.

$$a_4 = a_1 r^{4-1}$$

$$243 = a_1(-3)^3$$

$$-27a_1 = 243$$

$$a_1 = -9$$

Now find a_5 and a_n.

$$a_5 = a_1 r^{5-1}$$

$$a_5 = (-9)(-3)^4$$

$$a_5 = -729$$

$$a_n = a_1 r^{n-1}$$

$$a_n = (-9)(-3)^{n-1}$$

Note that

$$(-9)(-3)^{n-1} = -(-3)^2(-3)^{n-1}$$

$$= -(-3)^{n+1},$$

so $a_n = -(-3)^{n+1}$ is an equivalent formula for the nth term of this sequence.

11. -4, -12, -36, -108, ...

First find r.

$$r = \frac{-12}{-4} = 3$$

Now find a_5 and a_n.

$$a_5 = a_1 r^{5-1}$$

$$a_5 = (-4)(3)^4 = -324$$

$$a_n = a_1 r^{n-1}$$

$$a_n = -4(3)^{n-1}$$

13. $\frac{4}{5}$, 2, 5, $\frac{25}{2}$, ...

$$r = \frac{2}{\frac{4}{5}} = \frac{5}{2}$$

$$a_5 = a_1 r^{5-1}$$

$$a_5 = \left(\frac{4}{5}\right)\left(\frac{5}{2}\right)^4 = \frac{125}{4}$$

$$a_n = a_1 r^{n-1}$$

$$a_n = \left(\frac{4}{5}\right)\left(\frac{5}{2}\right)^{n-1}$$

Note that

$$\left(\frac{4}{5}\right)\left(\frac{5}{2}\right)^{n-1} = \frac{2^2}{5^1} \cdot \frac{5^{n-1}}{2^{n-1}}$$

$$= \frac{5^{n-2}}{2^{n-2}},$$

so $a_n = \dfrac{5^{n-2}}{2^{n-2}}$ is an equivalent formula for the nth term of this se-
quence.

15. 5, x, .6

If the sequence is arithmetic,

$$x - 5 = .6 - x.$$

16. $x - 5 = .6 - x$

$$2x = 5.6$$

$$x = 2.8$$

The solution is 2.8.
The sequence is 5, 2.8, .6.

17. 5, x, .6

If the sequence is geometric,

$$\frac{x}{5} = \frac{.6}{x}.$$

18. $\dfrac{x}{5} = \dfrac{.6}{x}$

$$x^2 = 3$$

$$x = \pm\sqrt{3}$$

The positive solution is $\sqrt{3}$.
The sequence is 5, $\sqrt{3}$, .6.

19. $a_3 = 5$ and $a_8 = \dfrac{1}{625}$

Using $a_n = a_1 r^{n-1}$, we write the
following system of equations:

$$a_3 = a_1 r^2 \qquad (1)$$
$$a_8 = a_1 r^7. \qquad (2)$$

Substituting the given values for a_3
and a_8, we have

$$a_1 r^2 = 5 \qquad (1)$$
$$a_1 r^7 = \frac{1}{625}. \qquad (2)$$

Dividing equation (2) by equation
(1), we obtain

$$\frac{a_1 r^7}{a_1 r^2} = \frac{\frac{1}{625}}{5} = \frac{1}{625} \cdot \frac{1}{5}$$

$$r^5 = \frac{1}{3125}$$

$$r = \frac{1}{5}.$$

Substituting this value into equation (1), we obtain

$$a_1 \left(\frac{1}{5}\right)^2 = 5$$

$$\frac{a_1}{25} = 5$$

$$a_1 = 125.$$

21. $a_4 = -\dfrac{1}{4}$ and $a_9 = -\dfrac{1}{128}$

Using $a_n = a_1 r^{n-1}$, we write the
following system of equations:

$$a_4 = a_1 r^3 \qquad (1)$$
$$a_9 = a_1 r^8. \qquad (2)$$

Substituting the given values for a_4
and a_9, we have

$$a_1 r^3 = -\frac{1}{4} \qquad (1)$$
$$a_1 r^8 = -\frac{1}{128}. \qquad (2)$$

Dividing equation (2) by equation
(1), we obtain

$$\frac{a_1 r^8}{a_1 r^3} = \frac{-\frac{1}{128}}{-\frac{1}{4}} = \left(-\frac{1}{128}\right)\left(-\frac{4}{1}\right)$$

$$r^5 = \frac{1}{32}$$

$$r = \frac{1}{2}.$$

Substituting this value into equation (1), we obtain

$$a_1\left(\frac{1}{2}\right)^3 = -\frac{1}{4}$$

$$\frac{a_1}{8} = -\frac{1}{4}$$

$$a_1 = -2.$$

23. 2, 8, 32, 128, ...

This geometric sequence has

$$r = \frac{8}{2} = 4 \text{ and } a_1 = 2.$$

Use the formula

$$S_n = \frac{a_1(1 - r^n)}{1 - r}$$

with $n = 5$, $a_1 = 2$, and $r = 4$.

$$S_5 = \frac{2[1 - 4^5]}{1 - 4}$$

$$= \frac{2(1 - 1024)}{-3}$$

$$= \frac{2(-1023)}{-3}$$

$$= 682$$

25. $18, -9, \frac{9}{2}, -\frac{9}{4}, \ldots$

This geometric sequence has

$$r = \frac{-9}{18} = -\frac{1}{2} \text{ and } a_1 = 18.$$

$$S_n = \frac{a_1(1 - r^n)}{1 - r}$$

$$S_5 = \frac{18\left[1 - \left(-\frac{1}{2}\right)^5\right]}{1 - \left(-\frac{1}{2}\right)}$$

$$= \frac{18\left(1 + \frac{1}{32}\right)}{\frac{3}{2}}$$

$$= 18\left(\frac{33}{32}\right)\left(\frac{2}{3}\right)$$

$$= \frac{99}{8}$$

27. $a_1 = 8.423$, $r = 2.859$

$$S_n = \frac{a_1(1 - r^n)}{1 - r}$$

$$S_5 = \frac{8.423[1 - (2.859)^5]}{1 - 2.859}$$

$$\approx \frac{8.423(-190.016)}{-1.859}$$

$$\approx 860.95$$

29. $\displaystyle\sum_{i=1}^{5} 3^i$

For this geometric series, $a_1 = 3$, $r = 3$, and $n = 5$.

$$S_n = \frac{a_1(1 - r^n)}{1 - r}$$

$$S_5 = \frac{3[1 - 3^5]}{1 - 3}$$

$$= \frac{3(1 - 243)}{-2}$$

$$= 363$$

31. $\displaystyle\sum_{j=1}^{6} 48\left(\frac{1}{2}\right)^j$

For this geometric series, $a_1 = 24$, $r = 1/2$, and $n = 6$.

$$S_n = \frac{a_1(1 - r^n)}{1 - r}$$

$$S_6 = \frac{24\left[1 - \left(\frac{1}{2}\right)^6\right]}{1 - \frac{1}{2}}$$

$$= \frac{24\left(1 - \frac{1}{64}\right)}{\frac{1}{2}}$$

$$= 24\left(\frac{63}{64}\right)\left(\frac{2}{1}\right)$$

$$= \frac{189}{4}$$

33. $\displaystyle\sum_{k=4}^{10} 2^k$

For this geometric series, $a_1 = 2^4 = 16$ and $r = 2$.

The number of terms is

$$n = 10 - 4 + 1 = 7.$$

$$S_n = \frac{a_1(1 - r^n)}{1 - r}$$

$$S_7 = \frac{16[1 - 2^7]}{1 - 2}$$

$$= \frac{16(1 - 128)}{-1}$$

$$= 2032$$

35. The sum of an infinite geometric series exists if $|r| < 1$.

37. 12, 24, 48, 96, ...

$$r = \frac{24}{12} = 2$$

The sum of the terms of this infinite geometric sequence would not converge since $r = 2$ is not between -1 and 1.

39. $-48, -24, -12, -6, \ldots$

$$r = \frac{-24}{-48} = \frac{1}{2}$$

41. $16 + 2 + \dfrac{1}{4} + \dfrac{1}{32} + \ldots$

This infinite geometric series has $a_1 = 16$ and

$$r = \frac{2}{16} = \frac{1}{8}.$$

The sum is

$$S_\infty = \frac{a_1}{1 - r}$$

$$= \frac{16}{1 - \frac{1}{8}}$$

$$= \frac{16}{\frac{7}{8}} = 16 \cdot \frac{8}{7} = \frac{128}{7}.$$

43. $100 + 10 + 1 + \ldots$

This series has $a_1 = 100$ and

$$r = \frac{10}{100} = \frac{1}{10}.$$

The sum is

$$S_\infty = \frac{a_1}{1 - r}$$

$$= \frac{100}{1 - \frac{1}{10}}$$

$$= \frac{100}{\frac{9}{10}} = 100 \cdot \frac{10}{9} = \frac{1000}{9}.$$

45. $\dfrac{4}{3} + \dfrac{2}{3} + \dfrac{1}{3} + \ldots$

This series has $a_1 = \dfrac{4}{3}$ and

$$r = \frac{2/3}{4/3} = \frac{1}{2}.$$

The sum is

$$S_\infty = \frac{a_1}{1 - r}$$

$$= \frac{\frac{4}{3}}{1 - \frac{1}{2}}$$

$$= \frac{\frac{4}{3}}{\frac{1}{2}} = \frac{4}{3} \cdot \frac{2}{1} = \frac{8}{3}.$$

47. $\displaystyle\sum_{i=1}^{\infty} 3\left(\frac{1}{4}\right)^{i-1}$

This series has

$$a_1 = 3\left(\frac{1}{4}\right)^{1-1} = 3\left(\frac{1}{4}\right)^{0} = 3 \cdot 1 = 3$$

and $r = 1/4$.
The sum is

$$\sum_{i=1}^{\infty} 3\left(\frac{1}{4}\right)^{i-1} = S_\infty = \frac{a_1}{1 - r}$$

$$= \frac{3}{1 - \frac{1}{4}}$$

$$= \frac{3}{\frac{3}{4}}$$

$$= 3 \cdot \frac{4}{3}$$

$$= 4.$$

49. $\displaystyle\sum_{k=1}^{\infty} (.3)^k$

This series has

$$a_1 = (.3)^1 = .3 \text{ and } r = .3.$$

The sum is

$$\sum_{k=1}^{\infty} (.3)^k = S_\infty = \frac{a_1}{1 - r}$$

$$= \frac{.3}{1 - .3}$$

$$= \frac{.3}{.7}$$

$$= \frac{3}{7}.$$

51. $g(x) = ab^x$

$g(1) = ab^1 = ab$

$g(2) = ab^2$

$g(3) = ab^3$

52. The sequence $g(1)$, $g(2)$, $g(3)$ is a geometric sequence because each term after the first is a constant multiple of the preceding term. The common ratio is $\dfrac{ab^2}{ab} = b$.

53. From Exercise 51, $a_1 = ab$. From Exercise 52, $r = b$. Therefore,

$$a_n = a_1 r^{n-1}$$

$$= ab(b)^{n-1}$$

$$= ab^n.$$

55. $\displaystyle\sum_{i=1}^{10} (1.4)^i$

Using the sequence feature of a graphing calculator, we obtain

$$S_{10} \approx 97.739.$$

57. $\displaystyle\sum_{j=3}^{8} 2(.4)^j$

Using the sequence feature of a graphing calculator, the sum is approximately .212.

59. **(a)** $a_n = a_1 \cdot 2^{n-1}$

(b) If $a_1 = 100$, we have

$$a_n = 100 \cdot 2^{n-1}.$$

Since $100 = 10^2$ and $1,000,000 = 10^6$, we need to solve the equation

$$10^2 \cdot 2^{n-1} = 10^6.$$

Divide both sides by 10^2.

$$2^{n-1} = 10^4$$

Take common logs (base 10) on both sides.

$$\log 2^{n-1} = \log 10^4$$
$$(n - 1) \log 2 = 4$$
$$n - 1 = \frac{4}{\log 2}$$
$$n = \frac{4}{\log 2} + 1 \approx 14.28$$

Since the number of bacteria is increasing, the first value of n where $a_n > 1,000,000$ is 15.

(c) Since a_n represents the number of bacteria after $40(n - 1)$ minutes, a_{15} represents the number after

$$40(15 - 1) = 40 \cdot 14$$
$$= 560 \text{ minutes}$$
$$\text{or } 9 \text{ hours, 20 minutes.}$$

61. This situation may be represented as a geometric sequence with $a_1 = 100$. Also, $r = .80$, since the strength of the mixture after each draining and replacing is

$$\frac{100 - 20}{100} = \frac{80}{100} = 80\%$$

of the previous strength.
Initially the percentage of chemical in the mixture is

$$100(.80)^0 = 100 \cdot 1 = 100.$$

After 1 draining, the percentage will be

$$100(.80)^1 = 100(.80) = 80.$$

After 9 drainings, the percentage will be

$$100(.80)^9 \approx 100(.134) = 13.4.$$

That is, after 9 drainings, the strength of the mixture will be about 13.4%.

63. A machine that loses 20% of its value yearly retains 80% of its value. At the end of 6 yr, the value of the machine will be

$$100,000(.80)^6 = \$26,214.40.$$

65. Use the formula for the sum of an infinite geometric sequence with $a_1 = 40$ and $r = .8$.

$$S_\infty = \frac{a_1}{1 - r}$$
$$= \frac{40}{1 - .8}$$
$$= \frac{40}{.2}$$
$$= 200$$

The pendulum bob will swing 200 cm.

67. Use the formula for the sum of the first n terms of a geometric sequence with $a_1 = 2$, $r = 2$, and $n = 5$.

$$S_n = \frac{a_1(1 - r^n)}{1 - r}$$

$$S_5 = \frac{2[1 - 2^5]}{1 - 2}$$

$$= \frac{2(1 - 32)}{-1}$$

$$= 62$$

Going back five generations, the total number of ancestors is 62. Next, use the same formula with $a_1 = 2$, $r = 2$, and $n = 10$.

$$S_{10} = \frac{2[1 - 2^{10}]}{1 - 2}$$

$$= \frac{2(1 - 1024)}{-1}$$

$$= 2046$$

Going back ten generations, the total number of ancestors is 2046.

69. When the midpoints of the sides of an equilateral triangle are connected, the length of a side of the new triangle is one-half the length of a side of the original triangle. Use the formula for the nth term of a geometric sequence with $a_1 = 2$, $r = 1/2$, and $n = 8$.

$$a_n = a_1 r^{n-1}$$

$$a_8 = (2)\left(\frac{1}{2}\right)^{8-1}$$

$$= 2\left(\frac{1}{128}\right)$$

$$= \frac{1}{64}$$

The eighth triangle has sides of length 1/64 m.

71. The future value of an annuity uses the formula $S_n = \dfrac{a_1(1 - r^n)}{1 - r}$, where $r = 1 + \text{interest rate}$.
The payments are \$1000 for 9 yr at 8% compounded annually, so $a_1 = 1000$, $r = 1.08$, and $n = 9$.

$$S_9 = \frac{1000[1 - (1.08)^9]}{1 - 1.08}$$

$$\approx 12{,}487.56$$

The future value is \$12,487.56.

73. The future value of an annuity uses the formula $S_n = \dfrac{a_1(1 - r^n)}{1 - r}$ where $r = 1 + \text{interest rate}$.
The payments are \$2430 for 10 yr at 6% compounded annually, so $a_1 = 2430$, $r = 1.06$, and $n = 10$.

$$S_{10} = \frac{2430[1 - (1.06)^{10}]}{1 - 1.06}$$

$$\approx 32{,}029.33.$$

The future value is \$32,029.33.

75. Since a_1, a_2, a_3, ... forms a geometric sequence, it has a common ratio

$$\frac{a_{n+1}}{a_n} = r_1.$$

Since b_1, b_2, b_3, ... forms a geometric sequence, it has a common ratio

$$\frac{b_{n+1}}{b_n} = r_2.$$

Show d_1, d_2, d_3, ... forms a geometric sequence, where

$$d_n = ca_n b_n.$$

d_1, d_2, d_3, ... has a common ratio $r = r_1 r_2$, since

$$\frac{d_{n+1}}{d_n} = \frac{ca_{n+1}b_{n+1}}{ca_n b_n} = r_1 r_2.$$

Thus, it forms a geometric sequence.

Section 8.4

1. $\dfrac{6!}{3!3!} = \dfrac{6 \cdot 5 \cdot 4 \cdot 3 \cdot 2 \cdot 1}{3 \cdot 2 \cdot 1 \cdot 3 \cdot 2 \cdot 1}$

$= \dfrac{6 \cdot 5 \cdot 4}{3 \cdot 2 \cdot 1}$

$= 20$

3. $\dfrac{7!}{3!4!} = \dfrac{7 \cdot 6 \cdot 5 \cdot 4 \cdot 3 \cdot 2 \cdot 1}{3 \cdot 2 \cdot 1 \cdot 4 \cdot 3 \cdot 2 \cdot 1}$

$= \dfrac{7 \cdot 6 \cdot 5}{3 \cdot 2 \cdot 1}$

$= 35$

5. $\dbinom{8}{3} = \dfrac{8!}{3!5!}$

$= \dfrac{8 \cdot 7 \cdot 6 \cdot 5 \cdot 4 \cdot 3 \cdot 2 \cdot 1}{3 \cdot 2 \cdot 1 \cdot 5 \cdot 4 \cdot 3 \cdot 2 \cdot 1}$

$= \dfrac{8 \cdot 7 \cdot 6}{3 \cdot 2 \cdot 1}$

$= 56$

7. $\dbinom{10}{8} = \dfrac{10!}{8!2!}$

$= \dfrac{10 \cdot 9 \cdot 8!}{8! \cdot 2 \cdot 1}$

$= \dfrac{10 \cdot 9}{2 \cdot 1}$

$= 45$

9. $\dbinom{13}{13} = \dfrac{13!}{13!1!}$

$= \dfrac{13!}{13! \cdot 1}$

$= 1$

11. $\dbinom{n}{n-1} = \dfrac{n!}{(n-1)![n-(n-1)]!}$

$= \dfrac{n \cdot (n-1)!}{(n-1)!1!}$

$= \dfrac{n}{1} = n$

15. $(x + y)^6$

$= x^6 + \dbinom{6}{1}x^5 y + \dbinom{6}{2}x^4 y^2 + \dbinom{6}{3}x^3 y^3$

$\quad + \dbinom{6}{4}x^2 y^4 + \dbinom{6}{5}xy^5 + y^6$

$= x^6 + \dfrac{6!}{1!5!}x^5 y + \dfrac{6!}{2!4!}x^4 y^2$

$\quad + \dfrac{6!}{3!3!}x^3 y^3 + \dfrac{6!}{4!2!}x^2 y^4$

$\quad + \dfrac{6!}{5!1!}xy^5 + y^6$

$= x^6 + 6x^5 y + 15x^4 y^2 + 20x^3 y^3$

$\quad + 15x^2 y^4 + 6xy^5 + y^6$

17. $(p - q)^5$

$= p^5 + \dbinom{5}{1}p^4(-q) + \dbinom{5}{2}p^3(-q)^2$

$\quad + \dbinom{5}{3}p^2(-q)^3 + \dbinom{5}{4}p(-q)^4 + (-q)^5$

$= p^5 + \dfrac{5!}{1!4!}p^4(-q) + \dfrac{5!}{2!3!}p^3 q^2$

$\quad + \dfrac{5!}{3!2!}p^2(-q^3) + \dfrac{5!}{4!1!}pq^4 - q^5$

$= p^5 - 5p^4 q + 10p^3 q^2 - 10p^2 q^3$

$\quad + 5pq^4 - q^5$

19. $(r^2 + s)^5$

$= (r^2)^5 + \binom{5}{1}(r^2)^4 s + \binom{5}{2}(r^2)^3 s^2$

$\quad + \binom{5}{3}(r^2)^2 s^3 + \binom{5}{4}(r^2)s^4 + s^5$

$= r^{10} + 5r^8 s + 10r^6 s^2 + 10r^4 s^3$

$\quad + 5r^2 s^4 + s^5$

21. $(p + 2q)^4$

$= p^4 + \binom{4}{1}p^3(2q) + \binom{4}{2}p^2(2q)^2$

$\quad + \binom{4}{3}p(2q)^3 + (2q)^4$

$= p^4 + 4p^3(2q) + 6p^2(4q^2)$

$\quad + 4p(8q^3) + 16q^4$

$= p^4 + 8p^3 q + 24p^2 q^2 + 32pq^3$

$\quad + 16q^4$

23. $(7p + 2q)^4$

$= (7p)^4 + \binom{4}{1}(7p)^3(2q)$

$\quad + \binom{4}{2}(7p)^2(2q)^2 + \binom{4}{3}(7p)(2q)^3$

$\quad + (2q)^4$

$= 2401p^4 + 4(686p^3 q)$

$\quad + 6(49p^2)(4q^2) + 4(7q)(8q^3)$

$= 2401p^4 + 2744p^3 q + 1176p^2 q^2$

$\quad + 224pq^3 + 16q^4$

25. $(3x - 2y)^6$

$= (3x)^6 + \binom{6}{1}(3x)^5(-2y)$

$\quad + \binom{6}{2}(3x)^4(-2y)^2 + \binom{6}{3}(3x)^3(-2y)^3$

$\quad + \binom{6}{4}(3x)^2(-2y)^4 + \binom{6}{5}(3x)(-2y)^5$

$\quad + (-2y)^6$

$= 729x^6 + 6(243x^5)(-2y)$

$\quad + 15(81x^4)(4y^2) + 20(27x^3)(-8y^3)$

$\quad + 15(9x^2)(16y^4) + 6(3x)(-32y^5)$

$\quad + 64y^6$

$= 729x^6 - 2916x^5 y + 4860x^4 y^2$

$\quad - 4320x^3 y^3 + 2160x^2 y^4$

$\quad - 576xy^5 + 64y^6$

27. $\left(\dfrac{m}{2} - 1\right)^6$

$= \left(\dfrac{m}{2}\right)^6 + \binom{6}{1}\left(\dfrac{m}{2}\right)^5(-1) + \binom{6}{2}\left(\dfrac{m}{2}\right)(-1)^2$

$\quad + \binom{6}{3}\left(\dfrac{m}{2}\right)^3(-1)^3 + \binom{6}{4}\left(\dfrac{m}{2}\right)^2(-1)^4$

$\quad + \binom{6}{5}\left(\dfrac{m}{2}\right)(-1)^5 + (-1)^6$

$= \dfrac{m^6}{64} - 6\left(\dfrac{m^5}{32}\right) + 15\left(\dfrac{m^4}{16}\right) - 20\left(\dfrac{m^3}{8}\right)$

$\quad + 15\left(\dfrac{m^2}{4}\right) - 6\left(\dfrac{m}{2}\right) + 1$

$= \dfrac{m^6}{64} - \dfrac{3m^5}{16} + \dfrac{15m^4}{16} - \dfrac{5m^3}{2} + \dfrac{15m^2}{4}$

$\quad - 3m + 1$

29. $\left(\sqrt{2}r + \dfrac{1}{m}\right)^4$

$= (\sqrt{2}r)^4 + \binom{4}{1}(\sqrt{2}r)^3\left(\dfrac{1}{m}\right)$

$\quad + \binom{4}{2}(\sqrt{2}r)^2\left(\dfrac{1}{m}\right)^2 + \binom{4}{3}(\sqrt{2}r)\left(\dfrac{1}{m}\right)^3$

$\quad + \left(\dfrac{1}{m}\right)^4$

$= (\sqrt{2})^4 r^4 + 4\left[(\sqrt{2})^3 r^3\left(\dfrac{1}{m}\right)\right]$

$\quad + 6\left[(\sqrt{2})^2 r^2\left(\dfrac{1}{m}\right)^2\right] + 4(\sqrt{2}r)\left(\dfrac{1}{m}\right)^3$

$\quad + \left(\dfrac{1}{m}\right)^4$

$= 4r^4 + \dfrac{8\sqrt{2}r^3}{m} + \dfrac{12r^2}{m^2} + \dfrac{4\sqrt{2}r}{m^3} + \dfrac{1}{m^4}$

31. $(4h - j)^8$

Use the formula

$$\binom{n}{k - 1}x^{n-(k-1)}y^{k-1}$$

with $n = 8$, $k = 6$, $k - 1 = 5$, and $n - (k - 1) = 3$.

The sixth term of the expansion is

$$\binom{8}{5}(4h)^3(-j)^5$$

$$= \frac{8!}{5!3!}(64h^3)(-j^5)$$

$$= 56(-64h^3j^5)$$

$$= -3584h^3j^5.$$

33. $(a^2 + b)^{22}$

Here $n = 22$, $k = 15$, $k - 1 = 14$, and $n - (k - 1) = 8$.
The fifteenth term of the expansion is

$$\binom{22}{14}(a^2)^8(b)^{14}$$

$$= \frac{22!}{14!8!}(a^{16})(b^{14})$$

$$= 319{,}770a^{16}b^{14}.$$

35. $(x - y^3)^{20}$

Here $n = 20$, $k = 15$, $k - 1 = 14$, and $n - (k - 1) = 6$.
The fifteenth term of the expansion is

$$\binom{20}{14}(x)^6(-y^3)^{14}$$

$$= \frac{20!}{14!6!}(x^6)(y^{42})$$

$$= 38{,}760x^6y^{42}.$$

37. $(3x^7 + 2y^3)^8$

This expansion has 9 terms, so the middle term is the fifth term, which is

$$\binom{8}{4}(3x^7)^4(2y^3)^4$$

$$= (70)(81x^{28})(16y^{12})$$

$$= 90{,}720x^{28}y^{12}.$$

39. If the coefficients of the fifth and eighth terms in the expansion of $(x + y)^n$ are the same, then the symmetry of the expansion can be used to determine n. There are four terms before the fifth term, so there must be four terms after the eighth term. This means that the last term of the expansion is the twelfth term. This in turn means that $n = 11$, since $(x + y)^{11}$ is the expansion that has twelve terms.

41. $(1.02)^{-3}$

$$= (1 + .02)^{-3}$$

$$= 1 + (-3)(.02) + \frac{(-3)(-4)}{2!}(.02)^2$$

$$\quad + \frac{(-3)(-4)(-5)}{3!}(.02)^3 + \ldots$$

$$= 1 - .06 + .0024 - \ldots$$

$$\approx .942$$

43. $(1.01)^{3/2}$

$$= (1 + .01)^{1.5}$$

$$= 1 + (1.5)(.01) + \frac{(1.5)(.5)}{2!}(.01)^2$$

$$\quad + \frac{(1.5)(.5)(-.5)}{3!}(.01)^3 + \ldots$$

$$= 1 + .015 + .0000375 + \ldots$$

$$\approx 1.015$$

45. $(1 + x)^{-1}$

$$= 1 + (-1)x + \frac{(-1)(-2)}{2!}x^2$$

$$\quad + \frac{(-1)(-2)(-3)}{3!}x^3$$

$$\quad + \frac{(-1)(-2)(-3)(-4)}{4!}x^4 + \ldots$$

$$= 1 - x + x^2 - x^3 + x^4 - \ldots$$

49. Using a calculator, we obtain the exact value

$$10! = 3,628,800.$$

Using Stirling's formula, we obtain the approximate value

$$10! \approx \sqrt{2\pi(10)} \cdot 10^{10} \cdot e^{-10}$$

$$\approx 3,598,695.619.$$

50. $\dfrac{3,628,800 - 3,598,695.619}{3,628,800} \approx .00830$

As a percent, this is .830%.

51. Using a calculator, we obtain the exact value

$$12! = 479,001,600.$$

Using Stirling's formula, we obtain the approximate value

$$12! \approx \sqrt{2\pi(12)} \cdot 12^{12} \cdot e^{-12}$$

$$\approx 475,687,486.5.$$

Find the percent error.

$$\dfrac{479,001,600 - 475,687,486.5}{479,001,600} \approx .00692$$

As a percent, this is .692%.

52. Using a calculator, we obtain the exact value

$$13! = 6,227,020,800.$$

Using Stirling's formula, we obtain the approximate value

$$13! \approx \sqrt{2\pi(13)} \cdot 13^{13} \cdot e^{-13}$$

$$\approx 6,187,239,475.$$

Find the percent error.

$$\dfrac{6,227,020,800 - 6,187,239,475}{6,227,020,800}$$

$$\approx .00639$$

As a percent, this is .639%.
As n gets larger, the percent error decreases.

53. To find the sum of the coefficients in the expansion of $(4x - 5)^7$, let $x = 1$.

$$[4(1) - 5]^7 = (4 - 5)^7 = (-1)^7 = -1$$

The sum $a_7 + a_6 + \ldots + a_1 + a_0$ is -1.

Section 8.5

1. Prove that S_n is true for every positive integer n.
Let S_n be the statement

$$2 + 4 + 6 + \ldots + 2n = n(n + 1).$$

S_1: $2 = 1(1 + 1)$

   $2 = 2$

S_2: $2 + 4 = 2(2 + 1)$

      $6 = 6$

S_3: $2 + 4 + 6 = 3(3 + 1)$

         $12 = 12$

S_4: $2 + 4 + 6 + 8 = 4(4 + 1)$

            $20 = 20$

S_5: $2 + 4 + 6 + 8 + 10 = 5(5 + 1)$

               $30 = 30$

Step 1 Show that the statement is true for n = 1. S_1 is the statement

$$2 = 1(1 + 1),$$

which is true.

Step 2 Show that if S_k is true, then S_{k+1} is also true.

S_k is the statement

$$2 + 4 + 6 + \ldots + 2k = k(k + 1)$$

and S_{k+1} is the statement

$$2 + 4 + 6 + \ldots + 2k + 2(k + 1)$$
$$= (k + 1)(k + 2).$$

Start with S_k:

$$2 + 4 + 6 + \ldots + 2k = k(k + 1).$$

Add the (k + 1)st term, 2(k + 1), to both sides:

$$2 + 4 + 6 + \ldots + 2k + 2(k + 1)$$
$$= k(k + 1) + 2(k + 1).$$

Now factor out the common factor on the right to get

$$2 + 4 + 6 + \ldots + 2k + 2(k + 1)$$
$$= (k + 1)(k + 2).$$

This result is the statement S_{k+1}. Thus, we have shown that if S_k is true, S_{k+1} is also true. The two steps required for a proof by mathematical induction have been completed, so the statement

$$2 + 4 + 6 + \ldots + 2n = n(n + 1)$$

is true for every positive integer n.

3. Let S_n be the statement

$$3 + 6 + 9 + \ldots + 3n = \frac{3n(n + 1)}{2}.$$

Prove that S_n is true for every positive integer n.

Step 1 S_1 is the statement

$$3 = \frac{3 \cdot 1(1 + 1)}{2}$$

$$3 = \frac{6}{2},$$

which is true.

Step 2 Show that if S_k is true, then S_{k+1} is also true.

S_k is the statement

$$3 + 6 + 9 + \ldots + 3k = \frac{3k(k + 1)}{2}.$$

Add the (k + 1)st term, 3(k + 1), to both sides.

$$3 + 6 + 9 + \ldots + 3k + 3(k + 1)$$
$$= \frac{3k(k + 1)}{2} + 3(k + 1)$$
$$= \frac{3k(k + 1)}{2} + \frac{6(k + 1)}{2}$$
$$= \frac{(k + 1)(3k + 6)}{2}$$

$$3 + 6 + 9 + \ldots + 3k + 3(k + 1)$$
$$= \frac{3(k + 1)(k + 2)}{2}$$

The final equation is the statement S_{k+1}. Thus, if S_k is true, S_{k+1} is also true.

Step 1 and 2 have been completed, so S_n is true for all positive integers n.

5. Let S_n be the statement

$$2 + 4 + 8 + \ldots + 2^n = 2^{n+1} - 2.$$

Prove that S_n is true for every positive integer n.

Step 1 S_1 is the statement

$$2 = 1^{1+1} - 2$$
$$2 = 4 - 2,$$

which is true.

Step 2   Show that if S_k is true, then S_{k+1} is also true. S_k is the statement.

$$2 + 4 + 8 + \ldots + 2^k = 2^{k+1} - 2.$$

Add the (k + 1)st term, 2^{k+1}, to both sides.

$$2 + 4 + 8 + \ldots + 2^k + 2^{k+1}$$
$$= (2^{k+1} - 2) + 2^{k+1}$$
$$= 2 \cdot 2^{k+1} - 2$$
$$= 2^{k+2} - 2$$

$$2 + 4 + 8 + \ldots + 2^k + 2^{k+1}$$
$$= 2^{(k+1)+1} - 2$$

The final equation is the statement S_{k+1}. Thus, if S_k is true, S_{k+1} is also true.

Therefore, by mathematical induction, S_n is true for every positive integer n.

7. Let S_n be the statement

$$1^2 + 2^2 + 3^2 + \ldots + n^2$$
$$= \frac{n(n + 1)(2n + 1)}{6}.$$

Step 1 S_1 is the statement

$$1^2 = \frac{1(2)(3)}{6},$$
$$1 = \frac{6}{6},$$

which is true.

Step 2   Show that if S_k is true, S_{k+1} is also true. S_k is the statement

$$1^2 + 2^2 = 3^2 + \ldots + k^2$$
$$= \frac{k(k + 1)(2k + 1)}{6}.$$

Add the (k + 1)st term, $(k + 1)^2$, to both sides.

$$1^2 + 2^2 + 3^2 + \ldots + k^2 + (k + 1)^2$$
$$= \frac{k(k + 1)(2k + 1)}{6} + (k + 1)^2$$
$$= \frac{k(2k^2 + 3k + 1)}{6} + k^2 + 2k + 1$$
$$= \frac{k(2k^2 + 3k + 1)}{6} + \frac{6(k^2 + 2k + 1)}{6}$$
$$= \frac{2k^3 + 3k^2 + k + 6k^2 + 12k + 6}{6}$$
$$= \frac{2k^3 + 9k^2 + 13k + 6}{6}$$
$$= \frac{(k + 1)(2k^2 + 7k + 6)}{6}$$
$$= \frac{(k + 1)(k + 2)(2k + 3)}{6}$$
$$= \frac{(k + 1)(k + 2)[2(k + 1) + 1]}{6}$$

The final result is the statement S_{k+1}.

Thus, if S_k is true, S_{k+1} is also true. Therefore, by mathematical induction, S_n is true for every positive integer n.

9. Let S_n be the statement

$$5 \cdot 6 + 5 \cdot 6^2 + 5 \cdot 6^3 + \ldots + 5 \cdot 6^n$$
$$= 6(6^n - 1).$$

Step 1 S_1 is the statement

$$5 \cdot 6 = 6(6^1 - 1)$$
$$30 = 6 \cdot 5,$$

which is true.

Step 2 Show that if S_k is true, S_{k+1} is also true.

S_k is the statement

$$5 \cdot 6 + 5 \cdot 6^2 = 5 \cdot 6^3 + \ldots + 5 \cdot 6^k$$
$$= 6(6^k - 1).$$

Add the $(k + 1)$st term, $5 \cdot 6^{k+1}$, to both sides.

$$5 \cdot 6 + 5 \cdot 6^2 + \ldots + 5 \cdot 6^k + 5 \cdot 6^{k+1}$$
$$= 6(6^k - 1) + 5 \cdot 6^{k+1}$$
$$= 6 \cdot 6^k - 6 + 5 \cdot 6^{k+1}$$
$$= 6^{k+1} + 5 \cdot 6^{k+1} - 6$$
$$= 1 \cdot 6^{k+1} + 5 \cdot 6^{k+1} - 6$$
$$= 6 \cdot 6^{k+1} - 6$$
$$= 6(6^{k+1} - 1)$$

The final result is the statement S_{k+1}. Thus, if S_k is true, S_{k+1} is also true. Therefore, by mathematical induction, S_n is true for every positive integer n.

11. Let S_n be the statement

$$\frac{1}{1 \cdot 2} + \frac{1}{2 \cdot 3} + \frac{1}{3 \cdot 4} + \ldots + \frac{1}{n(n + 1)}$$
$$= \frac{n}{n + 1}.$$

Step 1 S_1 is the statement

$$\frac{1}{1 \cdot 2} = \frac{1}{1 + 1},$$

which is true.

Step 2 Show that if S_k is true, S_{k+1} is also true. S_k is the statement

$$\frac{1}{1 \cdot 2} + \frac{1}{2 \cdot 3} + \frac{1}{3 \cdot 4} + \ldots + \frac{1}{k(k + 1)}$$
$$= \frac{k}{k + 1}.$$

Add the $(k + 1)$st term,

$$\frac{1}{(k + 1)(k + 2)}, \text{ to both sides.}$$

$$\frac{1}{1 \cdot 2} + \frac{1}{2 \cdot 3} + \ldots$$
$$+ \frac{1}{k(k + 1)} + \frac{1}{(k + 1)(k + 2)}$$
$$= \frac{k}{k + 1} + \frac{1}{(k + 1)(k + 2)}$$
$$= \frac{k(k + 2) + 1}{(k + 1)(k + 2)}$$
$$= \frac{k^2 + 2k + 1}{(k + 1)(k + 2)}$$
$$= \frac{(k + 1)^2}{(k + 1)(k + 2)}$$
$$= \frac{k + 1}{k + 2}$$

The final result is the statement S_{k+1}. Thus, if S_k is true, S_{k+1} is also true. Therefore, by mathematical induction, S_n is true for every positive integer n.

13. Let S_n be the statement

$$\frac{1}{2} + \frac{1}{2^2} + \frac{1}{2^3} + \ldots + \frac{1}{2^n} = 1 - \frac{1}{2^n}.$$

Step 1 S_1 is the statement

$$\frac{1}{2} = 1 - \frac{1}{2},$$

which is true.

Step 2   Show that if S_k is true, S_{k+1} is also true.

S_k is the statement

$$\frac{1}{2} + \frac{1}{2^2} + \frac{1}{2^3} + \ldots + \frac{1}{2^k} = 1 - \frac{1}{2^k}.$$

Add the $(k + 1)$st term, $\frac{1}{2^{k+1}}$, to both sides.

$$\frac{1}{2} + \frac{1}{2^2} + \frac{1}{2^3} + \ldots + \frac{1}{2^k} + \frac{1}{2^{k+1}}$$

$$= 1 - \frac{1}{2^k} + \frac{1}{2^{k+1}}$$

$$= 1 - \frac{1}{2^k} + \frac{1}{2^1 2^k}$$

$$= 1 - \frac{1}{2^k} + \frac{1}{2}\left(\frac{1}{2^k}\right)$$

$$= 1 - \frac{1}{2}\left(\frac{1}{2^k}\right)$$

$$= 1 - \frac{1}{2^{k+1}}$$

The final result is the statement S_{k+1}. Thus, if S_k is true, S_{k+1} is also true. Therefore, by mathematical induction, S_n is true for every positive integer n.

15. $2^n > 2n$

If $n = 1$, we have $2^1 > 2(1)$ or $2 > 2$, which is false.
If $n = 2$, we have $2^2 > 2(2)$ or $4 > 4$, which is false.
If $n = 3$, we have $2^3 > 2(3)$ or $8 > 6$, which is true.
For $n \geq 3$, the statement is true. The statement is false for $n = 1$ or 2.

17. $2^n > n^2$

If $n = 1$, we have $2^1 > 1^2$ or $2 > 1$, which is true.
If $n = 2$, we have $2^2 > 2^2$ or $4 > 4$, which is false.
If $n = 3$, we have $2^3 > 3^2$ or $8 > 9$, which is false.
If $n = 4$, we have $2^4 > 4^2$ or $16 > 16$, which is false.
If $n = 5$, we have $2^5 > 5^2$ or $32 > 25$, which is true.
For $n \geq 5$, the statement is true. The statement is false for $n = 2, 3,$ or 4.

19. Let S_n be the statement

$(a^m)^n = a^{mn}.$ (Assume that a and m are constant.)

Step 1 S_1 is the statement

$$(a^m)^1 = a^{m \cdot 1}$$
$$a^m = a^m.$$

Thus, S_1 is true.

Step 2   Show that if S_k is true, then S_{k+1} is also true.

S_k is the statement

$(a^m)^k = a^{mk}$, and

S_{k+1} is the statement

$(a^m)^{k+1} = a^{m(k+1)}.$

$(a^m)^{k+1}$

$= (a^m)^k (a^m)^1$ *Using the property $a^m \cdot a^n = a^{m+n}$*

$= a^{mk} \cdot a^m$ *Using the assumption that S_k is true*

$= a^{mk+m}$ *Using the property $a^m \cdot a^n = a^{m+n}$*

$= a^{m(k+1)}$ *Factor mk + m*

Thus, we have shown that S_1 is true, and that if S_k is true, S_{k+1} is also true. Therefore, by mathematical induction, S_n is true for every positive integer n.

21. Let S_n be the following statement: If $n \geq 3$, $2^n > 2n$.

Step 1 S_3 is the statement

$$2^3 > 2(3)$$
$$8 > 6,$$

which is true.

Step 2 Show that if S_k is true, then S_{k+1} is also true.
Since S_k is true,

$$2^k > 2k.$$

Multiply both sides by 2.

$$2(2^k) > 2(2k)$$
$$2^{k+1} > 2k + 2k$$

However, since $k \geq 3$, $2k > 2$.

$$2^{k+1} > 2k + 2k > 2k + 2$$
$$2^{k+1} > 2(k + 1)$$

Therefore, S_{k+1} is true.
By Steps 1 and 2, S_n is true for every positive integer $n \geq 3$.

23. Let S_n be the statement:

If $a > 1$, then $a^n > 1$.

Step 1 S_1 is the statement:

If $a > 1$, then $a > 1$,
which is obviously true.

Step 2 Show that if S_k is true, then S_{k+1} is also true.
S_k is the statement:

If $a > 1$, then $a^k > 1$.
Multiply both sides of the inequality

$$a^k > 1 \text{ by } a.$$

Since $a > 1 > 0$, the direction of the inequality symbol will not be changed.

$$a \cdot a^k > a \cdot 1$$
$$a^1 \cdot a^k > a$$
$$a^{k+1} > a \quad \textit{Using the property} \\ a^m \cdot a^n = a^{m+n}$$

Since $a > 1$, we have

$$a^{k+1} > a > 1.$$

Therefore, if $a > 1$, $a^{k+1} > 1$, and S_{k+1} is true.

By Steps 1 and 2, S_n is true for every positive integer n.

25. Let S_n be the statement:
If $0 < a < 1$, then $a^n < a^{n-1}$.

Step 1 S_1 is the statement:

If $0 < a < 1$, then $a < a^0$.
Since $a^0 = 1$, this is equivalent to:

If $0 < a < 1$, then $a < 1$, which is obviously true.

Step 2 Show that S_k if true, then S_{k+1} is also true.
S_k is the statement:

If $0 < a < 1$, then $a^k < a^{k-1}$.

S_{k+1} is the statement:

If $0 < a < 1$, then

$$a^{k+1} < a^{(k+1)-1}$$

or $a^{k+1} < a^k$.

Multiply both sides of the inequality

$$a^k < a^{k-1} \text{ by } a.$$

Since $a > 0$, the direction of the inequality symbol will not change.

$$a \cdot a^k < a \cdot a^{k-1}$$
$$a^{k+1} < a^{1+(k-1)} \quad \textit{Using the property } a^m \cdot a^n = a^{m+n}$$
$$a^{k+1} < a^{(k+1)-1}$$
$$a^{k+1} < a^k$$

The final result is the statement S_{k+1}. Thus, if S_k is true, S_{k+1} is also true. By Steps 1 and 2, S_n is true for every positive integer n.

27. Let S_n be the following statement: If $n \geq 4$, then $n! > 2^n$.

Step 1 S_4 is the statement

$$4! > 2^4$$
$$24 > 16$$

which is true.

Step 2 Show that if S_k is true, then S_{k+1} is also true.
Since S_k is true,

$$k! > 2^k.$$

Multiply both sides by $k + 1$. (Since $k + 1 > 0$, the direction of the inequality symbol is not changed.)

$$(k + 1)k! > (k + 1)2^k$$

By the definition of $n!$, we have $(k + 1) \cdot k! = (k + 1)!$.

$$(k + 1)! > (k + 1)2^k > 2 \cdot 2^k$$
$$k + 1 > 2$$
$$(k + 1)! > 2^{1+k}$$
$$(k + 1)! > 2^{k+1}$$

Therefore, S_{k+1} is true.
By Steps 1 and 2, S_n is true for every positive integer $n \geq 4$.

29. Let S_n be the statement that the number of points of intersection of n lines is $\dfrac{n^2 - n}{2}$.

Since 2 is the smallest number of lines that can intersect, we need to prove this statement for every positive integer $n \geq 2$.

Step 1 S_2 is the statement that 2 nonparallel lines intersect in one point, since

$$\frac{2^2 - 2}{2} = \frac{2}{2} = 1.$$

It is obvious that S_2 is true.

Step 2 Show that if S_k is true, S_{k+1} is also true.

S_k is the statement:

k lines in a plane (no lines parallel, no three lines passing through the same point) intersect in $\dfrac{k^2 - k}{2}$ points.

If one more line is added to the set of k lines, this new (k + 1)st line will intersect each of the previous k lines in one point. Thus, there will be k additional intersection points. Thus, the number of intersection points for k + 1 lines is

$$\frac{k^2 - k}{2} + k$$

$$= \frac{k^2 - k}{2} + \frac{2k}{2}$$

$$= \frac{k^2 + k}{2}$$

$$= \frac{(k^2 + 2k + 1) - (k + 1)}{2}$$

$$= \frac{(k + 1)^2 - (k + 1)}{2}.$$

This result is the statement S_{k+1}. By Steps 1 and 2, S_n is true for every positive integer $n \geq 2$.

31. The number of sides of the nth figure is $3 \cdot 4^{n-1}$ (from Exercise 26). To see this, let a_n = the number of sides of the nth figure.

$a_1 = 3$

$a_2 = 3 \cdot 4$ since each side of the first figure will develop into 4 sides.

$a_3 = 3 \cdot 4^2$, and so on.

This gives a geometric sequence with $a_1 = 3$ and $r = 4$, so

$$a_n = 3 \cdot 4^{n-1}.$$

To find the perimeter of each figure, multiply the number of sides by the length of each side. In each figure, the lengths of the sides are 1/3 the lengths of the sides in the preceding figure.

Thus, if P_n = perimeter of nth figure,

$$P_1 = 3(1) = 3$$

$$P_2 = 3 \cdot 4\left(\frac{1}{3}\right) = 4$$

$$P_3 = 3 \cdot 4^2\left(\frac{1}{9}\right) = \frac{16}{3}, \text{ and so on.}$$

This gives a geometric sequence with

$$P_1 = 3 \text{ and } r = \frac{4}{3}.$$

Thus, $P_n = a_1 r^{n-1}$

$$P_n = 3\left(\frac{4}{3}\right)^{n-1}.$$

The result may also be written as

$$P_n = \frac{3^1 \cdot 4^{n-1}}{3^{n-1}}$$

$$= \frac{4^{n-1}}{3^{-1} \cdot 3^{n-1}}$$

$$P_n = \frac{4^{n-1}}{3^{n-2}}.$$

33. With 1 ring, 1 move is required. With 2 rings, 3 moves are required. Note that

$$3 = 2 + 1.$$

With 3 rings, 7 moves are required. Note that

$$7 = 2^2 + 2 + 1.$$

With n rings,

$$2^{n-1} + 2^{n-2} + \ldots + 2^1 + 1$$

moves are required.

Let S_n be the following statement:
For n rings, the number of required
moves is

$$2^{n-1} + 2^{n-2} + \ldots + 2^1 + 1.$$

We will prove that S_n is true for
every positive integer n.

Step 1 S_1 is the following statement:
For 1 ring, the number of required moves is 2^{1-1} or 2^0, which
is 1. The statement is true.

Step 2 Show that if S_k is true, then
S_{k+1} is also true.
Assume k + 1 rings are on the
first peg. Since S_k is true, the
top k rings can be moved to the
second peg in

$$2^{k-1} + 2^{k-2} + \ldots + 2^k + 1$$

moves.
Now move the bottom ring to the
third peg.
Since S_k is true, move the k rings
on the second peg on top of the
largest ring on the third peg in

$$2^{k-1} + 2^{k-2} + \ldots + 2^k + 1$$

moves.
The total number of moves is

$$2(2^{k-1} + 2^{k-2} + \ldots + 2 + 1) + 1$$

or

$$2^k + 2^{k-1} + \ldots + 2^2 + 2 + 1.$$

Therefore, S_{k+1} is true.
By Steps 1 and 2, S_n is true for
every positive integer n.

Section 8.6

1. $P(12, 8)$

$$= \frac{12!}{(12 - 8)!}$$

$$= \frac{12!}{4!}$$

$$= \frac{12 \cdot 11 \cdot 10 \cdot 9 \cdot 8 \cdot 7 \cdot 6 \cdot 5 \cdot 4!}{4!}$$

$$= 12 \cdot 11 \cdot 10 \cdot 9 \cdot 8 \cdot 7 \cdot 6 \cdot 5$$

$$= 19,958,400$$

3. $P(9, 2) = \dfrac{9!}{(9 - 2)!} = \dfrac{9!}{7!} = \dfrac{9 \cdot 8 \cdot 7!}{7!}$

$$= 9 \cdot 8 = 72$$

5. $P(5, 1) = \dfrac{5!}{(5 - 1)!} = \dfrac{5!}{4!} = \dfrac{5 \cdot 4!}{4!} = 5$

7. $\dbinom{4}{2} = \dfrac{4!}{(4 - 2)!2!} = \dfrac{4!}{2!2!}$

$$= \dfrac{4 \cdot 3 \cdot 2!}{2!2!} = 6$$

9. $\dbinom{6}{0} = \dfrac{6!}{(6 - 0)!0!} = \dfrac{6!}{6! \cdot 1} = 1$

11. $\dbinom{12}{4} = \dfrac{12!}{(12 - 4)!4!} = \dfrac{12!}{8!4!}$

$$= \dfrac{12 \cdot 11 \cdot 10 \cdot 9 \cdot 8!}{8!4 \cdot 3 \cdot 2 \cdot 1}$$

$$= \dfrac{12 \cdot 11 \cdot 10 \cdot 9}{4 \cdot 3 \cdot 2 \cdot 1} = 495$$

13. **(a)** Since the order of digits in a
telephone number does matter, this
involves a permutation.

(b) Since the order of digits in a
social security number does matter,
this involves a permutation.

(c) Since the order of the cards in a poker hand does not matter, this involves a combination.

(d) Since the order of members on a committee of politicians does not matter, this involves a combination.

(e) Since the order of numbers of the "combination" on a combination lock does matter, this involves a permutation.

(f) Since the order does not matter, the lottery choice of six numbers involves a combination.

(g) Since the order of digits and/or letters on a license plate does matter, this involves a permutation.

15. $5 \cdot 3 \cdot 2 = 30$

There are 30 different homes available if a builder offers a choice of 5 basic plans, 3 roof styles, and 2 exterior finishes.

17. **(a)** The first letter can be one of 2.
The second letter can be one of 25.
The third letter can be one of 24.
The fourth letter can be one of 23.

Since $2 \cdot 25 \cdot 24 \cdot 23 = 27,600$, there are 27,600 different call letters without repeats.

(b) With repeats, the count is

$$2 \cdot 26 \cdot 26 \cdot 26 = 35,152.$$

(c) The first letter can be one of 2. The second letter can be one of 24, since it cannot repeat the first letter or be R.
The third letter can be one of 23, since it cannot repeat either of the first two letters or be R.
The fourth letter can be one of 1, since it must be R.
Since $2 \cdot 24 \cdot 23 \cdot 1 = 1104$, there are 1104 different such call letters.

19. Use the multiplication principle of counting.

$$3 \cdot 5 = 15$$

There are 15 different first- and middle-name arrangements.

21. **(a)** The first three positions could each be any one of 26 letters, and the second three positions could each be any one of 10 numbers.

$$26 \cdot 26 \cdot 26 \cdot 10 \cdot 10 \cdot 10$$
$$= 17,576,000$$

17,576,000 license plates were possible.

(b) $10 \cdot 10 \cdot 10 \cdot 26 \cdot 26 \cdot 26$
$$= 17,576,000$$

17,576,000 additional license plates were made possible by the reversal.

(c) $26 \cdot 10 \cdot 10 \cdot 10 \cdot 26 \cdot 26 \cdot 26$
$$= 456,976,000$$

456,976,000 plates were provided by prefixing the previous pattern with an additional letter.

23. The number of ways in which 6 people can be seated in 6 seats in a row is given by

$$P(6, 6) = \frac{6!}{(6 - 6)!}$$

$$= \frac{6!}{0!}$$

$$= 6!$$

$$= 720.$$

25. He has 6 choices for the first course, 5 choices for the second, and 4 choices for the third.

$$6 \cdot 5 \cdot 4 = 120$$

27. The number of ways in which the 3 officers can be chosen from the 15 members is given by

$$P(15, 3) = \frac{15!}{(15 - 3)!}$$

$$= \frac{15!}{12!}$$

$$= \frac{15 \cdot 14 \cdot 13 \cdot 12!}{12!}$$

$$= 15 \cdot 14 \cdot 13$$

$$= 2730.$$

29. 5 players can be assigned the 5 positions in

$$P(5, 5) = 5! = 120 \text{ ways.}$$

10 players can be assigned 5 positions in

$$P(10, 5) = \frac{10!}{(10 - 5)!}$$

$$= \frac{10!}{5!}$$

$$= 10 \cdot 9 \cdot 8 \cdot 7 \cdot 6$$

$$= 30,240 \text{ ways.}$$

31. We want to choose 4 committee members out of 30 and the order is not important. The number of possible committees is

$$\binom{30}{4} = \frac{30!}{(30 - 4)!4!}$$

$$= \frac{30!}{26!4!}$$

$$= \frac{30 \cdot 29 \cdot 28 \cdot 27 \cdot 26!}{4!26!}$$

$$= \frac{30 \cdot 29 \cdot 28 \cdot 27}{24}$$

$$= 27,405.$$

33. $\binom{6}{3} = \dfrac{6!}{(6 - 3)!3!}$

$$= \frac{6!}{3!3!}$$

$$= \frac{6 \cdot 5 \cdot 4 \cdot 3!}{3! \cdot 3 \cdot 2 \cdot 1}$$

$$= 20$$

20 different kinds of hamburgers can be made.

35. This problem involves choosing 2 members from a set of 5 members. There are $\binom{5}{2}$ such subsets.

$$\binom{5}{2} = \frac{5!}{3!2!}$$

$$= \frac{5 \cdot 4 \cdot 3!}{3! \cdot 2}$$

$$= \frac{5 \cdot 4}{2}$$

$$= 10$$

There are 10 different 2-card combinations.

37. Since 2 blue marbles are to be chosen and there are 8 blue marbles, this problem involves choosing 2 members from a set of 8 members. There are $\binom{8}{2}$ ways of doing this.

$$\binom{8}{2} = \frac{8!}{(8-2)!2!}$$

$$= \frac{8!}{6!2!}$$

$$= \frac{8 \cdot 7 \cdot 6!}{6!2!}$$

$$= \frac{8 \cdot 7}{2}$$

$$= 28$$

28 samples of 2 marbles can be drawn in which both marbles are blue.

39. There are 5 liberals and 4 conservatives, giving a total of 9 members. Three members are chosen as delegates to a convention.

(a) There are $\binom{9}{3}$ ways of doing this.

$$\binom{9}{3} = \frac{9!}{(9-3)!3!}$$

$$= \frac{9!}{6!3!}$$

$$= \frac{9 \cdot 8 \cdot 7 \cdot 6!}{6!3!}$$

$$= \frac{9 \cdot 8 \cdot 7}{6}$$

$$= 84$$

84 delegations are possible.

(b) To get all liberals, we must choose 3 members from a set of 5, which can be done $\binom{5}{3}$ ways.

$$\binom{5}{3} = \frac{5!}{(5-3)!3!}$$

$$= \frac{5!}{2!3!}$$

$$= \frac{5 \cdot 4 \cdot 3!}{2!3!}$$

$$= \frac{5 \cdot 4}{2}$$

$$= 10$$

10 delegations could have all liberals.

(c) To get 2 liberals and 1 conservative involves two independent events. First select the liberals. The number of ways to do this is

$$\binom{5}{2} = \frac{5!}{3!2!}$$

$$= \frac{5 \cdot 4 \cdot 3!}{3! \cdot 2 \cdot 1}$$

$$= 10.$$

Now select the conservative. The number of ways to do this is

$$\binom{4}{1} = \frac{4!}{(4-1)!1!}$$

$$= \frac{4!}{3!}$$

$$= 4.$$

To find the number of delegations, use the fundamental principle of counting. The number of delegations with 2 liberals and 1 conservative is

$$10 \cdot 4 = 40.$$

(d) If one particular person must be on the delegation, then there are 2 people left to choose from a set consisting of 8 members.

$$\binom{8}{2} = \frac{8!}{(8-2)!2!}$$

$$= \frac{8!}{6!2!}$$

$$= \frac{8 \cdot 7 \cdot 6!}{6!2!}$$

$$= \frac{8 \cdot 7}{2}$$

$$= 28$$

28 delegations are possible which include the mayor.

41.   The problem asks how many ways can Matthew arrange his schedule. Therefore, order is important, and this is a permutation problem. There are

$$P(8, 4) = \frac{8!}{(8-4)!} = \frac{8!}{4!} = 1680$$

ways to arrange his schedule.

43.   The order of the vegetables in the soup is not important, so this is a combination problem. There are

$$\binom{6}{4} = \frac{6!}{(6-4)!4!} = \frac{6!}{2!4!} = 15$$

different soups she can make.

45.   Order is important in seatings, so this is a permutation problem. All twelve children will have a specific location; the first eleven will sit down and the twelfth will be left standing.

$$P(12, 12) = \frac{12!}{(12-12)!}$$

$$= \frac{12!}{0!}$$

$$= 12!$$

$$= 479,001,600$$

There are 479,001,600 seatings possible.

47.   A club has 8 men and 11 women members. There are a total of 8 + 11 = 19 members, and 5 of them are to be chosen. Order is not important, so this is a combination problem.

(a) Choose all men.
That is, of the 8 men, choose 5.

$$\binom{8}{5} = \frac{8!}{(8-5)!5!}$$

$$= \frac{8!}{3!5!}$$

$$= \frac{8 \cdot 7 \cdot 6 \cdot 5!}{3!5!}$$

$$= \frac{8 \cdot 7 \cdot 6}{6}$$

$$= 56$$

56 committees having 5 men can be chosen.

(b) Choose all women.
That is, of the 11 women, choose 5.

$$\binom{11}{5} = \frac{11!}{(11-5)!5!}$$

$$= \frac{11!}{6!5!}$$

$$= \frac{11 \cdot 10 \cdot 9 \cdot 8 \cdot 7 \cdot 6!}{6!5!}$$

$$= \frac{11 \cdot 10 \cdot 9 \cdot 8 \cdot 7}{5 \cdot 4 \cdot 3 \cdot 2 \cdot 1}$$

$$= 462$$

462 committees having 5 women can be chosen.

(c) Choose 3 men and 2 women. Since choosing the men and choosing the women are independent events, we can use the multiplication principle of counting to find the number of committees with 3 men and 2 women.

$$\binom{8}{3} \cdot \binom{11}{2}$$

$$= \frac{8!}{(8-3)!3!} \cdot \frac{11!}{(11-2)!2!}$$

$$= \frac{8!}{5!3!} \cdot \frac{11!}{9!2!}$$

$$= \frac{8 \cdot 7 \cdot 6 \cdot 5!}{5!3!} \cdot \frac{11 \cdot 10 \cdot 9!}{9!2!}$$

$$= \frac{8 \cdot 7 \cdot 6}{6} \cdot \frac{11 \cdot 10}{2}$$

$$= 56 \cdot 55$$

$$= 3080$$

3080 committees having 3 men and 2 women can be chosen.

(d) Choose no more than 3 women. This means choose

0 women (and 5 men) or choose

1 women (and 4 men) or choose

2 women (and 3 men) or choose

3 women (and 2 men).

Thus, the number of possible committees with no more than 3 women is

$$\binom{11}{0} \cdot \binom{8}{5} + \binom{11}{1} \cdot \binom{8}{4} + \binom{11}{2} \cdot \binom{8}{3} + \binom{11}{3} \cdot \binom{8}{2}$$

$$= 1 \cdot 56 + 11 \cdot 70 + 55 \cdot 56 + 165 \cdot 28$$

$$= 56 + 770 + 3080 + 4620$$

$$= 8526.$$

8526 committees having no more than 3 women can be chosen.

49. Order is important in arrangements, so this is a permutation problem.

$$P(9, 5) = \frac{9!}{(9-5)!}$$

$$= \frac{9!}{4!}$$

$$= \frac{9 \cdot 8 \cdot 7 \cdot 6 \cdot 5 \cdot 4!}{4!}$$

$$= 9 \cdot 8 \cdot 7 \cdot 6 \cdot 5$$

$$= 15,120$$

The plants can be arranged in 15,120 ways.

51. $P(n, n-1) = \dfrac{n!}{[n-(n-1)]!}$

$$= \frac{n!}{1!} = \frac{n!}{1} = n!$$

$$P(n, n) = \frac{n!}{(n-n)!} = \frac{n!}{0!}$$

$$= \frac{n!}{1} = n!$$

Therefore, $P(n, n-1) = P(n, n)$.

53. $P(n, 0) = \dfrac{n!}{(n-0)!} = \dfrac{n!}{n!} = 1$

55. $\dbinom{n}{0} = \dfrac{n!}{(n-0)!0!} = \dfrac{n!}{n!0!}$

$$= \frac{n!}{n! \cdot 1} = \frac{n!}{n!} = 1$$

57. $\dbinom{n}{n-r} = \dfrac{n!}{[n-(n-r)!] \cdot (n-r)!}$

$$= \frac{n!}{r!(n-r)!}$$

$$= \frac{n!}{(n-r)!r!}$$

$$= \binom{n}{r}$$

59. (a) $\log 50! = \log 1 + \log 2 + \log 3 + \dots + \log 50$

Using a sum and sequence utility on a calculator, we obtain

log 50! ≈ 64.48307487

$\quad$ 50! ≈ $10^{64.48307487}$

$\quad$ 50! ≈ $10^{.48307487} \times 10^{64}$

$\quad$ 50! ≈ $3.04140932 \times 10^{64}$.

Computing the value directly, we obtain

$\quad$ 50! ≈ $3.04140932 \times 10^{64}$.

(b) log 60! = log 1 + log 2 + log 3

$\qquad\qquad$ + ... + log 60

Using a sum and sequence utility on a calculator, we obtain

log 60! ≈ 81.92017485

$\quad$ 60! ≈ $10^{81.92017485}$

$\quad$ 60! ≈ $10^{.92017485} \times 10^{81}$

$\quad$ 60! ≈ $8.320987113 \times 10^{81}$.

Computing the value directly, we obtain

$\quad$ 60! ≈ $8.320987113 \times 10^{81}$.

(c) log 65! = log 1 + log 2 + log 3

$\qquad\qquad$ + ... + log 65

Using a sum and sequence utility on a calculator, we obtain

log 65! ≈ 90.91633025

$\quad$ 65! ≈ $10^{90.91633025}$

$\quad$ 65! ≈ $10^{.91633025} \times 10^{90}$

$\quad$ 65! ≈ $8.247650592 \times 10^{90}$.

Computing the value directly, we obtain

$\quad$ 65! ≈ $8.247650592 \times 10^{90}$.

Section 8.7

1. Let h = heads, t = tails.

    The only possible outcome is a head. Hence, the sample space is

    $$S = \{h\}.$$

3. Since each coin can be a head or a tail and there are 3 coins, the sample space is

    $$S = \{(h, h, h), (h, h, t), (h, t, h),$$
    $$(t, h, h), (h, t, t), (t, h, t),$$
    $$(t, t, h), (t, t, t)\}.$$

5. Let c = correct answer, w = wrong answer.

    There are 2 possible answers for each question and 3 questions, so the sample space is

    $$S = \{(c, c, c), (c, c, w), (c, w, c),$$
    $$(w, c, c), (w, w, c), (w, c, w),$$
    $$(c, w, w), (w, w, w)\}.$$

7. **(a)** "The result is heads" is the event $E_1 = \{h\}$. This event is certain to occur, so $P(E_1) = 1$.

    (b) "The result is tails" is the event $E_2 = \emptyset$. This event is an impossible event, so $P(E_2) = 0$.

9. The sample space is

    $$S = \{(c, c, c), (c, c, w), (c, w, c),$$
    $$(w, c, c), (w, w, c), (w, c, w),$$
    $$(c, w, w), (w, w, w)\}.$$

$n(S) = 8$

(a) The event, all three answers correct, is

$$E = \{(c, c, c)\}.$$
$$n(E) = 1$$
$$P(e) = \frac{n(E)}{n(S)} = \frac{1}{8}$$

(b) The event, all three answers wrong, is

$$E = \{(w, w, w)\}.$$
$$n(E) = 1$$
$$P(E) = \frac{1}{8}$$

(c) The event, exactly two answers correct, is

$$E = \{(c, c, w), (c, w, c),$$
$$(w, c, c)\}.$$
$$n(E) = 3$$
$$P(E) = \frac{3}{8}$$

(d) The event, at least one answer correct, is

$$E = \{(c, c, c), (c, c, w), (c, w, c),$$
$$(w, c, c), (w, w, c), (w, c, w,),$$
$$(c, w, w)\}.$$
$$n(E) = 7$$
$$P(E) = \frac{7}{8}$$

13. There are 15 marbles, so $n(S) = 15$.

(a) E_1: yellow marble is drawn
There are 3 yellow marbles, so $n(E_1) = 3$.

$$P(E_1) = \frac{3}{15} = \frac{1}{5}$$

(b) E_2: black marble is drawn
There are no black marbles, so $n(E_2) = 0$.

$$P(E_2) = \frac{0}{15} = 0$$

(c) E_3: yellow or white marble is drawn
There are 3 yellow and 4 white marbles, so $n(E_3) = 3 + 4 = 7$.

$$P(E_3) = \frac{7}{15}$$

(d) E_4: yellow marble is drawn
There are 3 yellow marbles, so $n(E_4) = 3$.

$$P(E_4) = \frac{3}{15} = \frac{1}{5}$$

E_4': yellow marble is not drawn
There are 12 non-yellow marbles, so $n(E_4') = 12$.

$$P(E_4') = \frac{12}{15} = \frac{4}{5}$$

The odds in favor of drawing a yellow marble are

$$= \frac{P(E_4)}{P(E_4')}$$

$$= \frac{\frac{1}{5}}{\frac{4}{5}}$$

$$= \frac{1}{4} \quad \text{or} \quad 1 \text{ to } 4.$$

(e) From part (b), the probability that a blue marble is drawn is

$$P(E_2) = \frac{8}{15}.$$

The probability that a blue marble is not drawn is

$$P(E_2') = 1 - P(E_2) = \frac{7}{15}.$$

The odds against drawing a blue marble are

$$= \frac{P(E_2')}{P(E_2)}$$

$$= \frac{\frac{7}{15}}{\frac{8}{15}}$$

$$= \frac{7}{8} \quad \text{or} \quad 7 \text{ to } 8.$$

15. E: sum of numbers is 5

$$= \{1 \text{ and } 4, \ 2 \text{ and } 3\}$$

From Exercise 4 or Exercise 9, $n(S) = 10$.
Therefore,

$$P(E) = \frac{2}{10} = \frac{1}{5}$$

and $P(E') = 1 - P(E) = \frac{4}{5}.$

Odds in favor of sum being 5

$$= \frac{P(E)}{P(E')} = \frac{\frac{1}{5}}{\frac{4}{5}} = \frac{1}{4} \quad \text{or} \quad 1 \text{ to } 4.$$

17. Let E be the event "candidate wins." Then E′ is the event "candidate loses."
If the odds in favor of the candidate winning are 3 to 2, then

$$P(E) = \frac{3}{3 + 2} = \frac{3}{5}.$$

Therefore,

$$P(E') = 1 - P(E) = 1 - \frac{3}{5} = \frac{2}{5}.$$

The probability that the candidate will lose is $\frac{2}{5}$.

19. The sample space is S represented by $\{M, U, U, B, B, B, C, C, C, C\}$, so $n(S) = 10$.
The probability of the union of events E and F is represented by

$$P(E \text{ or } F) = P(E) + P(F) - P(E \cap F).$$

(a) $P(U) = \dfrac{2}{10} = \dfrac{1}{5},$

$$P(B) = \frac{3}{10},$$

$$P(U \cap B) = 0, \text{ so}$$

$$P(U \text{ or } B) = \frac{1}{5} + \frac{3}{10} - 0$$

$$= \frac{5}{10} = \frac{1}{2}.$$

(b) $P(B) = \dfrac{3}{10},$

$$P(C) = \frac{4}{10}$$

$$= \frac{2}{5},$$

$$P(B \cap C) = 0, \text{ so}$$

$$P(B \text{ or } C) = \frac{3}{10} + \frac{4}{10} - 0$$

$$= \frac{7}{10}.$$

(c) $P(B) = \dfrac{3}{10},$

$$P(M) = \frac{1}{10},$$

$$P(B \cap M) = 0, \text{ so}$$

$$P(B \text{ or } M) = \frac{3}{10} + \frac{1}{10} - 0$$

$$= \frac{4}{10} = \frac{2}{5}.$$

21. **(a)** P(less than $20)

$\qquad$ = P($5–$19.99 or below $5)

$\qquad$ = P($5–$19.99) + P(below $5)

$\qquad$ = .37 + .25

$\qquad$ = .62

(b) P($40 or more)

$\qquad$ = P($40–$69.99 or $70–$99.99

$\qquad$ or $100–$149.99 or $150 or

$\qquad$ more)

$\qquad$ = .09 + .07 + .08 + .03

$\qquad$ = .27

(c) P(more than $99.99)

$\qquad$ = P($100–$149.99 or $150 or

$\qquad$ more)

$\qquad$ = .08 + .03

$\qquad$ = .11

(d) P(less than $100)

$\qquad$ = P(below $5 or $5–$19.99 or

$\qquad$ $20–$39.99 or $40–$69.99

$\qquad$ or $70–$99.99)

$\qquad$ = .25 + .37 + .11 + .09 + .07

$\qquad$ = .89

23. There are four choices for the incorrect suit. The probability of picking the incorrect card in a given suit is 12/13. The probability of picking the correct card in each of the other three suits is 1/13. Therefore, the probability of getting three picks correct and winning $200 is

$$4 \cdot \frac{12}{13} \cdot \frac{1}{13} \cdot \frac{1}{13} \cdot \frac{1}{13}$$

$$= \frac{48}{28,561} \approx .001681.$$

25. P(5 or more years) = .30

P(female) = .28

P(retirement plan contributor) = .65

P(retirement plan contributor and

female) = $\frac{1}{2}$(.28) = .14

(a) P(male) = 1 – P(female)

$\qquad$ = 1 – .28

$\qquad$ = .72

(b) P(less than 5 years)

$\qquad$ = 1 – P(5 or more years)

$\qquad$ = 1 – .30

$\qquad$ = .70

(c) P(retirement plan contributor or

female)

$\qquad$ = P(contributor) + P(female)

$\qquad$ – P(contributor and female)

$\qquad$ = .65 + .28 – .14

$\qquad$ = .79

27. In this binomial experiment, we call having a girl a success.

Then, n = 5, r = 2, and p = 1/2.

$$P(2 \text{ girls}) = \binom{5}{2}\left(\frac{1}{2}\right)^2\left(1 - \frac{1}{2}\right)^3$$

$$= 10\left(\frac{1}{2}\right)^2\left(\frac{1}{2}\right)^3$$

$$= \frac{10}{32} = \frac{5}{16}$$

$$\approx .313$$

29. In this binomial experiment, we call having a girl a success.

Then, n = 5, r = 0, and p = 1/2.

$$P(\text{No girls}) = \binom{5}{0}\left(\tfrac{1}{2}\right)^{0}\left(1 - \tfrac{1}{2}\right)^{5}$$

$$= 1 \cdot 1 \cdot \tfrac{1}{32}$$

$$= \tfrac{1}{32}$$

$$\approx .031$$

31. In this binomial experiment, we call having a girl a success.

Then, n = 5; r = 0, 1, or 2; and p = 1/2.

P(At least 3 boys)

= P(0 girls) + P(1 girl)

+ P(2 girls)

$$= \binom{5}{0}\left(\tfrac{1}{2}\right)^{0}\left(1 - \tfrac{1}{2}\right)^{5} + \binom{5}{1}\left(\tfrac{1}{2}\right)^{1}\left(1 - \tfrac{1}{2}\right)^{4}$$

$$+ \binom{5}{2}\left(\tfrac{1}{2}\right)^{2}\left(1 - \tfrac{1}{2}\right)^{3}$$

$$= 1 \cdot 1 \cdot \tfrac{1}{32} + 5 \cdot \tfrac{1}{2} \cdot \tfrac{1}{16} + 10 \cdot \tfrac{1}{4} \cdot \tfrac{1}{8}$$

$$= \tfrac{1}{32} + \tfrac{5}{32} + \tfrac{10}{32}$$

$$= \tfrac{16}{32} = .5$$

33. In this binomial experiment, we call rolling a one a success.

Then n = 12, r = 12, and p = 1/6.

P(Exactly 12 ones)

$$= \binom{12}{12}\left(\tfrac{1}{6}\right)^{12}\left(1 - \tfrac{1}{6}\right)^{0}$$

$$= 1 \cdot \tfrac{1}{6^{12}} \cdot 1$$

$$= 6^{-12}$$

$$\approx 4.6 \times 10^{-10}$$

35. In this binomial experiment, we call rolling a one a success.

Then n = 12; r = 0, 1, 2, or 3; and p = 1/6.

P(No more than 3 ones)

= P(0 ones) + P(1 one) + P(2 ones)

+ P(3 ones)

$$= \binom{12}{0}\left(\tfrac{1}{6}\right)^{0}\left(1 - \tfrac{1}{6}\right)^{12} + \binom{12}{1}\left(\tfrac{1}{6}\right)^{1}\left(1 - \tfrac{1}{6}\right)^{11}$$

$$+ \binom{12}{2}\left(\tfrac{1}{6}\right)^{2}\left(1 - \tfrac{1}{6}\right)^{10} + \binom{12}{3}\left(\tfrac{1}{6}\right)^{3}\left(1 - \tfrac{1}{6}\right)^{9}$$

$$= 1 \cdot 1 \cdot \tfrac{5^{12}}{6^{12}} + 12 \cdot \tfrac{1}{6} \cdot \tfrac{5^{11}}{6^{11}} + 66 \cdot \tfrac{1}{6^{2}} \cdot \tfrac{5^{10}}{6^{10}}$$

$$+ 220 \cdot \tfrac{1}{6^{3}} \cdot \tfrac{5^{9}}{6^{9}}$$

$$\approx .875$$

37. Using the TABLE feature of a graphing calculator, the probabilities, in order, are .125, .375, .375, and .125.

39. **(a)** First compute

$$q = (1 - p)^{I} = (1 - .1)^{2} = .81.$$

Then, with S = 4, k = 3, and q = .81,

$$P = \binom{S}{k}q^{k}(1 - q)^{S-k} = \binom{4}{3}.81^{3}(1 - .81)^{4-3}$$

$$= 4 \times .81^{3} \times .19^{1} \approx .404.$$

There is about a 40.4% chance of exactly 3 people not becoming infected.

(b) Compute

$$q = (1 - p)^{I} = (1 - .5)^{2} = .25.$$

Then,

$$P = \binom{S}{k}q^{k}(1 - q)^{S-k} = \binom{4}{3}.25^{3}(1 - .25)^{4-3}$$

$$= 4 \times .25^{3} \times .75^{1} \approx .047.$$

There is about 4.7% chance of this occurring when the disease is highly infectious.

(c) Compute

$$q = (1 - p)^I = (1 - .5)^1 = .5.$$

Then, with $S = 9$, $K = 0$, and $p = .5$,

$$P = \binom{S}{k} q^k (1 - q)^{S-k} = \binom{9}{0} .5^0 (1 - .5)^9$$

$$= 1 \times 1 \times .5^9 \approx .002.$$

There is about a .2% chance of everyone becoming infected. This means that in a large family or group of people, it is highly un-likely that everyone will become sick even though the disease is highly infectious.

41. (a) $P_{i,j} = \dfrac{\binom{2i}{j}\binom{4-2i}{2-j}}{\binom{4}{2}}$

$$P_{00} = \frac{\binom{0}{0}\binom{4}{2}}{\binom{4}{2}} = 1$$

$$P_{01} = \frac{\binom{0}{1}\binom{4}{1}}{\binom{4}{2}} = 0$$

$$P_{02} = \frac{\binom{0}{2}\binom{4}{0}}{\binom{4}{2}} = 0$$

$$P_{10} = \frac{\binom{2}{0}\binom{2}{2}}{\binom{4}{2}} = \frac{1}{6}$$

$$P_{11} = \frac{\binom{2}{1}\binom{2}{1}}{\binom{4}{2}} = \frac{2}{3}$$

$$P_{12} = \frac{\binom{2}{2}\binom{2}{0}}{\binom{4}{2}} = \frac{1}{6}$$

$$P_{20} = \frac{\binom{4}{0}\binom{0}{2}}{\binom{4}{2}} = 0$$

$$P_{21} = \frac{\binom{4}{1}\binom{0}{1}}{\binom{4}{2}} = 0$$

$$P_{22} = \frac{\binom{4}{2}\binom{0}{0}}{\binom{4}{2}} = 1$$

(b) $P = \begin{bmatrix} P_{00} & P_{01} & P_{02} \\ P_{10} & P_{11} & P_{12} \\ P_{20} & P_{21} & P_{22} \end{bmatrix} = \begin{bmatrix} 1 & 0 & 0 \\ \frac{1}{6} & \frac{2}{3} & \frac{1}{6} \\ 0 & 0 & 1 \end{bmatrix}$

(c) The matrix is symmetric. The sum of the probabilities in each row is equal to 1. The greatest prob-abilities lie along the diagonal. This means that a mother cell is most likely to produce a daughter cell like itself.

Chapter 8 Review Exercises

1. $a_n = \dfrac{n}{n+1}$

$$a_1 = \frac{1}{1+1} = \frac{1}{2}$$

$$a_2 = \frac{2}{2+1} = \frac{2}{3}$$

$$a_3 = \frac{3}{3+1} = \frac{3}{4}$$

$$a_4 = \frac{4}{4+1} = \frac{4}{5}$$

$$a_5 = \frac{5}{5+1} = \frac{5}{6}$$

The first five terms are

$$\frac{1}{2}, \frac{2}{3}, \frac{3}{4}, \frac{4}{5}, \text{ and } \frac{5}{6}.$$

This sequence does not have a common difference or a common ratio, so the sequence is neither arithmetic nor geometric.

3. $a_n = 2(n + 3)$

$a_1 = 2(1 + 3) = 8$

$a_2 = 2(2 + 3) = 10$

$a_3 = 2(3 + 3) = 12$

$a_4 = 2(4 + 3) = 14$

$a_5 = 2(5 + 3) = 16$

The first five terms are

$$8, 10, 12, 14, \text{ and } 16.$$

There is a common difference, $d = 2$, so the sequence is arithmetic.

5. $a_1 = 5$; for $n \geq 2$, $a_n = a_{n-1} - 3$

$a_2 = a_{2-1} - 3 = a_1 - 3 = 5 - 3 = 2$

$a_3 = a_2 - 3 = 2 - 3 = -1$

$a_4 = a_3 - 3 = -1 - 3 = -4$

$a_5 = a_4 - 3 = -4 - 3 = -7$

The first five terms are

$$5, 2, -1, -4, \text{ and } -7.$$

There is a common difference, $d = -3$, so the sequence is arithmetic.

7. Arithmetic, $a_3 = \pi$, $a_4 = 1$

$d = a_4 - a_3 = 1 - \pi$

$a_3 = a_1 + 2d$

$\pi = a_1 + 2(1 - \pi)$

$\pi = a_1 + 2 - 2\pi$

$3\pi - 2 = a_1$

$a_2 = a_1 + d = (3\pi - 2) + (1 - \pi)$

$\qquad = 2\pi - 1$

$a_5 = a_1 + 4d = (3\pi - 2) + 4(1 - \pi)$

$\qquad = 3\pi - 2 + 4 - 4\pi$

$\qquad = -\pi + 2$

The first five terms are

$$3\pi - 2, \ 2\pi - 1, \ \pi, \ 1, \text{ and } -\pi + 2.$$

9. Geometric, $a_1 = -5$, $a_2 = -1$

$r = \dfrac{a_2}{a_1} = \dfrac{-1}{-5} = \dfrac{1}{5}$

$a_3 = a_1 r^2 = -5\left(\dfrac{1}{5}\right)^2 = -5 \cdot \dfrac{1}{25} = -\dfrac{1}{5}$

$a_4 = a_1 r^3 = -5\left(\dfrac{1}{5}\right)^3 = -5 \cdot \dfrac{1}{125}$

$\qquad = -\dfrac{1}{25}$

$a_5 = a_1 r^4 = -5\left(\dfrac{1}{5}\right)^4 = -5 \cdot \dfrac{1}{625}$

$\qquad = -\dfrac{1}{125}$

The first five terms are

$$-5, \ -1, \ -\frac{1}{5}, \ -\frac{1}{25}, \text{ and } -\frac{1}{125}.$$

11. $a_1 = -8$ and $a_7 = -\dfrac{1}{8}$

Use the formula for the nth term of a geometric sequence to solve for r.

$a_7 = a_1 r^{7-1}$

$-\dfrac{1}{8} = -8r^6$

$r^6 = \dfrac{1}{64}$

$r = \pm\dfrac{1}{2}$

There are two geometric sequences that satisfy the given conditions. If $r = 1/2$,

$$a_4 = (-8)\left(\frac{1}{2}\right)^3 = -1$$

and

$$a_n = -8\left(\frac{1}{2}\right)^{n-1}$$

or

$$a_n = -2^3\left(\frac{1}{2}\right)^{n-1} = -\left(\frac{1}{2}\right)^{n-4}.$$

If $r = -1/2$,

$$a_4 = (-8)\left(-\frac{1}{2}\right)^3 = 1$$

and

$$a_n = -8\left(-\frac{1}{2}\right)^{n-1}$$

or

$$a_n = (-2)^3\left(-\frac{1}{2}\right)^{n-1} = \left(-\frac{1}{2}\right)^{n-4}.$$

13. $a_1 = 6x - 9$, $a_2 = 5x + 1$

$d = a_2 - a_1 = (5x + 1) - (6x - 9)$

$\quad = -x + 10$

$a_n = a_1 + (n - 1)d$

$a_8 = a_1 + (8 - 1)d$

$\quad = (6x - 9) + 7(-x + 10)$

$\quad = 6x - 9 - 7x + 70$

$\quad = -x + 61$

15. $a_2 = 6$, $d = 10$

First, find a_1.

$$a_2 = a_1 + d$$
$$6 = a_1 + 10$$
$$a_1 = -4$$

Now we can find S_{12}.

$$S_n = \frac{n}{2}[2a_1 + (n - 1)d]$$

$$S_{12} = \frac{12}{2}[2(-4) + (12 - 1)(10)]$$

$$= 6[-8 + 110]$$

$$= 612$$

17. Geometric, $a_3 = 4$, $r = \frac{1}{5}$

$$a_4 = a_3 r = 4 \cdot \frac{1}{5} = \frac{4}{5}$$

$$a_5 = a_4 r = \frac{4}{5} \cdot \frac{1}{5} = \frac{4}{25}$$

19. Geometric, $a_1 = -1$, $r = 3$

Use the formula

$$S_n = \frac{a_1(1 - r^n)}{1 - r},$$

with $n = 4$.

$$S_4 = \frac{-1(1 - 3^4)}{1 - 3}$$

$$= \frac{-1(1 - 81)}{-2}$$

$$= \frac{-1(-80)}{-2}$$

$$= -40$$

21. $\displaystyle\sum_{i=1}^{7} (-1)^{i-1}$

This is a geometric series with $a_1 = 1$ and $r = -1$.

$$S_n = \frac{a_1(1 - r^n)}{1 - r}$$

$$S_7 = \frac{a_1(1 - r^7)}{1 - r}$$

$$= \frac{1[1 - (-1)^7]}{1 - (-1)}$$

$$= \frac{1 \cdot 2}{2} = 1$$

23. $\sum_{i=1}^{4} \frac{i+1}{i} = \frac{2}{1} + \frac{3}{2} + \frac{4}{3} + \frac{5}{4}$

$= \frac{24}{12} + \frac{18}{12} + \frac{16}{12} + \frac{15}{12}$

$= \frac{73}{12}$

25. $\sum_{j=1}^{2500} j = \frac{2500(2500+1)}{2}$

$= 1250(2501)$

$= 3,126,250$

27. $\sum_{i=1}^{\infty} \left(\frac{4}{7}\right)^i$

This is an infinite geometric series with $a_1 = 4/7$ and $r = 4/7$.

$S_\infty = \frac{a_1}{1-r}$

$= \frac{\frac{4}{7}}{1-\frac{4}{7}}$

$= \frac{\frac{4}{7}}{\frac{3}{7}}$

$= \frac{4}{3}$

29. $24 + 8 + \frac{8}{3} + \frac{8}{9} + \dots$

This is an infinite geometric series with $a_1 = 24$ and $r = \frac{8}{24} = \frac{1}{3}$.

$S_\infty = \frac{a_1}{1-r}$

$= \frac{24}{1-\frac{1}{3}}$

$= \frac{24}{\frac{2}{3}}$

$= 36$

31. $\frac{1}{12} + \frac{1}{6} + \frac{1}{3} + \frac{2}{3} + \dots$

This is an infinite geometric series with $a_1 = \frac{1}{12}$ and

$r = \frac{\frac{1}{6}}{\frac{1}{12}} = 2.$

Since $|r| > 1$, the series diverges.

33. $\sum_{i=1}^{4} (x_i{}^2 - 6)$

$= (x_1{}^2 - 6) + (x_2{}^2 - 6)$
$\quad + (x_3{}^2 - 6) + (x_4{}^2 - 6)$

$= (0^2 - 6) + (1^2 - 6) + (2^2 - 6)$
$\quad + (3^2 - 6)$

$= -6 + (-5) + (-2) + 3$

$= -10$

35. $4 - 1 - 6 - \dots - 66$

This series is the sum of an arithmetic sequence with $a_1 = 4$ and $d = -1 - 4 = -5$. Therefore, the nth term is

$a_n = a_1 + (n-1)d$

$= 4 + (n-1)(-5)$

$= 4 - 5n + 5$

$= -5n + 9,$

or, equivalently,

$a_i = -5i + 9.$

The last term of the series is -66, so

$a_i = -5i + 9$

becomes

$$-66 = -5i + 9$$

$$-75 = -5i$$

$$i = 15.$$

This indicates that the series consists of 15 terms.

$$4 - 1 - 6 - \ldots - 66 = \sum_{i=1}^{15} (-5i + 9)$$

37. $4 + 12 + 36 + \ldots + 972$

This series is the sum of a geometric sequence with $a_1 = 4$ and $r = 12/4 = 3$.

Find the nth term.

$$a_n = a_1 r^{n-1}$$

$$a_n = 4(3)^{n-1},$$

or, equivalently,

$$a_i = 4(3)^{i-1}.$$

Now find the number of terms.

$$972 = 4(3)^{i-1}$$

$$243 = 3^{i-1}$$

$$3^5 = 3^{i-1}$$

$$5 = i - 1$$

$$i = 6$$

Therefore,

$$4 + 12 + 36 + \ldots + 972$$

$$= \sum_{i=1}^{6} 4(3)^{i-1}.$$

39. $(x + 2y)^4$

$$= x^4 + \binom{4}{3} x^{4-1}(2y)^1$$

$$+ \binom{4}{2} x^{4-2}(2y)^2$$

$$+ \binom{4}{1} x^{4-3}(2y)^3 + (2y)^4$$

$(x + 2y)^4$

$$= x^4 + \frac{4!}{3!1!} x^3(2y) + \frac{4!}{2!2!} x^2(2y)^2$$

$$+ \frac{4!}{1!3!} x(2y)^3 + (2y)^4$$

$$= x^4 + 4x^3(2y) + 6x^2(4y^2)$$

$$+ 4x(8y^3) + 16y^4$$

$$= x^4 + 8x^3y + 24x^2y^2 + 32xy^3$$

$$+ 16y^4$$

41. $\left(3\sqrt{x} - \dfrac{1}{\sqrt{x}}\right)^5$

$$= [3x^{1/2} + (-x^{-1/2})]^5$$

$$= (3x^{1/2})^5 + \binom{5}{1}(3x^{1/2})^4(-x^{-1/2})$$

$$+ \binom{5}{2}(3x^{1/2})^3(-x^{-1/2})^2$$

$$+ \binom{5}{3}(3x^{1/2})^2(-x^{-1/2})^3$$

$$+ \binom{5}{4}(3x^{1/2})(-x^{-1/2})^4$$

$$+ (-x^{-1/2})^5$$

$$= 243x^{5/2} + (5)(81x^2)(-x^{-1/2})$$

$$+ (10)(27x^{3/2})(x^{-1})$$

$$+ (10)(9x)(-x^{-3/2})$$

$$+ (5)(3x^{1/2})(x^{-2}) + (-x^{-5/2})$$

$$= 243x^{5/2} - 405x^{3/2} + 270x^{1/2}$$

$$- 90x^{-1/2} + 15x^{-3/2} - x^{-5/2}$$

43. $(4x - y)^8$

The sixth term of the expansion is

$$\binom{8}{5}(4x)^3(-y)^5$$

$$= \frac{8!}{3!5!}(64x^3)(-y^5)$$

$$= 56(-64x^3y^5)$$

$$= -3584x^3y^5.$$

45. $(x + 2)^{12}$

The first four terms of this expansion are as follows.

$$(x)^{12} + \binom{12}{1}(x)^{11}(2)^1 + \binom{12}{2}(x)^{10}(2)^2$$

$$+ \binom{12}{3}(x)^9(2)^3$$

$$= x^{12} + \frac{12!}{11!1!}(x^{11})(2)$$

$$+ \frac{12!}{10!2!}(x^{10})(4) + \frac{12!}{9!3!}(x^9)(8)$$

$$= x^{12} + 12(2x^{11}) + 66(4x^{10})$$

$$+ 220(8x^9)$$

$$= x^{12} + 24x^{11} + 264x^{10} + 1760x^9$$

49. Let S_n be the statement

$$1 + 3 + 5 + 7 + \cdots + (2n - 1) = n^2.$$

Step 1 S_1 is $1 = 1^2$ or

$$1 = 1.$$

The statement is true for $n = 1$.

Step 2   Show that S_k implies S_{k+1}, where S_k is the statement

$$1 + 3 + 5 + 7 + \cdots + (2k - 1) = k^2$$

and S_{k+1} is the statement

$$1 + 3 + 5 + 7 + \cdots + [2(k + 1) - 1]$$

$$= (k + 1)^2.$$

Adding $[2(k + 1) - 1]$ to both sides of S_k gives

$$1 + 3 + 5 + 7 + \cdots + [2(k + 1) - 1]$$

$$= k^2 + [2(k + 1) - 1]$$

$$= k^2 + 2k + 2 - 1$$

$$= k^2 + 2k + 1$$

$$= (k + 1)^2.$$

The final result is the statement for $n = k + 1$; it has been shown that S_k implies S_{k+1}.
Therefore, by mathematical induction, S_n is true for every positive integer n.

51. Let S_n be the statement

$$2 + 2^2 + 2^3 + \ldots + 2^n = 2(2^n - 1).$$

Step 1 S_1 is the statement

$$2 = 2(2^1 - 1)$$

$$2 = 2 \cdot 1,$$

which is true.

Step 2   Show that if S_k is true, then S_{k+1} is also true, where S_k is the statement

$$2 + 2^2 + 2^3 + \ldots + 2^k = 2(2^k - 1),$$

and S_{k+1} is the statement

$$2 + 2^2 + 2^3 + \ldots + 2^k + 2^{k+1}$$

$$= 2(2^{k+1} - 1).$$

Start with S_k.   Add 2^{k+1} to both sides.

$$2 + 2^2 + 2^3 + \ldots + 2^k + 2^{k+1}$$

$$= 2(2^k - 1) + 2^{k+1}$$

$$= 2^{k+1} - 2 + 2^{k+1}$$

$$= 2 \cdot 2^{k+1} - 2 \cdot 1$$

$$= 2(2^{k+1} - 1)$$

Thus, $2 + 2^2 + 2^3 + \ldots + 2^n$
$= 2(2^n - 1)$ is true for every
positive integer n.

55. $P(6, 0) = \dfrac{6!}{(6 - 0)!}$

$= \dfrac{6!}{6!} = 1$

57. $9! = 9 \cdot 8 \cdot 7 \cdot 6 \cdot 5 \cdot 4 \cdot 3 \cdot 2 \cdot 1$
$= 362,880$

59. $2 \cdot 4 \cdot 3 \cdot 2 = 48$

48 different wedding arrangements
are possible.

61. There are 4 choices for the first
job, 3 choices for the second job,
and so on.

$$4 \cdot 3 \cdot 2 \cdot 1 = 24$$

There are 24 ways in which the jobs
can be assigned.

63. Order is important, so this is a
permutation problem.

$$P(9, 3) = \dfrac{9!}{6!}$$

$$= 9 \cdot 8 \cdot 7$$

$$= 504$$

The winners can be determined 504
ways.

65. (a) Let E be the event "picking a
green marble."

$n(E) = 4$

$n(S) = 4 + 5 + 6 = 15$
(total number of marbles)

$P(E) = \dfrac{P(E)}{n} = \dfrac{4}{15}$

(b) Let E be the next event "picking
a black marble."

$n(E) = 5$

$n(S) = 15$ (from part (a))

$P(E) = \dfrac{5}{15} = \dfrac{1}{3}$

The probability that the marble is
not black is given by

$$P(E') = 1 - P(E)$$

$$= 1 - \dfrac{1}{3} = \dfrac{2}{3}.$$

(c) Let E be the event "picking a
blue marble."

$n(E) = 0$ since there are no blue
marbles.

$n(S) = 15$ (from part (a))

$P(E) = \dfrac{n(E)}{n} = \dfrac{0}{15} = 0$

67. $P(\text{black king}) = \dfrac{n(\text{black kings})}{n(\text{cards in deck})}$

$= \dfrac{2}{52} = \dfrac{1}{26}$

69. $P(\text{ace or diamond})$

$= P(\text{ace}) + P(\text{diamond})$

$- P(\text{diamond and ace})$

$= \dfrac{4}{52} + \dfrac{13}{52} - \dfrac{1}{52}$

$= \dfrac{16}{52} = \dfrac{4}{13}$

71. "No more than 3" means "0 or 1 or 2
or 3."

$P(\text{no more than 3})$

$= P(0 \text{ or } 1 \text{ or } 2 \text{ or } 3)$

$= P(0) + P(1) + P(2) + P(3)$

$= .31 + .25 + .18 + .12$

$= .86$

73. In a sample of 5 filters, it is impossible for the number of defective filters to be more than 5. The probability of an impossible event is 0.

75. In this binomial experiment, we call tossing a tail a success.
Then, $n = 10$, $r = 4$, and $p = 1/2$.

$$P(\text{Exactly 4 tails})$$
$$= \binom{10}{4}\left(\frac{1}{2}\right)^4\left(1 - \frac{1}{2}\right)^6$$
$$= 210 \cdot \frac{1}{2^4} \cdot \frac{1}{2^6}$$
$$= \frac{210}{1024}$$
$$\approx .205$$

Chapter 8   Test

1. $a_n = (-1)^n(n^2 + 2)$

$n = 1$: $a_1 = (-1)^1(1^2 + 2) = -3$
$n = 2$: $a_2 = (-1)^2(2^2 + 2) = 6$
$n = 3$: $a_3 = (-1)^3(3^2 + 2) = -11$
$n = 4$: $a_4 = (-1)^4(4^2 + 2) = 18$
$n = 5$: $a_5 = (-1)^5(5^2 + 2) = -27$

The first five terms are

$$-3,\ 6,\ -11,\ 18,\ \text{and } -27.$$

This sequence does not have either a common difference or a common ratio, so the sequence is neither arithmetic nor geometric.

2. $a_n = -3 \cdot \left(\frac{1}{2}\right)^n$

$n = 1$: $a_1 = -3\left(\frac{1}{2}\right)^1 = -\frac{3}{2}$
$n = 2$: $a_2 = -3\left(\frac{1}{2}\right)^2 = -\frac{3}{4}$
$n = 3$: $a_3 = -3\left(\frac{1}{2}\right)^3 = -\frac{3}{8}$
$n = 4$: $a_4 = -3\left(\frac{1}{2}\right)^4 = -\frac{3}{16}$
$n = 5$: $a_5 = -3\left(\frac{1}{2}\right)^5 = -\frac{3}{32}$

The first five terms are

$$-\frac{3}{2},\ -\frac{3}{4},\ -\frac{3}{8},\ -\frac{3}{16},\ \text{and } -\frac{3}{32}.$$

This sequence has a common ratio, $r = 1/2$, so the sequence is geometric.

3. $a_1 = 2$, $a_2 = 3$, $a_n = a_{n-1} + 2a_{n-2}$, for $n \geq 3$

$n = 3$: $a_3 = a_2 + 2a_1 = 3 + 2(2) = 7$
$n = 4$: $a_4 = a_3 + 2a_2 = 7 + 2(3) = 13$
$n = 5$: $a_5 = a_4 + 2a_3 = 13 + 2(7)$
$$= 27$$

The first five terms are

$$2,\ 3,\ 7,\ 13,\ \text{and } 27.$$

There is no common difference or common ratio, so the sequence is neither arithmetic nor geometric.

4. $a_1 = 1$ and $a_3 = 25$

$$a_n = a_1 + (n - 1)d$$
$$a_3 = a_1 + 2d$$
$$25 = 1 + 2d$$
$$2d = 24$$
$$d = 12$$

$$a_5 = a_1 + 4d$$
$$= 1 + 4(12)$$
$$= 1 + 48$$
$$= 49$$

5. $a_1 = 81$ and $r = -\dfrac{2}{3}$

$$a_n = a_1 r^{n-1}$$
$$a_6 = 81\left(-\dfrac{2}{3}\right)^5$$
$$= 81\left(-\dfrac{32}{243}\right)$$
$$= -\dfrac{32}{3}$$

6. Arithmetic, with $a_1 = -43$ and $d = 12$

$$S_n = \dfrac{n}{2}[2a_1 + (n-1)d]$$
$$S_{10} = \dfrac{10}{2}[2(-43) + 9(12)]$$
$$= 5(-86 + 108)$$
$$= 5(22)$$
$$= 110$$

7. Geometric, with $a_1 = 5$ and $r = -2$

$$S_n = \dfrac{a_1(1 - r^n)}{1 - r}$$
$$S_{10} = \dfrac{5[1 - (-2)^{10}]}{1 - (-2)}$$
$$= \dfrac{5(1 - 1024)}{3}$$
$$= \dfrac{5(-1023)}{3}$$
$$= -1705$$

8. $\displaystyle\sum_{i=1}^{30} (5i + 2)$

This sum represents the sum of the first 30 terms of the arithmetic sequence having

$a_1 = 5 \cdot 1 + 2 = 7$ and

$a_n = a_{30} = 5 \cdot 30 + 2 = 152.$

$$\sum_{i=1}^{30} (5i + 2) = S_{30}$$
$$= \dfrac{n}{2}(a_1 + a_n)$$
$$= \dfrac{30}{2}(a_1 + a_{30})$$
$$= 15(7 + 152)$$
$$= 2385$$

9. $\displaystyle\sum_{i=1}^{5} (-3 \cdot 2^i)$

This sum represents the sum of the first five terms of the geometric sequence having

$a_1 = -3 \cdot 2^1 = -6$ and $r = 2.$

$$\sum_{i=1}^{5} -3 \cdot 2^i = S_5$$
$$= \dfrac{a_1(1 - r^n)}{1 - r}$$
$$= \dfrac{-6(1 - 2^5)}{1 - 2}$$
$$= \dfrac{-6(1 - 32)}{-1}$$
$$= 6(-31)$$
$$= -186$$

10. $\displaystyle\sum_{i=1}^{\infty} (2^i) \cdot 4$

This is the sum of an infinite geometric sequence with $r = 2$. Since $|r| > 1$, the sum does not exist.

11. $\displaystyle\sum_{i=1}^{\infty} 54\left(\frac{2}{9}\right)^i$

This is the sum of the infinite geometric sequence with

$a_1 = 54\left(\frac{2}{9}\right)^1 = 12$ and $r = \frac{2}{9}$.

$\displaystyle\sum_{i=1}^{\infty} 54\left(\frac{2}{9}\right)^i = S_\infty = \frac{a_1}{1 - r}$

$\qquad = \dfrac{12}{1 - \dfrac{2}{9}}$

$\qquad = \dfrac{12}{\dfrac{7}{9}}$

$\qquad = \dfrac{9 \cdot 12}{7}$

$\qquad = \dfrac{108}{7}$

12. $(x + y)^6$

$= x^6 + \binom{6}{1}x^5y + \binom{6}{2}x^4y^2 + \binom{6}{3}x^3y^3$

$\quad + \binom{6}{4}x^2y^4 + \binom{6}{5}xy^5 + y^6$

$= x^6 + \dfrac{6!}{1!5!}x^5y + \dfrac{6!}{2!4!}x^4y^2$

$\quad + \dfrac{6!}{3!3!}x^3y^3 + \dfrac{6!}{4!2!}x^2y^4$

$\quad + \dfrac{6!}{5!1!}xy^5 + y^6$

$= x^6 + 6x^5y + 15x^4y^2 + 20x^3y^3$

$\quad + 15x^2y^4 + 6xy^5 + y^6$

13. $(2x - 3y)^4$

$= (2x)^4 + \binom{4}{1}(2x)^3(-3y)$

$\quad + \binom{4}{2}(2x)^2(-3y)^2 + \binom{4}{3}(2x)(-3y)^3$

$\quad + (-3y)^4$

$= 16x^4 + 4(8x^3)(-3y) + 6(4x^2)(9x^2)$

$\quad + 4(2x)(-27y^3) + 81y^4$

$= 16x^4 - 96x^3y + 216x^2y^2 - 216xy^3$

$\quad + 81y^4$

14.   To find the third term in the expansion of $(w - 2y)^6$, use the formula

$$\binom{n}{k - 1}x^{n-(k-1)}y^{k-1}$$

with $n = 6$ and $k = 3$.
Then $k - 1 = 2$ and $n - (k - 1) = 4$.
Thus, the third term is

$\binom{6}{2}w^4(-2y)^2 = 15w^4(4y^2)$

$\qquad\qquad = 60w^4y^2.$

15. $C(10, 2) = \dfrac{10!}{(10 - 2)!2!}$

$\qquad = \dfrac{10 \cdot 9 \cdot 8!}{8! \cdot 2 \cdot 1}$

$\qquad = 45$

16. $\binom{7}{3} = \dfrac{7!}{(7 - 3)!3!}$

$\qquad = \dfrac{7 \cdot 6 \cdot 5 \cdot 4!}{4! \cdot 3 \cdot 2 \cdot 1}$

$\qquad = 35$

17. $P(11, 3) = \dfrac{11!}{(11 - 3)!}$

$\qquad = \dfrac{11!}{8!}$

$\qquad = \dfrac{11 \cdot 10 \cdot 9 \cdot 8!}{8!}$

$\qquad = 990$

18. $8! = 8 \cdot 7 \cdot 6 \cdot 5 \cdot 4 \cdot 3 \cdot 2 \cdot 1$

$\quad = 40,320$

19.   Prove that

$$8 + 14 + 20 + 26 + \ldots + (6n + 2)$$

$$= 3n^2 + 5n$$

is true for every positive integer n.

Step 1 S_1 is the statement

$$8 = 3(1^2) + 5(1),$$

which is true.

Step 2 Show that if S_k is true, then S_{k+1} is also true. S_k is the statement

$$8 + 14 + 20 + 26 + \ldots + (6k + 2)$$
$$= 3k^2 + 5k,$$

and S_{k+1} is the statement

$$8 + 14 + 20 + 26 + \ldots + (6k + 2) + [6(k + 1) + 2]$$
$$= 3(k + 1)^2 + 5(k + 1).$$

Start with S_k.

$$8 + 14 + 20 + 26 + \ldots + (6k + 2)$$
$$= 3k^2 + 5k$$

Add the $(k + 1)$st term, $6(k + 1) + 2$, to both sides.

$$8 + 14 + 20 + 26 + \ldots + (6k + 2) + [6(k + 1) + 2]$$
$$= (3k^2 + 5k) + [6(k + 1) + 2]$$
$$= 3k^2 + 5k + 5(k + 1) + (k + 1) + 2$$
$$= (3k^2 + 6k + 3) + 5(k + 1)$$
$$= 3(k^2 + 2k + 1) + 5(k + 1)$$
$$3 + 6 + 9 + \ldots + 3k + 3(k + 1)$$
$$= 3(k + 1)^2 + 5(k + 1)$$

The final equation is the statement S_{k+1}. Thus, we have shown that if S_k is true, then S_{k+1} is also true. The two steps required for a proof by mathematical induction have been completed, so the statement

$$8 + 14 + 20 + 26 + \ldots + (6n + 2)$$
$$= 3n^2 + 5n$$

is true for every positive integer n.

20. Using the fundamental principle of counting, the number of different types of shoes is

$$4 \cdot 3 \cdot 2 = 24.$$

21. Since order does matter in this situation, the number of ways the three offices can be filled is given by

$$P(20, 3) = \frac{20!}{17!}$$
$$= 20 \cdot 19 \cdot 18$$
$$= 6840.$$

22. The number of ways to select the men is $\binom{8}{2}$.

The number of ways to select the women is $\binom{12}{3}$.

The number of ways to choose the five people to attend the conference is given by

$$\binom{8}{2} \cdot \binom{12}{3} = 28 \cdot 220$$
$$= 6160.$$

For Exercises 24–27, the sample space is the set of all cards in a standard deck, so $n(S) = 52$.

24. Consider the event E: "drawing a red three." There are 2 red threes in a deck, so $n(E) = 2$.

$$P(E) = \frac{n(E)}{n(S)}$$
$$= \frac{2}{52}$$
$$= \frac{1}{26}$$

The probability of drawing a red three is $\frac{1}{26}$.

25. Consider the event E: "draw a face card." Each suit contains 3 face cards (jack, queen, and king), so the deck contains 12 face cards. Thus $n(E) = 12$

    and $P(E) = \frac{12}{52} = \frac{3}{13}$.

    The probability of drawing a card that is not a face card is

    $$P(E') = 1 - P(E)$$
    $$= 1 - \frac{3}{13} = \frac{10}{13}.$$

26. The events E: "draw a king" and F: "draw a spade" are not mutually exclusive, since it is possible to draw the king of spades, an out-come satisfying both events.

    $$P(E \text{ or } F) = P(E \cup F)$$
    $$= P(E) + P(F) - P(E \cap F)$$

    $P(\text{king or spade})$

    $= P(\text{king}) + P(\text{spade})$

      $- P(\text{king and spade})$

    $= \frac{4}{52} + \frac{13}{52} - \frac{1}{52}$

    $= \frac{16}{52} = \frac{4}{13}$

    The probability of drawing a king or a spade is $\frac{4}{13}$.

27. Consider the event E: "draw a face card." As shown in the solution to Problem 25,

    $$P(E) = \frac{3}{13} \quad \text{and} \quad P(E') = \frac{10}{13}.$$

    The odds in favor of drawing a face card are

    $$\frac{P(E)}{P(E')} = \frac{\frac{3}{13}}{\frac{10}{13}} = \frac{3}{10} \quad \text{or} \quad 3 \text{ to } 10.$$

28. "At most 2" means "0 or 1 or 2."

    $P(\text{at most } 2)$

    $= P(0 \text{ or } 1 \text{ or } 2)$

    $= P(0) + P(1) + P(2)$
    (since the events are mututally exclusive)

    $= .19 + .43 + .30$

    $= .92$

    The probability that at most 2 filters are defective is .92.

29. In this binomial experiment, we call rolling a four a success.
    Then, $n = 8$, $r = 3$, and $p = 1/6$.

    $P(\text{Exactly 3 fours}) = \binom{8}{3}\left(\frac{1}{6}\right)^3\left(1 - \frac{1}{6}\right)^5$

    $$= 56 \cdot \frac{1}{6^3} \cdot \frac{5^5}{6^5}$$

    $$\approx .104$$

30. In this binomial experiment, we call rolling a six a success.
    Then, $n = 8$, $r = 8$, and $p = 1/6$.

    $P(8 \text{ sixes}) = \binom{8}{8}\left(\frac{1}{6}\right)^8\left(1 - \frac{1}{6}\right)^0$

    $$= 1 \cdot \frac{1}{6^8} \cdot 1$$

    $$\approx .000000595$$

Cumulative Review Exercises (Chapter 1-8)

1. Write "the distance from x to 3 is at most 4" as an inequality involving an absolute value sign.

2. Use the properties of exponents to simplify $\left(\dfrac{x^3}{\sqrt{y}}\right)^4$.

3. Write out the binomial expansion of $(x - 2y)^4$.

4. Factor $(2a + 4)^2 + 2(2a + 4) - 15$.

5. Write $\dfrac{x^2 + 2x - 3}{x^2 + 4x - 5}$ in lowest terms.

6. Multiply $(x^{2/3} y^{1/4})^2 \cdot (xy^{-1/4})^{2/3}$.

7. Rationalize the denominator of $\dfrac{2\sqrt{3} + \sqrt{2}}{5 - \sqrt{6}}$.

8. Find $\dfrac{-5 + 6i}{6 - 5i}$.

9. Solve $y = \dfrac{1 - x}{1 + x}$ for x in terms of y.

10. A grocer creates a mixture of peanuts and raisins. Peanuts sell for $1.20 per pound and raisins sell for $1.70 per pound. How many pounds of each ingredient should be used to create a 50-pound mixture selling for $1.40 per pound?

11. Solve $2x^2 + 6x - 8 = 0$ by completing the square.

12. A rectangular garden is located next to a house. The remaining three sides are enclosed by 40 feet of fencing. If the area of the garden is 200 square feet, what are the dimensions of the garden?

13. Solve $x = (9x^2 - 20)^{1/4}$.

14. Solve $\dfrac{x}{x + 3} > 5$ for x.

15. Find an equation of the circle with center (−4, 5) and containing the point (2, 13).

16. Find the slope–intercept form of the equation of the line through the point (10, −2) and parallel to the line 4x + 5y = −3.

17. If $f(x) = \dfrac{1}{x}$, find and simplify $\dfrac{f(x + h) - f(x)}{h}$.

18. Without graphing, determine whether the graph of the equation $y = \dfrac{1}{1 + x^2} - 3x^4$ is symmetric with respect to the x–axis, the y–axis, the origin, or none of these.

19. Graph the parabola $f(x) = 3x^2 - 18x + 10$. Give the vertex, axis, domain, and range.

20. The polynomial $f(x) = 3x^3 + 13x^2 - 18x - 40$ has 2 as a zero. Find the other zeros.

21. Sketch the graph of the rational function $f(x) = \dfrac{2x}{x - 2}$.

22. Graph the polynomial function $f(x) = x^2(x + 3)(x - 4)$.

23. Find the inverse of $f(x) = \sqrt{8 - x}$.

24. Find the value of $\log_3 \left(\dfrac{1}{81}\right)$.

25. Solve $\ln x - \ln (2x + 3) = \ln 4$.

26. The radioactive element strontium 90 has a half–life of 28 years. If you begin with 5 grams of strontium 90, how much will you have after 10 years?

27. A cup of uncooked rice contains 15 g of protein and 810 calories. A cup of uncooked soybeans contains 22.5 g of protein and 270 calories. How many cups of each should be used for a meal containing 9.5 g of protein and 324 calories?

28. Solve the system
$$2xy + 1 = 0$$
$$x + 16y = 2.$$

29. Find two numbers which differ by 3 and whose squares have a sum of 29.

30. Use the Gauss–Jordan method to solve the following system of equations.

$$2x - 4y + 2z = 0$$
$$x + 3y - z = 5$$
$$3x + 2y + z = 12$$

31. Let $A = \begin{bmatrix} 3 & \frac{1}{2} & -1 \\ -4 & 2 & 0 \end{bmatrix}$, $B = \begin{bmatrix} -2 & 5 \\ 4 & -2 \\ 6 & 1 \end{bmatrix}$.   Find AB and BA.

32. In a certain town, the percentages of voters voting Democratic and Republican in certain election by various age groups is given by the following matrix.

$$\begin{array}{cc} & \text{Dem.} \quad \text{Rep.} \\ A = & \begin{bmatrix} .70 & .30 \\ .50 & .50 \\ .45 & .55 \end{bmatrix} \begin{array}{l} \text{Under 40} \\ \text{40–60} \\ \text{Over 60} \end{array} \end{array}$$

The population of voters in the town by age group is given by the matrix

$$\begin{array}{ccc} \text{Under 40} & \text{40–60} & \text{Over 60} \\ B = [12{,}000 & 10{,}000 & 8000]. \end{array}$$

Compute and interpret the entries of the matrix BA.

33. Evaluate $\begin{vmatrix} 3 & 6 & -3 \\ 0 & 5 & 4 \\ 1 & -2 & 7 \end{vmatrix}$.

34. Use Cramer's rule to solve the system of linear equations.

$$2x - 3y + z = 0$$
$$4x + y + 3z = 17$$
$$-6x + 5y - z = 2$$

35. Solve the system of linear equations by using the inverse of the coefficient matrix.

$$-2x + 2y - z = 3$$
$$3x - 5y + 4z = 4$$
$$5x - 6y + 4z = 5$$

36. Graph the solution set of the system of nonlinear inequalities.

$$x^2 - 4x + y \le 1$$
$$e^{.5x} - y \le 0$$

37. A machine shop manufactures two types of bolts. Each can be made on any of three groups of machines, but the time required on each group differs, as shown in the table.

Machine groups	I	II	III
Bolts Type 1	.4 hour	.5 hour	.2 hour
Type 2	.3 hour	.2 hour	.4 hour

Production schedules are made up for one week at a time. In this period there are 1200 hours of machine time available for each machine group. Type 1 bolts sell for 10¢ and type bolts for 12¢. How many of each type of bolt should be manufactured per week to maximize revenue? What is the maximum revenue?

38. Give the focus, directrix, and axis for the parabola $(x + 3)^2 = 20(y - 8)$.

39. Find an equation for the ellipse with center at $(6, 7)$, a vertex at $(6, 11)$, and a focus at $(6, 10)$.

40. Find the equation of the hyperbola with center $(5, -2)$, vertex $(2, -2)$, and asymptotes $y + 2 = \pm 2(x - 5)$.

41. Identify the type of conic section with the equation $9x^2 - 54x + 4y^2 + 40y = -145$.

42. Evaluate $\displaystyle\sum_{k=1}^{4} (1 + k)^{k-1}$.

43. Find a_1 for the arithmetic sequence with $a_6 = 29$ and $a_{20} = 99$.

44. Evaluate $\displaystyle\sum_{i=1}^{10} (-2 + 4i + 3 \cdot 2^i)$.

45. Find a_1 and r for the geometric sequence with $a_4 = 1$ and $a_6 = .0625$.

46. Evaluate $\displaystyle\sum_{k=1}^{\infty} 3 \cdot (.8)^k$.

47. Find the ninth term of the binomial expansion of $(2x - y)^{12}$.

48. Use mathematical induction to prove that $4 + 4^2 + 4^3 + \ldots + 4^n = \dfrac{4(4^n - 1)}{3}$ for every positive integer n.

49. In a batch of 100 computer diskettes, seven are defective. A sample of three diskettes is to be selected from the batch. How many samples are possible? How many of the samples consist of all defective diskettes?

50. If nine percent of all candy bars sold in the United States are Yummy bars, what are the odds that a randomly-selected candy bar is a Yummy bar?

51. A couple decides to have four children. What is the probability that among the children will be at least one boy and at least one girl?

Solutions to Cumulative Review Exercises (Chapters 1–8)

1. "The distance from x to 3 is at most 4" is written as $|x - 3| \le 4$.

2. $\left(\dfrac{x^3}{\sqrt{y}}\right)^4 = \left(\dfrac{x^3}{y^{1/2}}\right)^4$

   $= \dfrac{(x^3)^4}{(y^{1/2})^4}$

   $= \dfrac{x^{12}}{y^2}$

3. To expand $(x - 2y)^4$, use the coefficients in row 4 of Pascal's triangle.

   $(x - 2y)^4 = x^4 + 4x^3(-2y) + 6x^2(-2y)^2$
   $\qquad\qquad\quad + 4x(-2y)^3 + (-2y)^4$
   $\qquad\quad = x^4 + 4x^3(-2y) + 6x^2(4y^2)$
   $\qquad\qquad\quad + 4x(-8y^3) + 16y^4$
   $\qquad\quad = x^4 - 8x^3y + 24x^2y^2 - 32xy^3$
   $\qquad\qquad\quad + 16y^4$

4. $(2a + 4)^2 + 2(2a + 4) - 15$

   Let $m = 2a + 4$. With this substitution, the trinomial becomes

   $m^2 + 2m - 15 = (m + 5)(m - 3)$.

   Now replace m with $2a + 4$.

   $(m + 5)(m - 3)$
   $\quad = [(2a + 4) + 5][(2a + 4) - 3]$

   Thus,

   $(2a + 4)^2 + 2(2a + 4) - 15$
   $\quad = (2a + 9)(2a + 1)$.

5. $\dfrac{x^2 + 2x - 3}{x^2 + 4x - 5} = \dfrac{(x + 3)(x - 1)}{(x + 5)(x - 1)}$

   $= \dfrac{x + 3}{x + 5}$

6. $(x^{2/3}y^{1/4})^2 \cdot (xy^{-1/4})^{2/3}$

   $= x^{4/3}y^{1/2} \cdot x^{2/3}y^{-1/6}$

   $= x^{6/3}y^{2/6}$

   $= x^2y^{1/3}$

7. $\dfrac{2\sqrt{3} + \sqrt{2}}{5 - \sqrt{6}} = \dfrac{2\sqrt{3} + \sqrt{2}}{5 - \sqrt{6}} \cdot \dfrac{5 + \sqrt{6}}{5 + \sqrt{6}}$

   $= \dfrac{10\sqrt{3} + 2\sqrt{18} + 5\sqrt{2} + \sqrt{12}}{25 - 6}$

   $= \dfrac{10\sqrt{3} + 6\sqrt{2} + 5\sqrt{2} + 2\sqrt{3}}{19}$

   $= \dfrac{12\sqrt{3} + 11\sqrt{2}}{19}$

8. $\dfrac{-5 + 6i}{6 - 5i} = \dfrac{-5 + 6i}{6 - 5i} \cdot \dfrac{6 + 5i}{6 + 5i}$

   $= \dfrac{-30 - 25i + 36i + 30i^2}{36 - 25i^2}$

   $= \dfrac{-30 + 11i - 30}{36 + 25}$

   $= \dfrac{-60 + 11i}{61}$

   $= -\dfrac{60}{61} + \dfrac{11}{61}i$

9. $y = \dfrac{1 - x}{1 + x}$

   $y(1 + x) = \dfrac{(1 - x)}{(1 + x)} \cdot (1 + x)$

   $y + xy = 1 - x$

   $x + xy = 1 - y$

   $x(1 + y) = 1 - y$

   $x = \dfrac{1 - y}{1 + y}$

10. Let x = the number of pounds of raisins;

$50 - x$ = the number of pounds of peanuts.

$$1.70x + 1.20(50 - x) = 1.40(50)$$
$$1.7x + 60 - 1.2x = 70$$
$$.5x = 10$$
$$x = 20$$

The grocer should use 20 pounds of raisins and 30 pounds of peanuts.

11. $2x^2 + 6x - 8 = 0$

$$x^2 + 3x - 4 = 0$$
$$x^2 + 3x = 4$$
$$x^2 + 3x + \frac{9}{4} = \frac{16}{4} + \frac{9}{4}$$
$$\left(x + \frac{3}{2}\right)^2 = \frac{25}{4}$$
$$x + \frac{3}{2} = \pm\sqrt{\frac{25}{4}}$$
$$x + \frac{3}{2} = \pm\frac{5}{2}$$
$$x = -\frac{3}{2} \pm \frac{5}{2}$$
$$x = -\frac{3}{2} + \frac{5}{2} = \frac{2}{2} = 1$$
or $\quad x = -\frac{3}{2} - \frac{5}{2} = -\frac{8}{2} = -4$

Solution set: $\{1, -4\}$

12. Let x = length of the two sides of equal length;

$40 - 2x$ = length of the third side.

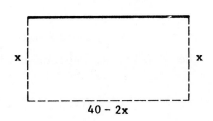

Use the formula for the area of a rectangle, $L \cdot W = A$.

$$x(40 - 2x) = 200$$
$$40x - 2x^2 = 200$$
$$0 = 2x^2 - 40x + 200$$

Divide both sides by 2.

$$0 = x^2 - 20x + 100$$
$$0 = (x - 10)^2$$
$$x = 10$$

The dimensions of the garden are 10 feet by 20 feet.

13. $x = (9x^2 - 20)^{1/4}$

Raise both sides to the fourth power.

$$x^4 = [(9x^2 - 20)^{1/4}]^4$$
$$x^4 = 9x^2 - 20$$
$$x^4 - 9x^2 + 20 = 0$$
$$(x^2 - 4)(x^2 - 5) = 0$$
$$x^2 - 4 = 0 \quad \text{or} \quad x^2 - 5 = 0$$
$$x^2 = 4 \quad \text{or} \quad x^2 = 5$$
$$x = \pm 2 \quad \text{or} \quad x = \pm\sqrt{5}$$

The potential solutions $x = -2$ and $x = -\sqrt{5}$ do not check in the original equation.

Solution set: $\{2, \sqrt{5}\}$

14.
$$\frac{x}{x + 3} > 5$$
$$\frac{x}{x + 3} - 5 > 0$$
$$\frac{x}{x + 3} - \frac{5(x + 3)}{x + 3} > 0$$
$$\frac{x - 5x - 15}{x + 3} > 0$$
$$\frac{-4x - 15}{x + 3} > 0$$

The numerator is zero when $x = -15/4$. The denominator is zero when $x = -3$.

$$-4x - 15 \quad ++\ \circ\ -\ \ -\ -\ -\ -\ -$$
$$x + 3 \quad -\ -\ \ -\ \circ\ +\ +\ +\ +\ +\ +$$

Sign of $-\dfrac{15}{4}$ -3 0

$$\dfrac{-4x - 15}{x + 3} \quad -\quad +\qquad -$$

The sign graph shows that the quotient is positive in the open interval $(-15/4, -3)$.

Therefore, the solution set is the open interval $\left(-\dfrac{15}{4}, -3\right)$.

15. Label the points $C(-4, 5)$ and $P(2, 13)$.

The radius is the distance from the center to any point on the circle.

$$r = d(C, P)$$
$$= \sqrt{[2 - (-4)]^2 + (13 - 5)^2}$$
$$= \sqrt{6^2 + 8^2} = \sqrt{36 + 64}$$
$$= \sqrt{100} = 10$$

The radius is 10 and the center is $C(-4, 5)$. An equation of the circle is

$$[x - (-4)]^2 + (y - 5)^2 = 10^2$$
$$(x + 4)^2 + (y - 5)^2 = 100.$$

16. Write the equation $4x + 5y = -3$ in slope–intercept form.

$$4x + 5y = -3$$
$$5y = -4x - 3$$
$$y = -\dfrac{4}{5}x - \dfrac{3}{5}$$

The slope of the line is $m = -4/5$. The line whose equation is to be written is parallel to the given line and therefore has the same slope. Use the point–slope form; then change the equation to slope–intercept form.

$$y - y_1 = m(x - x_1)$$
$$y - (-2) = -\dfrac{4}{5}(x - 10)$$
$$y + 2 = -\dfrac{4}{5}x + 8$$
$$y = -\dfrac{4}{5}x + 6$$

17. $f(x) = \dfrac{1}{x}$

$$\dfrac{f(x + h) - f(x)}{h}$$
$$= \dfrac{\dfrac{1}{x + h} - \dfrac{1}{x}}{h}$$
$$= \dfrac{x(x + h) \cdot \dfrac{1}{x + h} - x(x + h) \cdot \dfrac{1}{x}}{x(x + h)h}$$
$$= \dfrac{x - (x + h)}{x(x + h)h}$$
$$= \dfrac{x - x - h}{x(x + h)h}$$
$$= \dfrac{-h}{x(x + h)h}$$
$$= \dfrac{-1}{x(x + h)}$$

18. $y = \dfrac{1}{1 + x^2} - 3x^4$

To test for symmetry with respect to the y-axis, we replace x with $-x$.

$$y = \dfrac{1}{1 + (-x)^2} - 3(-x)^4$$
$$y = \dfrac{1}{1 + x^2} - 3x^4$$

We obtain an equivalent equation.
The graph is symmetric with respect
to the y-axis.
To test for symmetry with respect to
the x-axis, we replace y with $-y$.

$$-y = \frac{1}{1 + x^2} - 3x^4$$

$$y = \frac{1}{1 + x^2} + 3x^4$$

We do not obtain an equivalent equation. The graph is not symmetric
with respect to the x-axis.
To test for symmetry with respect to
the origin, we replace x with $-x$ and
y with $-y$.

$$-y = \frac{1}{1 + (-x)^2} - 3(-x)^4$$

$$-y = \frac{1}{1 + x^2} - 3x^4$$

$$y = -\frac{1}{1 + x^2} + 3x^4$$

We do not obtain an equivalent equation. The graph is not symmetric
with respect to the origin.
Therefore, the graph is symmetric
with respect to the y-axis only.

19. $f(x) = 3x^2 - 18x + 10$
$$= 3(x^2 - 6x) + 10$$
$$= 3(x^2 - 6x + 9) + 10 - 27$$
$$= 3(x - 3)^2 - 17$$

The coordinates of the vertex are
(3, -17). The equation of the axis
is x = 3.
The domain is $(-\infty, \infty)$.
Since the parabola opens upward, the
smallest y-value occurs at the vertex, so the range is $[-17, \infty)$.

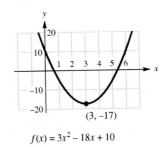

$$f(x) = 3x^2 - 18x + 10$$

20. $f(x) = 3x^3 + 13x^2 - 18x - 40$

Since 2 is a zero of $f(x)$, $x - 2$ is
a factor.

```
2 | 3   13   -18   -40
  |      6    38    40
  --------------------
    3   19    20     0
```

$$f(x) = (x - 2)(3x^2 + 19x + 20)$$

The remaining zeros are the zeros
of the quotient polynomial,
$3x^2 + 19x + 20$.

$$3x^2 + 19x + 20 = 0$$
$$(3x + 4)(x + 5) = 0$$
$$3x + 4 = 0 \quad \text{or} \quad x + 5 = 0$$
$$x = -\frac{4}{3} \quad \text{or} \quad x = -5$$

The other zeros are $-4/3$ and -5.

21. $f(x) = \dfrac{2x}{x - 2}$

The line x = 2 is the only vertical
asymptote. Since the degree of the
numerator is the same as the degree
of the denominator, the graph has a
horizontal asymptote at

$$y = \frac{a_n}{b_n} = \frac{2}{1} = 2.$$

Since

$$f(0) = \frac{2 \cdot 0}{0 - 2} = 0,$$

the y-intercept is 0.

The numerator, 2x, is equal to 0 only when x = 0, so the only x-intercept is 0.

Plot some additional points to complete the graph.

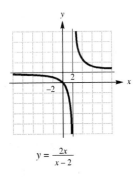

$$y = \frac{2x}{x - 2}$$

22. $f(x) = x^2(x + 3)(x - 4)$

The function has the following zeros: 0 (multiplicity 2), −3, and 4.

These zeros divide the x-axis into four regions.

Region	Test point	Value of f(x)	Sign of f(x)
$(-\infty, -3)$	−3.5	45.9375	Positive
$(-3, 0)$	−1	−10	Negative
$(0, 4)$	1	−12	Negative
$(4, \infty)$	4.5	75.9375	Positive

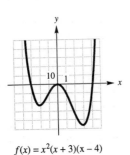

$$f(x) = x^2(x + 3)(x - 4)$$

23. $f(x) = \sqrt{8 - x}$

This function is one-to-one and thus has an inverse. Note that the range of f is [0, ∞); therefore, the domain of f^{-1} is [0, ∞).

Let f(x) = y and solve for x.

$$y = \sqrt{8 - x}$$
$$y^2 = 8 - x$$
$$x = 8 - y^2 = f^{-1}(y)$$

Now exchange x and y to get

$$y = 8 - x^2 = f^{-1}(x)$$

or $f^{-1}(x) = 8 - x^2$.

Since the domain of f^{-1} is [0, ∞), the inverse of function is

$$f^{-1}(x) = 8 - x^2, \; x \geq 0.$$

24.   Let $y = \log_3 \left(\frac{1}{81}\right)$.

Then,

$$3^y = \frac{1}{81}$$
$$3^y = \frac{1}{3^4}$$
$$3^y = 3^{-4}$$
$$y = -4.$$

Therefore,

$$\log_3 \left(\frac{1}{81}\right) = -4.$$

25. $\ln x - \ln (2x + 3) = \ln 4$

$$\ln \left(\frac{x}{2x + 3}\right) = \ln 4$$

Since the natural logarithm function is one-to-one, we have

$$\frac{x}{2x + 3} = 4$$

$$x = 8x + 12$$

$$-7x = 12$$

$$x = -\frac{12}{7}.$$

Solution set: $\left\{-\frac{12}{7}\right\}$

26. We use the equation $y = Ce^{kx}$ where C is the initial amount and x is the number of years. Since the half-life is 28 years,

$$\frac{1}{2}C = Ce^{k(28)}$$

$$.5 = e^{28k}$$

$$28k = \ln .5$$

$$k = \frac{\ln .5}{28} \approx -.0248.$$

Therefore, $y = 5e^{-.0248x}$.
To find the amount remaining after 10 years, let $x = 10$.

$$y = 5e^{-.0248(10)}$$

$$y = 5e^{-.248}$$

$$y \approx 3.9.$$

After 10 years, about 3.9 grams will be left.

27. Let x = the number of cups of un-cooked rice;

y = the number of cups of un-cooked soybeans.

From the information given in the exercise, we can write the system

$$15x + 22.5y = 9.5 \quad (1)$$

$$810x + 270y = 324. \quad (2)$$

We may solve the system by any of the methods discussed in the textbook. We will use the elimination method.
Multiply equation (1) by -54 and add the result to equation (2).

$$-810x - 1215y = -513$$

$$\underline{810x + 270y = 324}$$

$$-945y = -189$$

$$y = \frac{-189}{-945} = \frac{1}{5}$$

Substitute this value into equation (2).

$$810x + 270\left(\frac{1}{5}\right) = 324$$

$$810x + 54 = 324$$

$$810x = 270$$

$$x = \frac{270}{810} = \frac{1}{3}$$

Use 1/3 cup of uncooked rice and 1/5 cup of uncooked soybeans.

28. $2xy + 1 = 0 \quad (1)$
    $x + 16y = 2 \quad (2)$

This is a nonlinear system because equation (1) is nonlinear. We will use the substitution method.
Solve equation (2) for x.

$$x = 2 - 16y \quad (3)$$

Substitute this expression for x in equation (1).

$$2(2 - 16y)y + 1 = 0$$

$$4y - 32y^2 + 1 = 0$$

$$0 = 32y^2 - 4y - 1$$

$$0 = (8y + 1)(4y - 1)$$

$$y = -\frac{1}{8} \quad \text{or} \quad y = \frac{1}{4}$$

For each value of y, substitute into equation (3) to find the corresponding value of x.

If $y = -\frac{1}{8}$, then

$$x = 2 - 16\left(-\frac{1}{8}\right) = 4.$$

If $y = \frac{1}{4}$, then

$$x = 2 - 16\left(\frac{1}{4}\right) = -2.$$

Solution set: $\left\{\left(4, -\frac{1}{8}\right), \left(-2, \frac{1}{4}\right)\right\}$

29. Let x = the larger number;
y = the smaller number.

From the given information, we have the nonlinear system

$$x - y = 3 \qquad (1)$$
$$x^2 + y^2 = 29. \qquad (2)$$

We will use the substitution method to solve the system.

Solve equation (1) for x.

$$x = 3 + y \qquad (3)$$

Substitute this expression into equation (2).

$$(3 + y)^2 + y^2 = 29$$
$$9 + 6y + y^2 + y^2 = 29$$
$$2y^2 + 6y - 20 = 0$$

Divide by 2, then solve the resulting equation by factoring.

$$y^2 + 3y - 10 = 0$$
$$(y + 5)(y - 2) = 0$$
$$y = -5 \quad \text{or} \quad y = 2$$

For each value of y, substitute into equation (3) to find the corresponding value of x.

If $y = -5$, then

$$x = 3 + (-5) = -2.$$

If $y = 2$, then

$$x = 3 + 2 = 5.$$

The numbers are −2 and −5, or 5 and 2.

30. $2x - 4y + 2z = 0$
$x + 3y - z = 5$
$3x + 2y + z = 12$

$$\begin{bmatrix} 2 & -4 & 2 & | & 0 \\ 1 & 3 & -1 & | & 5 \\ 3 & 2 & 1 & | & 12 \end{bmatrix}$$

$$\begin{bmatrix} 1 & 3 & -1 & | & 5 \\ 2 & -4 & 2 & | & 0 \\ 3 & 2 & 1 & | & 12 \end{bmatrix} \quad \text{R1} \leftrightarrow \text{R2}$$

$$\begin{bmatrix} 1 & 3 & -1 & | & 5 \\ 0 & -10 & 4 & | & -10 \\ 0 & -7 & 4 & | & -3 \end{bmatrix} \quad \begin{matrix} -2\text{R1} + \text{R2} \\ -3\text{R1} + \text{R3} \end{matrix}$$

$$\begin{bmatrix} 1 & 3 & -1 & | & 5 \\ 0 & 1 & -\frac{2}{5} & | & 1 \\ 0 & -7 & 4 & | & -3 \end{bmatrix} \quad -\frac{1}{10}\text{R2}$$

$$\begin{bmatrix} 1 & 0 & \frac{1}{5} & | & 2 \\ 0 & 1 & -\frac{2}{5} & | & 1 \\ 0 & 0 & \frac{6}{5} & | & 4 \end{bmatrix} \quad \begin{matrix} -3\text{R2} + \text{R1} \\ \\ 7\text{R2} + \text{R3} \end{matrix}$$

$$\begin{bmatrix} 1 & 0 & \frac{1}{5} & | & 2 \\ 0 & 1 & -\frac{2}{5} & | & 1 \\ 0 & 0 & 1 & | & \frac{10}{3} \end{bmatrix} \quad \frac{5}{6}\text{R3}$$

$$\begin{bmatrix} 1 & 0 & 0 & \bigm| & \frac{4}{3} \\ 0 & 1 & 0 & \bigm| & \frac{7}{3} \\ 0 & 0 & 1 & \bigm| & \frac{10}{3} \end{bmatrix} \qquad \begin{array}{l} -\frac{1}{5}R3 + R1 \\[1em] \frac{2}{5}R3 + R2 \end{array}$$

Solution set: $\left\{ \left(\frac{4}{3}, \frac{7}{3}, \frac{10}{3} \right) \right\}$

31. $AB = \begin{bmatrix} 3 & \frac{1}{2} & -1 \\ -4 & 2 & 0 \end{bmatrix} \cdot \begin{bmatrix} -2 & 5 \\ 4 & -2 \\ 6 & 1 \end{bmatrix} = \begin{bmatrix} -6+2-6 & 15-1-1 \\ 8+8+0 & -20-4+0 \end{bmatrix} = \begin{bmatrix} -10 & 13 \\ 16 & -24 \end{bmatrix}$

$BA = \begin{bmatrix} -2 & 5 \\ 4 & -2 \\ 6 & 1 \end{bmatrix} \begin{bmatrix} 3 & \frac{1}{2} & -1 \\ -4 & 2 & 0 \end{bmatrix} = \begin{bmatrix} -6-20 & -1+10 & 2+0 \\ 12+8 & 2-4 & -4+0 \\ 18-4 & 3+2 & -6+0 \end{bmatrix}$

$= \begin{bmatrix} -26 & 9 & 2 \\ 20 & -2 & -4 \\ 14 & 5 & -6 \end{bmatrix}$

32.

$$\begin{array}{cccc} \text{Under} & & \text{Over} & \text{Dem.} \quad \text{Rep.} \\ 40 & 40\text{--}60 & 60 & \\ BA = [12{,}000 & 10{,}000 & 8000] & \begin{bmatrix} .70 & .30 \\ .50 & .50 \\ .45 & .55 \end{bmatrix} \begin{array}{l} \text{Under } 40 \\ 40\text{--}60 \\ \text{Over } 60 \end{array} \end{array}$$

$= [8400 + 5000 + 3600 \quad 3600 + 5000 + 4400]$

$= [17{,}000 \quad 13{,}000]$

The matrix BA shows that 17,000 voted Democratic and 13,000 voted Republican.

33. We will evaluate the determinant by expansion about the first column.

$$\begin{vmatrix} 3 & 6 & -3 \\ 0 & 5 & 4 \\ 1 & -2 & 7 \end{vmatrix} = 3 \begin{vmatrix} 5 & 4 \\ -2 & 7 \end{vmatrix} - 0 \begin{vmatrix} 6 & -3 \\ -2 & 7 \end{vmatrix} + 1 \begin{vmatrix} 6 & -3 \\ 5 & 4 \end{vmatrix}$$

$$= 3[35 - (-8)] - 0 + [24 - (-15)]$$

$$= 3(43) + 39 = 129 + 39 = 168$$

34. $2x - 3y + z = 0$

$4x + y + 3z = 17$

$-6x + 5y - z = 2$

$$D = \begin{vmatrix} 2 & -3 & 1 \\ 4 & 1 & 3 \\ -6 & 5 & -1 \end{vmatrix} = 36$$

$$D_x = \begin{vmatrix} 0 & -3 & 1 \\ 17 & 1 & 3 \\ 2 & 5 & -1 \end{vmatrix} = 14$$

$$D_y = \begin{vmatrix} 2 & 0 & 1 \\ 4 & 17 & 3 \\ -6 & 2 & -1 \end{vmatrix} = 64$$

$$D_z = \begin{vmatrix} 2 & -3 & 0 \\ 4 & 1 & 17 \\ -6 & 5 & 2 \end{vmatrix} = 164$$

$x = \dfrac{D_x}{D} = \dfrac{14}{36} = \dfrac{7}{18}$

$y = \dfrac{D_y}{D} = \dfrac{64}{36} = \dfrac{16}{9}$

$z = \dfrac{D_z}{D} = \dfrac{164}{36} = \dfrac{41}{9}$

Solution set: $\left\{ \left(\dfrac{7}{18}, \dfrac{16}{9}, \dfrac{41}{9} \right) \right\}$

35. $-2x + 2y - z = 3$

$3x - 5y + 4z = 4$

$5x - 6y + 4z = 5$

Let $A = \begin{bmatrix} -2 & 2 & -1 \\ 3 & -5 & 4 \\ 5 & -6 & 4 \end{bmatrix}$, $X = \begin{bmatrix} x \\ y \\ z \end{bmatrix}$,

and $B = \begin{bmatrix} 3 \\ 4 \\ 5 \end{bmatrix}$.

Find A^{-1}.

$[A \mid I_3] = \begin{bmatrix} -2 & 2 & -1 & 1 & 0 & 0 \\ 3 & -5 & 4 & 0 & 1 & 0 \\ 5 & -6 & 4 & 0 & 0 & 1 \end{bmatrix}$

$\begin{bmatrix} 1 & -1 & \frac{1}{2} & -\frac{1}{2} & 0 & 0 \\ 3 & -5 & 4 & 0 & 1 & 0 \\ 5 & -6 & 4 & 0 & 0 & 1 \end{bmatrix}$ $-\frac{1}{2}R1$

$\begin{bmatrix} 1 & -1 & \frac{1}{2} & -\frac{1}{2} & 0 & 0 \\ 0 & -2 & \frac{5}{2} & \frac{3}{2} & 1 & 0 \\ 0 & -1 & \frac{3}{2} & \frac{5}{2} & 0 & 1 \end{bmatrix}$ $\begin{matrix} -3R1 + R2 \\ \\ -5R1 + R3 \end{matrix}$

$\begin{bmatrix} 1 & -1 & \frac{1}{2} & -\frac{1}{2} & 0 & 0 \\ 0 & 1 & -\frac{5}{4} & -\frac{3}{4} & -\frac{1}{2} & 0 \\ 0 & -1 & \frac{3}{2} & \frac{5}{2} & 0 & 1 \end{bmatrix}$ $-\frac{1}{2}R2$

$\begin{bmatrix} 1 & -1 & \frac{1}{2} & -\frac{1}{2} & 0 & 0 \\ 0 & 1 & -\frac{5}{4} & -\frac{3}{4} & -\frac{1}{2} & 0 \\ 0 & 0 & \frac{1}{4} & \frac{7}{4} & -\frac{1}{2} & 1 \end{bmatrix}$ $R2 + R3$

$\begin{bmatrix} 1 & -1 & \frac{1}{2} & -\frac{1}{2} & 0 & 0 \\ 0 & 1 & -\frac{5}{4} & -\frac{3}{4} & -\frac{1}{2} & 0 \\ 0 & 0 & 1 & 7 & -2 & 4 \end{bmatrix}$ $4R3$

$\begin{bmatrix} 1 & 0 & -\frac{3}{4} & -\frac{5}{4} & -\frac{1}{2} & 0 \\ 0 & 1 & 0 & 8 & -3 & 5 \\ 0 & 0 & 1 & 7 & -2 & 4 \end{bmatrix}$ $\begin{matrix} R2 + R1 \\ \frac{5}{4}R3 + R2 \end{matrix}$

$\begin{bmatrix} 1 & 0 & 0 & 4 & -2 & 3 \\ 0 & 1 & 0 & 8 & -3 & 5 \\ 0 & 0 & 1 & 7 & -2 & 4 \end{bmatrix}$ $\frac{3}{4}R3 + R1$

$A^{-1} = \begin{bmatrix} 4 & -2 & 3 \\ 8 & -3 & 5 \\ 7 & -2 & 4 \end{bmatrix}$

(A graphing calculator can also be used to find A^{-1}.)

$X = A^{-1}B$

$= \begin{bmatrix} 4 & -2 & 3 \\ 8 & -3 & 5 \\ 7 & -2 & 4 \end{bmatrix} \begin{bmatrix} 3 \\ 4 \\ 5 \end{bmatrix}$

$= \begin{bmatrix} 19 \\ 37 \\ 33 \end{bmatrix}$

Solution set: $\{(19, 37, 33)\}$

36. $x^2 - 4x + y \le 1$

$e^{.5x} - y \le 0$

To graph the solution set of $x^2 - 4x + y \le 1$, write the inequality as

$$y \le -x^2 + 4x + 1.$$

Graph the parabola $y = -x^2 + 4x + 1$, which has vertex (2, 5) and opens downward, as a solid curve. Shade the region below the parabola.
To graph the solution of $e^{.5x} - y \le 0$ write the inequality as

$$y \ge e^{.5x}.$$

Graph the exponential curve $y = e^{.5x}$ as a solid curve.
Shade the region above the curve.
The solution set of the system is the intersection of the two regions.

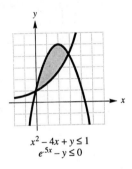

$x^2 - 4x + y \le 1$
$e^{.5x} - y \le 0$

37. Let x = the number of Type 1 bolts;
y = the number of Type 2 bolts.

We want to maximize the revenue,

$.10x + .12y,$

subject to the constraints

$.4x + .3y \le 1200$

$.5x + .2y \le 1200$

$.2x + .4y \le 1200$

$x \ge 0,\ y \ge 0.$

We find the region of feasible solutions by graphing the solution set of the above system of inequalities.

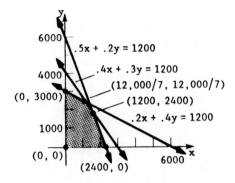

The corner points of this region are (0, 0), (0, 3000), (1200, 2400), (12,000/7, 12,000/7), and (2400, 0).
The point (12,000/7, 12,000/7) is the intersection of the lines

$.5x + .2y = 1200$

and $.4x + .3y = 1200.$

The point (1200, 2400) is the intersection of the lines

$.2x + .4y = 1200$

and $.4x + .3y = 1200.$

The coordinates of these points can be found by solving each of these systems of linear equations using any of the methods discussed in Chapter 6 of the textbook.
We find the value of $.10x + .12y$ at each corner point.

Point	Value of $.10x + .12y$
(0, 0)	0
(0, 3000)	360
(1200, 2400)	408
$\left(\dfrac{12,000}{7}, \dfrac{12,000}{7}\right)$	$\dfrac{2640}{7} \approx 377.14$
(2400, 0)	240

The maximum revenue is \$408 when 1200 Type 1 and 2400 Type 2 bolts are manufactured.

38. $(x + 3)^2 = 20(y - 8)$

The graph of the equation is a parabola with vertex at $(-3, 8)$ and opening upward.

$$4p = 20$$
$$p = 5$$

The focus is 5 units above the vertex. The coordinates of the focus are $(-3, 13)$.

The directrix is a horizontal line 5 units below the vertex. The equation of the directrix is $y = 3$.

The axis of the parabola is the vertical line through the vertex. The equation of the axis is $x = -3$.

39. Since the center, a vertex, and a focus all lie on the vertical line $x = 6$, this ellipse has a vertical major axis.

The form of the equation is

$$\frac{(x - 6)^2}{b^2} + \frac{(y - 7)^2}{a^2} = 1.$$

Since $(6, 7)$ is the center and $(6, 11)$ is a vertex,

$$a = 11 - 7 = 4.$$

Since $(6, 7)$ is the center and $(6, 10)$ is a focus,

$$c = 10 - 7 = 3.$$

Since

$$c^2 = a^2 - b^2,$$

we have

$$9 = 16 - b^2$$
$$b^2 = 7.$$

The equation is

$$\frac{(x - 6)^2}{7} + \frac{(y - 7)^2}{16} = 1.$$

40. Since the center and a vertex lie on the horizontal line $y = -2$, this is a horizontal hyperbola.

The form of the equation is

$$\frac{(x - 5)^2}{a^2} - \frac{(y + 2)^2}{b^2} = 1.$$

Since $(5, -2)$ is the center and $(2, -2)$ is a vertex, $a = 5 - 2 = 3$, and $a^2 = 9$.

The slopes of the asymptotes are ± 2. Therefore,

$$\frac{b}{a} = 2$$
$$\frac{b}{3} = 2$$
$$b = 6$$
$$b^2 = 36.$$

The equation is

$$\frac{(x - 5)^2}{9} - \frac{(y + 2)^2}{36} = 1.$$

41. $9x^2 - 54x + 4y^2 + 40y = -145$

Complete the square on x and on y.

$$9(x^2 - 6x) + 4(y^2 + 10y) = -145$$
$$9(x^2 - 6x + 9) + 4(y^2 + 10y + 25)$$
$$= -145 + 81 + 100$$
$$9(x + 3)^2 + 4(y + 5)^2 = 36$$
$$\frac{(x + 3)^2}{4} + \frac{(y + 5)^2}{9} = 1$$

The conic section with this equation is an ellipse.

42. $\sum\limits_{k=1}^{4} (1 + k)^{k-1} = 2^0 + 3^1 + 4^2 + 5^3$
$$= 1 + 3 + 16 + 125$$
$$= 145$$

43. $a_6 = 29$ and $a_{20} = 99$

Using the formula
$$a_n = a_1 + (n - 1)d,$$
we write a system of equations:
$$a_6 = a_1 + 5d \quad (1)$$
$$a_{20} = a_1 + 19d. \quad (2)$$

Substituting $a_6 = 29$ and $a_{20} = 99$, we have
$$a_1 + 5d = 29 \quad (3)$$
$$a_1 + 19d = 99. \quad (4)$$

Multiply equation (3) by 19 and equation (4) by −5; then add the resulting equations.
$$19a_1 + 95d = 551$$
$$\underline{-5a_1 - 95d = -495}$$
$$14a_1 \qquad = 56$$
$$a_1 = 4$$

44. $\sum\limits_{i=1}^{10} (-2 + 4i + 3 \cdot 2^i)$
$$= \sum\limits_{i=1}^{10} (4i - 2) + \sum\limits_{i=1}^{10} (3 \cdot 2^i)$$

$\sum\limits_{i=1}^{10} (4i - 2)$ is the sum of an arithmetic sequence with $a_1 = 2$, $d = 4$, and $n = 10$. Use the formula for the sum of the first n terms of an arithmetic sequence.

$$S_n = \frac{n}{2}[2a_1 + (n - 1)d]$$
$$S_{10} = \frac{10}{2}[2(2) + 9(4)] = 5(40) = 200$$

$\sum\limits_{i=1}^{n} (3 \cdot 2^i)$ is the sum of a geometric sequence with $a_1 = 6$, $r = 2$, and $n = 10$. Use the formula for the sum of the first n terms of a geometric sequence.

$$S_n = \frac{a_1(1 - r^n)}{1 - r}$$
$$S_{10} = \frac{6(1 - 2^{10})}{1 - 2} = \frac{6(-1023)}{-1} = 6138$$

Thus,

$$\sum\limits_{i=1}^{10} (-2 + 4i + 3 \cdot 2^i) = 200 + 6138$$
$$= 6338.$$

45. $a_4 = 1$ and $a_6 = .0625$

Since $.0625 = \frac{1}{16}$, $a_6 = \frac{1}{16}$.

We use the formula $a_n = a_1 r^{n-1}$ to write a system of equations.
$$a_4 = a_1 r^3 \quad (1)$$
$$a_6 = a_1 r^5 \quad (2)$$

Substitute $a_4 = 1$ and $a_6 = \frac{1}{16}$ to obtain
$$a_1 r^3 = 1 \quad (3)$$
$$a_1 r^5 = \frac{1}{16} \quad (4)$$

Therefore,

$$\frac{a_1 r^5}{a_1 r^3} = \frac{\frac{1}{16}}{1}$$
$$r^2 = \frac{1}{16}$$
$$r = \pm\frac{1}{4}.$$

Substitute each value for r into equation (3) to find the corresponding value of a_1.

If $r = 1/4$,

$$a_1\left(\frac{1}{4}\right)^3 = 1$$

$$\frac{1}{64}a_1 = 1$$

$$a_1 = 64.$$

If $r = -1/4$,

$$a_1\left(-\frac{1}{4}\right)^3 = 1$$

$$-\frac{1}{64}a_1 = 1$$

$$a_1 = -64.$$

Thus, there are two geometric sequences that satisfy the given conditions: one sequence with $a_1 = 64$ and $r = 1/4$, and the other with $a_1 = -64$ and $r = -1/4$.

46. $\displaystyle\sum_{k=1}^{\infty} 3 \cdot (.8)^k$

This is the sum of an infinite geometric sequence with $a_1 = 3(.8) = 2.4$ and $r = .8$.

$$S_\infty = \frac{a_1}{1 - r}$$

$$= \frac{2.4}{1 - .8}$$

$$= \frac{2.4}{.2}$$

$$= 12$$

47. Use the formula

$$\binom{n}{k-1}x^{n-(k-1)}y^{k-1}$$

with $n = 12$, $k = 9$, $k - 1 = 8$, and $n - (k - 1) = 4$.

The ninth term of the expansion of $(2x - y)^{12}$ is given by

$$\binom{12}{8}(2x)^4(-y)^8 = 495 \cdot 16x^4y^8$$

$$= 7920x^4y^8.$$

48. Let S_n be the statement

$$4 + 4^2 + 4^3 + \ldots + 4^n = \frac{4(4^n - 1)}{3}.$$

Step 1 S_1 is the statement

$$4 = \frac{4(4^1 - 1)}{3}$$

$$4 = 4.$$

S_1 is true.

Step 2 Show that if S_k is true, then S_{k+1} is also true. S_k is the statement.

$$4 + 4^2 + 4^3 + \ldots + 4^k = \frac{4(4^k - 1)}{3}.$$

Add the $(k + 1)$st term, 4^{k+1}, to both sides.

$$4 + 4^2 + 4^3 + \ldots + 4^k + 4^{k+1}$$

$$= \frac{4(4^k - 1)}{3} + 4^{k+1}$$

$$= \frac{4^{k+1} - 4}{3} + \frac{3 \cdot 4^{k+1}}{3}$$

$$= \frac{4 \cdot 4^{k+1} - 4}{3}$$

$$4 + 4^2 + 4^3 + \ldots + 4^k + 4^{k+1}$$

$$= \frac{4(4^{k+1} - 4)}{3}$$

This is the statement S_{k+1}.

By Steps 1 and 2, S_n is true for every positive integer n.

49. In choosing samples, order is not imporuant, so we use combinations. The number of possible samples is

$$\binom{100}{3} = 161,700.$$

The number of possible samples consisting of all defective diskettes is

$$\binom{7}{3} = 35.$$

50. If A is the event that a candy bar is a Yummy bar,

$$P(A) = .09,$$

so

$$P(A') = .91.$$

The odds in favor of A are given by

$$\frac{P(A)}{P(A')} = \frac{.09}{.91} = \frac{9}{91}$$

or 9 to 91.

51. In this binomial experiment, we define a success as having a boy. Then n = 4; r = 1, 2, or 3; and p = 1/2.

P(at least one boy and at least one girl)
= P(1 boy) + P(2 boys) + P(3 boys)

$$= \binom{4}{1}\left(\frac{1}{2}\right)^1\left(1-\frac{1}{2}\right)^3 + \binom{4}{2}\left(\frac{1}{2}\right)^2\left(1-\frac{1}{2}\right)^2$$

$$+ \binom{4}{3}\left(\frac{1}{2}\right)^3\left(1-\frac{1}{2}\right)^1$$

$$= 4\cdot\frac{1}{16} + 6\cdot\frac{1}{16} + 4\cdot\frac{1}{16}$$

$$= \frac{14}{16} = \frac{7}{8}$$

APPENDIX   A   SETS

1. The elements of the set $\{12, 13, 14, \ldots, 20\}$ are all the natural numbers from 12 to 20 inclusive. These numbers are

    12, 13, 14, 15, 16, 17, 18, 19, and 20.

3. The elements of the set $\{1, 1/2, 1/4, \ldots, 1/32\}$ form a geometric sequence with $a_1 = 1$ and $r = 1/2$, that is, the first number is 1, and each number after the first is found by multiplying the preceding number by $1/2$. There are 6 elements:

    1, 1/2, 1/4, 1/8, 1/16, and 1/32.

5. To find the elements of the set $\{17, 22, 27, \ldots, 47\}$, start with 17 and add 5 to find the next number. The elements of this set form an arithmetic sequence with $a_1 = 17$ and $d = 5$. There are 7 elements.

    17, 22, 27, 32, 37, 42, and 47.

7. The elements of the set $\{$all natural numbers greater than 7 and less than 15$\}$ are

    8, 9, 10, 11, 12, 13 and 14.

9. The set $\{4, 5, 6, \ldots, 15\}$ has a limited number of elements, so it is a finite set.

11. The set $\{1, 1/2, 1/4, 1/8, \ldots\}$ has an unlimited number of elements, so it is an infinite set.

13. The set $\{x \mid x$ is a natural number larger than 5$\}$, which can also be written as $\{6, 7, 8, 9, \ldots\}$, has an unlimited number of elements, so it is an infinite set.

15. There are an infinite number of fractions between 0 and 1, so $\{x \mid x$ is a fraction between 0 and 1$\}$ is an infinite set.

17. 6 is an element of the set $\{3, 4, 5, 6\}$, so we write $6 \in \{3, 4, 5, 6\}$.

19. -4 is not an element of $\{4, 6, 8, 10\}$, so we write $-4 \notin \{4, 6, 8, 10\}$.

21. 0 is an element of $\{2, 0, 3, 4\}$, so we write $0 \in \{2, 0, 3, 4\}$.

23. $\{3\}$ is a subset of $\{2, 3, 4, 5)$, not an element of $\{2, 3, 4, 5\}$, so we write $\{3\} \notin \{2, 3, 4, 5\}$.

25. $\{0\}$ is a subset of $\{0, 1, 2, 5\}$, not an element of $\{0, 1, 2, 5\}$, so we write $\{0\} \notin \{0, 1, 2, 5\}$.

27. 0 is not an element of $\emptyset$, since the empty set contains no elements. Thus, $0 \notin \emptyset$.

29. $3 \in \{2, 5, 6, 8\}$

3 is not one of the elements in $\{2, 5, 6, 8\}$, so this statement is false.

31. $1 \in \{3, 4, 5, 11, 1\}$

Since 1 is one of the elements of $\{3, 4, 5, 11, 1\}$, the statement is true.

33. $9 \notin \{2, 1, 5, 8\}$

Since 9 is not one of the elements of $\{2, 1, 5, 8\}$, the statement is true.

35. $\{2, 5, 8, 9\} = \{2, 5, 9, 8\}$

This statement is true because both sets contain exactly the same elements.

37. $\{5, 8, 9\} = \{5, 8, 9, 0\}$

These two sets are not equal because $\{5, 8, 9, 0\}$ contains the element 0, which is not an element of $\{5, 8, 9\}$. Therefore, the statement is false.

39. $\{x | x \text{ is a natural number less than } 3\} = \{1, 2\}$

Since 1 and 2 are the only natural numbers less than 2, this statement is true.

41. $\{5, 7, 9, 19\} \cap \{7, 9, 11, 15\} = \{7, 9\}$

The symbol "$\cap$" means the intersection of the two sets, which is the set of all elements that belong to both sets. Since 7 and 9 are the only elements belonging to both sets, the statement is true.

43. $\{2,\ 1,\ 7\} \cup \{1,\ 5,\ 9\} = \{1\}$

The symbol "$\cup$" means the union of two sets, which is the set of all elements that belong to either one of the sets or to both sets.

$$\{2,\ 1,\ 7\} \cup \{1,\ 5,\ 9\} = \{1,\ 2,\ 5,\ 7,\ 9\},$$

while

$$\{2,\ 1,\ 7\} \cap \{1,\ 5,\ 9\} = \{1\}.$$

Therefore, the statement is false.

45. $\{3,\ 2,\ 5,\ 9\} \cap \{2,\ 7,\ 8,\ 10\} = \{2\}$

Since 2 is the only element belonging to both sets, the statement is true.

47. $\{3,\ 5,\ 9,\ 10\} \cap \emptyset = \{3,\ 5,\ 9,\ 10\}$

In order to belong to the intersection of two sets, and element must belong to both sets. Since the empty set contains no elements, $\{3,\ 5,\ 9,\ 10\} \cap \emptyset = \emptyset$, so the statement is false.

49. $\{1,\ 2,\ 4\} \cup \{1,\ 2,\ 4\} = \{1,\ 2,\ 4\}$

Since the two sets are equal, their union contains the same elements, 1, 2, and 4. Thus, the statement is true.

51. $\emptyset \cup \emptyset = \emptyset$

Since the empty set contains no elements, the statement is true.

For Exercises 53–63,

$$A = \{2,\ 4,\ 6,\ 8,\ 10,\ 12\},$$
$$B = \{2,\ 4,\ 8,\ 10\},$$
$$C = \{4,\ 10,\ 12\},$$
$$D = \{2,\ 10\},$$
$$\text{and}\quad U = \{2,\ 4,\ 6,\ 8,\ 10,\ 12,\ 14\}.$$

53. $A \subseteq U$

This statement says "A is a subset of U." Since every element of A is also an element of U, the statement is true,

55. D ⊆ B

Since both elements of D, 2 and 10, are also elements of B, D is a subset of B. The statement is true.

57. A ⊆ B

Set A contains a two elements, 6 and 12, that are not elements of B. Thus, A is not a subset of B. The statement is false.

59. ∅ ⊆ A

The empty set is a subset of every set, so the statement is true.

61. {4, 8, 10} ⊆ B

Since 4, 8, and 10 are all elements of B, {4, 8, 10} is a subset of B. The statement is true.

63. B ⊆ D

Since B contains two elements, 4 and 8, that are not elements of D, B is not a subset of D. The statement is false.

65. Every element of {2, 4, 6} is also an element of {3, 2, 5, 4, 6}, so {2, 4, 6} is a subset of {3, 2, 5, 4, 6}. We write

$$\{2,\ 4,\ 6\} \subseteq \{3,\ 2,\ 5,\ 4,\ 6\}.$$

67. Since 0 is an element of {0, 1, 2}, but is not an element of {1, 2, 3, 4, 5}, {0, 1, 2} is not a subset of {1, 2, 3, 4, 5}. We write

$$\{0,\ 1,\ 2\} \not\subseteq \{1,\ 2,\ 3,\ 4,\ 5\}.$$

69. The empty set is a subset of every set, so ∅ ⊆ {1, 4, 6, 8}.

For Exercises 71–93,

$$U = \{0,\ 1,\ 2,\ 3,\ 4,\ 5,\ 6,\ 7,\ 8,\ 9,\ 10,\ 11,\ 12,\ 13\},$$
$$M = \{0,\ 2,\ 4,\ 6,\ 8\},$$
$$N = \{1,\ 3,\ 5,\ 7,\ 9,\ 11,\ 13\},$$
$$Q = \{0,\ 2,\ 4,\ 6,\ 8,\ 10,\ 12\},$$
and $R = \{0,\ 1,\ 2,\ 3,\ 4\}.$

71. $M \cap R$

The only elements belonging to both M and R are 0, 2, and 4, so

$$M \cap R = \{0,\ 2,\ 4\}.$$

73. $M \cup N$

The union of two sets contains all elements that belong to either set or to both sets.

$$M \cup N = \{0,\ 1,\ 2,\ 3,\ 4,\ 5,\ 6,\ 7,\ 8,\ 9,\ 11,\ 13\}$$

75. $M \cap U$

Since $M \subseteq U$, the intersection of M and U will contain the same elements as M.

$$M \cap U = M \quad \text{or} \quad \{0,\ 2,\ 4,\ 6,\ 8\}$$

77. $N \cup R = \{0,\ 1,\ 2,\ 3,\ 4,\ 5,\ 7,\ 9,\ 11,\ 13\}$

79. N'

The set N′ is the complement of set N, which means the set of all elements in the universal set U that do not belong to N.

$$N' = Q \quad \text{or} \quad \{0,\ 2,\ 4,\ 6,\ 8,\ 10,\ 12\}$$

81. $M' \cap Q$

First form M′, the complement of M. M′ contains all elements of U that are not elements of M.

$$M' = \{1,\ 3,\ 5,\ 7,\ 9,\ 10,\ 11,\ 12,\ 13\}$$

Now form the intersection of M' and Q.

$$M' \cap Q = \{10,\ 12\}$$

83. $\emptyset \cap R$

Since the empty set contains no elements, there are no elements belonging to both $\emptyset$ and R.   Thus, $\emptyset$ and R are disjoint sets, and $\emptyset \cap R = \emptyset$.

85. $N \cup \emptyset$

Since $\emptyset$ contains no elements, the only elements belonging to N or $\emptyset$ are the elements of N.   Thus,

$$N \cup \emptyset = N \quad \text{or} \quad \{1,\ 3,\ 5,\ 7,\ 9,\ 11,\ 13\}.$$

87. $(M \cap N) \cup R$

First form the intersection of M and N.   Since M and N have no common elements, $M \cap N = \emptyset$.
Thus,

$$(M \cap N) \cup R = \emptyset \cup R$$
$$= R \quad \text{or} \quad \{0,\ 1,\ 2,\ 3,\ 4\}.$$

89. $(Q \cap M) \cup R$

First form the intersection of Q and M.

$$Q \cap M = \{0,\ 2,\ 4,\ 6,\ 8\} = M$$

Now form the union of this set with R.

$$(Q \cap M) \cup R = M \cup R$$
$$= \{0,\ 1,\ 2,\ 3,\ 4,\ 6,\ 8\}$$

91. $(M' \cup Q) \cap R$

First, find M', the complement of M.

$$M' = \{1,\ 3,\ 5,\ 7,\ 9,\ 10,\ 11,\ 12,\ 13\}$$

Next, form the union of M' and Q.

$$M' \cup Q = \{0,\ 1,\ 2,\ 3,\ 4,\ 5,\ 6,\ 7,\ 8,\ 9,\ 10,\ 11,\ 12,\ 13\}$$
$$= U$$

Thus,

$$(M' \cup Q) \cap R = U \cap R$$
$$= R \quad \text{or} \quad \{1, 2, 3, 4\}.$$

93. $Q' \cap (N' \cap U)$

First, find Q', the complement of Q.

$$Q' = \{1, 3, 5, 7, 9, 11, 13\} = N$$

Now find N', the complement of P.

$$N' = \{0, 2, 4, 6, 8, 10, 12\} = Q$$

Next, form the union of N' and U.

$$N' \cup U = Q \cup U = Q$$

Finally, we have

$$Q' \cap (N' \cap U) = Q' \cap Q = \emptyset$$

Since the intersection of Q' and $(N' \cap U)$ is $\emptyset$, Q' and $(N' \cap U)$ are disjoint sets.

95. M' is the set of all students in this school who are not taking this course.

97. $N \cap P$ is the set of all students in this school who are taking both calculus and history.

99. $M \cup P$ is the set of all students in this school who are taking this course or history or both.